FOURTH EDITION

ALTERNATIVE SWEETENERS

FOURTH EDITION

ALTERNATIVE SWEETENERS

EDITED BY LYN O'BRIEN NABORS

CRC Press
Taylor & Francis Group
Boca Raton   London   New York

CRC Press is an imprint of the
Taylor & Francis Group, an **Informa** business

CRC Press
Taylor & Francis Group
6000 Broken Sound Parkway NW, Suite 300
Boca Raton, FL 33487-2742

First issued in paperback 2016

Version Date: 20110816

ISBN 13: 978-1-138-19856-2 (pbk)
ISBN 13: 978-1-4398-4614-8 (hbk)

Library of Congress Cataloging-in-Publication Data

Alternative sweeteners / editor, Lyn O'Brien Nabors. -- 4th ed.
　　p. ; cm.
　Includes bibliographical references and index.
　ISBN 978-1-4398-4614-8 (hardback : alk. paper)
　I. Nabors, Lyn O'Brien, 1943- II. Title.
　[DNLM: 1. Sweetening Agents. WA 712]

　LC classification not assigned
　664'.6--dc23
　　　　　　　　　　　　　　　　　　　　　　　　　　　2011033775

Visit the Taylor & Francis Web site at
http://www.taylorandfrancis.com

and the CRC Press Web site at
http://www.crcpress.com

This book is dedicated to all authors who have made *Alternative Sweeteners*, editions one through four, possible.

Contents

PART II REDUCED-CALORIE SWEETENERS

PART III CALORIC ALTERNATIVES

PART IV   MULTIPLE INGREDIENT APPROACH

Preface to the Fourth Edition

Sweeteners are forever in the news. Whether it's information about a new sweetener or questions about one that has been on the market for years, interest in sweeteners and sweetness continues. This fourth edition of *Alternative Sweeteners* provides information on new (e.g., advantame), recently evaluated (e.g., stevia), and numerous other alternatives to sucrose. Many are successfully marketed today, others may find a place in the market should consumers' desire for "natural" continue, while some are of academic interest only. The chapters have been provided by those most familiar with the sweeteners, those who have developed, make and/or use the sweeteners.

A new chapter, "The Benefits of Reduced Calorie Foods and Beverages in Weight Management," has been added to this volume. With the obesity epidemic, all "tools" for weight control are being scrutinized, with low-calorie sweeteners and the products in which they are used at the forefront. This new chapter examines numerous studies related to this issue.

All chapters are extensively referenced for those who want to learn more.

Contributors

Eyassu G. Abegaz
Ajinomoto Corporate Services LLC
Washington, DC

Sue E. Andress
The NutraSweet Company
Chicago, Illinois

Michael H. Auerbach
Danisco A/S
Tarrytown, New York

Abraham I. Bakal
ABIC International Consultants, Inc.
Fairfield, New Jersey

Jörg Bernard
Südzucker AG
Obrigheim, Germany

Hans Bertelsen
MD Foods Ingredients amba
Videbaek, Denmark

Ihab E. Bishay
Ajinomoto Food Ingredients LLC
Chicago, Illinois

Navroz Boghani
Cadbury plc
Whippany, New Jersey

Barbara A. Bopp
TAP Pharmaceutical Products, Inc.
Deerfield, Illinois

Francisco Borrego
Ferrer Grupo
Beniel-Murcia, Spain

Saskia Brokx
PURAC biochem bv.
Gorinchem, The Netherlands

Allan W. Buck
Archer Daniels Midland Company
Decatur, Illinois

Robert G. Bursey
Ajinomoto Corporate Services LLC
Washington, DC

Harriett H. Butchko
The NutraSweet Company
Chicago, Illinois

Michael Carakostas
ToxStrategies, Inc.
St. Helena Island, South Carolina

C. Phil Comer
The NutraSweet Company
Chicago, Illinois

Cesar M. Compadre
University of Arkansas for Medical Sciences
Little Rock, Arkansas

Peter de Cock
Cargill R&D Center Europe
Vilvoorde, Belgium

Ronald C. Deis
Corn Products International
Newark, Delaware

Lee B. Dexter (deceased)
Lee B. Dexter & Associates
Austin, Texas

Milda E. Embuscado
Cerestar USA, Inc.
Hammond, Indiana

Kristian Eriknauer
MD Foods Ingredients amba
Viby J, Denmark

John C. Fry
Connect Consulting
Horsham, United Kingdom

V. Lee Grotz
McNeil Nutritionals
Fort Washington, Pennsylvania

Søren Juhl Hansen
MD Foods Ingredients amba
Videbaek, Denmark

Michael E. Hendrick (deceased)
Pfizer Inc.
Groton, Connecticut

Beth Hubrich
Calorie Control Council
Atlanta, Georgia

Frances Hunt
International Sweeteners Association
Brussels, Belgium

Peter R. Jamieson
Corn Products International
Newark, Delaware

Malcolm W. Kearsley
Cadbury plc
Whippany, New Jersey

A. Douglas Kinghorn
College of Pharmacy
The Ohio State University
Columbus, Ohio

Christian Klug
Nutrinova Nutrition Specialties and Food
    Ingredients GmbH
Kelsterbach, Germany

Rene Soegaard Laursen
MD Foods Ingredients amba
Brabrand, Denmark

Anh S. Le
SPI Polyols, Inc.
New Castle, Delaware

Graeme Locke
Cultor Food Science,
Surrey, United Kingdom

Dale A. Mayhew
The NutraSweet Company
Chicago, Illinois

Paul H. J. Mesters
PURAC biochem bv.
Gorinchem, The Netherlands

Brad I. Meyers
The NutraSweet Company
Chicago, Illinois

Helen Mitchell
Danisco (UK) Ltd.
Surrey, United Kingdom

Samuel Molinary
Consultant, Scientific & Regulatory Affairs
Beaufort, South Carolina

Frances K. Moppett
Pfizer, Inc.
Groton, Connecticut

Kathleen Bowe Mulderrig
SPI Polyols, Inc.
New Castle, Delaware

Lyn O'Brien Nabors
Calorie Control Council
Atlanta, Georgia

Philip M. Olinger
Polyol Innovations, Inc.
Reno, Nevada

Sakharam K. Patil
Cerestar USA, Inc.
Hammond, Indiana

Joan Patton
Calorie Control Council
Atlanta, Georgia

Tammy Pepper
Danisco (UK) Ltd.
Surrey, United Kingdom

Robert C. Peterson
Tate & Lyle
Decatur, Illinois

Indra Prakash
The Coca-Cola Company
Atlanta, Georgia

Paul Price
Price International
Libertyville, Illinois

Mary E. Quinlan
Tate & Lyle
Decatur, Illinois

Richard Reo
McNeil Nutritionals
Fort Washington, Pennsylvania

Alan B. Richards
Hayashibara International Inc.
Westminster, Colorado

James Saunders
Biospherics Incorporated
Beltsville, Maryland

Anke Sentko
Beneo Group
Mannheim, Germany

Djaja Djendoel Soejarto
PCRPS, Department of Medicinal Chemistry
    and Pharmacognosy
College of Pharmacy
University of Illinois at Chicago
Chicago, Illinois

W. Wayne Stargel
The NutraSweet Company
Chicago, Illinois

Julian Stowell
Danisco (UK) Ltd.
Surrey, United Kingdom

John A. Van Velthuijsen
PURAC biochem bv.
Gorinchem, The Netherlands

Christian M. Vastenavond
NUTRILAB NV
Heusden Zolder, Belgium

Gert-Wolfhard von Rymon Lipinski
MK Food Management Consulting GmbH
Bad Vilbel, Germany

John S. White
White Technical Research
Argenta, Illinois

Christine D. Wu
Department of Pediatric Dentistry
College of Dentistry
University of Illinois at Chicago
Chicago, Illinois

Christos Zacharis
Danisco (UK) Ltd.
Surrey, United Kingdom

Chapter 1

Alternative Sweeteners: An Overview

Lyn O'Brien Nabors

Contents

Introduction

Alternative sweeteners to sucrose continue to be of great interest to the food industry, health professionals, consumers, and the media. They may assist the industry in reducing the calorie content of good-tasting foods and beverages, which is also of interest to health professionals. Perhaps one reason for media and consumer interest in alternative sweeteners is that their role is more easily understood than that of, for example, antioxidants. This understanding has led to controversy, which will be discussed in later chapters.

Scientists and food technologists have been researching sweeteners and sweetness for more than 100 years. The number of approved sweeteners has increased substantially in the last three decades. Food product developers now have a number of sweeteners from which to choose in order to provide more product choices to meet the increasing demand for good-tasting products that have reduced calories. Variety is important as no sweetener, including sucrose, is perfect for all uses.

Ideal Sweetener

The ideal sweetener does not exist. Even sucrose, the gold standard, is unsuitable for some pharmaceuticals and chewing gums. Alternative sweeteners provide food and beverage choices to control caloric, carbohydrate, and/or sugar intake; assist in weight maintenance or reduction; aid in the management of diabetes; assist in the control of dental caries; enhance the usability of pharmaceuticals and cosmetics; provide sweetness in times of sugar shortage; and assist in the cost-effective use of limited resources.

The ideal sweetener should be at least as sweet as sucrose, colorless, odorless, and noncariogenic. It should have a clean, pleasant taste with immediate onset, without lingering. The more a sweetener tastes and functions like sucrose, the greater consumer acceptability. If it can be processed similarly to sucrose with existing equipment, it is more desirable to the industry.

The ideal sweetener should be water soluble and stable in both acidic and basic conditions and over a wide range of temperatures. Length of stability and consequently the shelf life of the final product are also important. The final food product should taste similar to the traditional one. A sweetener must be compatible with a wide range of food ingredients as sweetness is only one component of complex flavor systems.

Safety is essential. The sweetener must be nontoxic and metabolized normally or excreted unchanged, and studies verifying its safety should be in the public domain.

To be successful, a sweetener should be priced competitively in relation to sucrose and comparable sweeteners. It should be easily produced, stored, and transported.

Relative Sweetness

Perceived sweetness is dependent upon and can be modified by a number of factors. The chemical and physical composition of the medium in which the sweetener is dispersed impacts the taste and intensity. The concentration of the sweetener, the temperature at which the product is consumed, the pH, other ingredients in the product, and the sensitivity of the taster are all important. Sucrose is the usual standard and taste is king.

The intensity of the sweetness of a given substance in relation to sucrose is determined on a weight basis. Table 1.1 provides approximate relative sweetness values for many of the alternative sweeteners discussed in this volume.

Multiple Ingredient Approach

The development and approval of a variety of low-calorie sweeteners provide industry with the opportunity to meet the ever-growing consumer demand for light products. Having a variety of alternative sweeteners from which to choose enables manufacturers to use the ingredient, or combination of ingredients, best suited for a given product. This is known as the multiple sweetener approach to calorie control.

A variety of approved sweeteners is essential because no sweetener is perfect for all uses. The advantages of the multiple sweetener approach have long been known. Sweeteners vary not only in sweetness intensity but also in mouthfeel, onset and duration of sweetness, perceived aftertaste, solubility and stability at various levels of pH, and temperature. As noted above, the availability of a variety of sweeteners is important because no sweetener, including sucrose, is perfect for all uses.

Table 1.1   Relative Sweetness of Alternatives to Sucrose

	Approximate Sweetness (Sucrose = 1)
Lactitol	0.4
Polyglycitol, maltitol syrups	0.4–0.9
Teahouse	0.45
Isomalt	0.45–0.65
Isomaltulose	0.48
Sorbitol	0.6
Erythritol	0.7
Mannitol	0.7
Maltitol	0.9
Tagatose	0.9
Xylitol	1.0
High-fructose corn syrup, 55%	1.0
High-fructose corn syrup, 90%	1.0+
Crystalline fructose	1.2–1.7
Cyclamate	30
Glycyrrhizin	50–100
Aspartame	180
Lo han guo fruit extract	180
Acesulfame K	200
Saccharin	300–500
Steviol glycosides	300
Sucralose	600
Hernandulcin	1,000
Monellin	1,500–2,000
Neohesperidine dihydrochalcone	1,800
Alitame	2,000
Thaumatin	2,000–3,000
Neotame	8,000
Advantame	20,000

With the availability of several sweeteners, each can be used in the applications for which it is best suited and manufacturers can overcome limitations of individual sweeteners by using them in blends. Blends can improve taste, solubility, and stability. For example, blends of aspartame and saccharin or aspartame and acesulfame potassium (acesulfame K) in beverages increase stability and maintain sweetness for extended shelf life.

During the 1960s, cyclamate and saccharin were blended together in a variety of popular diet soft drinks and other products. This was really the first practical application of the multiple sweetener approach. The primary advantage of this sweetener blend was that saccharin (300 times sweeter than sucrose) boosted the sweetening power of cyclamate (30 times sweeter than sucrose), while cyclamate masked the aftertaste that can be associated with saccharin.

The two sweeteners, when combined, have a synergistic effect, that is, the sweetness of the combination is greater than the sum of the individual parts. This is true for most sweetener blends. Cyclamate was the major factor in launching the diet segment of the carbonated beverage industry. By the time cyclamate was banned in the United States in 1970, products and trademarks had been well established. Such a large market for diet beverages provided a tremendous incentive to develop new sweeteners.

After cyclamate was taken off the market in 1970, saccharin was the only low-calorie alternative to sugar available in the United States for more than a decade, although cyclamate remained available in many other countries.

Now, with the availability of acesulfame K, aspartame, cyclamate, neotame, steviol glycosides, sucralose, and saccharin in most regions of the world, and the recognition of the reduced caloric value of the polyols, a multiple sweetener approach is being frequently utilized, providing better taste and increasing consumer choices.

Because of the synergistic effect when using sweetener blends, less total sweetener is usually required. Also, since the sweeteners used do not chemically react with one another and the individual sweeteners do not break down into unknown or unsafe ingredients, the use of sweeteners in combination assures manufacturers and consumers of continued safe new product and taste choices.

The polyols are also important adjuncts to sugar-free product development. There are eight polyols or sugar alcohols generally available. They are erythritol, polyglycitol or maltitol syrups, isomalt, lactitol, maltitol, mannitol, sorbitol, and xylitol. These sweeteners, with the exception of xylitol, are not as sweet as sugar, and are therefore often used in combination with low-calorie sweeteners to enhance their sweetness while providing the bulk that low-calorie sweeteners lack. The polyols, which are reduced in calories, combine well and are synergistic with low-calorie sweeteners, resulting in good-tasting, reduced-calorie products that are very similar to their traditional counterparts.

The US FDA allows the use of the following caloric values for the polyols: 1.6 for mannitol; 2.0 for isomalt and lactitol; 2.1 for maltitol; 2.4 for xylitol; 2.6 for sorbitol; and 3.0 for polyglycitols, hydrogenated starch hydrolysates, or maltitol syrups. The EU has assigned a caloric value of 2.4 to all polyols except erythritol, which has zero calories.

In the United States, fountain soft drinks generally contain a combination of saccharin and aspartame, and bottled drinks are available with combinations of aspartame and acesulfame K as well as sucralose and aspartame. Triple blends, such as acesulfame K, aspartame, and saccharin, and aspartame, cyclamate, and acesulfame K are being used in some parts of the world. Chewing gums are a good example in which a combination of low-calorie sweetener(s) and polyol(s) is used.

The taste profile of sweeteners varies. The sweet taste of some sweeteners is perceived immediately, while that of others is delayed. The duration of sweetness can also vary. Again, chewing gum is a good example for the use of multiple sweeteners. By combining, for example, three or more sweeteners, you would incorporate the benefits of all to have an immediate-onset, long-lasting, tapering-off sweetness.

Sweetener Safety

All sweeteners are carefully studied and reviewed by numerous scientific and regulatory bodies around the world before they are used in foods and beverages. Alternative sweeteners, individually and as a group, are the most extensively researched ingredients in the food supply. The safety of the approved sweeteners has been evaluated and confirmed by numerous regulatory and scientific bodies repeatedly.

International Regulatory Groups

Food ingredients are evaluated and/or regulated by numerous national and international bodies. International groups include the Joint Food and Agriculture Organization/World Health Organization (FAO/WHO) Expert Committee on Food Additives (JECFA) and the Codex Alimentarius Commission.

The objective of the FAO/WHO program on food additives is to make systematic evaluations of food additives and provide advice to member states of FAO and WHO on the control of additives and related health aspects. The two groups responsible for implementing the program are the JECFA and the Committee on Food Additives of the Joint FAO/WHO Codex Alimentarius Commission (JECFA 1974).

The JECFA is made up of an international group of experts who serve without remuneration in their personal capacities rather than as representatives of their governments or other bodies. Members are selected primarily for their ability and technical experience, with consideration given to adequate geographical distribution. Their reports contain the collective views of the group and do not necessarily represent the decision or the stated policy of the WHO or FAO. The experts convene to give advice on technical and scientific matters, establishing specifications for identity and purity for food additives, evaluating the toxicological data, and recommending, where appropriate, acceptable daily intakes for humans. The committee also acts in an advisory capacity for the Codex Committee on Food Additives and Contaminants, now two committees, the Codex Committee on Food Additives and the Codex Committee on Contaminants in Food (JECFA 1974).

The Codex Alimentarius Commission was established in 1962 to implement the Joint FAO/WHO Food Standards Program. Membership is comprised of those member nations and associate members of FAO and WHO that have notified the director general of FAO or WHO of their wish to be members.

The Codex Alimentarius is a collection of internationally adopted food standards and related texts presented in a uniform manner which aim to protect consumers' health and ensure fair practices in the food trade. The publication of the Codex Alimentarius is intended to guide and promote the elaboration and establishment of definitions and requirements for foods to assist in their harmonization and in doing so to facilitate international trade (CAC 2008).

The Codex Alimentarius includes provisions for food additives. The Codex Committee on Food Additives (CCFA) is charged with establishing or endorsing acceptable maximum levels for individual food additives; preparing priority lists of food additives for risk assessment by the Joint FAO/WHO Expert Committee on Food Additives; assigning functional classes to individual food additives; recommending specifications of identity and purity for food additives for adoption by the Commission; considering methods of analysis for the determination of additives in food; and considering and elaborating standards or codes for related subjects such as the labeling of food additives when sold as such (CAC 2008).

CCFA is developing a General Standard for Food Additives (GSFA) that lists food additives reviewed and assigned an acceptable daily intake (ADI) (either numerical or "not specified") by the JECFA. The GSFA provides a list of food categories for which an additive may be used and the levels of use for each category (GSFA 2010). Countries that do not have the review capability, such as the United States, those in the EU, Canada, and Japan, depend upon JECFA for scientific evaluation of food additives and the categories of use and use level guidelines provided by the CCFA in the GSFA.

The World Trade Organization (WTO) encourages countries to harmonize food standards on the basis of Codex standards and uses its decisions to settle trade disputes. In addition, the WTO recognizes JECFA specifications for food additives in international trade, increasing the importance of both the Codex and JECFA.

The European Food Safety Authority (EFSA) was created in January 2002 as an independent source of scientific advice and communication on risks associated with the food chain. EFSA's Scientific Committee and Panels carry out EFSA's scientific risk assessment work. They are composed of highly qualified risk assessment experts from across Europe with a range of expertise. Sweetener issues are addressed by EFSA's Panel on Food Additives and Nutrient Sources Added to Food (ANS) (EFSA 2010).

In the European Union, the harmonization of legislation to ensure that all member states have similar laws and regulations is an ongoing process. The Sweeteners Directive, Directive 94/35/EC, of the European Parliament and Council was adopted June 30, 1994. This directive has been amended four times. The articles of this legislation contain explanations and special provisions for the use of sweeteners in foods and beverages; the annex of this directive indicates the maximum levels of listed low-calorie sweeteners in a given food category (ISA 2010). More recently, the European Parliament and the Council adopted a framework regulation (Regulation 1333/2008), which consolidates all current authorizations for food additives, including sweeteners, into one legal text, as of January 2011 (ISA 2010).

As of November 30, 2010, the following low-calorie sweeteners are authorized in the EU: acesulfame K, aspartame, aspartame-acesulfame salt, cyclamate, neohesperidine DC, neotame, saccharin, sucralose, and thaumatin. Steviol glycosides are under review.

Specifications for the identity and purity of food additives, including sweeteners, are provided in a number of compendia. For example, Codex Alimentarius provides specifications for food additives reviewed by JECFA, European Commission Directive 95/31/EC provides specifications for sweeteners approved in the EU, and *Food Chemicals Codex* (*FCC*) provides specifications for quality and purity of more than 1000 food-grade substances. *FCC* is recognized internationally (*FCC* 2010). *FCC* specifications are cited, by reference, in the US Code of Federal Regulations. In Canada, *FCC* is officially recognized in the Canadian Food and Drug Regulations as the reference for specifications for food additives. Under New Zealand food regulations, a food additive is defined as being of appropriate quality if it complies with *FCC* monographs and, similarly, the National Food Authority of Australia frequently refers to the *FCC* specifications to define food additives.

US Regulation of Sweeteners

Food additives were first subjected to regulation in the United States under the Food and Drug Act of 1906. Section 402(a)(1) of the act states that a food shall be deemed adulterated:

> If it bears or contains any poisonous or deleterious substance which may render it injurious to health; but in case the substance is not an added substance, such food shall not

be considered adulterated under this clause if the quantity of such substance in such food does not ordinarily render it injurious to health. (FD&C 2004)

"Added" is not defined but is generally understood to mean a substance not present in a food in its natural state. The intent of this section is to prohibit any level of added food substance inconsistent with public health. The Federal Food, Drug, and Cosmetic Act of 1938 contains food safety provisions similar to those in the 1906 Act.

The basic Food, Drug, and Cosmetic Act was last updated in 1958. Section 201(s) defines a "food additive" as:

> any substance the intended use of which results or may reasonably be expected to result, directly or indirectly, in its becoming a component or otherwise affecting the characteristics of any food (including any substance intended for use in producing, manufacturing, packing, processing, preparing, treating, packaging, transporting, or holding food; and including any source of radiation intended for any such use). (FD&C 2004)

Congress passed the Food Additives Amendment, Section 409 of the Food, Drug, and Cosmetic Act, in 1958. This amendment exempts two important groups of substances from the food additive definition. Those exempted are (a) substances generally recognized as safe (GRAS) among experts qualified by scientific training and experience to evaluate safety, and (b) substances that either the United States FDA or the US Department of Agriculture (USDA) had sanctioned for use in food prior to 1958 (so-called "prior sanction" substances). The amendment does not pertain to pesticide chemicals in or on raw agricultural commodities.

The "Delaney Clause" is part of the 1958 Food Additives Amendment. The clause states that "no additive shall be deemed to be safe if it is found to induce cancer when ingested by man or animal, or if it is found after tests which are appropriate for the evaluation of the safety of food additives, to induce cancer in man or animal" (FD&C 2004). The Delaney Clause is often referred to but rarely used. Debate frequently centers on the undefined phrases "induce cancer" and "tests which are appropriate."

The 1958 Food Additives Amendment forbids the use of any food additive not approved by the US FDA, and the agency may only approve additives shown to be "safe." Section 409 of the Act outlines the requirements for requesting approval for a food additive (i.e., "Petition to establish safety") and details the action to be taken by the US FDA in dealing with such a petition.

A petitioner, requesting the issuance of a food additive regulation, must provide, in addition to any explanatory or supporting data:

> (2)(A) the name and all pertinent information concerning such food additive, including, where available, its chemical identity and composition; (B) a statement of the conditions of proposed use of such additive, including all directions, recommendations, and suggestions proposed for the use of such additive, and including samples of its proposed labeling; (C) all relevant data bearing on the physical or other technical effect such additive is intended to produce, and the quantity of the additive required to produce such effect; (D) a description of practicable methods of determining the quantity of such additive in or on food, and any substance formed in or on food, because of its use; and (E) full reports of investigations made with respect to the safety of such additive, including full information as to the methods and controls used in conducting such investigations. (FD&C 2004, Section 409(b))

The Federal Food, Drug, and Cosmetic Act does not describe the safety investigations to be conducted on the proposed food additive. The US FDA, therefore, issued a document entitled "Toxicological Principles for the Safety Assessment of Direct Food Additives and Color Additives Used in Food" (referred to as the "Redbook") in 1982 (Redbook 2000). Redbook II was issued in 1993 and is periodically updated. It is intended to provide guidance on criteria used for the safety assessment of direct food additives and color additives used in food and to assist petitioners in developing and submitting toxicological safety data for US FDA review. Although the Redbook is not legally binding and the US FDA notes that a petitioner may follow the guidelines and protocols in the Redbook or choose to use alternative procedures, the agency suggests that alternative procedures be discussed informally with the agency before proceeding (Redbook 2000).

According to federal law and regulations, any substance that is "Generally Recognized As Safe" (GRAS) for a particular use(s) may be used in food(s) for that purpose without premarket approval from the US FDA (CFR 2010). The GRAS exemption has been greatly misunderstood and misinterpreted over the years. It has been associated with a "second" tier safety protection, based on a less rigorous standard than that for food additives, when in reality the safety standard applicable to GRAS ingredients is the same as for food additives, "reasonable certainty of no harm" (Rulis and Levitt 2009, 22).

The real distinction between GRAS ingredients and food additives is information about the safety of the GRAS ingredient for its intended use that is common knowledge and the general availability and acceptance of that information "across the scientific community of food ingredient safety experts" (Rulis and Levitt 2009, 26).

The US FDA is now using a "GRAS notification system," under which manufacturers may notify the US FDA that the company has determined that their substance is GRAS. In notifying the US FDA of their GRAS determination, manufacturers must provide evidence supporting their decisions. Such data includes generally available and accepted scientific data, information, methods, or principles. Under certain circumstances, other scientific data, as well as analytical methods, methods of manufacture, and/or accepted scientific principles could be relied upon as part of the technical information. The US FDA notes that the quantity and quality of scientific evidence required to demonstrate the safety may vary depending on the estimated dietary exposure and the chemical, physical, and physiological properties of the substance. The notice summary must consider the totality of the publicly available information and evidence about the safety of the substance for its intended use, including favorable and potentially unfavorable information.

After evaluating the notification, and if the US FDA agrees that the evidence submitted supports a GRAS determination, the agency issues a letter notifying the manufacturer that the US FDA has no questions at the time concerning the company's conclusion that the substance is GRAS under the intended conditions of use. A notification may be revisited if new information indicates a reason for concern (FR 1997).

Acceptable Daily Intake

As part of the evaluation of a food additive, many regulatory bodies establish an Acceptable Daily Intake (ADI) level. JECFA defines the ADI "for man, expressed on a body weight basis, is the amount of a food additive that can be taken daily in the diet, even over a lifetime, without risk" (JECFA 1974, 10). The ADI may be used as a benchmark to evaluate the actual intake of a substance and as an aid in reviewing possible additional uses for a food ingredient. The ADI is expressed in milligrams per kilogram of body weight.

The ADI is a conservative estimate that incorporates a considerable safety factor. It is established from toxicological testing in animals, and sometimes humans, and is usually estimated by applying an intentionally conservative safety factor (generally a 100-fold safety factor). Animal tests are used to determine the maximum dietary level of an additive demonstrating no toxic effects, a "no observable effect level" or NOEL. The NOEL is then used to determine the ADI. For example, if safety evaluation studies of a given substance demonstrate a NOEL of 1000 mg/kg, using a 100-fold safety factor the ADI would be 10 mg/kg body weight per day for humans.

The ADI does not represent a maximum allowable daily intake level. It should not be regarded as a specific point at which safety ends and possible health concerns begin. In fact, the US FDA has said it is not concerned that consumption occasionally may exceed the ADI. The agency has stressed that because the ADI has a built-in safety margin and is based on a chronic lifetime exposure, occasional consumption in amounts greater than the ADI "would not cause adverse effects" (Young 1985, 4).

Conclusion

The demand for low-calorie products continues, as does the demand for ingredients that make them possible. Taste remains king and the approval of multiple alternative sweeteners facilitates the development of an increasing number of good-tasting products. As obesity increases in many parts of the world and as consumers become increasingly aware that "calories still count," the number of successful light products, containing alternative sweeteners, should soar.

References

Code of Federal Regulations (CFR). 2010. Title 21, Sections 170.30, 170.35, US Government Printing Office, Washington, DC, April 1.

Codex Alimentarius Commission (CAC). 2008. Procedural Manual. 18th ed. Joint FAO/WHO Food Standards Programme. Food and Agricultural Organization of the United Nations, Rome.

European Food Safety Authority (EFSA). 2010. http://www.efsa.europa.eu/en/aboutefsa.htm. Accessed November 16, 2010.

Federal Food, Drug, and Cosmetic (FD&C) Act, as amended. 2004. Food. Unsafe food additives (Chapter 4). http://www.fda.gov/RegulatoryInformation/Legislation/FederalFoodDrugandCosmeticActFDCAct/FDCActChapterIVFood/ucm107843.htm. Accessed November 16, 2010.

Federal Register (FR). 1984. Food additives permitted for direct addition to food for human consumption; Aspartame; Denial of Requests for Hearing; Final Rule [Docket 75F-0355 and 82F-0305], February 22, pp. 6672–6682.

Federal Register (FR). 1997. Substances Generally Recognized As Safe [Docket No. 97N-0103], April 17, pp. 10937–10964.

Food Chemicals Codex (*FCC*). 2010. http://www.usp.org/fcc/. Accessed November 16, 2010.

General Standard for Food Additives (GSFA). 2010. Available online at http://www.codexalimentarius.net/gsfaonline/index.html. Accessed November 17, 2010.

International Sweeteners Association (ISA). 2010. http://www.isabru.org/EN/about_sweeteners_sweeteners_directive.asp. Accessed November 15, 2010.

Joint FAO/WHO Expert Committee on Food Additives (JECFA). 1974. Toxicological Evaluation of Certain Food Additives with a Review of General Principles and Specifications, World Heath Organization, Geneva.

Redbook. 2000. Toxicological principles for the safety assessment of direct food additives and color additives used in food. US Food and Drug Administration, Center for Food Safety and Applied Nutrition. http://www.fda.gov/Food/GuidanceComplianceRegulatoryInformation/GuidanceDocuments/FoodIngredientsandPackaging/Redbook/ucm078409.htm. Accessed November 16, 2010.

Rulis, A.M., and Levitt, J.A. 2009. FDA's food ingredient approval process. Safety assurance based on scientific assessment. *Regul Toxicol Pharmacol* 53(1):20–31.

Young, F.E. 1985. Report in response to questions on the sweetener, aspartame. (Report from US Commissioner of Food and Drug to Senator John Heinz).

LOW-CALORIE
SWEETENERS

Chapter 2

Acesulfame Potassium

Christian Klug and Gert-Wolfhard von Rymon Lipinski

Contents

Introduction

Clauss and Jensen in 1967 incidentally discovered a sweet-tasting compound, 5,6-dimethyl-1,2,3-oxathiazin-4(3H)-one 2,2-dioxide, which had a new ring system that had not been previously synthesized (Clauss and Jensen 1973). Systematic research on dihydrooxathiazinone dioxides revealed many sweet-tasting compounds. Variations of substitutions in positions five and six of the ring system noticeably influenced the intensity and purity of the sweetness. All synthesized substances exhibited some sweetness, even those without substitutions on the ring system. The maximum sweetness was found in compounds with short-chain alkyl groups.

Sensory evaluations of the different substances showed that the substitutions on the ring system not only influenced the intensity but also the purity of the sweetness.

In addition to variations of substitutions on the ring system itself, structurally similar compounds were synthesized to investigate whether other variations within the ring system would influence the sweet taste. Evaluations revealed no new sweet-tasting compounds. Even methylation on the nitrogen in the ring furnishes a compound without sweetness.

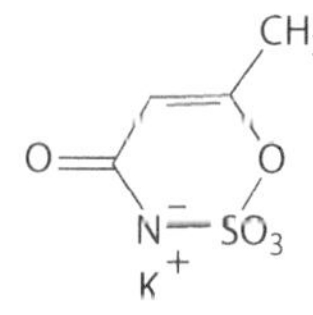

Figure 2.1   Acesulfame K.

Of the compounds, 6-methyl-1,2,3-oxathiazin-4(3H)one-2,2-dioxide exhibited the most favorable taste properties. As production of this compound was less difficult than that of related substances, it was chosen for systematic evaluation of suitability as an intense sweetener for use in foods and beverages (Figure 2.1).

In 1978, the World Health Organization registered acesulfame potassium salt (acesulfame K) as the generic name for this compound.

Production

Dihydrooxathiazinone dioxides can be synthesized from different raw materials using different products and routes. Suitable starting materials are ketones, β-diketones, derivatives of β-oxocarbonic acids, and alkynes that may be reacted with halogen sulfonyl isocyanates. The compounds formed from such reactions are transformed into N-halogen sulfonyl acetoacetic acid amides. In the presence of potassium hydroxide, these compounds cyclize to the dihydrooxathiazinone dioxide ring system. Because dihydrooxathiazinone dioxides are highly acidic compounds, salts of the ring system are formed. The production of acesulfame K requires potassium hydroxide; however, sodium hydroxide or calcium hydroxide can also be used to obtain the respective salts (Clauss and Jensen 1973; Clauss et al. 1976; Arpe 1978; von Rymon Lipinski and Huddart 1983). As starting materials for these production routes are difficult to handle, they were abandoned at an early stage.

Acetoacetamide-N-sulfonic acid, another suitable starting material, cyclizes in the presence of sulfur trioxide to form the dihydrooxathiazinone dioxide ring system, which may be reacted with potassium hydroxide to yield acesulfame K. Again, the production of salts other than the potassium salt, seems possible (Linkies and Reuschling 1990). Continuous production of acesulfame K is possible using this route of synthesis, which allows large-scale production.

A number of patent applications claiming other possible production routes have been published, but it is questionable whether any of these are of commercial importance.

Properties

Acesulfame K is a white, crystalline powder. The originally analyzed crystals are monoclinic of the $P2_1/c$ order. X-ray diffraction demonstrated that the ring system is almost plane, whereas the distances between the single atoms are less than those calculated from the theoretical values. Acesulfame K seems to exist in two polymorphic forms (Paulus 1975; Velaga et al. 2010). The specific gravity of acesulfame K is 1.83 g/cm³.

The shelf life of pure, solid acesulfame K appears to be almost unlimited at room temperature. Samples kept at room temperature for more than six years and either exposed to or protected from

Table 2.1 Solubilities of Acesulfame K

Solvent	Temperature (°C)	g/100 ml Solvent
Water	0	15
Water	20	27
Water	100	~130
Anhydrous ethanol	20	~0.1
Aqueous ethanol 50%	20	~10
Aqueous ethanol 15%	20	~220
Glacial acetic acid	20	30
Sucrose syrup (dry matter 62.5%)	20	≥100
Invert sugar syrup (62.5% dry matter)	20	≥160
Fructose syrup (50% dry matter)	20	≥150
Sorbitol syrup (70% dry matter)	20	≥75
Maltitol syrup (80% dry matter)	20	≥100
Isomalt syrup (25% dry matter)	20	≥250

light showed no signs of decomposition or differences in analytical data compared with freshly produced material (von Rymon Lipinski et al. 1983).

Acesulfame K does not show a definitive melting point. When the product is heated under conditions used for melting point determination, decomposition is normally observed at temperatures well above 200°C. The decomposition limit appears to depend on the heating rate. No decomposition of acesulfame K has been observed under conditions of temperature exposure normally found for food additives. Upon thermal decomposition, acesulfame K yields potassium sulfate as the solid residue (de Carvalho et al. 2009). In contrast to acesulfame K, acesulfame acid has a sharp and definitive melting point at 123°C (Clauss et al. 1976).

Even at room temperature, acesulfame K dissolves readily in water. The solubility at 20°C is approximately 27 g/100 ml of water. This increases to approximately 130 g/100 ml at 100°C. In alcohols, however, acesulfame K is much less soluble. At 20°C, approximately 0.1 g/100 ml dissolves in anhydrous ethanol. In mixtures of alcohol and water, the solubility increases with rising water content. Approximately 10 g/100 ml dissolves in 50% ethanol (v/v) (Clauss et al. 1976). Aqueous solutions of acesulfame K are almost neutral (Table 2.1).

In view of the temperature/solubility ratio of acesulfame K solutions in water, the product can be easily purified by recrystallization. High-purity acesulfame K can therefore be produced on a technical scale while meeting the purity requirements for food additives.

Sensory Properties

Acesulfame K exhibits approximately 200 times the sweetness of a 3% sucrose solution, although slightly higher values have been reported (Hoppe and Gassmann 1985a). The sweetness intensity depends on the concentrations of the sucrose solution to which it is compared. At the threshold level the intensity is much greater and decreases with increasing sucrose concentrations to values from 130 per 100 times the sucrose value. Normally, acesulfame K can be considered to be approximately half as sweet as sodium saccharin, similarly sweet as aspartame, and four to five times sweeter than sodium cyclamate. In acid foods and beverages for the same concentrations, a slightly higher sweetness may be perceived compared to neutral solutions.

The sweet taste of acesulfame K is perceived quickly and without unpleasant delay. Compared with other intense sweeteners (aspartame, alitame) acesulfame K has a faster sweetness onset (Ott et al. 1991). The sweet taste is not lingering and does not persist longer than the intrinsic taste of food. In aqueous solutions with high concentrations of acesulfame K, a bitter taste can sometimes be detected, which increases with increasing concentration (Schiffman et al. 1995). This results from activation not only of the T1R3 receptor responsible for sweetness perception but also of the TRPV1, hTAS2R43, and hTAS2R44 receptors mediating bitterness (Riera et al. 2007; Riera et al. 2008; Kuhn et al. 2004). In foodstuffs with lower concentrations of acesulfame K, this effect is not of great importance (Hoppe and Gassmann 1985b). As for all intense sweeteners, however, assessments of different taste characteristics depend on the product in which the sweeteners are used (Paulus and Braun 1988). In addition, recent studies revealed antagonists to the bitter taste receptors which may be able to reduce bitterness, should it be perceived at all (Slack et al. 2010). It was further observed that the sweetness of acesulfame K solutions does not decrease with rising temperatures to the extent of other intense sweeteners (Hoppe and Gassmann 1980).

A strong synergistic taste enhancement was noted in mixtures of acesulfame K and aspartame, sodium cyclamate or sucralose, whereas only slight taste enhancement was perceived in mixtures of acesulfame K and saccharin (Ayya and Lawless 1992). Acesulfame K and aspartame taste approximately 100 times sweeter or less than sucrose in a solution containing 100 g/l. A blend of the sweeteners, however, tastes more than 300 times sweeter than the same sucrose solution, which means a quantitative synergism in the range of 40%–50%. In tertiary blends of acesulfame K, aspartame, and cyclamate the synergistic sweetness enhancement may be as high as 90%. This effect cannot be explained from the individual sweetness intensity curves and should therefore be attributable to synergism (Frank et al. 1989; von Rymon Lipinski and Klein 1988). Although the strongest taste enhancement was found for blends of acesulfame K and aspartame of around 1:1 (w/w), pronounced synergism was also observed for other blend ratios. Acesulfame K exhibits synergism when combined with many of the other intense or nutritive sweeteners, including sucralose, high-fructose corn syrup, thaumatin, and fructose.

Although acesulfame K can be used as an intense sweetener by itself and does not show particular taste problems if used in appropriate concentrations, mixtures with other sweetening agents are of much practical interest (Portman and Kilcast 1998; Hanger et al. 1996; Gelardi 1987). Studies showed that, in particular, mixtures of acesulfame K with various other intense sweeteners had especially desirable taste properties. This qualitative improvement of taste seems to be caused by an addition of the time-intensity profiles of the individual sweeteners. Thus, combinations balance the characteristics of the sweeteners used in such blends (van Tournot et al. 1985). Mixtures of acesulfame K and aspartame, acesulfame K and sodium cyclamate, acesulfame K and/or sucralose, and sometimes other mixtures in which both sweetening agents almost equally contributed to the sweetness of the mixture were favorably viewed. Mixtures of intense sweeteners containing acesulfame K are closer in taste to sucrose than the single sweeteners. In particular, acesulfame K/aspartame and acesulfame K/aspartame/saccharin/cyclamate mixtures are similar to sucrose for taste attributes known to cause differences among sweeteners (sweet or bitter side tastes or aftertaste). The lingering sweetness of aspartame and sucralose was substantially reduced when blending either of the sweeteners with acesulfame K. Blends consisting of two or three different sweeteners on repeated presentation exhibited less reduction in sweetness intensity over four repeated sips than a single sweetener at an equivalent sweetness level. Tertiary combinations tended to be slightly more effective than binary combinations at lessening the effect of repeated exposure to a given sweet stimulus. These findings suggest that the decline in sweetness intensity experienced over repeated exposure to a sweet stimulus could be reduced by the blending of sweeteners (Schiffman et al. 2003).

More recent studies on interactions of sweeteners with flavors show that the sweetness quality can substantially differ from flavor to flavor. For acesulfame K such data were generated for cola, orange, peach, strawberry, and lemon-lime flavors as well as for coffee and tea (Saelzer 2004; Rathjen 2003; Meyer 2000; Meyer 2001).

Mixtures of acesulfame K with polyols or sugars produced favorable taste reports. Mixtures of acesulfame K with sugar alcohols have a full and well-balanced sweetness and are therefore particularly suitable for sugarless confectionery, fruit preparations, and other foods that require a bulking agent. The rounding effect seems attributable to the fast onset of the sweetness of acesulfame K. When acesulfame K is combined with bulk sweeteners (e.g., maltitol), the mixture provides a taste closer to sucrose, because nonsweet taste and aftertaste versus sucrose are significantly reduced. Suitable blend ratios of acesulfame K with the bulk sweeteners are approximately 1:100 in a mixture with xylitol, 1:150 in a mixture with maltitol, 1:150–200 in a mixture with sorbitol, and 1:250 in a mixture with isomalt (von Rymon Lipinski 1985).

Blending acesulfame K with other intense or bulk sweeteners can therefore help optimize the flavor profile of a food or beverage as blends taste more like sucrose. Blends of acesulfame K and fructose or high-fructose corn syrup in particular are found to be similar to sucrose in sweetened beverages (Rathjen 2005). In combination with complex carbohydrates like oligofructose or inulin, not only the sweetness of blends of acesulfame K with other sweeteners but also the mouthfeel are improved (Wiedmann and Jager 1997).

Toxicology

Because toxicological evaluations of intense sweeteners are of crucial importance for approval and subsequent use, a full range of toxicological studies was carried out with acesulfame K.

The acute oral toxicity of acesulfame K is so low that it can be regarded as practically nontoxic. The LD_{50} was orally determined at 6.9–8.0 g/kg body weight. The intraperitoneal LD_{50} is 2.2 g/kg body weight (Anonymous 1991). Subchronic toxic effects were investigated in a 90-day study with rats. The animals were fed concentrations of 0%–10% acesulfame K in the diet. Potential carcinogenicity and chronic toxicity were studied in rats fed up to 3% acesulfame K in the diet. A carcinogenicity study was conducted in mice fed concentrations of up to 3% acesulfame K and chronic toxicity effects of acesulfame K were studied in beagle dogs for 2 years (Anonymous 1991). Feeding studies in genetically modified mice intended to be used as a model in carcinogenicity studies with up to 3% acesulfame K in the diet had no effect on survival, mean body weights, or feed consumption. It was concluded that under the conditions of the studies, there were no neoplasms, nonneoplastic lesions, or evidence of carcinogenic activity attributed to exposure to acesulfame K (Anonymous 2005).

No mutagenicity was found in several respective studies. Among such studies, there were a dominant lethal test, a micronucleus test, bone marrow investigations in hamsters, tests for malignant transformation, DNA binding, and others (Anonymous 1991). A claimed dose-response relationship for chromosome aberrations could not be confirmed in a cross-check of the slides of the study or in a repeated study under the same conditions (Selzer Rasmussen 1999; Voelkner 1998a, 1998b; Mukherjee and Chakrabarti 1997).

The toxicological studies on acesulfame K demonstrated that the compound would be safe for use as an intense sweetener. This view was confirmed by the Joint Expert Committee on Food Additives (JECFA) of the Food and Agriculture Organization (FAO) and World Health Oraganization (WHO), which concluded that the data showed acesulfame K to be neither mutagenic nor carcinogenic. Therefore, an acceptable daily intake (ADI) of 0–9 mg/kg of body weight was first allocated and later increased to 0–15 mg/kg (Anonymous 1991). On the basis of a detailed evaluation of the available animal studies, the United States Food and Drug Administration (US FDA) also allocated an ADI of up to 15 mg/kg of body weight (Anonymous 1998). The Scientific Committee for Foods of the EU published an assessment stating that long-term studies did not show any dose-related increase in specific tumors or any treatment-related pathological changes of significance. Therefore, an ADI of 0–9 mg/kg of body weight was allocated by the EU (Anonymous 1985) using a dog study with a safety factor of 100 and not using a rat study with a safety factor of 100, as done by JECFA and the US FDA.

Metabolism and Physiological Characteristics

Acesulfame K is not metabolized by the human body. To investigate possible metabolic transformations, ^{14}C-labeled acesulfame K was studied in rats, dogs, and pigs. Because animal studies did not show any metabolism, human volunteers were also given labeled acesulfame K. The different animal species, as well as the human volunteers, excreted the original compound. No activity attributable to metabolites was found (Anonymous 1991).

Because acesulfame K is excreted completely unmetabolized, it does not have any caloric value. In conjunction with the metabolic studies, the pharmacokinetics of acesulfame K were also investigated. These studies were carried out in rats, dogs, pigs, and later in human volunteers. All animal species, as well as humans, quickly absorbed acesulfame K, but there was rapid excretion of the compound, mainly in the urine. A multiple-dose study showed no accumulation in tissues (Anonymous 1991). Serum determination of acesulfame K can be performed by high-performance liquid chromatographic (HPLC) analysis. No activity attributable to metabolites was found. After prolonged exposure to acesulfame K, animals did not show any sign of induced metabolism. Again, after administration of ^{14}C-labeled acesulfame K, only the original substance was found in the excreta (Anonymous 1991).

The secretion of insulin and the blood glucose levels are apparently not influenced by acesulfame K under conditions of normal food consumption (Steiniger et al. 1995; Haertel and Graubaum 1993). Therefore studies carried out in isolated pancreas cells or under laboratory conditions may not be of practical importance, especially as concentrations used in these studies are higher than those resulting from oral ingestion (Fujita et al. 2009; Malaisse et al. 1998; Liang et al. 1987a, 1987b).

Acesulfame K is not metabolized by bacteria including *Streptococcus mutans* and other microorganisms that may contribute to the formation of caries. Acesulfame K was tested in several studies. Although an inhibition of dental plaque microorganisms or *S. mutans* was not always reported, in other test systems a clear inhibition was demonstrated (Linke 1983). Synergism in the inhibition of bacteria was observed in mixtures of intense sweeteners (Siebert et al. 1987) or mixtures of acesulfame K, saccharin, and fluoride (Brown and Best 1988). Although in these studies concentrations higher than those for customary sweetness levels were used, lack of acidogenicity was also demonstrated at levels of practical importance (Park et al. 1995).

Stability and Reactions in Foods

Long-term and heat stability are important factors for the use of intense sweeteners in many food products and in beverages. In this regard, various conditions have to be met. In foods and beverages, pH levels vary from neutral to the acid range and may, in extreme cases like certain soft drinks, decrease to values of pH 3.0 or less. In this wide pH range, even after prolonged storage, no decrease of sweetness intensity is detected.

In aqueous media, acesulfame K is distinguished by very good stability. After several months of storage at room temperature, virtually no change in acesulfame K concentration was found in the pH range common for beverages. Prolonged continuous exposure to 30°C, a condition that will hardly be found in practice, does not cause losses exceeding 10%, the threshold for recognition of sweetness differences (Hoppe and Gassmann 1985). Even at temperatures of 40°C, the threshold for detection of sweetness differences is exceeded after several months only for products having pH 3.0 or less (von Rymon Lipinski 1988a).

Extensive studies were performed with buffered aqueous solutions. Results for pH levels and storage conditions commonly found for soft drinks are given in Table 2.2. After 10 years of storage of a solution buffered to pH 7.5 at room temperature, no significant loss of acesulfame K was detected. The half-lives at 20°C were determined to be 11.5 years at pH 5.77 and 6.95 years at pH 3.22 (Coiffard et al. 1997).

Acesulfame K-containing beverages can be pasteurized under normal pasteurization conditions without loss of sweetness. Pasteurizing for longer periods at lower temperatures is possible, as is

Table 2.2 Stability of Acesulfame K in Buffered Aqueous Solutions

	Recovery (%)	
Storage Time (Weeks)	pH 3.0	pH 3.5
20°C		
16	98	98
30	98	99
50	98	99
100	95	98
30°C		
16	97	100
30	95	97
50	91	96

Note: Acesulfame K is considered to be highly temperature stable (Coiffard et al. 1997).

short-term pasteurization for a few seconds at high temperatures. Sterilization is possible without losses under normal conditions (i.e., temperatures at approximately 100°C for products having lower pH levels and 121°C for products around and greater than 4). In a solution of pH 4.0, which was heated to 120°C for 1 hour, no loss of acesulfame could be measured. Half-life values determined at 100°C demonstrate that the common treatments of foods and beverages should not cause any substantial decomposition of acesulfame K (von Rymon Lipinski 1989).

Under Ultra high temperature (UHT) (Lotz et al. 1992) and microwave (Korb et al. 1992) treatment, acesulfame K is stable. In baking studies, no indication of decomposition of acesulfame K was found even when biscuits with low water content were baked at high oven temperatures for short periods (Klug et al. 1992). This corresponds to the observation that acesulfame K decomposes at temperatures well above 200°C.

Potential decomposition products of acesulfame K can therefore only be found under extreme conditions. Under such conditions, compounds of hydrolytic decomposition are mainly acetone, CO_2, ammonium salts, sulfate, and amidosulfonate. In the hydrolytic decomposition, the ring system is initially opened, which quickly yields the end products of hydrolysis. Only occasionally, and then in very acidic media, can traces of derivatives of acetoacetic acid be detected (Arpe 1978).

Applications

Acesulfame K can be used as a sweetening agent in a wide range of products, for instance in low-calorie products, diabetic foods, sugarless products, oral hygiene preparations, pharmaceuticals, and animal feeds. Many of the product groups acesulfame K has been used in are shown in Table 2.3.

Acesulfame K is suitable for low-calorie and diet beverages because of its good stability in aqueous solutions even at low pH typical of diet soft drinks. If used by itself, acesulfame K can impart sweetness comparable to 8%–10% sucrose, but mixtures of acesulfame K with other intense sweeteners are more predominantly used because of the sucrose-like taste these blends provide. In countries where mixtures of intense sweeteners with other sweetening agents are permitted, mixtures of acesulfame K with fructose, glucose, high-fructose corn syrup, or sucrose can be used. Generally, beverages containing such mixtures of acesulfame K with bulk sweeteners are rated to be fuller bodied because of the slightly higher viscosity and the different taste profiles of sugars and acesulfame K. A substantial number of low-calorie beverages, however, are sweetened with mixtures of acesulfame K and aspartame or other intense sweeteners. These beverages benefit from the synergism and the improved taste characteristics provided by such blends. For example, a sweetness level equivalent to approximately 10% of sucrose is imparted by concentrations in a range of 500–600 mg/l of acesulfame K or aspartame, whereas the same sweetness level can be achieved by using a blend of only 160–180 mg/l of each of the sweeteners, depending on the flavor. Blend ratios between acesulfame K and aspartame, in particular, may be different in beverages having different flavors to match flavor and sweetness profiles (Saelzer 2004; Rathjen 2003; Meyer 2000, 2001). Of particular importance in beverages is sweetener stability, and the stability of acesulfame K at lower pH increases sweetness retention in beverages versus sweeteners that have less stability (e.g., aspartame). Stable sweetness with good taste quality is achieved in acesulfame K/sucralose blends.

When blending acesulfame K with other intense sweeteners for beverage applications, the blend ratio may depend on different factors, including the flavor or flavor type. Blends of acesulfame K and aspartame (40:60) in orange-flavored beverages have been noted to have time intensity curves of sweetness and fruitiness similar to sucrose-sweetened beverages (Matysiak and Noble 1991).

Table 2.3 Applications of Acesulfame K

Tabletop sweeteners
Carbonated soft drinks
Noncarbonated soft drinks
Squashes and dilutables
Fruit nectars
Ice teas
Herbal infusions
Ciders
Powdered beverages
Instant coffees and teas
Flavored milk and whey
Cocoas
Yoghurts
Desserts
Rice puddings
Ice creams
Cakes
Cookies
Jams
Fruit purees
Hard candies
Soft candies
Gum confections
Chocolate confections
Marzipans
Breath mints
Chewing gums
Pickled vegetables
Marinated fish
Slimming diets
Dietary supplements
Toothpastes
Mouthwashes
Pharmaceuticals

In raspberry-flavored beverages containing natural flavors, 40/60 to 25/75 (acesulfame K/aspartame) blend ratios are considered optimum, whereas beverages with artificial raspberry flavors were found best with blend ratios of 50/50 to 20/80 (Baron and Hanger 1998). Optimal taste profiles when using intense sweeteners are considered to have high fruit flavor and minimum side tastes or aftertastes, and sweetener mixtures containing acesulfame K exhibit these properties. It was also demonstrated that green and leafy flavors showed a greater volatility in beverages sweetened with an acesulfame K/aspartame blend compared to sucrose-based beverages (King et al. 2006). When only single sweeteners are substituted for sucrose in beverages, flavor problems are encountered and it has been noted that mixtures of sweeteners can minimize these flavor problems (Meyer 2001; Meyer 2000; Nahon et al. 1996). Most notable are mixtures including acesulfame K, which produce taste profiles similar to sucrose. Some recommended blend ratios for sugar-free and reduced-sugar beverages are given in Tables 2.4 and 2.5. Optimal blend ratios complement the taste profile of the specific flavor of the beverage and are, however, best determined by sensory testing of various blends.

Developments in beverages over recent years include replacement of sugar in only sugar-containing beverages with intense sweeteners like acesulfame K. In regions of the world where sugar is expensive or has inconsistent quality, intense sweeteners are used in combination with

Table 2.4 Recommended Acesulfame K Levels in Sugar-Free Beverages (10% Sucrose Equivalence)

Flavor	ACK (mg/l)	APM (mg/l)	SUC (mg/l)	SAC (mg/l)	CYC (mg/l)
Cola	110	260	—	—	—
	97	—	146	—	—
	194	97	36	—	—
	100	100	—	—	280
Lemon-lime	110	257	—	—	—
	168	168	—	—	—
	175	—	100	—	—
	115	125	55	—	—
	100	100	—	—	280
	65	65	—	40	280
Orange	109	258	—	—	—
	175	—	100	—	—
	110	121	53	—	—
	100	100	—	—	280
Strawberry	179	179	—	—	—
	175	—	100	—	—

Note: *ACK* Acesulfame K; *APM* Aspartame; *SUC* Sucralose; *SAC* Sodium saccharin; *CYC* Sodium cyclamate.

Table 2.5 Recommended Acesulfame K/Aspartame Blend Levels in Sugar-Free Beverages (10% Sucrose Equivalence)

Sugar replacement (%)	50	50	65	65	80	80
High-fructose corn syrup (solids, g/l)	50	—	35	—	20	—
Sucrose (g/l)	—	50	—	35	—	35
Acesulfame K (mg/l)	80	76	98	88	115	108
Aspartame (mg/l)	80	76	98	88	115	108

sucrose to sweeten many different beverages. Stability in warm climates for longer storage conditions and stability at low pH are necessary for sweeteners that can replace sugar. Acesulfame K exhibits excellent properties for use in this type of application.

Producing jams and marmalades with intense sweeteners only is difficult. Bulking agents are important to improve the texture and shelf life of such products. In trials for the production of jams with acesulfame K, highly acceptable preparations were made with sorbitol as the bulking material. Other sugar alcohols and polydextrose also proved to be suitable. In combinations with acesulfame K, the concentration of sorbitol or other bulking agents can be reduced, yielding products with a noticeable reduction in caloric values compared to sucrose-containing products. These jams and marmalades are less protected from microbial spoilage compared to sucrose-containing products. In view of the low concentration of osmotically active compounds, the addition of preservatives can help to avoid such microbial spoilage. The flavor stability of jams and marmalades containing low levels of dry solids, however, is normally lower than the storage stability of standard products (Hoerlein et al. 1995).

Acesulfame K is suitable for canned fruit with higher pH values, which are normally sterilized at 121°C, as well as for products with lower pH values, which are normally sterilized at lower temperatures. Owing to their good temperature stability, blends of acesulfame K and sucralose are especially suitable for the production of sugar-free and sugar-reduced canned fruit (Saelzer 2005a).

Confectionery items can be made with acesulfame K if suitable bulking ingredients or bulk sweeteners are added to give the necessary volume. As the sweetness intensity of most sugar alcohols and bulking ingredients is lower than the sweetness of sucrose, acesulfame K imparts the desired sweetness, whereas sugar alcohols or low-calorie bulking ingredients provide the necessary bulk and texture. Combinations of acesulfame K with sorbitol can have a good taste pattern. Gum confections containing sorbitol instead of sucrose and, in addition, 1000 mg/kg acesulfame K were produced without difficulties and showed a good, fruity, sweet taste. Hard-boiled candies can be manufactured using acesulfame K as the intense sweetener and isomalt, maltitol, or lactitol as the bulking ingredient. Acesulfame K rounds the sweetness of these sugar alcohols and brings the taste to that of standard, sugar-containing products. Such combinations are especially compatible with fruit flavors. Because of the good temperature stability of acesulfame K, it can be added before cooking if no acids are added. Alternatively, it can be added together with acids, flavorings, and colorants after cooking. In soft candies, acesulfame K shows good shelf stability. Again, suitable blends of polyols and acesulfame K allow the production of sugarless soft candy coming close to standard products in taste and texture. Sugarless marzipan having good taste, texture, and shelf stability can be manufactured using acesulfame K as the intense sweetener and isomalt as a bulking ingredient, with some addition of sorbitol to soften the texture. In chocolate and related products, acesulfame K can be added at the beginning of the production process (e.g., before rolling). It withstands all treatments including conching without detectable decomposition (von Rymon Lipinski and Gorstelle 1989). Starch-based confectionery items also benefit from the addition of acesulfame K because they withstand the extrusion temperatures while adding sweetness, as these types of products are limited in the amount of sugar that can be incorporated.

Bulking agents with a sweet taste like sorbitol are generally used for the production of sugarless chewing gums. The low sweetness of most polyols is normally enhanced by intense sweeteners. The fast onset of sweetness of acesulfame K is beneficial in forming the initial taste. Therefore, acesulfame K-containing chewing gum has a pleasant, sweet taste from the beginning. Because of its good solubility, acesulfame K may be dissolved fairly quickly by the saliva. Prolonged sweetness may be achieved by encapsulation of some of the acesulfame K or acesulfame K in combinations with other sweetening ingredients. In sugar-based chewing gum, acesulfame K can enhance some flavors and contribute to a longer flavor perception at levels of a few hundred milligrams per kilogram. In recipes using acesulfame K, it can either be blended into the chewing gum mass as fine crystals or dissolved in sorbitol syrup.

In reduced-calorie baked goods, bulking agents like polydextrose substitute for sugar and flour and may help to reduce the level of fats. Acesulfame K combines well with suitable bulking ingredients and bulk sweeteners and therefore allows the production of sweet-tasting baked goods with fewer calories. In diabetic products, combinations of acesulfame K and polyols like isomalt, lactitol, maltitol, or sorbitol can provide volume and sweetness. Texture and sweetness intensity can be similar to those of sucrose-containing products. Although lower sweetness for baked compared with unbaked goods was reported in one study (Redlinger and Setser 1987), extensive work on acesulfame K in baked goods did not give any indication of losses during baking. Even after prolonged exposure to elevated temperatures or elevated oven temperatures, no statistically significant deviation from the added concentrations was analyzed (Klug et al. 1992). Sensory studies showed good and pleasant sweetness for baked goods containing acesulfame K alone or in combination with bulking ingredients or bulk sweeteners. Reduced-calorie baked goods (muffins and frostings) sweetened partially with acesulfame K were found to be equal to full calorie products in sensory testing. Also, sweetener combinations of acesulfame K (with sucralose, but also with aspartame/saccharin or aspartame alone) in shortbread cookies were found to have sweetness time intensity profiles similar to those of sucrose (Lim et al. 1989).

Acesulfame K can easily be used in fruit-flavored dairy products. It withstands pasteurization of fruit preparations and pasteurization of the product itself (Lotz et al. 1992). For nonfermented products like flavored milk, cocoa beverages, and similar products, the use of acesulfame K as the single sweetener or in combination with sucralose is advisable because good heat and shelf life stability is required. In certain countries where cyclamate is approved, blends of acesulfame K and cyclamate are used in this application, too. Whenever temperature stability is not very important, particularly well-rounded taste profiles are obtained from blends of acesulfame K and aspartame, for example, in fruit-flavored yogurt and yogurt analog products. Acesulfame K is apparently not attacked by lactic acid bacteria and other microorganisms used in fermented milk products (von Rymon Lipinski 1988). In ice cream and analog products, the modification of freezing and crystallization of water has to be achieved by the use of appropriate levels of sugar substitutes. Again, acesulfame K can round and enhance the sweetness because it combines particularly well with the taste of polyols (von Rymon Lipinski 1988b).

In addition to the food categories discussed in detail, many other products can be sweetened with acesulfame K; examples are shown in Table 2.3.

Acesulfame K finds several applications in delicatessen products like pickles, ketchup, sauces, dressings, and delicatessen salads. Production without preservatives requires low pH levels and acesulfame K can mellow the harsh taste of vinegar at the concentrations normally used. One more advantage compared to sugar is that microorganisms cannot use acesulfame K for energy and, therefore, not for growth. Combinations of acesulfame K or blends of acesulfame K and sucralose together with polydextrose, modified starches, or other thickeners allow production of calorie-reduced sauces and ketchup (Saelzer 2005b).

The production of tabletop preparations containing acesulfame K is similar to those containing other intense sweeteners. Effervescent tablets, granules, powder, and solutions have been produced. Because of the good solubility of acesulfame K in water, highly concentrated solutions suitable for household use can be manufactured. Similarly, no problems have been reported in the dissolution of tablets or powders. Solid tabletop preparations are generally used in hot beverages and the solubility of acesulfame K is extremely good at such elevated temperatures. Depending on the field of application, it may be reasonable to develop specific blend ratios. Acesulfame K/aspartame blends (either 50/50 or 30/70) are well suited for tea, as they provide a round, well-balanced tea flavor, while an acesulfame K/sucralose blend (55/45) is recommended for coffee (Saelzer 2004).

Outside the food sector, acesulfame K can be used to sweeten oral hygiene products, pharmaceuticals, tobacco products, and animal feedstuffs. In cosmetics, such as toothpaste and mouthwash preparations, and in similar products, some ingredients (e.g., surfactants) tend to impart a bitter taste. Ordinarily, special flavors and sweeteners are needed to mask such taste and to produce an initial pleasant flavor. Because the sweetness of acesulfame K is perceived quickly, it is particularly suitable for these oral hygiene products. It is compatible with commonly used flavoring agents in toothpastes and mouthwash preparations. If the flavor concentration is not crucial for the taste character of the cosmetic product, the fast onset of the sweetness of acesulfame K requires low concentrations of flavoring ingredients. As mentioned earlier, acesulfame K is highly compatible with sorbitol, which is often used in toothpaste. If glycerol is used as a humectant and brightener, no solubility problems are incurred. Problems have not been encountered when acesulfame K is dissolved in mouthwash formulations containing alcohol, despite the low solubility of acesulfame K in pure ethanol. In mixtures of ethanol and water, the solubility of acesulfame K is normally greater than concentrations used in mouthwash preparations (Schmidt et al. 1998; von Rymon Lipinski and Lueck 1981, von Rymon Lipinski et al. 1976).

Pharmaceuticals sometimes have unpleasant flavor or taste characteristics; again, acesulfame K can mask such bitter taste patterns. Acesulfame K-containing silver and platinum complexes were tested for cytotoxicity with interesting results. The platinum complex showed good activity against two types of the dengue virus. The silver complex presented activity against *Mycobacterium tuberculosis*, and Gram-negative (*Escherichia coli* and *Pseudomonas aeruginosa*) and Gram-positive (*Enterococcus faecalis*) microorganisms (Cavicchioli et al. 2010).

Outside the food sector acesulfame K is used as an experimental anion in ionic liquid. Its use would have the advantage that the anion would be derived from a safe food additive (Carter et al. 2004).

Analytical Methods

Thin-layer chromatographic detection of acesulfame K is possible (von Rymon Lipinski and Brixius 1979).

Quantitative determinations can be carried out by liquid chromatographic methods and iso-tachophoresis. Because the volatility of acesulfame K is low and methylation produces differing ratios of methyl derivatives, the quantitative determination by gas chromatography is impossible. Apart from the direct spectroscopic determination of acesulfame K at 227 nm, no spectroscopic methods for the determination are available.

Liquid chromatographic determination of acesulfame K in foods is simple, as beverages or aqueous extracts from foods often can be injected onto the columns immediately after filtration. In case of difficulties, clarification by addition of zinc sulfate and potassium is advisable. The methods described are basically reversed-phase separations with detection in ultraviolet (UV) light (Hagenauer-Hener et al. 1986) coupled with electrospray ionization mass spectrometric detection (ESI-MS) (Yang and Chen 2009) or with evaporative light scattering detection (HPLC-ELSD) (Wasik and Buchgraber 2007). Publications on liquid chromatographic analyses of acesulfame K are numerous. The same liquid chromatographic procedures can be used in assaying feeds, which should also be clarified by adding zinc sulfate and potassium hexacyanoferrate(II).

Acesulfame K and other sweeteners including cyclamate can be detected by a conductivity detector (Biemer 1989). Because conductivity detectors are used in isotachophoresis, they can also be used for separation of acesulfame K from other food ingredients and intense sweeten-ers. Quantitative determination of acesulfame K in the presence of other sweeteners has been described (Hermannova et al. 2006).

Most methods described allow the separation of acesulfame K from other intense sweeteners that may be used in combination.

A simple determination of acesulfame K and aspartame in sweetener tablets is possible by Fourier transform middle-infrared (FTIR) spectrometry (Armenta et al. 2004).

Regulatory Status

The completion of a comprehensive safety evaluation program paved the way for the approval of acesulfame K as a food additive.

The evaluation of the data by the Joint Expert Committee for Food Additives of the WHO and FAO (JECFA) resulted in approval for food use with the allocation of an ADI of 0–15 mg/kg of body weight (Anonymous 1991). In addition, specifications were published that were revised later and are now published in JECFA's Online Compendium of Food Additive Specifications (JECFA 2010). The Scientific Committee for Foods of the EU performed another evaluation of the safety of acesulfame K. Again, acesulfame K was approved for food use, and an ADI of 0–9 mg/kg of body weight was allocated and later confirmed (Anonymous 1985).

After the favorable assessments published by the WHO and FAO and the Scientific Committee for Foods of the EU, more than 100 countries have approved the use of ace-sulfame K in at least some products. They include the United States, all member states of the European Union, Canada, Japan, Australia, and New Zealand. In the United States former approvals for certain product groups have been replaced by approval as a "gen-eral purpose sweetener" for use following Good Manufacturing Practice (Anonymous 2003). The EU approves use of acesulfame K in 43 product categories with different maxi-mum levels (Anonymous 1994). Acesulfame K monographs are included in the European Pharmacopoeia and the US National Formulary.

It is sometimes claimed that the intake of acesulfame K may exceed the ADI. While such claims are based on calculations assuming use of acesulfame K at the approved maximum in all product categories for all food consumed, several studies are based on food diaries. One especially reliable study used the fact that acesulfame K is excreted unchanged to monitor intake over 24 hours in children. It showed intake of a fraction of the ADI even in small children (Wilson et al. 1999). All studies with a reliable view of the consumption of acesulfame K show intakes well below the ADI in all groups of the population (Renwick 2006).

References

Anonymous. 1985. *Report of the Scientific Committee for Food No. 16 - Sweeteners.* Luxembourg: Commission of the European Communities.

Anonymous. 1991. Acesulfame K. In *Toxicological evaluation of certain food additives and contaminants*, 183–218. Geneva: WHO Food Additives Series No. 28.

Anonymous. 1994. Directive 94/35/EC of 30 June 1994 on sweeteners for use in foodstuffs. *Official Journal L* 237:3–12 (with subsequent amendments).

Anonymous. 2003. Food additives permitted for direct addition to food for human consumption; acesulfame potassium. *Fed Regist* 68(250):75411–75413.

Anonymous. 2005. *NTP Report on the Toxicity Studies of Acesulfame Potassium (CAS No. 55589-62-3) In FVB/N-TgN(v-Ha-ras)Led (Tg.AC) Hemizygous Mice and Carcinogenicity Studies of Acesulfame Potassium in B6.129-Trp53tm1Brd (N5) Haploinsufficient Mice.* Research Triangle Park: National Toxicology Program NTP GMM 2, NIH Publication No. 06-4460.

Armenta, S., Garrigues, S., and de la Guardia, V. 2004. FTIR determination of aspartame and acesulfame-K in tabletop sweeteners. *J Agric Food Chem* 52(26):7798–7803.

Arpe, H.-J. 1978. Acesulfame K, a new noncaloric sweetener. In *Health and sugar substitutes. Proceeding of the ERGOB conference, Geneva 1978*, ed. B Guggenheim, 178–183. Basel: Karger.

Ayya, N., and Lawless, V. 1992. Quantitative and qualitative evaluation of high intensity sweeteners and sweetener mixtures. *Chem Senses* 17:245–259.

Baron, R., and Hanger, V. 1998. Using acid level, acesulfame potassium/aspartame blend ratio and flavor type to determine optimum flavor profiles of fruit flavoured beverages. *J Sens Studies* 13:269–283.

Biemer, T.A. 1989. Analysis of saccharin, acesulfame K and sodium cyclamate by high-performance ion chromatography. *J Chromatog* 463:463–468.

Brown, A.T., and Best, V. 1988. Apparent synergism between the interaction of saccharin, acesulfame K and fluoride with hexitol metabolism by *Streptococcus mutans*. *Caries Res* 22:2–6.

Carter, E., Culver, S., Fox, P., Goode, R.D., Ntai, I., Tickell, M.D., Traylor, R.K., Hoffman, N.W., and Davis, J.H. 2004. Sweet success: Ionic liquids derived from non-nutritive sweeteners. *Chem Commun* 2004:630–631.

Cavicchioli, M., Massabni, A.C., Heinrich, T.A., Costa-Neto, C.M., Abrao, E.P., Fonseca, B.A., Castellano, E.E., Corbi, P.P., Lustri, W.R., and Leite, C.Q. 2010. Pt(II) and Ag(I) complexes with acesulfame: Crystal structure and a study of their antitumoral, antimicrobial and antiviral activities. *J Inorg Biochem* 104(5):533–540.

Clauss, K., Lück, E., and von Rymon Lipinski, G.-W. 1976. Acetosulfam, ein neuer Suessstoff. *Z Lebensm Unters Forsch* 162:37–40.

Clauss, K., and Jensen, H. 1973. Oxathiazinon dioxides - a new group of sweetening agents. *Angew Chem* 85:965–973; *Angew Chem (Int Ed)* 12:869–876.

Coiffard, C., Coiffard, L.M.J., and de Roeck-Holtzbauer, Y. 1997. Influence of pH on the thermodegradation of acesulfame K in aqueous diluted solutions. *S T P Pharma Sciences* 7(5):382–385.

de Carvalho, L.C., Pinotti Segato, M., Spezia Nunes, R., Novak, C., Tadeu, E., and Cavalheiro, G. 2009. Thermoanalytical studies of some sweeteners. *J Therm Anal Cal* 97(1):359–365.

Frank R.A., Mize, S.J., and Carter, R. 1989. An assessment of binary mixture interactions for nine sweeteners. *Chem Senses* 14:621–632.

Fujita, Y., Wideman, R.D., Speck, M., Asadi, A., King, D.S., Webber, T.D., Haneda, M., Timothy J., and Kieffer, T.J. 2009. Incretin release from gut is acutely enhanced by sugar but not by sweeteners in vivo. *Am J Physiol* 296(3, Pt. 1):E473–479.

Gelardi, R.C. 1987. The multiple sweetener approach and new sweeteners on the horizon. *Food Technol* 41(1):123–124.

Haertel, B., and Graubaum, H.-J. 1993. Einfluss von Suessstoff-Loesungen auf die Insulinsekretion und den Blutglukosespiegel. *Ernaehrungsumschau* 4:152–155.

Hagenauer-Hener, U., Frank, C., Hener, U., and Mosandl, A. 1986. Bestimmung von Aspartame, Acesulfame K, Saccharin, Coffein, Sorbinsaeure und Benzoesaeure in Lebensmitteln mittels HPLC. *Dtsch Lebensm Rdsch* 86:348–351.

Hanger, L.Y., Lotz, A., and Lepeniotis, S. 1996. Descriptive profiles of selected high intensity sweeteners (HIS), HIS blends and sucrose. *J Food Sci* 61:456–458.

Hermannova, M., Krivankova, L., Bartos, M., and Vytras, K. 2006. Direct simultaneous determination of eight sweeteners in foods by capillary isotachophoresis. *J Separation Sci* 29(8):1132–1137.

Hoerlein, L., Lotz, A., and Gierschner, K. 1995. Herstellung von brennwertverminderter diabetikergeeigneter Erdbeerkonfitüre unter Verwendung der beiden Suessstoffe Acesulfam-K und Aspartam. *Ind Obst-u. Gemueseverwertung* 80(1):3–7.

Hoppe, K., and Gassmann, B. 1980. Neue Aspekte der Suesskraftbeurteilung und des Einsatzes von Suessungsmitteln. *Nahrung* 24:423–431.

Hoppe, K., and Gassmann, B. 1985a. Vergleichstabellen zur Suesseintensitaet von 16 Suessungsmitteln. *Lebensmittelindustrie* 32:227–231.

Hoppe, K., and Gassmann, B. 1985b. Bestimmung der Missgeschmacksschwellen von Saccharin, Cyclamat, Acesulfam und Aspartam. *Nahrung* 29:417–420.

Joint FAO/WHO Expert Committee on Food Additives (JECFA). 2010. Combined Compendium of Food Additive Specifications. http://www.fao.org/ag/agn/jecfa-additives/details.html?id=841. Accessed June 8, 2011.

King, B.M., Arents, P., Bouter, N., Duineveld, C.A.A., Meyners, M., Schroff, S.I., and Soekhai, S.T. 2006. Sweetener/sweetness-induced changes in flavor perception and flavor release of fruity and green character in beverages. *J Agric Food Chem* 54(7):2671–2677.

Klug, C., von Rymon Lipinski, G.-W., and Boettger, D. 1992. Baking stability of acesulfame K. *Z Lebensm Unters Forsch* 194(5):476–478.

Korb, M., Kniel, B., and Meyer, E. 1992. Mikrowellenstabilität der Suessstoffe Acesulfam und Aspartam. *ZFL* 43(9):494–498.

Kuhn C., Bufe, B., Winnig, M., Hofmann, T., Frank, O., Behrens, M., Lewtschenko, T., Slack, J.P., Ward, C.D., and Meyerhof, W. 2004. Bitter taste receptors for saccharin and acesulfame K. *J Neurosci* 24(45):10260–10265.

Liang, Y., Maier, V., Steinbach, G., Lalić, L., and Pfeiffer, E.F. 1987a. The effect of artificial sweetener on insulin secretion. II. Stimulation of insulin release from isolated rat islets by Acesulfame K (in vitro experiments). *Horm Metab Res* 7:285–289.

Liang, Y., Steinbach, G., Maier, V., and Pfeiffer, E.F. 1987b. The effect of artificial sweetener on insulin secretion. 1. The effect of acesulfame K on insulin secretion in the rat (studies in vivo). *Horm Metab Res* 6:233–238.

Lim, H., Setser, C.S., and Kim, S.S. 1989. Sensory studies of high potency multiple sweetener systems for shortbread cookies with and without polydextrose. *J Food Sci* 54:625–628.

Linke H.A. 1983. Adherence of Streptococcus mutans to smooth surface in the presence of artificial sweeteners. *Microbios* 36:41–45.

Linkies, A., and Reuschling, D. 1990. Ein neues Verfahren zur Herstellung von 6-Methyl-1,2,3-oxathiazin-4(3H)-on-2,2dioxid Kaliumsalz. *Synthesis* 405–406.

Lotz, A., Klug, C., and von Rymon Lipinski, G.-W. 1992. Stability of acesulfame K during high temperature processing under conditions relevant for dairy products. *Z Lebensmtechnol Verfahrenst* 43(5):EFS 21–23.

Malaisse, W.J., Vanonderbergen, A., Louchami, K., Jijakli, H., and Malaisse-Lagae, F. 1998. Effects of artificial sweeteners on insulin release and cationic fluxes in rat pancreatic islets. *Cell Signal* 10(10):727–733.

Matysiak, N.L., and Noble, A.C. 1991. Comparison of temporal perception of fruitiness in model systems sweetened with aspartame, an aspartame+acesulfame K blend, or sucrose. *J Food Sci* 56:823–826.

Meyer, S. 2001. Custom-tailored sweetness for fruit flavours. *Soft Drinks Intl* 2001(9):38–40.

Meyer, S. 2000. Taste interactions—adjusting sweeteners to flavours. *World Food Ingred* 2000(12):38–40.

Mukherjee, A., and Chakrabarti, J. 1997. In vivo cytogenetic studies on mice exposed to acesulfame K – a non-nutritive sweetener. *Food Chemical Toxicol* 35:1177–1179.

Nahon, D.F., Roozen, J.P., and de Graaf, C. 1996. Sweetness flavor interactions in soft drinks. *Food Chem* 56:283–289.

Ott, D.B., Edwards, C.L., and Palmer, S.J. 1991. Perceived taste intensity and duration of nutritive and non-nutritive sweeteners in water using time-intensity (T-I) evaluations. *J Food Sci* 56:535–542.

Park, K.K., Schemehorn, B.R., Stookey, G.K., Butchko, H.H., and Sanders, V. 1995. Acidogenicity of high-intensity sweeteners and polyols. *Am J Dent* 8(1):23–26.

Paulus, E.F. 1975. 6-Methyl-1,2,3-oxathiazin-4(3H)-on 2,2-dioxide. *Acta Cryst (B)* 31:1191–1192.

Paulus, K., and Braun, M. 1988. Suesskraft und Geschmacksprofil von Suessstoffen. *Ernaehrungs-Umschau* 35:384–391.

Portman, M.O., and Kilcast, D. 1998. Descriptive profiles of synergistic mixtures of bulk and intense sweeteners. *Food Quality Preference* 9:221–229.

Rathjen, S. 2005. US Patent Application 20050037121 (to Nutrinova).

Rathjen, S. 2003. Fine-tuning sweetener blends in lemon-lime drinks. *Intl Food Ingredients* 2003(4–5):26–28.

Redlinger, P.E., and Setser, C.S. 1987. Sensory quality of selected sweeteners: Unbaked and baked flour doughs. *J Food Sci* 52:1391–1393, 1413.

Renwick, A. 2006. The intake of intense sweeteners—an update review. *Food Addit Contam* 23(4):327–338.

Riera, C.E., Vogel, H., Simon, S.A., Damak, S., and le Coutre, J. 2008. The capsaicin receptor participates in artificial sweetener aversion. *Biochem Biophys Res Comm* 376(4):653–657.

Riera, C.E., Vogel, H., Simon, S.A., and le Coutre, J. 2007. Artificial sweeteners and salts producing a metallic taste sensation activate TRPV1 receptors. *Am J Physiol* 293(2, Pt. 2):R626–634.

Saelzer, K. 2005a. The world of fruits in cans. *Fruit Processing* 2005(5/6):169–167.

Saelzer, K. 2005b. More ketchup—less calories. Sunett presents new opportunities for low calorie sauces. *Innovations in Food Technology* 2004(2):46–47.

Saelzer, K. 2004. New customised sweetening systems for tea and coffee. *Innovations in Food Technology* 2004(2):2–3.

Schiffman, S.S., Sattely-Miller, E.A., Graham, B.G., Zervakis, J., Butchko, H.H., and Stargel, W.W. 2003. Effect of repeated presentation on sweetness intensity of binary and ternary mixtures of sweeteners. *Chem Senses* 28(3):219–229.

Schiffman, S.S., Booth, B.J., Losee, M.L., Pecoreand, S.D., and Warwick, Z.S. 1995. Bitterness of sweeteners as a function of concentration. *Brain Res Bull* 36(5):505–513.

Schmidt, R., Janssen, E., Haeussler, O., Duriez, X., and Baron, R.F. 1998. Eine hochwertige Suessungsalternative fuer Zahnpasten. *SOEFW Journal* 124(3):148–150.

Selzer Rasmussen, E. 1999. Evaluation of the clastogenicity of Acesulfame K. *Pharmacol Toxicol* (Suppl. I) 60.

Siebert G., Ziesenitz, S.C., and Lotter, J. 1987. Marked caries inhibition in the sucrose-challenged rat by a mixture of nonnutritive sweeteners. *Caries Res* 21:141–148.

Slack, J.P., Brockhoff, A., Batram, C., Menzel, S., Sonnabend, C., Born, S., Galindo M.M., et al. 2010. Modulation of bitter taste perception by a small molecule hTAS2R antagonist. *Curr Biol* 20(12):1104–1109.

Steiniger J., Graubaum, H.-J., Steglich, H.-D., Schneider, A., and Metzner, C. 1995. Gewichtsreduktion mit saccharose-und suessstoffhaltiger Reduktionsdiaet. *Ernaehrungsumschau* 42:430–437.

van Tournot, P., Pelgroms, J., and van der Meeren, J. 1985. Sweetness evaluation of mixtures of fructose with saccharin, aspartame or acesulfame K. *J Food Sci* 50:469–472.

Velaga, S.P., Vangala, V.R., Basavoju, S., and Bostroem, D. 2010. Polymorphism in acesulfame sweetener: Structure-property and stability relationships of bending and brittle crystals. *Chem Commun* 46(20):3562–3564.

Voelkner, W. 1998a. Chromosome aberration assay in bone marrow cells of the mouse with Acesulfame K. RCC-CCR Project 609900.

Voelkner, W. 1998b. Microscopic evaluation of cytogenetic preparations. RCC-CCR Project 621100.

von Rymon Lipinski, G.-W. 1988a. Einsatz von Sunett® in Erfrischungsgetränken. *Swiss Food* 10(5):25–27.

von Rymon Lipinski, G.-W. 1988b. Einsatz von Sunett® in Joghurt und anderen Milcherzeugnissen. *Swiss Food* 10(6):29–32.

von Rymon Lipinski, G.-W. 1989. Stability and synergism-important characteristics for the application of Sunett® In *Food Ingredients Conference Proceedings 1989*, 249–251. Maarsen: Expoconsult Publishers.

von Rymon Lipinski, G.-W. 1985. The new Intense sweetener Acesulfame K. *Food Chem* 16:259–269.

von Rymon Lipinski, G.-W., and Brixius, C. 1979. Duennschichtchromatographischer Nachweis von Acesulfam, Saccharin und Cyclamat. *Z Lebensm Unters Forsch* 168:212–213.

von Rymon Lipinski, G.-W., and Gorstelle, E. 1989. The application of Sunett® in confectionery. *Confect Prod* 55:597–599.

von Rymon Lipinski, G.-W., and Huddart, B.E. Acesulfame K. *Chem Ind* 1983:427–432.

von Rymon Lipinski, G.-W., and Klein, E. 1988. In *Synergistic effects in blends of Acesulfame K and Aspartame*. ACS Conference "Sweeteners-Carbohydrate and Low Calorie." Los Angeles, 1988, No. 27.

von Rymon Lipinski G.-W., and Lueck, E. 1981. Acesulfame K: A new sweetener for oral cosmetics. *Manuf Chemist* 52:37.

Wasik, A., and Buchgraber, M. 2007. Simultaneous determination of nine intense sweeteners by HPLC-ELSCD. European Commission-Joint Research Centre validated method. http://publications.jrc.ec.europa.eu/repository/bitstream/111111111/12460/1/6968%20-%20EUR%2022727%20Report%20SOP%20Sweeteners.pdf. Accessed June 9, 2011.

Wiedmann, M., and Jager, M. 1997. Innovative sweetening systems: Synergies, functional and health benefits. *Food Ingred Anal Intl* 1997(12):51–56.

Wilson, L.A., Wilkinson, K., Crews, H.M., Davies, A.M., Dick, C.S., and Dumsday, V.L. 1999. Urinary monitoring of saccharin and acesulfame-K as biomarkers of exposure to these additives. *Food Addit Contam* 16:227–238.

Yang, D.J., and Chen, B. 2009. Simultaneous Determination of Nonnutritive Sweeteners in Foods by HPLC/ESI-MS. *J Agr Food Chem* 57(8):3022–3027.

Chapter 3

Advantame

Ihab E. Bishay and Robert G. Bursey

Contents

Introduction

Advantame is a new ultrahigh potency sweetener and flavor enhancer developed by Ajinomoto. It is derived from aspartame, a dipeptide sweetener composed of aspartic acid and phenylalanine, and vanillin. Advantame is approximately 20,000 times sweeter than sugar and 100 times sweeter than aspartame. Advantame provides zero calories and has a clean, sweet, sugar-like taste with no undesirable taste characteristics. It is functional in a wide array of beverages and foods and is most suitable for use in blends with other high-potency or carbohydrate sweeteners. Advantame is stable under dry conditions. In aqueous food systems, its stability is similar to aspartame with greater stability predicted at higher and neutral pH, as well as higher temperature conditions (e.g., baking and other prolonged heating processes), and in yogurt.

The results of numerous safety studies confirm that advantame is safe for use by the general population and people with diabetes. In addition, no special labeling for phenylketonuric (PKU) individuals is expected to be required.

Advantame is currently under review by the United States Food and Drug Administration (US FDA) for dry use applications and is also under review in Australia and New Zealand and the European Union (EU). Its unique properties provide the food technologist with an additional tool to produce innovative new foods and beverages to meet the demand of consumers to have great tasting foods without the calories of sugar.

Discovery

Advantame was the result of a long-term research program at Ajinomoto to discover new high-potency sweeteners with desirable taste characteristics. The goal was to explore new aspartyl-based sweet molecules based on the following strategy (Amino et al. 2008):

■ Select lead compounds
■ Conduct lead optimization and structure-activity studies
■ Apply computer aided molecular modeling
■ Synthesize compounds and screen potential new sweeteners

This research program advanced as the field of research of sweet taste itself advanced, starting with Schallenberger and Acree's "AH-B" model (Shallenberger 1996; Eggers et al. 2000) and Ariyoshi's proposed model for aspartame derivatives (Ariyoshi 1976) to the hypothesis of the interaction between sweet peptides and the sweet taste receptor. This included the use of computational chemistry to analyze the structure of sweet molecules and develop models for sweet molecules to the identification of the sweet taste receptor by various groups (Hoon et al. 1999; Kitagawa et al. 2001; Li et al. 2001, 2002; Max et al. 2001; Nelson et al. 2001; Sainz et al. 2001; Zhao et al. 2003). Initial screening of candidates was based on sweet potency, taste quality, and availability of synthesis. The initial screening reduced the candidates from hundreds of compounds to approximately ten. A second more rigorous screening was based on the following criteria: sweet potency, detailed taste profile, physical-chemical properties, feasibility of industrial production and estimation of metabolic dynamics in the human body by *in vitro* assay.

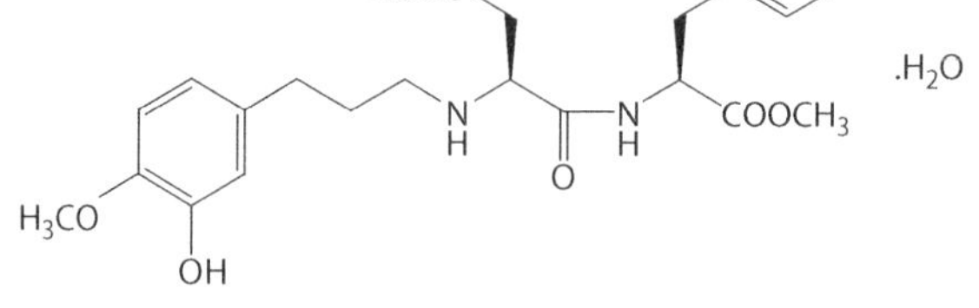

Figure 3.1 Structure of advantame.

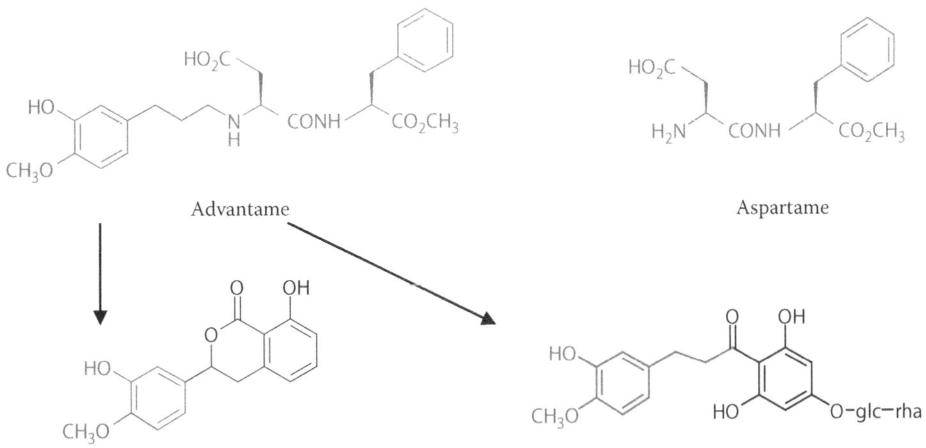

Phyllodulcin
SP(3)=400
Japanese traditional sweetener
obtained from *Saxifragaceae amateur* leaf

Neohesperidin dihydrochalcone
SP(5)=1430
Obtained from citrus fruits

Figure 3.2 Structural similarities between advantame and sweeteners of natural origin.

The result of this process is a novel sweetener, N-[*N*-[3-(3-hydroxy-4-methoxyphenyl) propyl]-α-aspartyl]-L-phenylalanine 1-methyl ester monohydrate (initially code named ANS9801 but now with the common name advantame, molecular formula: $C_{24}H_{30}N_2O_7 \cdot H_2O$, CAS no. 714229-20-6) (Figure 3.1). It is interesting to note that advantame has a structural similarity to natural sweeteners, for example the 3-hydroxy-4-methoxyphenyl group exists in phyllodulcin, a sweetener found in the leaves of *Hydrangea serrata* (Hydrangeaceae) (Ujihara et al. 1995) and neohesperidin dihydrochalcone, a sweetener derived from the hydrogenation of the flavanone neohesperidin, which is found in citrus fruits (Horowitz and Gentili 1961, 1963a, 1963b, 1969; Horowitz 1964) (Figure 3.2).

Physical Characteristics and Chemistry

The starting materials of advantame are aspartame and vanillin. It is synthesized from aspartame and (3-hydroxy-4-methoxyphenyl) propylaldehyde (HMPA) in a one step process by reductive N-alkylation, carried out by treating aspartame with HMPA with hydrogen in the presence of a platinum catalyst. HMPA is derived from vanillin by a four-step synthesis (Figure 3.3) (Amino et al. 2008).

Advantame has a molecular weight of 476.52 g/mol and the monohydrate has a melting point of 101.5°C. Its specifications are listed in Table 3.1. Furthermore, the solubility of advantame in water, ethanol and ethyl acetate has been measured, and the results are presented in Table 3.2. Considering the ultrahigh potency of advantame, its solubility in these solvents is more than sufficient for the required functionality.

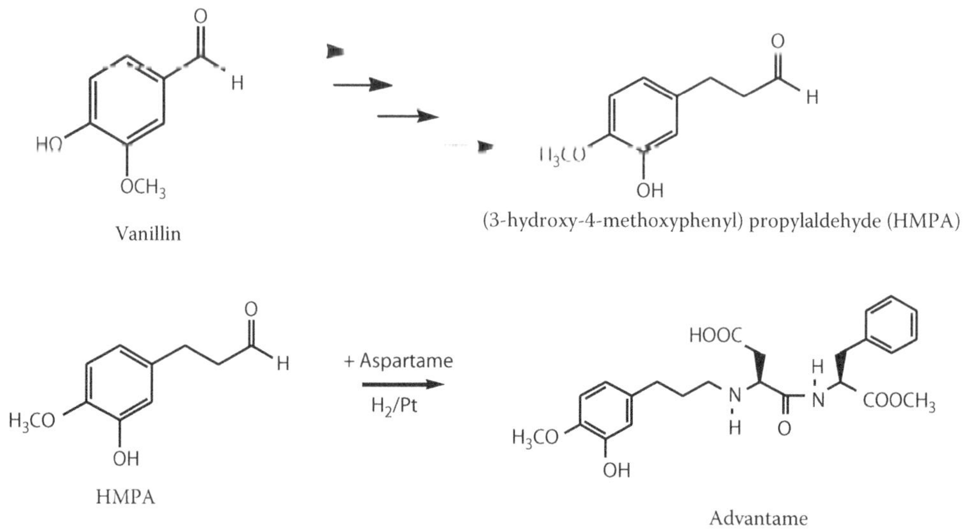

(3-hydroxy-4-methoxyphenyl) propylaldehyde (HMPA)

Vanillin

HMPA

Advantame

Figure 3.3 Synthesis of advantame.

Stability

The stability of advantame is dependent upon pH, moisture, and temperature. Advantame in dry applications, such as tabletop or powdered soft drinks, is very stable and maintains its functionality during normal storage and handling conditions. Figures 3.4 and 3.5 show the stability of advantame in tabletop and powdered soft drinks, respectively, when stored at 25°C and

Table 3.1   Advantame Specifications

Specification Parameter	*Specification Value*
Identification	
Description	White to yellow powder
IR absorption spectrum	Same as reference standard
Purity	
Assay	Not less than 97.0% and not more than 102.0% on anhydrous basis
N-[N-[3-(3-hydroxy-4-methoxyphenyl) propyl]-α-aspartyl]-L-phenylalanine	Not more than 1.0%
Total other related substances	Not more than 1.5%
Water	Not more than 5.0%
Residue on ignition	Not more than 0.2%
Lead	Not more than 1 ppm
Specific rotation [α]20D	Between –45° and –38°

Table 3.2   Advantame Solubility After 30 min at Various Temperatures in Water, Ethanol, and Ethyl Acetate

The Solubility of Advantame in Water, Ethyl Acetate, and Ethanol After 30 Minutes			
	Solubility of Advantame (g/100 ml of Solvent)		
Temperature (°C)	Water	Ethanol	Ethyl Acetate
15	0.076	0.798	0.165
25	0.099	1.358	0.279
40	0.210	3.827	0.796
50	0.310	9.868	1.600
60	0.586	32.277	2.271

60% relative humidity (RH). Figure 3.6 shows the stability of advantame in a model carbonated soft drink at pH 3.2 and stored at 25°C. Under these conditions, advantame displays the same stability profile as aspartame indicating that both sweeteners follow the same degradation mechanism.

Research is still being conducted on advantame stability in other matrices and under different conditions. It can be theorized that advantame would have the same stability as aspartame under relatively low pH and low temperature environments, since under these conditions the primary mechanism of degradation is the de-esterification of the methyl ester (Homler 1984; Pattanaargson et al. 2001). However, under relatively higher pH and/or temperature conditions, advantame is expected to demonstrate better stability than aspartame. Under these conditions, the primary mechanism for degradation of aspartame is the formation of diketopiperazine (DKP) (Homler 1984; Pattanaargson et al. 2001). However, the formation of DKP in advantame is blocked due to steric hindrance by the bulky vanillyl functional group. Furthermore since the dipeptide linkage of advantame is not recognized as such by microorganisms, it is further expected that it would have excellent stability in yogurt, even when added prior to the fermentation process.

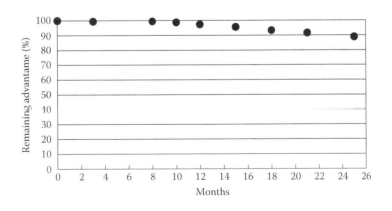

Figure 3.4   Advantame stability in tabletop mix. 25°C and 60% RH.

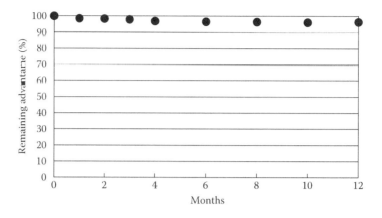

Figure 3.5   Advantame stability in powdered soft drink mix. 25°C and 60% RH.

Taste

Potency

In sweetener research, sucrose is the standard against which other compounds are compared. Sucrose equivalence (SE) is the standardized sweetness intensity scale established for comparing sweet compounds. An x% SE is equivalent in sweetness to an x% sucrose in water solution. Advantame has a concentration-response graph with a curvilinear shape similar to other high potency sweeteners. Figure 3.7 and Table 3.3 show that advantame in water has a potency range from 7,000 times that of sugar at high concentrations (14% SE) to almost 50,000 times that of sugar at low concentrations (3% SE), and although it is a derivative of aspartame, advantame is 70 to 120 times more potent than aspartame. Like other high-potency sweeteners, the actual sweetness potency of advantame is dependent on its concentration as well as the matrix in which it is used. Based on these data, advantame can reach an extrapolated maximum sweetness intensity (plateau) of 15.8% SE in water. Sweeteners such as aspartame, acesulfame potassium (acesulfame K), sodium cyclamate, and sodium saccharin attain their maximum sweetness intensity in water

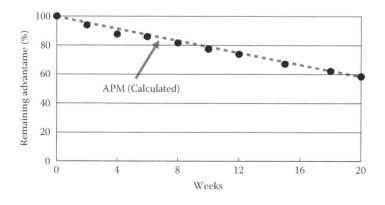

Figure 3.6   Advantame stability in a model carbonated soft drink with calculated aspartame stability for comparison (pH 3.2, 25°C).

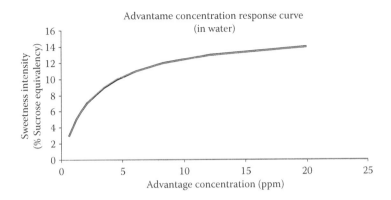

Figure 3.7   Concentration-response curve for advantame in water.

at approximately 16.0% SE, 11.6% SE, 11.3% SE, and 9.0% SE, respectively (DuBois et al. 1992). This extrapolated maximum sweetness intensity is an important measure of the "cleanliness" of the sweetness, indirectly measuring the presence or lack of competing tastes and/or flavors. A high extrapolated maximum sweetness intensity indicates a very clean sweet profile, while a relatively low extrapolated maximum sweetness intensity indicates off-taste and off-flavors competing with the sweet taste and not allowing a sufficiently high extrapolated maximum sweetness intensity.

Table 3.3   Advantame Potency at Various Concentrations

Sweetness Intensity (% Sucrose Equivalency)	Advantame Relative Potency	
	Relative to Aspartame (Aspartame/Advantame)	Relative to Sucrose (Sucrose/Advantame)
3	120	47778
4	119	44074
5	118	40370
6	116	36667
7	114	32963
8	112	29259
9	109	25556
10	105	21852
11	100	18148
12	94	14444
13	85	10741
14	70	7037

Table 3.4 Estimated Use Levels of Advantame in Various Foods and Beverages

Application	Concentration of Advantame
Tabletop Sweetener	0.65 mg in 1.3 g sachet (as consumed. 2–4 ppm)
Powdered Soft Drink	250–600 ppm (as consumed: 3–7 ppm)
Carbonated Soft Drink	4–10 ppm
Hot Fill Beverage	3–4 ppm
Chewing Gum	100–400 ppm
Yogurt	4–7 ppm
Yellow Cake	10–18 ppm

Because of its remarkable sweetness potency, advantame is used in food and beverage products at considerably lower concentrations than other high-potency sweeteners. Table 3.4 lists some estimated use level ranges for advantame in various foods and beverages when using advantame as the sole sweetener.

Taste Profile

A descriptive analysis profile of advantame in water was carried out using QDA® methodology (Amino et al. 2008; Stone and Sidel 1998) and compared to that of aspartame. Advantame was tested at 5 ppm and compared to aspartame at 500 ppm. Results are shown in Figures 3.8 and 3.9. The only characteristic of advantame is sweetness with no other competing off-tastes or off-flavors and with a very similar taste profile to that of aspartame. The primary difference in profile between advantame and aspartame is in the aftertaste attributes, with advantame having a slightly longer sweet aftertaste than aspartame.

Flavor Enhancement

Advantame has very interesting properties as a flavor enhancer. In order to assess this functionality, it was necessary to test the functionality of advantame as a flavor enhancer below its sweetening levels. Initial testing was to determine the group difference threshold level (GDTL) of advantame, and then to establish whether advantame enhanced flavor at or below its GDTL and at what range of concentrations this flavor enhancement was observed (Carr et al. 1993; ASTM 2004). In testing for the functionality of advantame as a flavor enhancer in beverages, for example, the GDTL was initially established in water sweetened with 9.56% sucrose. The panelists were asked to choose the sweeter sample between the control (only sweetener with 9.56% sucrose) and the test (sweetened with 0.2, 0.5, 0.8, 1.1, 1.4 or 1.7 ppm advantame). Using this method and calculating the GDTL by logistic regression, the GDTL of advantame was determined to be 1.46 ppm. Strawberry flavored beverages were then sweetened with 9.56% sucrose and various levels of advantame were added, all below the GDTL (0.1, 0.4, 0.7, 1.0 and 1.46 ppm advantame). These sucrose/advantame strawberry beverages (test) were compared to sucrose-only strawberry beverages (control). The panelists were asked to choose the more flavorful sample between the two. Table 3.5 summarizes the results which show that advantame displays flavor enhancement at and below its GDTL

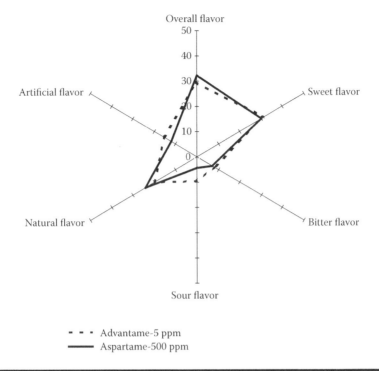

Figure 3.8 QDA® profile of advantame compared to aspartame.

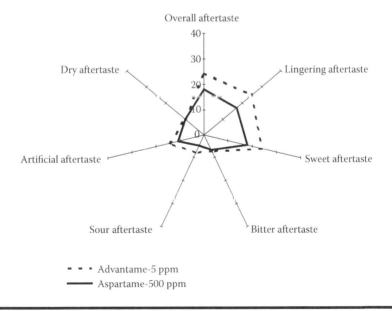

Figure 3.9 QDA® aftertaste profile of advantame compared to aspartame.

Table 3.5 Flavor Enhancement of Advantame in Strawberry Flavored Beverage. GDTL is 1.46 ppm

Advantame Concentration	Total Reads	Panelists Choosing the Sample with Advantame as More Flavorful		Probability of No Flavor Enhancement
(in beverage)	(N*2 replications)	Number	Percent	(p-value)
0.1 ppm	83	40	48.2%	0.5868
0.4 ppm	83	48	57.8%	0.0619
0.7 ppm	83	42	50.6%	0.4132
1.0 ppm	83	59	71.1%	0.0000*
1.46 ppm	83	51	62.2%	0.0099*

* Flavor enhancement.

at 1.0 ppm and 1.46 ppm. Similarly the GDTL was determined to be 0.92 ppm in unflavored yogurt sweetened with 10% sucrose. Advantame was then tested below the GDTL in vanilla flavored yogurt, also sweetened with 10% sucrose. Table 3.6 summarizes the results which show that advantame enhances the flavor of vanilla yogurt below its GDTL (0.4, 0.6 and 0.8 ppm). Finally the GDTL was investigated in unflavored chewing gum sweetened with an aspartame/acesulfame K blend with advantame concentrations ranging from 2.5 ppm to 40 ppm. Results of the GDTL analysis showed that all the concentrations tested were below the GDTL. Therefore a GDTL was not established, but it can be concluded that the GDTL of advantame in unflavored chewing gum is above 40 ppm. Flavor enhancement was then tested in mint flavored chewing gum, also sweetened with an aspartame/acesulfame K blend, with advantame concentrations ranging from 2.5 ppm to 30 ppm. The results are summarized in Table 3.7. Advantame shows flavor enhancement

Table 3.6 Flavor Enhancement of Advantame in Vanilla Flavored Yogurt. GDTL is 0.92 ppm

Advantame Concentration	Total Reads	Panelists Choosing the Sample with Advantame as More Flavorful		Probability of No Flavor Enhancement
(in yogurt)	(N*2 replications)	Number	Percent	(p-value)
0.2 ppm	84	50	59.5%	0.0506
0.4 ppm	83	51	61.4%	0.0238*
0.6 ppm	83	57	68.7%	0.0004*
0.8 ppm	84	54	64.3%	0.0058*
1.0 ppm	83	56	67.5%	0.0010+

* Flavor enhancement ; + above GDTL.

Table 3.7 Flavor Enhancement of Advantame in Mint Flavored Chewing Gum. GDTL is above 40 ppm

Advantame Concentration	Total Reads	Panelists Choosing the Sample with Advantame as More Flavorful		Probability of No Flavor Enhancement
(in chewing gum)	(N*2 replications)	Number	Percent	(p-value)
2.5 ppm	82	42	51.22%	0.046
5.0 ppm	82	50	60.98%	0.03*
7.5 ppm	81	51	62.96%	0.01*
10.0 ppm	83	50	60.24%	0.04*
12.5 ppm	81	43	53.09%	0.33
15.0 ppm	81	34	41.98%	0.09
17.5 ppm	84	49	58.33%	0.08
20.0 ppm	83	45	54.22%	0.25
25.0 ppm	83	44	53.01%	0.33
30.0 ppm	81	49	60.49%	0.04*

* Flavor enhancement.

at various concentrations, namely 5.0, 7.5, 10, and 30 ppm. Even though the chewing gum data is not as consistent as the beverage and yogurt data, there is still a clear indication of flavor enhancement of advantame in chewing gum. The functionality of advantame in chewing gum can provide flavor enhancement as well as an increase in sweet duration to deliver longer chew times depending on the concentrations used.

Application as a Sweetener Replacer

The clean sweet taste of advantame allows food technologists to substitute a portion of carbohydrate sweeteners such as sugar or high fructose corn syrup with advantame, while maintaining a taste that is indistinguishable from the 100% carbohydrate-sweetened product. Internal research using triangle testing with expert panels on various beverage formulations such as cola, orange and lemon-lime flavored carbonated soft drinks and fruit punch, lemonade, lemon flavored tea and lime flavored non-carbonated soft drinks, has shown, for example, that, depending on the formulation, up to 40% of a sweetener can be replaced with advantame without altering the taste. Typically advantame can replace between 20% and 30% of the sweetness in most applications without altering the taste profile. Using advantame to replace sugar or high fructose corn syrup will reduce sugar, calories, and cost.

Advantame can also be used to replace a portion of other high potency sweeteners while maintaining the same taste profile or, in some cases, improving the overall taste profile and enhancing the flavor. Generally the level of substitution that can be accomplished with advantame when replacing other high potency sweeteners, is greater than the level that can be accomplished when replacing carbohydrate sweeteners. Replacing a portion of high potency sweetener can also be accomplished while reducing cost, and in many cases also improving taste.

Safety and Regulatory Status

Safety Evaluation

The safety of advantame has been evaluated in a large number of studies conducted in the mouse, rat, rabbit, dog and human under Good Laboratory Practices (GLP) and included all those recommended in the US FDA's Redbook (US FDA 2000). For each of these key studies, a preliminary non-GLP study was performed to establish the appropriate dose-range to be used. For most of the subchronic, chronic and reproductive studies, advantame was fed at specific concentrations in standard animal chow diets. In addition, studies were carried out in rats, dogs and humans to evaluate the metabolic fate of advantame ingested orally. Two clinical studies were also performed in humans to assess the tolerability and safety of advantame in both normoglycemic and diabetic humans.

The toxicity studies conducted in animals provided extensive data that advantame exhibits no evidence of genotoxicity and causes neither adverse effects nor organ toxicity even when fed at extremely high doses. No observed adverse effect levels (NOAELs) determined in chronic animal studies ranged from 5700 to 7350 mg/kg body weight/day in mice, 2600 to 3450 mg/kg body weight/day in rats and 2000 mg/kg body weight/day in dogs. Doses of up to 0.5 mg of advantame produced no adverse effects in either normoglycemic or diabetic individuals. Because of the potency of sweetness exhibited by advantame, use levels of the sweetener by even heavy users are estimated to be in the microgram per kilogram of body weight range, providing a very large margin of safety associated with its use.

Toxicology and Carcinogenicity

Advantame was shown to be non-mutagenic in an Ames assay when incubated with various strains of *Salmonella typhimurium*, with or without metabolic activation. It similarly did not exhibit any mutagenic potential in the *in vitro* mouse lymphoma assay. No genotoxic potential was demonstrated in the mouse micronucleus assay when advantame was administered to mice by oral gavage.

Dietary studies of advantame fed at levels of up to 50,000 ppm to mice, rats and dogs for periods of 4 or 13 weeks clearly demonstrated the safety of the sweetener. In each case, no treatment related deaths, clinical signs of toxicity, organ toxicity or other toxicologically significant effects were seen and the NOAEL was considered to be 50,000 ppm in the diet (the highest level tested) representing the aforementioned NOAELs expressed on a milligram per kilogram body weight per day basis.

The toxicological effects and potential carcinogenicity of chronic exposure from the ingestion of varying doses (up to 50,000 ppm) of advantame were evaluated in mice and rats over 104 weeks and in dogs for 1 year. Changes in body weight gain in animals ingesting the highest doses of advantame (50,000 ppm) were sporadically reported but these differences could be attributed to palatability and/or nutrient dilution issues rather than toxicity of the sweetener. There was, however, no evidence of treatment-related carcinogenicity or organ toxicity in any of the species of animals tested at any of the levels of advantame fed. Once again, the NOAEL of advantame fed for periods of 1 year to dogs and 2 years to mice or rats is 50,000 ppm in the diet, equivalent to 5693 and 7351 mg/kg body weight/day in male and female mice, respectively,

2621 and 3454 mg/kg body weight/day in male and female rats, respectively, and 2058 and 2139 mg/kg body weight/day in male and female dogs, respectively.

Reproduction and Teratology

Advantame was fed to male and female rats (the F0 generation) at doses up to 50,000 ppm for 10 weeks prior to mating and then through gestation, lactation and until termination. Offspring of these F0 rats (the F1 generation) were fed similar levels of advantame prior to their mating, during gestation and during lactation. Detailed records of reproductive performance, histopathology of reproductive organs and the health and well-being of mothers, fathers and/or pups of the F0, F1 and F2 generations were evaluated. The NOAEL for reproductive toxicity was shown to be 50,000 ppm, the highest dose tested. No adverse effects of advantame were seen in any of the parameters measured in the F0, F1 or F2 generations.

In a study to evaluate the teratogenicity of advantame, the sweetener was added to the diets of female rats at doses up to 50,000 ppm. The females were then mated to stock (not receiving advantame) males and sacrificed at day 20 of gestation. Reproductive performance, fetal development and incidence and types of malformations were evaluated. Evidence of inappetence in rats consuming the highest concentrations of advantame was observed along with decreased body weight gain, but these effects were not considered to be adverse. The NOAEL for the developing rat fetus was established to be 50,000 ppm or 4828 mg/kg body weight/day.

In a teratology study conducted in New Zealand white rabbits, advantame was administered on gestation days 6–28 by oral gavage at doses up to 2000 mg/kg body weight/day. The maternal NOAEL was considered to be 500 mg/kg body weight/day based upon one death at the 1000 mg/kg body weight/day due to gastrointestinal (GI) tract distress (mediated through inappetence). The NOAEL for the embryo/fetal survival and development was considered to be 1000 mg/kg body weight/day based upon a slightly greater number of fetal deaths at the 2000 mg/kg body weight/day dose.

Metabolism and Human Studies

Ingested advantame is rapidly hydrolyzed in the GI tract to the corresponding acid. Only a small percentage of advantame or the corresponding acid is absorbed from the lumen of the GI tract with approximately 87%–93% excreted in the feces and the remainder excreted in urine as the corresponding acid. There is no tissue-specific retention of either advantame or the corresponding acid.

Two clinical studies of the tolerability of advantame in humans, including both normoglycemic and diabetic subjects, were conducted. Included were a single dose pharmacokinetic study, an escalating dose study in healthy males, a 4-week safety and tolerability study in healthy males, and a 12 week safety study in diabetics. Consumption of a single dose of advantame up to 0.5 mg/kg body weight/day was shown to be safe in both normoglycemic and diabetic individuals. No serious adverse events were reported during any of the human studies and advantame had no effect on plasma glucose or insulin levels in normoglycemic individuals and also had no effect on glucose tolerance or insulin resistance in diabetics.

Regulatory Status

Advantame has undergone an assessment of its safety for use as a flavor enhancer by a panel of experts convened by the Flavor and Extract Manufacturers Association and was deemed to be

Generally Recognized As Safe (GRAS) for use in non-alcoholic beverages (at levels of 1.5 ppm or less), chewing gum (at levels of 50 ppm or less), milk products (at levels of 1.0 ppm or less) and frozen dairy products (at levels of 1.0 ppm or less). Further, a Food Additive Petition was submitted in April 2009 to the US FDA seeking approval for use of advantame in powdered beverages and for tabletop use (US FDA 2009). Food additive petitions for use of advantame in various foods have also been filed with the Food Standards Australia New Zealand (July 2009) and more recently with the European Food Standards Authority.

Summary

Advantame is an ultrahigh potency sweetener derived from aspartame and vanillin. It is approximately 20,000 times sweeter than sucrose and 100 times sweeter than aspartame. Advantame has a clean sweet taste very similar to aspartame. Sweetness duration is only slightly longer. The stability of advantame is also similar to aspartame in liquid applications. Stability in other applications is currently being investigated, but in most applications is expected to be similar to aspartame, and may be better than aspartame in some applications involving higher pH or higher temperature conditions. Advantame is also a flavor enhancer, displaying flavor enhancement properties below its sweetness threshold concentrations. Furthermore, due to its taste profile, advantame is well suited for chewing gum where it can increase the flavor impact and increase the chew time. Due to its excellent taste and functionality, advantame can be used to partially replace sugar, high fructose corn syrup or other high potency sweeteners to reduce cost, calories and/or sugar content while maintaining the same taste profile or improving the taste and flavor profile of the products.

References

Amino, Y., Mori, K., Tomiyama, Y., Sakata, H., and Fujieda, T. 2008. Development of new, low calorie sweetener: New aspartame derivative. In *Sweetness and sweeteners: Biology, chemistry, and psychophysics, ACS symposium series 979,* eds. D.K. Weerasinghe and G.E. DuBois, 463–480. Washington, DC: American Chemical Society.

Ariyoshi, Y. 1976. The structure-taste relationships of aspartyl dipeptide esters. *Agric Biol Chem* 40:983–992.

ASTM Method E1432. 2004. Standard practice for defining and calculating individual and group sensory thresholds from forced-choice data sets of intermediate size. ASTM International West Conshohocken, PA.

Carr, B.T., Pecore, S.D., Gibes, K.M., and DuBois, G.E. 1993. Sensory methods for sweetener evaluation. In *Flavor measurement*, eds. C.-T. Ho and C.H. Manley. New York: Marcel Dekker.

DuBois, G., Walters, D., Schiffman, S., Warwick, Z., Booth, B., Pecore, S., Gibes, K., Carr, B., and Brands, L. 1992. Concentration-response relationships of sweeteners. In *Sweeteners discovery, molecular design, and chemoreception*, eds. D.E. Walters, F.T. Orthoefer, and G.E. DuBois, 261–276. Washington, DC: American Chemical Society.

Eggers, S.C., Acree, T.E., and Schallenberger, R.S. 2000. Sweetness chemoreception theory and sweetness transduction. *Food Chem* 68(1):45–49.

Homler B.E. 1984. Properties and stability of aspartame. *J Food Technol* 38(July):50–55.

Hoon, M., Adler, E., Lindemeier, J., Battey, J.F., Ryba, N.J.P., and Zuker, C.S. 1999. Putative mammalian taste receptors: a class of taste-specific GCPRs with distinct topographic selectivity. *Cell* 96:541–551.

Horowitz, R.M. 1964. Relations between the taste and structure of some phenolic glycosides, In *Biochemistry of Phenolic Compounds,* ed. J.B. Harborne, 545–571. New York: Academic Press.

Horowitz, R.M., and Gentili, B. 1961. Phenolic glycosides of grapefruit: A relation between bitterness and structure. *Arch Biochem Biophys* 92:191–192.

Horowitz, R.M., and Gentili, B. 1963a. Dihydrochalcone derivatives and their use as sweetening agents, US Patent 3,087,821 (April 30, 1963).

Horowitz, R.M., and Gentili, B. 1963b. Flavonoids of citrus VI. The structure of neohesperidose, *Tetrahedron* 19:773–782.

Horowitz, R.M., and Gentili, B. 1969. Taste and structure in phenolic glycosides, 1. *Agric Food Chem* 17:696–700.

Kitagawa, M., Kusakabe, Y., Miura, H., Ninomiya, Y., and Hino, A. 2001. Molecular genetic identification of a candidate receptor gene for sweet taste. *Biochem Biophys Res Commun* 283(1):236–242.

Li, X., Inoue, M., Reed, D.R., Huque, T., Puchalski, R.B., Tordoff, M.G., Ninomiya, Y., Beauchamp, G.K., and Bachmanov, A.A. 2001. High-resolution genetic mapping of the saccharin preference locus (*Sac*) and the putative sweet taste receptor (T1R1) gene (Gpr70) to mouse distal Chromosome 4. *Mamm Genome* 12:13–16.

Li, X., Staszewski, L., Xu, H., Durick, K., Zoller, M., and Adler, E. 2002. Human receptors for sweet and umami taste. *Proc Nat Acad Sci USA* 99(7):4692–4696.

Max, M., Shanker, Y.G., Huang, L., Rong, M., Liu, Z., Campagne, F., Weinstein, H., Damak, S., and Margolskee, R.F. 2001. Tas1r3, encoding a new candidate taste receptor, is allelic to the sweet responsiveness locus Sac. *Nat Genet* 28(1):58–63.

Nelson, G., Hoon, M.A., Chandrashekar, J., Zhang, Y., Ryba, N.J., and Zuker, C.S. 2001. Mammalian sweet taste receptors. *Cell* 106:381–390.

Pattanaargson, S., Chuapradit, C., and Srisukphonraruk, S. 2001. Aspartame degradation in solutions at various pH condition, *J Food Science* 66(6):808–809.

Sainz, E., Korley J.N., Battey, J.F., and Sullivan, S.L. 2001. Identification of a novel member of the T1R family of putative taste receptors. *J Neurochem* 77:896–903.

Shallenberger, R.S. 1996. The AH,B glycophore and general taste chemistry. *Food Chem* 56(3):209–214.

Stone, H., and Sidel, J.L. 1998. Quantitative descriptive analysis: development, applications and the future. *Food Technol* 52(8):48–52.

Ujihara, M., Shinozaki, M., and Kato, M. 1995. Accumulation of phyllodulcin in sweet-leaf plants of *Hydrangea serrata* and its neutrality in the defense against a specialist leafmining herbivore. *Res Popul Ecol* 37(2):249–257.

US FDA. 2000. Toxicological principles for the safety assessment of food ingredients: Redbook 2000. Food and Drug Administration, U.S. (FDA), Center for Food Safety and Applied Nutrition (CFSAN); Washington, DC.

US FDA. 2009. Ajinomoto Co.: Filing a food additive petition. Federal Register 74 (138)35871.

Zhao, G.Q., Zhang, Y., Hoon, M.A., Chandrashekar, J., Erlenbach I., Ryba, N.J., and Zuker, C.S. 2003. The receptors for mammalian sweet and umami taste. *Cell* 115:255–266.

Chapter 4

Alitame

Michael H. Auerbach, Graeme Locke, and Michael E. Hendrick[*]

Contents

Introduction

The field of high-potency sweeteners has long been an active area of research. Stimulated by the accidental discovery of the potent sweet taste of aspartame in 1965 (Ripper et al. 1986; Mazur et al. 1969), dipeptides became an area of increasing interest for chemists and food technologists alike. Starting in the early 1970s, an intensive, systematic program to develop high-potency sweeteners was carried out at Pfizer Central Research. This program, which involved the synthesis of a large number of dipeptides of diverse structural types, culminated in the preparation of alitame. Alitame and structurally related peptide sweeteners are the subject of a US patent (Brennan and Hendrick 1983).

---

[*] Deceased.

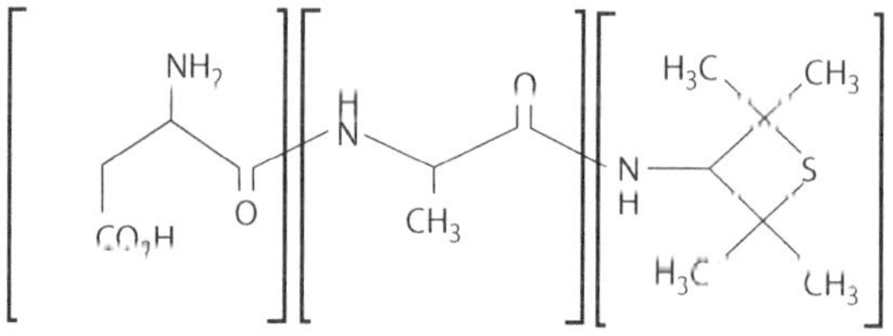

Development:
Discovered: 1979 Pfizer Central Research
Patented: 1983 (US 4,411,925)
US FDA filing: 1986 Food Additive Petition

Figure 4.1 Alitame: structure and development.

Structure

The structure of alitame is shown in Figure 4.1, which emphasizes the component parts of the molecule. Alitame is formed from the amino acids L-aspartic acid and D-alanine, with a novel C-terminal amide moiety. It is this novel amide (formed from 2,2,4,4-tetramethylthietanylamine) that is the key to the very high sweetness potency of alitame. The structure of alitame was developed by following leads from a number of synthesized model compounds. Within the series of L-aspartyl-D-alanine amides, those structural features that were found to be most conducive to high sweetness potency include small to moderate ring size, presence of small-chain branching α to the amine-bearing carbon, and the introduction of the sulfur atom into the carbocyclic ring.

Organoleptic Properties

Alitame is a crystalline, odorless, nonhygroscopic powder. Its sweetness potency, determined by comparison of the sweetness intensity of alitame solutions with concentrations in the range of 50 μg/ml to a 10% solution of sucrose, is approximately 2000 times that of sucrose. Compared with threshold concentrations of sucrose (typically 2%–3%), the potency of alitame increases to approximately 2900 times that of sucrose. This phenomenon is typical of high-potency sweeteners. Use of higher concentrations of alitame allows preparation of solutions with a sweetness intensity equivalent to that of sucrose solutions of 40% or greater.

The sweetness of alitame is of a high quality, sucrose-like and without accompanying bitter or metallic notes often typical of high-potency sweeteners. The sweetness of alitame develops rapidly in the mouth and lingers slightly, in a manner similar to that of aspartame.

Alitame has been found to exhibit synergy when combined with both acesulfame K and cyclamate. High-quality blends may be obtained with these and other sweeteners, including saccharin.

Solubility

At the isoelectric pH, alitame is very soluble in water. Excellent solubility is also found in other polar solvents, as is shown in Table 4.1. As expected from the molecule's polar structure,

Table 4.1 Solubility of Alitame in Various Solvents at 25°C

Solvent	Solubility (%w/v)
Water	13.1
Methanol	41.9
Ethanol	61.0
Propylene glycol	53.7
Chloroform	0.02
N-Heptane	0.001

alitame is virtually insoluble in lipophilic solvents. In aqueous solutions, the solubility rapidly increases with temperature and as the pH deviates from the isoelectric pH. This effect is shown in Table 4.2.

Decomposition Pathways

The principal pathways for the reaction of alitame with water are relatively simple compared with aspartame (Figure 4.2). The major pathway involves hydrolysis of the aspartylalanine dipeptide bond to give aspartic acid and alanyl-2,2,4,4-tetramethylthietane amide ("alanine amide"). The α,β-aspartic rearrangement, common to all peptides bearing N-terminal aspartic acid (Bodanszky and Martinez 1983), also occurs to give the β-aspartic isomer of alitame. This rearranged dipeptide hydrolyzes at a slower rate than alitame to give the same products as those arising from the parent compound (i.e., aspartic acid and alanine amide). No cyclization to diketopiperazine or hydrolysis of the alanine amide bond is detectable in solutions of alitame that have undergone up to 90% decomposition. All three major decomposition products are completely tasteless at levels that are typical in foods.

Table 4.2 Solubility[a] of Alitame as a Function of pH and Temperature

pH	5°C	20°C	30°C	40°C	50°C
2.0	41.7	48.7	56.4	50.3	54.0
3.0	32.2	39.2	46.5	50.9	53.9
4.0	12.9	13.9	17.3	20.4	37.6
5.0	11.7	12.8	14.9	16.8	29.2
6.0	11.6	13.2	14.9	19.5	32.8
7.0	11.8	14.3	17.6	29.5	51.8
8.0	14.8	24.9	46.8	56.2	52.1

[a] Water solubility (%w/v).

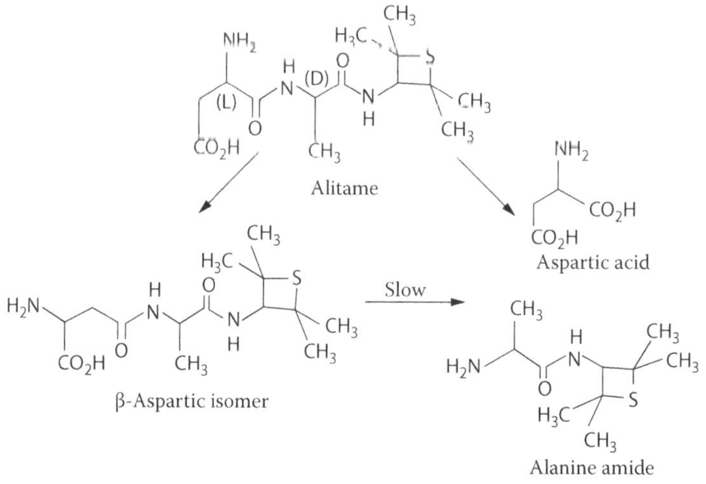

Figure 4.2   Main degradation pathways of alitame.

Stability

In Figure 4.3, the half-lives for alitame and aspartame (Homler 1984) in buffer solutions of various pH values are compared. As can be seen, the solution stability of alitame approaches the optimum for aspartic acid dipeptides. At acid pH (2–4), alitame solution half-lives are more than twice those of aspartame. As the pH increases, this stability advantage greatly increases. In particular in the neutral pH range (5–8) alitame is completely stable for more than 1 year at room temperature.

Alitame is sufficiently stable for use in hard and soft candies, heat-pasteurized foods, and in neutral pH foods processed at high temperatures, such as sweet baked goods. The dramatic heat stability advantage of alitame over aspartame under simulated baking conditions is illustrated in Figure 4.4. Actual half-life data for the two sweeteners under these conditions may be found in Table 4.3. This allows alitame to survive the thermal and pH conditions of the baking process with insignificant decomposition.

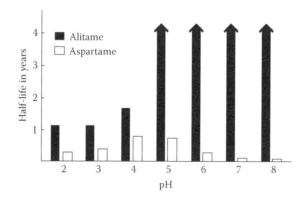

Figure 4.3   Alitame and aspartame stability in buffer solutions at 23°C.

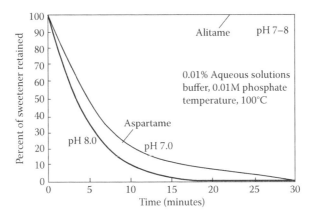

Figure 4.4   Thermal stability of alitame and aspartame.

Compatibility

Alitame exhibits excellent functionality and is compatible with a wide variety of freshly prepared foods. In accordance with its chemical structure, it can undergo chemical reaction with certain food components. In particular, high levels of reducing sugars, such as glucose and lactose, may react with alitame in heated liquid or semiliquid systems, such as baked goods, to form Maillard reaction products. High levels of flavor aldehydes can behave similarly. Such reactions have been reported for aspartame and other aspartyl dipeptides (Cha and Ho 1988; Ariyoshi and Sato 1972; Hussein et al. 1984; Stamp and Labuza 1983). Therefore, the compatibility of alitame with a given recipe will depend on the actual ingredients present and the thermal and pH exposure involved in the manufacturing process.

Prolonged storage of alitame in a few standard acidic liquid beverage recipes may result in an incompatibility as measured organoleptically (off-flavors). This is not reflected in storage stability as measured by chemical assay for alitame and its degradation products. Levels of off-flavorant(s) are below the limits of modern analytical detection. Substances which may produce off-flavors on storage with alitame in liquid products are hydrogen peroxide, sodium bisulfite, ascorbic acid, and some types of caramel color at pHs less than 4.0. Several solutions to overcome these problems have been discovered.

Table 4.3   Stability of Alitame and Aspartame at Elevated Temperature in Aqueous Solution

Sweetener	Temp (°C)	Half-Life (hr)		
		pH 6.0	pH 7.0	pH 8.0
Alitame	100	13.5	13.4	12.6
Alitame	115	2.1	2.1	2.1
Aspartame	100	0.4	0.1	0.04
Aspartame	115	0.1	0.03	0.02

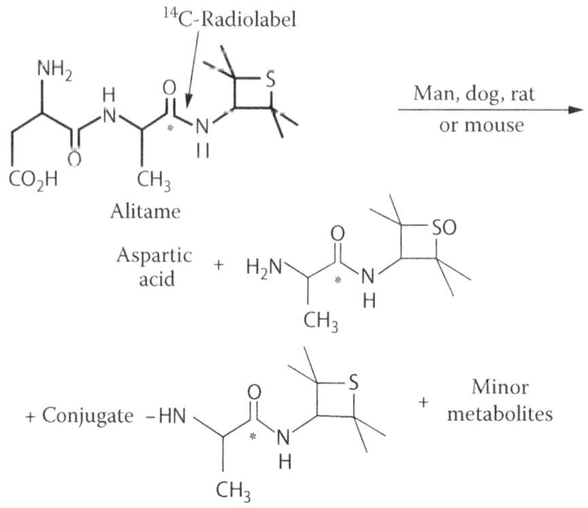

Figure 4.5 Metabolism of alitame.

Metabolism

Alitame is well absorbed after oral administration to the mouse, rat, dog, or man. Most of an oral dose (77%–96%) is excreted in urine as a mixture of metabolites (Figure 4.5). The remainder (7%–22%) is excreted in the feces, primarily as unchanged alitame. Radiochemical balances of 97%–105% were obtained in the four species.

In all four species the metabolism of alitame is characterized by the loss of aspartic acid followed by conjugation and/or oxidation at the sulfur atom of the alanine amide fragment yielding the corresponding sulfoxide isomers and the sulfone. Further hydrolysis of the alanine amide fragments does not take place. In the rat and dog the alanine amide is partially acetylated, and in man it is partially conjugated with glucuronic acid. No observable cleavage of the alanine amide bond or rupture of the thietane ring takes place.

Because the aspartic acid portion of the molecule is available for normal amino acid metabolism, alitame is partially caloric. The maximum caloric contribution of 1.4 calories per gram is clearly insignificant at the use levels in the diet.

Safety

Alitame has been evaluated in the extensive series of studies designed to establish its safety as a sweetener in the human diet. A list of the major studies completed is shown in Table 4.4. On the basis of substantial safety factors determined in these studies, it is concluded that alitame is safe for its intended use as a component of the diet of man.

The appropriateness of the animal studies is supported by the fact that the major metabolites of the compound are common to laboratory animals and man. The cumulative data amassed during the safety evaluation demonstrate that this sweetener has an extremely low order of toxicity that supports its estimated mean chronic daily human dietary intake of 0.34 mg/kg body weight, assuming that alitame is the sole sweetener in all food categories requested in the Food Additive

Table 4.4   Major Alitame Safety Studies Completed

Genetic toxicology (5 studies)
Three-month dog (0.5%, 1%, 2% of the diet)
Three-month rat (0.5%, 1%, 2% of the diet)
Rat teratology (100, 300, 1000 mg/kg/day)
Rabbit teratology (100, 300, 1000 mg/kg/day)
One-month mouse (1%, 2%, 5% of the diet)
Twelve-month rat (0.01%, 0.03%, 0.1%, 0.3%, 1% of the diet)
Eighteen-month dog (10, 30, 100, 500 mg/kg/day)
Rat reproduction (0.1%, 0.3%, 1% of the diet)
Twenty-four-month mouse oncogenicity (0.1%, 0.3%, 0.7% of the diet)
Twenty-four-month rat oncogenicity (0.1%, 0.3%, 1% of the diet; two studies in species)
Rat, whole body autoradiography and microhistoautoradiography
Rat, neurobehavioral (3 studies)
Man, metabolism
Man, no effects (15 mg/kg/day, 14 days)
Man, no effects (10 mg/kg/day, 90 days)
Diabetic men and women, no effects (10 mg/kg/day, 90 days; two studies)

Petition. In all the studies the no-effect level (NOEL) was consistently greater than 100 mg/kg, or >300 times the mean chronic human exposure.

The primary effect of alitame in animals treated with levels greater than the NOEL (>100 mg/kg) for periods ranging from five days to two years, was a dose-related increase in liver weight secondary to the induction of hepatic microsomal metabolizing enzymes, a common adaptive response of the liver to xenobiotics. This high-dose effect abated during chronic studies and after cessation of treatment. At a dose of 15 mg/kg/day in man (44 times the mean chronic intake estimate and equivalent to consumption of 18 liters of alitame-sweetened carbonated beverage per day by a 60 kg individual), no enzyme induction was observed over a 14-day period.

Furthermore, neither enzyme induction nor any other effects were observed over a 90-day period in a double-blind toleration study of 130 subjects of both sexes of three races (Caucasian, African-American, and Asian) receiving alitame 10 mg/kg/day, a dose equivalent to consumption of 12 liters of carbonated beverage per day.

No evidence of carcinogenic potential was noted in rats and mice treated for two years at doses as high as 564 and 1055 mg/kg/day, respectively. In doses as high as 1000 mg/kg/day during organogenesis, alitame was devoid of embryotoxic or teratogenic potential in rats and rabbits. In two-generation reproduction studies, there were no compound related effects on mating behavior, pregnancy rate, the course of gestation, litter size, or maternal or offspring survival.

Mutagenicity assays, both *in vitro* and *in vivo*, revealed no genotoxicity at gene or chromosomal levels in microbial and mammalian test systems. Alitame had no detectable activity in a battery of pharmacological systems used to assess autonomic, gastrointestinal, renal, and central nervous system functions. No effects were noted on fasting blood glucose or on the disposition of an oral glucose load.

Regulatory Status

In 1986 a Food Additive Petition was submitted to the US Food and Drug Administration requesting broad clearance for alitame (Federal Register 1986). The petition requested approval of alitame as a sweetener and flavoring in specified foods in amounts necessary to achieve the intended effect and in accordance with good manufacturing practice. Table 4.5 presents the 16 food product categories requested in this petition.

Table 4.5   Food Product Categories Requested in Alitame 1986 Food Additive Petition to US FDA

Baked goods and baking mixes (restricted to fruit-, custard-, and pudding-filled pies; cakes; cookies, and similar baked products)
Presweetened, ready-to-eat breakfast cereals
Milk products (restricted to flavored milk- or dairy-based beverages and mixes for their preparations, yogurt, and dietetic milk products)
Frozen desserts and mixes
Fruit and water ices and mixes
Fruit drinks, ades, and mixes, including diluted juice beverages and concentrates (frozen and non-frozen) for dilute juice beverages
Confections and frosting
Jams, jellies, preserves, sweet spreads
Sweet sauces, toppings, and mixes
Gelatins, puddings, custards, fillings, and mixes
Beverages, nonalcoholic and mixes
Dairy product analogs (restricted to toppings and topping mixes)
Sugar substitutes
Sweetened coffee and tea beverages, including mixes and concentrates
Candy (including soft and hard candies and cough drops)
Chewing gum

In addition to the United States, permission for the use of alitame in food has been requested from a number of other countries and regulatory agencies. Alitame was approved for use in Australia in December 1993; in Mexico in May 1994; in New Zealand in October 1994; in People's Republic of China in November 1994; in Indonesia in October 1995; in Colombia in April 1996, and in Chile in June 1997. Alitame was reviewed by the Joint FAO/WHO Expert Committee on Food Additives in 1995 (Abbott 1996a) and 1996 (Abbott 1996b) and allocated an acceptable daily intake (ADI) of 1 mg/kg of body weight (Abbott 1996b). A WHO/FAO specification monograph has been published (JECFA 1995). Alitame INS No. 956 is also listed in the Codex Alimentarius General Standard for Food Additives (GSFA 1997) and in the 7th edition of the *Food Chemicals Codex (FCC)* (USP 2010).

References

Abbott, P.J. 1996a. Alitame. In *Toxicological evaluation of certain food additives and contaminants*. 44th JECFA, WHO Food Additives Series 35, pp. 209–254.

Abbott, P.J. 1996b. Alitame. In *Toxicological evaluation of certain food additives*. 46th JECFA, WHO Food Additives Series 37, pp. 9–11.

Ariyoshi, Y., and Sato, N. 1972. The reaction of aspartyl dipeptide esters with ketones. *Bull Chem Soc Japan* 45:2015–2018.

Bodanszky, M., and Martinez, J. 1983. Side reactions in peptide synthesis. In *The peptides*, eds. E. Gross and J. Meienhofer, Vol. 5, 143–148. New York: Academic Press, Inc.

Brennan, T.M., and Hendrick, M.E. US Patent No. 4,411,925 (October 25, 1983).

Cha, A.S., and Ho, C.T. 1988. Studies of the interaction between aspartame and flavor vanillin by high performance liquid chromatography. *J Food Science* 53:562–564.

Federal Register 51(188):34503 (September 29,1986).

GSFA (General Standard for Food Additives), Draft Schedule 1 on Sweeteners—Additives with Numerical ADIs. In Report of the 29th Session of the Codex Committee on Food Additives and Contaminants (CCFAC), Alinorm 97/12A, March 1997, pp. 156–157, 67–78.

Homler, B.M. 1984. Properties and stability of aspartame. *Food Technology* (July) 38:50–55.

Hussein, M.M., D'Amelia, R.P., Manz, A.L., Jacin, H., and Chen, W.-T.C. 1984. Determination of reactivity of aspartame with flavor aldehydes by gas chromatography, HPLC and GPC. *J Food Science* 49:520–524.

JECFA, 44th meeting, 1995. Alitame. In *Compendium of food additive specifications*. FAO Food and Nutrition Paper 52 Addendum 3, pp. 9–13.

Mazur, R.H., Schlatter, J.M., and Goldkamp, A.H. 1969. Structure-taste relationships of some dipeptides. *J Am Chem Soc* 91:2684–2691.

Ripper, A., Homler, B.E., and Miller, G.A. 1986. Aspartame. In *Alternative sweeteners*, eds. L.O Nabors and R.C. Gelardi, 43–70. New York: Marcel Dekker, Inc.

Stamp, J.A., and Labuza, T.P. 1983. Kinetics of the Maillard Reaction Between Aspartame and Glucose in Solution at High Temperatures. *J Food Science* 48:543–544.

United States Pharmacopeia (USP). 2010. Alitame. In *Food Chemicals Codex (FCC)*, 7th edition, pp. 30 31. Rockville, MD: United States Pharmacopeial Convention.

Chapter 5

Aspartame

Eyassu G. Abegaz, Dale A. Mayhew, Harriett H. Butchko,
W. Wayne Stargel, C. Phil Comer, and Sue E. Andress

Contents

Introduction and History

More than four decades have elapsed since aspartame was discovered accidentally in 1965 by G.D. Searle and Co. chemist James Schlatter (Mazur 1976, Mazur and Ripper 1979). Currently, it is estimated that aspartame is used in approximately 6,000 different products worldwide.

The safety of aspartame has been tested extensively in animal and human studies. It is undoubtedly the most thoroughly studied of the high intensity sweeteners. The safety of aspartame has been affirmed by numerous scientific bodies and regulatory agencies, including the Joint FAO/WHO Expert Committee on Food Additives of the Codex Alimentarius (JECFA 1980), the Scientific Committee for Food of the Commission of European Communities (SCF 1989), the US Food and Drug Administration (US FDA 1981, 1983), and the regulatory agencies of more than 100 additional countries around the world.

Physical Characteristics and Chemistry

Structure

Aspartame is a dipeptide composed of two amino acids, L-aspartic acid and the methyl ester of L-phenylalanine. The chemical structure of aspartame is depicted in Figure 5.1. Aspartame sold for commercial use meets all requirements of the *Food Chemicals Codex* (*FCC* 2010).

Stability

In liquids and under certain conditions of moisture, temperature, and pH, the ester bond is hydrolyzed forming the dipeptide, aspartylphenylalanine, and methanol. Ultimately, aspartylphenylalanine can be hydrolyzed to its individual amino acids, aspartate and phenylalanine (Mazur et al. 1969; Mazur 1974). Alternatively, methanol may also be hydrolyzed by the cyclization of aspartame to form its diketopiperazine (DKP) (Figure 5.2).

The decomposition of aspartame is indicative of first-order kinetics, and stability is determined by time, moisture, temperature and pH. Under dry conditions, the stability of aspartame is excellent; it is, however, affected by extremely high temperatures which are not typical for the

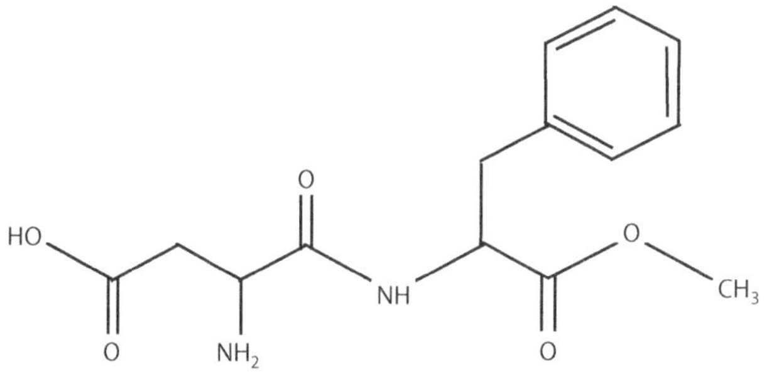

L-aspartyl-L-phenylalanine methyl ester

Figure 5.1   Chemical structure of aspartame.

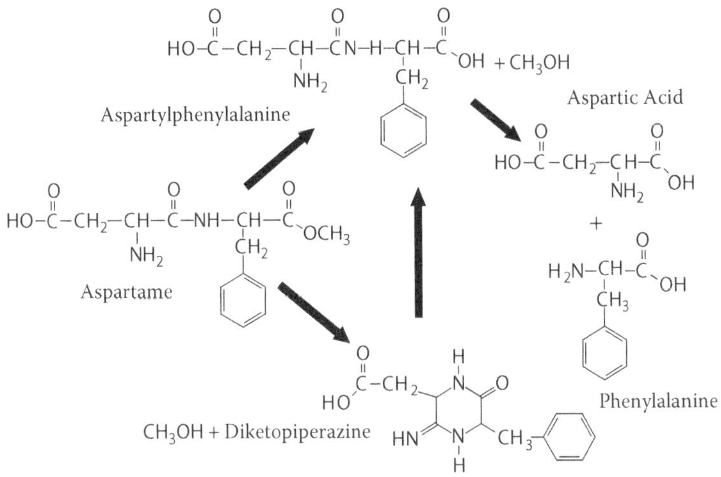

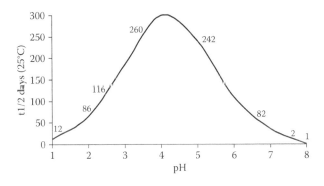

Figure 5.2   Principal conversion products of aspartame.

production of dry food products (Beck 1974, 1978). The combined effect of time, temperature, and pH on the stability of aspartame in solutions is shown in Figure 5.3.

At 25°C, the maximum stability is observed at pH ~4.3. Aspartame functions very well over a broad range of pH conditions but is most stable in the weak acidic range in which most foods exist (between pH 3 and pH 5). A frozen dairy dessert may have a pH ranging from 6.5 to more than 7.0 but, due to the frozen state, the rate of reaction is dramatically reduced. In addition, because of the lower free moisture, the shelf life stability of aspartame exceeds the predicted shelf life stability of these products.

Solubility in Food and Beverage Applications

Aspartame is slightly soluble in water (approximately 1.0% at 25°C) and is sparingly soluble in alcohol (Beck 1974, 1978). It is not soluble in fats or oils. An important consideration is the rate of dry mix dissolution, especially for many products prepared by mixing aspartame-sweetened products with water or with milk at different temperatures. Solubility is a function of both temperature and pH.

Figure 5.3   Stability of aspartame in aqueous buffers at 25°C.

Taste

The taste of aspartame is described as clean and sweet like sugar but without the bitter chemical or metallic aftertaste often associated with some other high-intensity sweeteners. Comparisons of the sweetness of aspartame and sucrose employing quantitative descriptive analyses reveal that the taste profile for aspartame closely resembles that for sucrose. Consumers as well as food industry and university-based studies have documented this sugar-like taste of aspartame (Mazur et al. 1969; Schiffman 1984).

Flavor-Enhancing Property

Various food and beverage flavors are enhanced or extended by aspartame, especially acid fruit flavors. Such flavor enhancements or flavor extensions are particularly evident with naturally derived flavors (Baldwin and Korschgen 1979). This flavor-enhancing property, as evident in chewing gum, may extend flavor up to four times longer (Bahoshy et al. 1976). Such a characteristic is important in many food applications.

Sweetness Intensity

Depending on the food or beverage system, the intensity (i.e., potency) of aspartame has been determined to be 160 to 220 times the sweetness of sucrose (Beck 1974, 1978). Generally, an inverse relationship exists between the intensity of aspartame and the concentrations of sucrose being replaced. Overall, the relative sweetness of aspartame may vary depending upon the flavor system, the pH, and the amount of sucrose or other sugars being replaced (Beck 1974, 1978; Homler et al. 1991).

Food and Beverage Applications

Aspartame is approved for general use in foods (US FDA 1996a), including carbonated soft drinks, powdered soft drinks, yogurt, hard candy, confectionery and so forth. The stability of aspartame is excellent in dry-product applications (e.g., tabletop sweetener, powdered drinks, dessert mixes). Aspartame can withstand the heat processing used for dairy products and juices, aseptic processing and other processes where high-temperature short-time and ultra-high temperature conditions are employed. The potential for aspartame to hydrolyze or cyclize may, under some conditions of excessive heat, limit some applications of aspartame.

There is a wide range over which aspartame sweetness levels are acceptable. The loss of aspartame due to certain combinations of pH, moisture and temperature can lead to a gradual loss of perceived sweetness, with no development of off-flavors since conversion products of aspartame are tasteless (Beck 1974, 1978).

Blends or combinations of sweeteners are often used to achieve the desired level of sweetness in food and beverage products that traditionally have been sweetened with single sweeteners. Aspartame works well in admixture with other sweeteners including sugar. The flavor enhancement quality of aspartame masks bitter flavors even at sub-sweetening levels and makes aspartame a desirable choice in blends with those sweeteners which possess potentially undesirable or more complex taste profiles.

Established Safety of Aspartame and Its Components

Aspartame has been proven to be a remarkably safe sweetener with more than 200 scientific studies in animals and humans confirming its safety. Vigilant postmarketing surveillance of anecdotal complaints from consumers revealed no consistent pattern of symptoms related to consumption of aspartame. An extensive postmarketing research program to evaluate these allegations in controlled, scientific studies further confirmed that aspartame is not associated with adverse health effects.

Acceptable Daily Intake versus Actual Intake

Based on the results of the comprehensive safety studies in animals (Molinary 1984; Kotsonis and Hjelle 1996), an Acceptable Daily Intake (ADI) of 40 mg/kg/day for aspartame was set by JECFA (1980). Based on both animal and human data, the US FDA set an ADI for aspartame of 50 mg/kg/day (US FDA 1984).

The ADI represents the amount of a food additive that can be consumed daily for a lifetime with no ill effects (Lu 1988; Renwick 1990, 1991). It is not a maximum amount that can be safely consumed on a given day. A person may occasionally consume a food additive in quantities exceeding the ADI without adverse effects. Aspartame is approximately 200 times sweeter than sugar, therefore an ADI of 50 mg/kg is the sweetness equivalent of approximately 600 grams (1.3 pounds) of sucrose consumed daily by a 60 kg person over a lifetime. If aspartame replaced all the sucrose in our diet, consumption would be well below the ADI at approximately 8.3 mg/kg/day (US FDA 1981).

Actual consumption levels of aspartame were monitored from 1984 to 1992 through dietary surveys in the United States (Butchko and Kotsonis 1991, 1994, 1996; Butchko et al. 1994). Average daily aspartame consumption at the 90th percentile ("eaters" only) in the general population ranged from approximately 2 to 3 mg/kg. Consumption by 2–5 year old children in these surveys ranged from approximately 2.5 to 5 mg/kg/day. Aspartame consumption has also been estimated in several other countries around the world. Although survey methodologies differed among these evaluations, aspartame consumption is remarkably consistent and is only a fraction of the ADI (Virtanen et al. 1988; Heybach and Ross 1989; Hinson and Nicol 1992; Bar and Biermann 1992; Bergsten 1993; Collerie de Borely and Renault 1994; MAFF 1995; Hulshof and Bouman 1995; Toledo and Ioshi 1995; NFAA 1995; Leclercq et al. 1999; Garnier-Sagne et al. 2001; Ilback et al. 2003; Arcella et al. 2004; FSANZ 2004; Chung et al. 2005; Renwick 2006; Leth et al. 2007; Husøy et al. 2008; Lino et al. 2008).

Extensive Safety Studies with Aspartame in Humans

In addition to the comprehensive battery of toxicology studies in animals, the safety of aspartame and its metabolic constituents has been exhaustively assessed in humans. In addition to healthy males and females, several human subgroups were studied, including infants, children, adolescents, obese individuals, diabetics, individuals with renal disease, individuals with liver disease, lactating women, and individuals heterozygous for the genetic disease, phenylketonuria (PKU), who have a decreased ability to metabolize phenylalanine. These and longer-term studies showed no untoward health consequences from aspartame (Hoffman 1972, 1973; Langlois 1972; Frey 1973, 1976; Knopp et al. 1976; Koch et al. 1976; Stern et al. 1976; Stegink et al. 1977, 1979a, 1979b, 1980, 1981a, 1981b; Nerhling et al. 1985; Horwitz et al. 1988; London, 1988; Gupta et al. 1989; Hertelendy et al. 1993).

The results of the human studies, along with the animal research, have provided convincing evidence that aspartame is safe for the general population, including pregnant women and children

Safety of Dietary Components of Aspartame

Following consumption, aspartame is rapidly metabolized into its components: aspartate, phenylalanine, and methanol. These components have sometimes been the objects of speculation regarding their potential for adverse effects. As discussed below, the components of aspartame are normal constituents of the diet, and there is no evidence for adverse effects associated with consumption of aspartame.

Aspartate

Plasma concentrations of aspartate are not altered by even enormous amounts of aspartame (Stegink et al. 1977, 1979a, 1979b, 1980, 1981a, 1988; Filer et al. 1983). Oral administration at a dosage of 34 mg/kg/day, more than ten times estimated 90th percentile consumption, does not change plasma aspartate concentrations in humans (Stegink et al. 1977). Chronic administration of even higher doses of aspartame (75 mg/kg/day for 24 weeks) did not change mean fasting plasma concentrations of aspartate (Leon et al. 1989). The addition of glutamate and aspartame (each at 34 mg/kg) to a meal providing 1 gram protein/kg did not increase plasma concentrations of either glutamate or aspartate beyond the changes due to the meal itself (Stegink et al. 1982; Stegink et al. 1987). As Figure 5.4 illustrates, at current levels of consumption only a small fraction of daily dietary intake of aspartate is derived from aspartame in adults or children (Butchko and Kotsonis 1996).

Phenylalanine

Consumption of aspartame-sweetened foods does not increase plasma phenylalanine concentrations beyond those which normally occur postprandially (Stegink et al. 1977, 1979a, 1979b, 1980). Doses of aspartame at approximately 30 mg/kg/day did not increase plasma phenylalanine concentrations above those observed after eating a protein-containing meal in normal adults,

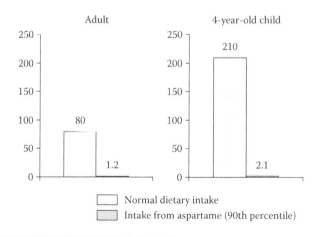

Figure 5.4   Dietary intake of aspartic acid vs. intake from aspartame in adults and children.

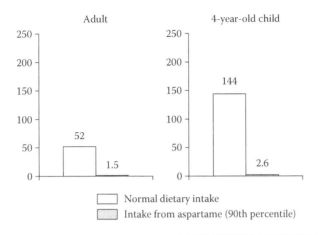

Figure 5.5   Dietary intake of phenylalanine vs. intake from aspartame in adults and children.

phenylketonuric heterozygotes, or non-insulin dependent diabetic populations (Filer and Stegink 1989). As Figure 5.5 illustrates, at current levels of consumption only a small fraction of daily dietary intake of the essential amino acid phenylalanine is derived from aspartame in adults or children (Butchko and Kotsonis 1996). The only individuals who must be concerned regarding aspartame's phenylalanine content are those with phenylketonuria, a rare genetic disease in which the body cannot properly metabolize phenylalanine. These individuals must severely restrict phenylalanine intake from all dietary sources, including aspartame.

Methanol

Aspartame yields approximately 10% methanol by weight. The amount of methanol released from aspartame is well below normal dietary exposures to methanol from fruits, vegetables, and juices (Figure 5.6) (US FDA 1983; Butchko and Kotsonis 1991). Aspartame consumption, even in amounts many times those consumed in foods and beverages, does not alter baseline blood

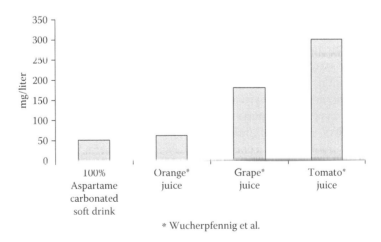

Figure 5.6   Methanol content of beverage sweetened with aspartame vs. fruit juices. (From Wucherpfennig et al., *Flussuges Obst.*, 8:348–354, 1983.)

concentrations of methanol or formate (Steginnk et al. 1981b, 1983). Whereas methanol exposure at the 90th percentile of chronic aspartame consumption is 0.3 mg/kg/day, the US FDA has established acceptable levels of exposure to methanol at 7.1 to 8.4 mg/kg/day for 60 kg adults (US FDA 1996b). Thus acceptable dietary exposures to methanol are approximately 25 times potential exposures to methanol following 90th percentile consumption of aspartame.

Trocho et al. (1998) concluded from a study in rats that aspartame may be hazardous because formaldehyde adducts in tissue proteins and nucleic acids from aspartame may accumulate. However, according to Tephly (1999), the doses of aspartame used in the study do not even yield blood methanol concentrations outside control values. Further, the amounts of aspartame equal to that in about 75 cans of beverage as a single bolus result in no detectable increase in blood formate concentrations in humans whereas increased urinary formate excretion shows that the body is well able to handle even excessive amounts of aspartame. In addition, there is no accumulation of blood or urinary methanol or formate with long term exposure to aspartame. Thus, Tephly (1999) concluded that "the normal flux of one-carbon moieties whether derived from pectin, aspartame, or fruit juices is a physiologic phenomenon and not a toxic event."

Postmarketing Surveillance of Anecdotal Health Reports

Aspartame has been the subject of extensive postmarketing surveillance. Shortly after its widespread marketing, there were a number of anecdotal reports of health effects which some consumers related to their consumption of aspartame-containing products. Not unexpectedly, negative media stories influenced the numbers and types of these reports. The NutraSweet Company took the unprecedented step of instituting a postmarketing surveillance system to document and evaluate these anecdotal reports (Butchko and Kotsonis 1994; Butchko et al. 1994, 1996). Data from this system were evaluated by the company and also shared with the US FDA, as discussed below. Following the approval of aspartame in carbonated beverages in 1983, an increase in the reporting of adverse health events allegedly associated with the consumption of aspartame-containing products led the US FDA to request the Centers for Disease Control (CDC) to evaluate these reports (CDC 1984; Bradstock et al. 1986). The CDC was charged with describing the types of complaints and with determining if there were any specific clusters or types of complaints that would indicate the need for further study. The CDC (1984) concluded, "Despite great variety overall, the majority of frequently reported symptoms were mild and are symptoms that are common in the general populace." The CDC could not identify any specific group of symptoms that were clearly related to aspartame but felt that focused clinical studies would be the best mechanism to address the issues raised by these reports.

In the mid-1980s, the US FDA Center for Food Safety and Applied Nutrition (CFSAN) established its own passive surveillance system, the Adverse Reaction Monitoring System (ARMS) to monitor and evaluate anecdotal reports of adverse health effects thought to be related to foods, food and color additives, and vitamin/mineral supplements (Tollefson 1988; Tollefson et al. 1988). Through this system, spontaneous reports of food-associated adverse health events received from consumers, physicians, and industry are documented, investigated, and evaluated.

Based upon reviews of anecdotal complaints for aspartame, the US FDA concluded that there is no "reasonable evidence of possible public health harm" and "no consistent or unique patterns of symptoms reported with respect to aspartame that can be causally linked to its use" (Tollefson 1988; Tollefson et al. 1988). In a recent evaluation, they further concluded that there was a gradual decrease in the number of reports regarding aspartame received over time and that the reports remained comparable to previous ones in terms of demographics, severity, strength of association, and symptoms (Wilcox and Gray 1995).

In considering the anecdotal reports, it is important to keep in mind that, since approximately 100 million people consume aspartame in the United States alone, it is not unlikely that a consumer may experience a medical event or ailment near the same time aspartame is consumed. The error of inclinations to associate causality to coincidence is perhaps best stated by one scientist who indicated, "As aspartame is estimated to be consumed by about half the US population, one need not be an epidemiologist to grasp the problem of establishing a cause-and-effect relationship. Half the headaches in America would be expected to occur in aspartame users, as would half the seizures and half the purchases of Chevrolets." (Raines 1987).

Research to Investigate Issues Raised in the Postmarketing Period

In order to expand the knowledge base about aspartame and further address the anecdotal reports and other scientific issues raised during the postmarketing period, a number of studies, including "focused" studies in humans as suggested by the CDC, were carried out. The results of a long-term clinical study with high doses of aspartame (75 mg/kg/day for 24 weeks, or approximately 25–30 times current consumption levels at the 90th percentile), for example, provided additional confirmatory evidence of aspartame's safety (Leon et al. 1989). Even after these enormous daily doses of aspartame, there were no changes in clinical or biochemical parameters or increased adverse experiences compared with a placebo. Clinical studies to evaluate whether aspartame causes headache, seizures, or allergic-type reactions were performed (Kulczycki 1986; Schiffman et al. 1987, 1995; Amery 1988; Koehler and Glaros 1988; Schiffman 1988; Garriga et al. 1991; Van den Eden et al. 1991; Camfield et al. 1992; Geha et al. 1993; Shaywitz and Novotny 1993; Shaywitz et al. 1994a; Levy et al. 1995; Rowan et al. 1995; Jacob and Stechschulte 2008; Abegaz and Bursey 2009). In some of these studies, individuals identified through the company's medical postmarketing surveillance system, who were convinced that aspartame caused their symptoms, were studied (Schiffman et al. 1987; Geha et al. 1993; Shaywitz et al. 1994a; Rowan et al. 1995). In addition, studies were performed to evaluate whether aspartame had any effect on brain neurotransmitter concentrations or neurotransmission or on indicators of brain function, such as memory, learning, mood, and behavior (Wurtman, 1983, Wolraich et al. 1985; Caballero et al. 1986; Ferguson et al. 1986; Milich and Pelham 1986; Goldman et al. 1986; Ryan-Harshman et al. 1987; Kruesi et al. 1987; Garattini et al. 1988, 1989, 1990; Saravis et al. 1990; Stokes et al. 1991; Walton et al. 1993; Butchko et al. 1994; Lajtha et al. 1994; Shaywitz et al. 1994a, 1994b; Stokes et al. 1994; Trefz et al. 1994; Wolraich et al. 1994; Spiers et al. 1998; Simintzi et al. 2007a, 2007b; Abegaz and Bursey 2008; Renwick 2008). The weight of evidence from the results of the research done with aspartame clearly demonstrates that, even in amounts many times greater than those typically consumed, aspartame is not associated with adverse health effects (Stegink and Filer 1984; AMA 1985; Stegink 1987a, 1987b; Janssen et al. 1988; Butchko and Kotsonis 1989; Fisher 1989; Sze 1989; Fernstrom 1991; Meldrum 1993; Jobe and Dailey 1993; Tschanz et al. 1996).

False Allegations of Brain Tumors

In 1996, a group led by long-time aspartame critic John Olney contended that the reported increase in the rate of brain tumors in the United States was related to marketing of aspartame (Olney et al. 1996). Olney and colleagues described what they termed a "surge in brain tumors in the mid 1980s" based on selective analysis of the US Surveillance Epidemiology and End Results (SEER) tumor data base.

The arguments of Olney et al. implicitly require two biologically indefensible assumptions: firstly, that a certain factor (aspartame) could cause an observed increase in the incidence of brain cancer in less than four years and secondly, that even more widespread exposure to this factor would cause no further increase in the incidence of that cancer in subsequent years. However, the trend of increased brain tumor rates started well before aspartame was approved, and overall brain tumor rates have actually been decelerating in recent years (Levy and Hedeker 1996).

Further, the pattern of increased brain tumor rates has been noted primarily in the very elderly (Muir et al. 1994; Greig et al. 1990; Werner et al. 1995; Davis et al. 1991), not the typical age group of aspartame consumers. In addition, it is widely thought that apparent increases in brain tumor rates in the mid-1980s may not reflect genuine increases in brain tumors but rather enhanced detection, largely resulting from the availability of sophisticated non-invasive diagnostic technology, such as CT and MRI (Boyle et al. 1990; Greig et al. 1990; Marshall 1990; Davis et al. 1991; La Vecchia et al. 1992; Modan et al. 1992; Muir et al. 1994; Werner et al. 1995).

Epidemiologists have criticized Olney and coworkers' attempted association between the introduction of aspartame and occurrence of brain tumors (Davies et al. 1996; Ross 1998). For example, Ross (1998) stated, "From an epidemiologic perspective, the conclusion of the report may well represent a classic example of 'ecologic fallacy'... . There is no information available regarding whether the individuals who developed brain tumors consumed aspartame. For example, one might also invoke (a) cellular phone, home computer, and VCR usage; (b) depletion of the ozone layer; or (c) increased use of stereo headphones as potentially causative agents ... some or all of these possibilities may or may not have any biological plausibility to the observed associations." Seife (1999) humorously chided that Olney and coworkers had neglected to consider the close statistical correlation that exists between increased brain tumor incidence and the rise of national debt driven by supply-side economics in the mid-1980s.

Further, a case-control study specifically evaluating aspartame consumption and the risk of childhood brain tumors was published by Gurney et al. (1997). The results of the study showed that children with brain tumors were no more likely to have consumed aspartame than control children and that there was no elevated risk from maternal consumption of aspartame during pregnancy. Gurney and coworkers concluded, "... it appears unlikely that any carcinogenic effect of aspartame ingestion could have accounted for the recent brain tumor trends as Olney et al. contend" (Gurney 1997).

Nonetheless, the allegations regarding aspartame and brain tumors have been carefully evaluated by scientists at regulatory agencies in the United States, Australia/New Zealand, the United Kingdom, and the European Union with the unanimous conclusion that aspartame does not cause cancer (US FDA 1996c; SCF 1997; ANZFA 1997).

European Ramazzini Foundation Aspartame Feeding Studies in Rats and Allegations of Cancers

In 2006, researchers from European Ramazzini Foundations (ERF) published a study where they alleged rats administered aspartame with their feed over a life-time (until natural death) had increased malignant tumors, increased incidence of lymphoma and leukemias, transitional cell carcinomas of the renal pelvis and urethra and an increased positive trend of malignant schwanomas of peripheral nerves (Soffritti et al. 2006). The investigators concluded aspartame is a "multipotential carcinogenic agent" and that "a reevaluation of the present guidelines on the use and consumption of aspartame is urgent and cannot be delayed" (Soffritti et al. 2006).

In 2007, the ERF published a second study in male and female rats that were administered aspartame with their feed from the 12th day of fetal life until natural death (Soffritti et al. 2007). The authors reported significant dose related increases in incidents of lymphomas and leukemias in male and female rats and significant dose related increases in mammary cancer in female rats. The authors alleged that their study confirmed that an increased carcinogenic effect is observed in rats exposed to aspartame over a life-time, starting from fetal life (Soffritti et al. 2007).

Following the publication of the ERF studies, the European Community requested the European Food Safety Authority to evaluate the ERF aspartame rat study and alleged health effects (EFSA 2006, 2009). After conducting a thorough review, the EFSA's expert panel expressed concern that:

1. Increased incidents of lymphoma/leukemia in ERF aspartame treated rats may be attributed to high background incidence of chronic inflammatory changes in the lung due to chronic respiratory disease infection in the rat colony.
2. The preneoplastic and neoplastic lesions of the renal pelvis, urethra and bladder in female rats in the high dose groups were most likely from an imbalance in calcium metabolism.
3. There is uncertainty and possible misdiagnosis of the reported malignant schwannoma tumors in aspartame treated rats.
4. An unorthodox approach was undertaken by the ERF investigators whereby all malignant tumor incidents/malignant tumor bearing animals were aggregated for statistical purposes (which was not justified).
5. The reported increased incidence of mammary carcinomas was highly variable within the ERF carcinogenic studies and was not reported in their first aspartame study in which rats were fed much higher doses of aspartame.

The panel concluded, "on the basis of all the evidence currently available from the ERF study, other recent studies and previous evaluations that there is no reason to revise the previously established ADI for aspartame of 40 mg/kg bw" (EFSA 2006).

It is important to note also that in 2005, the National Toxicology Program (NTP), a research arm of the US National Institute of Environmental Health Sciences (NIEHS), published the results of three carcinogenicity and toxicity studies involving aspartame feeding in three different transgenic mice models, sensitive for spontaneous lymphomas and sarcomas, brain carcinogens and forestomach tumors. It was concluded that there was no evidence of carcinogenicity observed in three transgenic models that consumed aspartame doses as high as 7500 mg/kg/day (NTP 2005).

After the publication of the ERF aspartame study and alleged cancer effects, Magnuson et al. (2007) also published one of the most comprehensive reviews on the safety of aspartame. In this publication, the authors, including leading experts in toxicology and other safety related fields, reviewed all the chronic toxicity studies with aspartame that were submitted to regulatory agencies for the pre-market market approval of aspartame and also the two ERF aspartame studies. The expert panel concluded that aspartame does not present a cancer risk at the current level of intakes (Magnuson et al. 2007). Other investigators also dismiss the ERF aspartame studies and concur with conclusions of the EFSA expert panel (Abegaz 2007; Magnuson and Williams 2008; Schoeb et al. 2009, 2010; Ward and Alden 2009).

Epidemiological studies on the effect of aspartame on cancer have been published since the two ERF studies publications. A 2006 US National Institutes of Health (NIH) prospective study looked at the association between aspartame beverage consumption and hematopoietic cancers (lymphomas and leukemias) and gliomas (brain cancer). The study recruited 285,079 men and 188,905 women between the ages of 50 and 71 years who participated in the US NIH-AARP Diet Health Study cohort from 1995–2000. The investigators collected data on daily intake of aspartame-containing beverages from base line using self administered food frequency question naires. Histologically confirmed incidences of cancer were identified from state cancer registries The results of the study indicated that even higher levels of aspartame intake were not associated with a risk of overall hematopoietic cancer and glioma in men and women (Lim et al. 2006).

Gallus et al. (2007) and Bosetti et al. (2009) also investigated an integrated network of case-control studies between 1991 and 2004 using histologically confirmed incidences of cancers. They studied the association of consumption of the low calorie sweetener saccharin and other sweeteners including aspartame-containing products and the risk of common neoplasms in Italy. The investigators reported no evidence of adverse effects of low-calorie sweeteners (aspartame) consumption and the risk of oral cavity, pharynx, esophagus, colon, rectum, larynx, breast, ovary, prostate, kidney, stomach, pancreas and endometrial cancers (Gallus et al. 2007; Bosetti et al. 2009).

Other major regulatory agencies such as the Health Canada, Food Standards Australia New Zealand, US FDA, and the UK Food Standards Agency have all reviewed the ERF aspartame studies, concluded that that there is no need to revise previously published ADIs and stand behind an overwhelming body of scientific evidence that exists to support the safety of aspartame (Health Canada 2006; FSANZ 2006; US FDA 2007; EFSA 2006, 2010; FSA 2008).

Internet Misinformation

Although the Internet is a revolutionary source of information, anyone can post information on it without being held accountable for accuracy. The Internet has unfortunately been exploited and provided an unwitting forum for dissemination of calculated misinformation, scientifically unfounded allegations, and speculations by a handful of individuals. Society has unfortunately not yet developed the critical evaluation skills necessary for distinguishing Internet fact from Internet fiction. A number of websites, for example, use pseudoscience to attribute any number of maladies to aspartame. Virtually all of the information on these websites is distorted, anecdotal, from anonymous sources, or is scientifically implausible. In a number of cases, responsible medical organizations have been compelled to respond to bogus Internet allegations. For example, the senior medical advisor of one organization so targeted, the Multiple Sclerosis Foundation, stated, "This campaign by the 'aspartame activists' is not innocent drum banging" as they have created a danger in that individuals who should seek appropriate medical treatment instead blame aspartame for their medical conditions. He further stated that whatever the ultimate agenda of 'aspartame activists,' "it is not public health." (Squillacote 1999).

Misinformation on the Internet has caught the attention of various regulatory agencies, such as the US FDA (Henkel 1999), the Ministry of Health in Brazil (Agenci Saude 1999) and the Ministry of Health in the United Kingdom (FSA 2000). These agencies have evaluated the allegations, including those regarding a number of serious diseases such as multiple sclerosis, lupus erythematosus, Gulf War syndrome, brain tumors, and a variety of other diseases. They have concluded that these allegations are anecdotal and that there is no reliable scientific evidence that aspartame is responsible for any of these conditions, thus reaffirming the safety of aspartame. Further, scientific organizations around the globe have also rebutted these attacks on the Internet and issued statements of support for aspartame's safety (CCC 2010).

Worldwide Regulatory Status

In addition to the United States, aspartame has been approved for food and beverage and/or table-top sweetener use in more than 100 countries.

Conclusion

The availability of aspartame to food manufacturers worldwide has been one of the major factors responsible for the growth of the "light" and "low-calorie" segments of the food industry. Aspartame provides many opportunities for formulating new products while lowering or limiting calories and sugar consumption. The clean, sugar-like taste and unique flavor-enhancing properties of aspartame, combined with the exhaustive documentation of its safety, have contributed to acceptance by consumers, the food industry, and health professionals worldwide.

References

Abegaz, E.G. 2007. Aspartame not linked to cancer. *Environ Health Perspect* 115:A16– A17.

Abegaz, E.G., and Bursey, R.G. 2008. Response to "The effect of aspartame on acetylcholinesterase activity in hippocampal homogenates of suckling rats" by Simmintzi et al. *Pharmacol Res* 57:87–88.

Abegaz, E.G., and Bursey, R.G. 2009. Formaldehyde, aspartame, migraines: A possible connection. *Dermatitis* 20:176–177.

Agencia Saude, Sao Paulo, Brazil. 1999. Forum of scientific discussion: Aspartame. http://anvsl.saude.gov.br/Alimento/aspartame.html.

American Medical Association (AMA). 1985. Council on scientific affairs. Aspartame: Review of safety issues. *JAMA* 254:400–402.

Amery, W.K. 1988. More on aspartame and headache. *Headache* 28:624.

Arcella, D., Le Donne, C., Piccinelli, R., and Leclercq, C. 2004. Dietary estimated intake of intense sweeteners by Italian teenagers. Present levels and projections derived from the INRAN-RM-2001 food survey. *Food Chem Toxicol* 42:677–685.

Australia New Zealand Food Authority (ANZFA). 1997. Information paper. Aspartame: Information for consumers.

Bahoshy, B.J., Klose, R.E., and Nordstrom, H.A. 1976. Chewing gums of longer lasting sweetness and flavor, US Patent 3,943,258, General Foods Corp., White Plains, New York, March, 9.

Baldwin, R.E., and Korschgen, B.M. 1979. Intensification of fruit-flavors by aspartame. *J Food Sci* 44:938–939.

Bar, A., and Biermann, C. 1992. Intake of intense sweeteners in Germany. *Z Ernahrungswiss* 31:25–39.

Beck, C.I. 1974. Sweetness, character and applications of aspartic acid-based sweeteners. In *ACS sweetener symposium*, ed. G. Inglett, 164–181. Westport: AVI Publishing.

Beck, C.I. 1978. Application potential for aspartame in low calorie and dietetic foods. In *Low calorie and special dietary foods*, ed. B.K. Dwivedi, 59–114. West Palm Beach: CRC Press.

Bergsten, C. 1993. Intake of acesulfame, aspartame, cyclamate and saccharin in Norway. SNT Norwegian Food Control Authority: Report 3.

Bosetti, C., Gallus, S., Talamini, R., Montella, M., Franceschi, S., Negri, E., and La Vecchia, C. 2009. Artificial sweeteners and the risk of gastric, pancreatic, and endometrial cancers in Italy. *Cancer Epidemiol Biomarkers Prev* 18:2235–2238.

Boyle, P., Maisonneuve, P., Saracci, R., and Muir, C.S. 1990. Is the increased incidence of primary malignant brain tumors in the elderly real? *J Natl Cancer Inst* 82:1594–1596.

Bradstock, M.K., Serdula, M.K., Marks, J.S., Barnard, R.J., Crane, N.T., Remington, L., and Trowbridge, F. L. 1986. Evaluation of reactions to food additives: The aspartame experience. *Am J Clin Nutr* 43:464–469.

Butchko, H.H., and Kotsonis, F.N. 1989. Aspartame: Review of recent research. *Comments Toxicol* 3:253–278.

Butchko, H.H., and Kotsonis, F.N. 1991. Acceptable daily intake vs. actual intake: The aspartame example. *J Am Coll Nutr* 10:258–266.

Butchko, H.H., Tschanz, C., and Kotsonis, F.N. 1994. Postmarketing surveillance of food additives. *Reg Toxicol Pharmacol* 20:105–118.

Butchko, H.H., and Kotsonis, F.N. 1994. Postmarketing surveillance in the food industry: The aspartame case study. In *Nutritional toxicology*, eds. F.N. Kotsonis, M. Mackey, and J. Hjelle, 235–250. New York: Raven Press.

Butchko, H.H. 1994. Adverse reactions to aspartame: Double-blind challenge in patients from a vulnerable population by Walton et al. *Biol Psychiatry* 36:206–207.

Butchko, H.H., and Kotsonis, F.N. 1996. Acceptable daily intake and estimation of consumption. In *The clinical evaluation of a food additive: Assessment of aspartame,* eds. C. Tschanz, H.H. Butchko, W.W. Stargel, and F.N. Kotsonis, 43–53. Boca Raton: CRC Press.

Butchko, H.H., Tschanz, C., and Kotsonis, F.N. 1996. Postmarketing surveillance of anecdotal medical complaints. In *The clinical evaluation of a food additive: Assessment of aspartame*, eds. C. Tschanz, H. H. Butchko, W.W. Stargel, and F.N. Kotsonis, 183–193. Boca Raton: CRC Press.

Caballero, B., Mahon, B.E., Rohr, F.J., Levy, H.L., and Wurtman, R.J. 1986. Plasma amino acid levels after single-dose aspartame consumption in phenylketonuria, mild hyperphenylalaninemia, and heterozygous state for phenylketonuria. *J Pediatr* 109:668–671.

Camfield, P.R., Camfield, C.S., Dooley, J., Gordon, K., Jollymore, S., and Weaver, D.F. 1992. Aspartame exacerbates EEG spike-wave discharge in children with generalized absence epilepsy: A double-blind controlled study. *Neurology* 42:1000–1003.

Calorie Control Council (CCC). http://www.aspartame.org. Accessed September 17, 2010.

Centers for Disease Control and Prevention (CDC). 1984. Evaluation of consumer complaints related to aspartame use. MMWR 33:605–607.

Collerie de Borely, A., and Renault C. 1994. Intake of intense sweeteners in France (1989–1992). CREDOC, Département Prospective de la Consommation, Paris.

Chung, M.S., Suh, H.J., Yoo, W., Choi, S.H., Cho, Y.J., Cho, Y.H., and Kim, C.J. 2005. Daily intake assessment of saccharin, stevioside, D-sorbitol and aspartame from various processed foods in Korea. *Food Addit Contam* 22:1087–1097.

Davies, S.M., Ross, J., and Woods, W.G. 1996. Aspartame and brain tumors: Junk food science. *Causes Childhood Cancer Newslett* 7(6).

Davis, D.L., Ahlbom, A., Hoel, D., and Percy, C. 1991. Is brain cancer mortality increasing in industrial countries? *Am J Ind Med* 19:421–431.

Dodge, R.E., Warner, D., Sangal, S., and O'Donnell, R.D. 1990. The effect of single and multiple doses of aspartame on higher cognitive performance in humans. Aerospace Medical Association 61st Annual Scientific Meeting, A30.

European Food Safety Authority (EFSA). 2006. Opinion of the scientific panel on food additives, flavorings, processing aids and materials in contact with food (AFC) on a request from the commission related to a new long-term carcinogenicity study on Aspartame. Question number EFSA-Q-2005–122. *Eur Food Safety Authority J* 365:1–44.

European Food Safety Authority (EFSA). 2009. Updated opinion on a request from the European Commission related to the 2nd ERF carcinogenicity study on aspartame, taking into consideration study date submitted by the Ramazzini Foundation in February 2009. Scientific opinion of the panel on foods additives and nutrients sources added to food. Question number EFSA-Q-2009-00474. *Eur Food Safety Authority J* 1015:1–18.

European Food Safety Authority (EFSA). 2010. Report of the meeting on aspartame with National Experts. Question Number: EFSA-Q-2009-00488.

Ferguson, H.B., Stoddart, C., and Simeon, J.G. 1986. Double-blind challenge studies of behavioral and cognitive effects of sucrose-aspartame ingestion in normal children. *Nutr Rev* 44:144–150.

Fernstrom, J.D. 1991. Central nervous system effects of aspartame. In *Sugars and sweeteners,* eds. N.D. Kretchmer and C.B. Hallenbeck, 151–173. Boca Raton: CRC Press.

Filer, L.J., Baker, G.L., and Stegink, L.D. 1983. Effect of aspartame loading on plasma and erythrocyte free amino acid concentrations in one-year-old infants. *J Nutr* 113:1591–1599.

Filer, L.J., and Stegink, L.D. 1989. Aspartame metabolism in normal adults, phenylketonuric heterozygotes, and diabetic subjects. *Diabetes Care* 12:67–74.

Fisher, R.S. 1989. Aspartame, neurotoxicity, and seizures: A review. *J Epilepsy* 2:55–64.

Food Chemical Codex (*FCC*). 2010. *Aspartame,* pp. 35–36. Baltimore: United Book Press, Inc.

Food Standards Agency (FSA). 2000. Food safety information bulletin No 117. http://archive.food.gov.uk/maff/archive/food/bulletin/2000/no117/asparte.htm. Accessed June 9, 2011.

Food Standard Agency (FSA). 2008. Aspartame. http://www.food.gov.uk/safereating/chemsafe/additives-branch/sweeteners/55174. Accessed January 3, 2009.

Food Standards Australia New Zealand (FSANZ). 2004. Consumption of intense sweeteners in Australia and New Zealand: Benchmark survey 2003. Ray Morgan Research Report Series 8:60, 62–63, 66–67, 87.

Food Standards Australia New Zealand (FSANZ). 2006. Aspartame. Fact Sheet. http://www.foodstandards.gov.au/scienceandeducation/factsheets/factsheets2006/aspartame2006.cfm. Accessed January 15, 2007.

Frey, G.H. 1973. Long term tolerance of aspartame by normal adults. GD Searle & Co. Research Report (E-60), Submitted to FDA February 9.

Frey, G.H. 1976. Use of aspartame by apparently healthy children and adolescents. *J Toxicol Environ Health* 2:401–415.

Gallus, S., Scotti, L., Negri, E., Talamini, R., Franceschi, S., Montella, M., Giacosa, A., Dal Maso, L., and La Vecchia, C. 2007. Artificial sweeteners and cancer risk in a network of case-control studies. *Ann Oncol* 18:40–44.

Garattini, S., Caccia, S., Romano, M., Diomede, L., Guiso, G., Vezzani, A., and Salmona, M. 1988. Studies on the susceptibility to convulsions in animals receiving abuse doses of aspartame. In *Dietary phenylalanine and brain function*, eds. R.J. Wurtman and W. E. Ritter, 131–143. Boston: Birkhauser.

Garnier-Sagne, I., Leblanc, J.C., and Verger, P. 2001. Calculation of the intake of three intense sweeteners in young insulin-dependent diabetics. *Food Chem Toxicol* 39:745–749.

Garriga, M., Berkebile, C., and Metcalfe, D. 1991. A combined single-blind, double-blind, placebo-controlled study to determine the reproducibility of hypersensitivity reactions to aspartame. *J Allergy Clin Immunol* 87:821–827.

Geha, R., Buckley, C.E., Greenberger, P., Patterson, R., Polmar, S., Saxon, A., Rohr, A., Yang, W., and Drouin, M. 1993. Aspartame is no more likely than placebo to cause urticaria/angioedema: Results of a multicenter, randomized, double-blind, placebo-controlled, crossover study. *J Allergy Clin Immunol* 92:513–520.

Goldman, J.A., Lerman, R.H., Contois, J.H., and Udall, J.N. 1986. Behavioral effects of sucrose on preschool children. *J Abnorm Child Psychol* 14:565–577.

Greig, N.H., Ries, L.G., Yancik, R., and Rapoport, S.I. 1990. Increasing annual incidence of primary malignant brain tumors in elderly. *J Natl Cancer Inst* 82:1621–1624.

Gupta, V., Cochran, C., Parker, T.F., Long, D.L., Ashby, J., Gorman, M.A., and Liepa, G.U. 1989. Effect of aspartame on plasma amino acid profiles of diabetic patients with chronic renal failure. *Am J Clin Nutr* 49:1302–1306.

Gurney, J.G., Pogoda, J.M., Holly, E.A., Hecht, S.S., and Preston-Martin, S. 1997. Aspartame consumption in relation to childhood brain tumor risk: Results from a case—control study. *J Natl Cancer Inst* 89:1072–1074.

Health Canada. 2006. Health Canada comments on the recent study related to the safety of aspartame. http://www.hc-sc.gc.ca/fn-an/securit/addit/sweeten-edulcor/aspartame_statement-eng.php. Accessed June 9, 2011.

Henkel, J. 1999. Sugar substitutes: Americans opt for sweetness and lite. FDA Consumer 12–16,

Hertelendy, Z.I., Mendenhall, C.L., Rouster, S.D., Marshall, L., and Weesner, R. 1993. Biochemical and clinical effects of aspartame in patients with chronic, stable alcoholic liver disease. *Am J Gastroenterol* 88(5):737–743.

Heybach, J.P., and Ross, C. 1989. Aspartame consumption in a representative sample of Canadians. *J Can Diet Assoc* 50:166–170.

Hinson, A.L. and Nicol, W.M. 1992. Monitoring sweetener consumption in Great Britain. *Food Addit Contam* 9:669–681.

Hoffman, R. 1972. Short term tolerance of aspartame by obese adults. GD Searle & Co. Research Report (E-24), Submitted to FDA November 30.

Hoffman, R. 1973. Long term tolerance of aspartame by obese adults. GD Searle & Co. Research Report (E-64), Submitted to FDA June 14.

Homler, B.E., Deis, R.C., and Shazer, W.H. 1991. Aspartame. In *Alternative sweeteners*, eds. L.O. Nabors and R.C. Gelardi, 39–69. New York: Marcel Dekker.

Horwitz, D.L., McLane, M., and Kobe, P. 1988. Response to single dose of aspartame or saccharin by NIDDM patients. *Diabetes Care* 11:230–234.

Hulshof, K.F.A.M., and Bouman, M. 1995. Use of various types of sweeteners in different population groups: 1992 Dutch National Food Consumption Survey. TNO Nutrition and Food Research Institute, Netherlands.

Husøy, T., Mangschou, B., Fotland, T.Ø., Kolset, S.O., Nøtvik Jakobsen, H., Tømmerberg, I., Bergsten, C., Alexander, J., and Frost Andersen, L. 2008. Reducing added sugar intake in Norway by replacing sugar sweetened beverages with beverages containing intense sweeteners—a risk benefit assessment. *Food Chem Toxicol* 46:3099–3105.

Ilbäck, N.G., Alzin, M., Jahrl, S., Enghardt-Barbieri, H., and Busk, L. 2003. Estimated intake of the artificial sweeteners acesulfame-K, aspartame, cyclamate and saccharin in a group of Swedish diabetics. *Food Addit Contam* 20:99–114.

Jacob, S.E., Stechschulte, S. 2008. Formaldehyde, aspartame, migraines: A possible connection. *Dermititis* 19:E10–11.

Janssen, P.J.C.M., and van der Heijden, C.A. 1988. Aspartame: Review of recent experimental and observational data. *Toxicology* 50:1–26.

Jobe, P., and Dailey, J. 1993. Aspartame and seizures. *Amino Acids* 4:197–235.

Joint FAO/WHO Expert Committee on Food Additives (JECFA). 1980. *Aspartame*. Toxicological evaluation of certain food additives. WHO Tech Rep Ser. 653. Rome: World Health Organization.

Knopp, R.H., Brandt, K., and Arky, R.A. 1976. Effects of aspartame in young persons during weight reduction. *J Toxicol Environ Health* 2:417–428.

Koch, R., Shaw, K.N.F., Williamson, M., and Haber, M. 1976. Use of aspartame in phenylketonuric heterozygous adults. *J Toxicol Environ Health* 2:453–457.

Koehler, S.M., and Glaros, A. 1988. The effect of aspartame on migraine headache. *Headache* 28:10–14.

Kotsonis, F.N., and Hjelle, J.J. 1996. The safety assessment of aspartame: Scientific and regulatory considerations. In *The clinical evaluation of a food additive: Assessment of aspartame,* eds. C. Tschanz, H.H. Butchko, W.W. Stargel and F.N. Kotsonis, 23–41. Boca Raton: CRC Press.

Kruesi, M.J.P., Rapoport, J.L., Cummings, E.M., Berg, C.J., Ismond, D.R., Flament, M., Yarrow, M., and Zahn-Waxler, C. 1987. Effects of sugar and aspartame on aggression and activity in children. *Am J Psychiatry* 144:1487–1490.

Kulczycki, A. 1986. Aspartame-induced urticaria. *Ann Intern Med* 104:207–208.

Langlois, K. 1972. Short term tolerance of aspartame by normal adults. GD Searle & Co. Research Report (E-23), Submitted to the FDA November 30.

Lajtha, A., Reilly, M.A., and Dunlop, D.S. 1994. Aspartame consumption: Lack of effects on neural function. *J Nutr Biochem* 5:266–283.

Lapierre, K.A., Greenblatt, D.J., Goddard, J.E., Harmatz, J.S., and Shader, R.I. 1990. The neuropsychiatric effects of aspartame in normal volunteers. *J Clin Pharmacol* 30:454–460.

La Vecchia, C., Lucchini, F., Negri, E., Boyle, P., Maisonneuve, P., and Levi, F. 1992. Trends of cancer mortality in Europe, 1955–1989: IV, urinary tract, eye, brain and nerves, and thyroid. *Eur J Cancer* 28A:1210–1281.

Leclercq, C., Berardi, D., Sorbillo, M.R., and Lambe, J. 1999. Intake of saccharin, aspartame, acesulfame K and cyclamate in Italian teenagers: Present levels and projections. *Food Addit Contam* 16:99–109.

Leon, A.S., Hunninghake, D.B., Bell, C., Rassin, D.K., and Tephly, T.R. 1989. Safety of long-term large doses of aspartame. *Arch Intern Med* 149:2318–2324.

Leth, T., Fabricius, N., and Fagt, S. 2007. Estimated intake of intense sweeteners from non-alcoholic beverages in Denmark. *Food Addit Contam* 24:227–235.

Levy, P.S., Hedeker, D., and Sanders, P.G. 1995. Letter to the Editor. *Neurology* 45:1631–1632.

Levy, P.S., and Hedeker, D. 1996. Letter to the Editor. *J Neuropathol Exp Neurol* 55:1280.

Lieberman, H.R., Caballero, B., Emde, G.G., and Bernstein, J.G. 1988. The effects of aspartame on human mood, performance, and plasma amino acid levels. In *Dietary phenylalanine and brain function,* eds. R.J. Wurtman and E. Ritter-Walker, 196–200. Boston: Birkhauser.

Lino, C.M., Costa, I.M., Pena, A., Ferreira, R., and Cardoso, S.M. 2008. Estimated intake of the sweeteners, acesulfame-K and aspartame, from soft drinks, soft drinks based on mineral waters and nectars for a group of Portuguese teenage students. *Food Addit Contam* 25:1291–1296.

Lim, U., Subar, A.F., Mouw, T., Hartge, P., Morton, L.M., Stolzenberg-Solomon, R., Campbell, D., Hollenbeck, A.R., and Schatzkin, A. 2006. Consumption of aspartame-containing beverages and incidence of hematopoietic and brain malignancies. *Cancer Epidemiol Biomarkers Prev* 15:1654–1659.

London, R.S. 1988. Saccharin and aspartame. Are they safe to consume during pregnancy? *J Reprod Med* 33:17–21.

Lu, F.C. 1988. Acceptable daily intake: Inception, evolution, and application. *Regul Toxicol Pharmacol* 8:45–60.

Magnuson, B.A., Burdock, G.A., Doull, J., Kroes, R.M., Marsh, G.M., Pariza, M.W., Spencer, P.S., Waddell, W.J., Walker, R., and Williams, G.M. 2007. Aspartame: A safety evaluation based on current use levels, regulations, and toxicological and epidemiological studies. *Crit Rev Toxicol* 37:629–727.

Magnuson, B., and Williams, G.M. 2008. Carcinogenicity of aspartame in rats not proven. *Environ Health Perspect* 116:A239–240.

Marshall, E. 1990. Experts clash over cancer data. *Science* 250:900–902.

Mazur R.H., Schlatter J., and Goldkamp, A.H. 1969. Structure-taste relationships of some dipeptides. *J Am Chem Soc* 91:2684–2691.

Mazur, R.H. 1974. Aspartic acid-based sweeteners In *Symposium: Sweeteners*, ed. G.E. Inglett. Westport: AVI Publishing.

Mazur R.H. 1976. Aspartame—a sweet surprise. *J Toxicol Environ Health* 2:243–249.

Mazur R.H., and Ripper, A. 1979. Peptide-based sweeteners. In *Developments in sweeteners,* eds. C.A.M Hough, K.J. Parker, and A.J. Vlitos, 125–134. London: Applied Science.

Meldrum, B. 1993. Amino-acids as dietary excitotoxins: A contribution to understanding neurodegenerative disorders. *Brain Res Rev* 18:293–314.

Milich, R., and Pelham, W.E. 1986. Effects of sugar ingestion on the classroom and playgroup behavior of attention deficit disordered boys. *J Consult Clin Psychol* 54:714–718.

Ministry of Agriculture, Fisheries and Food (MAFF). 1995. Survey of the intake of sweeteners by diabetics. Food Surveillance Info Sheet. London, UK.

Modan, B., Wagener, D.K., Feldman, J.J., Rosenberg, H.M., and Feinleib, M. 1992. Increased mortality from brain tumors: A combined outcome of diagnostic technology and change of attitude toward the elderly. *Am J Epidemiol* 135:1349–1357.

Molinary, S.V. 1984. Preclinical studies of aspartame in nonprimate animals. In *Aspartame physiology and biochemistry*, eds. L.D. Stegink and L.J. Filer, 289–306. New York: Marcel Dekker.

Muir, C.S., Storm, H.H., and Polednak, A. 1994. Brain and other nervous system tumours. *Cancer Surv* 19/20:369–392.

National Food Authority of Australia (NFAA). 1995. Survey of intense sweetener consumption in Australia: Final report, Canberra, Australia.

National Toxicology Program (NTP). 2005. NTP report on the toxicology studies of aspartame (CAS No. 22839-47-0) in genetically modified (FVB Tg.AC hemizygous) and B6.129-Cdkn2atm1Rdp (N2) deficient mice and carcinogenicity studies of aspartame in genetically modified [B6.129 Trp53tm1Brd (N5) haploinsufficient] mice (feed studies). National Toxicology Program. 1–222.

Nerhling, J.K., Kobe, P., McLane, M.P., Olson, R.E., Kamath, S., and Horwitz, D.L. 1985. Aspartame use by persons with diabetes. *Diabetes Care* 8:415–147.

Olney, J.W., Farber, N.B., Spitznagel, E., and Robins, L.N. 1996. Increasing brain tumor rates: Is there a link to aspartame? *J Neuropathol Exp Neurol* 55:1115–1123.

Perego, C., DeSimoni, M.G., Fodritto, F., Raimondi, L., Diomede, L., Salmona, M., Algeri, S., and Garattini, S. 1988. Aspartame and the rat brain monoaminergic system. *Toxicol Lett* 44:331–339.

Raines, A. 1987. Another side to aspartame (letter). *Washington Post,* June 2.

Reilly, M.A., Debler, E.A., Fleischer, A., and Lajtha, A. 1989. Lack of effect of chronic aspartame ingestion on aminergic receptors in rat brain. *Biochem Pharmacol* 38:4339–4341

Reilly, M.A., Debler, E.A., and Lajtha, A. 1990. Perinatal exposure to aspartame does not alter aminergic neurotransmitter systems in weanling rat brain. *Res Commun Psychol Psychiatry Behav* 15:141–159.

Renwick, A.G. 1990. Acceptable daily intake and the regulation of intense sweeteners. *Food Addit Contam* 7:463–475.

Renwick, A.G. 1991. Safety factors and establishment of acceptable daily intakes. *Food Addit Contam* 8:135–149.

Renwick, A.G. 2006. The intake of intense sweeteners—an update review. *Food Addit Contam* 23:327–338.

Renwick, A.G. 2008. Response to "The effect of aspartame metabolites on the suckling rat frontal cortex acetylcholinesterase. An *in vitro* study" by Simintzi et al. *Food Chem Toxicol* 46:1206–1207.

Ross, J.A. 1998. Brain tumors and artificial sweeteners? A lesson on not getting soured on epidemiology. *Med Pediatr Oncol* 30:7–8.

Rowan, J.A., Shaywitz, B.A., Tuchman, L., French, J.A., Luciano, D., and Sullivan, C.M. 1995. Aspartame and seizure susceptibility: Results of a clinical study in reportedly sensitive individuals. *Epilepsia* 36:270–275.

Ryan-Harshman, M., Leiter, L.A., and Anderson, G.H. 1987. Phenylalanine and aspartame fail to alter feeding behavior, mood and arousal in men. *Physiol Behav* 39:247–253.

Saravis, S., Schachar, R., Zlotkin, S., Leiter, L.A., and Anderson, G.H. 1990. Aspartame: Effects on learning, behavior, and mood. *Pediatrics* 86:75–83.

Schoeb, T.R., McConnell, E.E., Juliana, M.M., Davis, J.K., Davidson, M.K., and Lindsey, J.R. 2009. Mycoplasma pulmonis and lymphoma in bioassays in rats. *Vet Pathol* 46:952–959.

Schoeb, T.R., and McConnell, E. E. 2010. Commentary: Further comments on mycoplasma pulmonis and lymphoma in bioassay of rats. *Vet Pathol.* doi: 10.117/0300985810377183.

Schiffman, S.S. 1984. Comparison of taste properties of aspartame with other sweeteners. In *Aspartame: Physiology and biochemistry*, eds. L.D. Stegink and L.J. Filer, 207–246. New York: Marcel Dekker.

Schiffman, S.S., Buckley, C.E., Sampson, H.A., Massey, E.W., Baraniuk, J.N., Follett, J., and Warwick, Z.S. 1987. Aspartame and susceptibility to headache. *N Engl J Med* 317:1181–1185.

Schiffman, S.S. 1988. Aspartame and headache. *Headache* 28:370.

Schiffman, S.S., Buckley, C.E., Sampson, H.A., Massey, E.W., Baraniuk, J.N., and Schiffman, S. 1995. Letter to Editor. *Neurology* 45:1632.

Scientific Committee for Food (SCF). 1989. Aspartame. Food Science and Techniques. Reports of the Scientific Committee for Food, 21st series. Commission of the European Communities, Luxembourg, pp. 22–23.

Scientific Committee for Food (SCF). 1997. Aspartame: Extract of minutes of the 107th meeting of the Scientific Committee for Food held on 12–13 June in Brussels.

Seife, C. 1999. Increasing brain tumor rates: Is there a link to deficit spending? *J Neuropathol Exp Neurol* 58:404–405.

Shaywitz, B.A., and Novotny, E.J. 1993. Letter to Editor. *Neurology* 43:630.

Shaywitz, B.A., Anderson, G.M., Novotny, E.J., Ebersole, J., Sullivan, C.M., and Gillespie, S.M. 1994a. Aspartame has no effect on seizures or epileptiform discharges in epileptic children. *Ann Neurol* 35:98–103.

Shaywitz, B.A., Sullivan, C.M., Anderson, G.M., Gillespie, S.M., Sullivan, B., and Shaywitz, S.E. 1994b. Aspartame, behavior, and cognitive function in children with attention deficit disorder. *Pediatrics* 93:70–75.

Simintzi, I., Schulpis K.H., Angelogianni, P., Liapi, C., and Tsakaris, S. 2007a. The effect of aspartame metabolites on the suckling rat frontal cortex acetylcholinesterase. An *in vitro* study. *Food Chem Toxicol* 45:2397–2401.

Simintzi, I., Schulpis, K.H., Angelogianni, P., Liapi, C., and Tsakaris, S. 2007b. The effect of aspartame on acetylcholinesterase activity in hippocampal homogenates of suckling rats. *Pharmacol Res* 56:155–159.

Soffritti, M., Belpoggi, F., Esposti, D.D., Lambertini, L., Tibaldi, E., and Rigano, A. 2006. First experimental demonstration of the multipotential carcinogenic effects of aspartame administered in the feed of Sprague-Dawley rats. *Environ Health Perspect* 114:379–385.

Soffritti, M., Belpoggi, F., Tibaldi, E., Esposti, D.D., and Lauriola, M. 2007. Life-span exposure to low doses of aspartame beginning during prenatal life increases cancer effects in rats. *Environ Health Perspect* 115:1293–1297.

Spiers, P.A., Sabaounjian, L., Reiner, A., Myers, D.K., Wurtman, J., and Schomer, D.L. 1998. Aspartame: Neuropsychologic and neurophysiologic evaluation of acute and chronic effects. *Am J Clin Nutr* 68:531–537.

Squillacote, D. 1999. Aspartame (Nutrasweet): No danger. The Multiple Sclerosis Foundation. http://www.aspartame.net/opinion/Multiple_Sclerosis_Foundation.asp Accessed June 9, 2011.

Stegink, L.D., Filer, L.J., and Baker, G.L. 1977. Effect of aspartame and aspartate loading upon plasma and erythrocyte free amino acid levels in normal adult volunteers. *J Nutr* 107:1837–1845.

Stegink L.D., Filer, L.J., Baker, G.L., and McDonnell, J. 1979a. Effect of aspartame loading upon plasma and erythrocyte amino acid levels in phenylketonuric heterozygotes and normal adult subjects. *J Nutr* 109:708–717.

Stegink, L.D., Filer, L.J., and Baker, G.L. 1979b. Plasma, erythrocyte and human milk levels of free amino acids in lactating women administered aspartame or lactose. *J Nutr* 109:2173–2181.

Stegink, L.D., Filer, L.J., Baker, G.L., and McDonnell, J.E. 1980. Effect of an abuse dose of aspartame upon plasma and erythrocyte levels of amino acids in phenylketonuric heterozygous and normal adults. *J Nutr* 110:2216–2224.

Stegink, L.D., Filer, L.J., and Baker, G.L. 1981a. Plasma and erythrocyte concentrations of free amino acids in adult humans administered abuse doses of aspartame. *J Toxicol Environ Health* 7:291–305.

Stegink, L.D., Brummel, M.C., McMartin, K., Martin-Amat, G., Filer, L.J., Baker, G.L., and Tephly, T.R. 1981b. Blood methanol concentrations in normal adult subjects administered abuse doses of aspartame. *J Toxicol Environ Health* 7:281–290.

Stegink, L.D., Filer, L.J., and Baker, G.L. 1982. Effect of aspartame plus monosodium l-glutamate ingestion on plasma and erythrocyte amino acid levels in normal adult subjects fed a high protein meal. *Am J Clin Nutr* 36:1145–1152.

Stegink, L.D., Brummel, M.C., Filer, L.J., and Baker, G.L. 1983. Blood methanol concentrations in one-year-old infants administered graded doses of aspartame. *J Nutr* 13:1600–1606.

Stegink, L.D., and Filer, L.J. eds. 1984. *Aspartame: Physiology and Biochemistry*. New York: Marcel Dekker.

Stegink, L.D., Filer, L.J., and Baker, G.L. 1987. Plasma amino acid concentrations in normal adults ingesting aspartame and monosodium l-glutamate as part of a soup/beverage meal. *Metabolism* 36:1073–1079.

Stegink, L.D. 1987a. Aspartame: Review of the safety issues. *Food Technol* 41:119–121.

Stegink, L.D. 1987b. The aspartame story: A model for the clinical testing of a food additive. *Am J Clin Nutr* 46:204–215.

Stegink, L.D., Filer, L.J., and Baker, G.L. 1988. Repeated ingestion of aspartame-sweetened beverage: Effect on plasma amino acid concentrations in normal adults. *Metabolism* 37:246–251.

Stern, S.B., Bleicher, S.J., Flores, A., Gombos, G., Recitas, D., and Shu, J. 1976. Administration of aspartame in non-insulin-dependent diabetics. *J Toxicol Environ Health* 2:429–439.

Stokes, A.F., Belger, A., Banich, M.T., and Taylor, H. 1991. Effects of acute aspartame and acute alcohol ingestion upon the cognitive performance of pilots. *Aviat Space Environ Med* 62:648–653.

Stokes, A.F., Belger, A., Banich, M.T., and Bernadine, E. 1994. Effects of alcohol and chronic aspartame ingestion upon performance in aviation relevant cognitive tasks. *Aviat Space Environ Med* 65:7–15.

Sze, P.Y. 1989. Pharmacological effects of phenylalanine on seizure susceptibility: An overview. *Neurochem Res* 14:103–111.

Tephly, T.R. 1999. Comments on the purported generation of formaldehyde and adduct formation from the sweetener aspartame. *Life Sci* 65:157–160.

Toledo, M.C.F., and Ioshi, S.H. 1995. Potential intake of intense sweeteners in Brazil. *Food Addit Contam* 12:799–808.

Tollefson, L. 1988. Monitoring adverse reactions to food additives in the US Food and Drug Administration. *Regul Toxicol Pharmacol* 8:438–446.

Tollefson, L., Barnard, R.J., and Glinsmann, W.H. 1988. Monitoring of adverse reactions to aspartame reported to the US Food and Drug Administration. In *Dietary phenylalanine and brain function*, eds. R.J. Wurtman and E. Ritter-Walker, 317–337. Boston: Birkhauser.

Trefz, F., de Sonneville, L., Matthis, P., Benninger, C., Lanz-Englert, B., and Bickel, H. 1994. Neuropsychological and biochemical investigations in heterozygotes for phenylketonuria during ingestion of high dose aspartame (a sweetener containing phenylalanine). *Hum Genet* 93.369–374.

Trocho, C., Pardo, R., Rafecas, I., Virgili, J., Remesar, X., Fernandez-Lopez, J.A., and Alemany, M. 1998. Formaldehyde derived from dietary aspartame binds to tissue components *in vivo*. *Life Sci* 63:337–349.

Tschanz, C., Butchko, H.H., Stargel, W.W., and Kotsonis, F.N., eds. 1996. *The clinical evaluation of a food additive: assessment of aspartame*. New York: CRC Press.

US Food and Drug Administration (US FDA). 1981. Aspartame: Commissioner's final decision. *Fed Regist* 46:38285–38308.

US Food and Drug Administration (US FDA). 1983. Food additives permitted for direct addition to food for human consumption: Aspartame. *Fed Regist* 48:31376–31382.

US Food and Drug Administration (US FDA). 1984. Food additives permitted for direct addition to food for human consumption: Aspartame. *Fed Regist* 49:6672–6682.

US Food and Drug Administration (US FDA). 1996a. Food additives permitted for direct addition to food for human consumption: Aspartame. *Fed Regist* 61:33654–33656.

US Food and Drug Administration (US FDA). 1996b. Food additives permitted for direct addition to food for human consumption: Dimethyl dicarbonate. *Fed Regist* 61:26786–26788.

US Food and Drug Administration (US FDA). 1996c. FDA statement on aspartame. FDA Talk Paper. November 18, 1996.

US Food and Drug Administration (US FDA). 2007. FDA statements on European aspartame study CFSAN/Office of Food Additive Safety. http://www.fda.gov/Food/FoodIngredientsPackaging/FoodAdditives/ucm208580.htm. Accessed June 9, 2011.

Van den Eeden, S.K., Koepsell, T.D., van Belle, G., Longstreth, W.T., and McKnight, B. 1991. A randomized crossover trial of aspartame and headaches. *Am J Epidemiol* 134:789.

Virtanen, S.M., Rasanen, L., Paganus, A., Varo, P., and Akerblom, H.K. 1988. Intake of sugars and artificial sweeteners by adolescent diabetics. *Nutr Rep Int* 38:1211–1218.

Walton, R.G., Hudak, R., and Green-Waite, R.J. 1993. Adverse reactions to aspartame: Double-blind challenge in patients from a vulnerable population. *Biol Psychiatry* 34:13–17.

Ward, J.M., and Alden, C.L. 2009. Confidence in rodent carcinogenesis bioassays. *Vet Pathol* 46:790–791.

Werner, M.H., Phuphanich, S., and Lyman, G.H. 1995. The increasing incidence of malignant gliomas and primary central nervous system lymphoma in the elderly. *Cancer* 76:1634–1642.

Wilcox, T.G., and Gray, D.M. 1995. Summary of adverse reactions attributed to aspartame. FDA Health Hazard Evaluation Board.

Wolraich, M., Milich, R., Stumbo, P., and Schultz, F. 1985. Effects of sucrose ingestion on the behavior of hyperactive boys. *J Pediatr* 106:675–682.

Wolraich, M.L., Lindgren, S.D., Stumbo, P.H., Stegink, L.D., Appelbaum, M.I., and Kiritsy, M.C. 1994. Effects of diets high in sucrose or aspartame on the behavior and cognitive performance of children. *N Engl J Med* 330:301–307.

Wucherpfennig, K., Dietrich, H., and Bechtel, J. Alcohol actual, total and potential methyl alcohol of fruit juices. *Flussiges Obst* 8: 348–354.

Wurtman, R.J. 1983. Neurochemical changes following high-dose aspartame with dietary carbohydrates. *N Engl J Med* 309:429–430.

Chapter 6

Aspartame-Acesulfame

John C. Fry, Brad I. Meyers, and Dale A. Mayhew

Contents

Introduction

Aspartame-acesulfame is the first commercially viable member of a group of compounds called sweetener-sweetener salts. These salts owe their existence to the fact that some high-potency sweeteners form positively-charged ions in solution, while others are negative. It is thus theoretically possible to combine two oppositely-charged sweeteners to create a compound where each molecule contains both "parent" sweeteners. To look at this another way, many current permitted sweeteners are sold as their metal salts, for example sodium cyclamate, calcium saccharin and acesulfame potassium (acesulfame K). In a sweetener-sweetener salt, the positively-charged metal ion (sodium, calcium or potassium) is replaced by another sweetener which itself carries a positive charge.

Aspartame was probably the first realistic candidate for this role of positively-charged sweetener, although there have since been others, such as alitame and neotame. Despite the availability of aspartame for decades, the practical difficulties of preparing sweetener-sweetener salts seem to have defeated most researchers. The patent literature records only one attempt at laboratory synthesis (Palomo Coll 1986), via a route involving dissolution of an unstable form of the negatively-charged sweetener in a toxic organic solvent, a procedure which produces only small quantities of poor crystals. Realistic commercial manufacture requires a synthesis which can be carried out in an aqueous medium, and which satisfies both food industry demands for purity and commercial requirements for economic yield. This was achieved in 1995, when Fry and Van Soolingen invented such a process (Fry 1996) and used it to produce a range of high-potency sweetener salts, including the first, unique crystals of aspartame-acesulfame. Patent applications on the Fry-Van Soolingen method have been filed widely, including in Europe (Fry and Van Soolingen 1997) and the United States (Fry and Van Soolingen 1998). Also the subject of patents is the aspartame-acesulfame compound itself. From all the possible sweetener-sweetener salts, this then became the leading candidate for full commercial production. Aspartame-acesulfame is now marketed by The NutraSweet Company under their trademark Twinsweet™.

The reasons for choosing this particular salt are clear. Firstly, there are the well-known advantages of blending aspartame with acesulfame K. These sweeteners exhibit quantitative synergy which means that, when used jointly, they are more potent than would have been expected based on their properties when used independently. Furthermore, the quality of sweetness is improved by blending the two sweeteners. The best features of the sweetness profiles of each come to the fore when they are combined, and a favored blend to achieve this is 60:40 by weight of aspartame and acesulfame K, respectively. This ratio happens to contain approximately equal numbers of molecules of each. In the aspartame-acesulfame salt, of course, chemistry dictates that the sweeteners are exactly equimolar.

Another benefit of using aspartame with acesulfame K in liquid products is that the mixture offers greater sweetness stability and longer shelf life in comparison to aspartame alone. This arises because acesulfame is more stable than aspartame and mitigates the sweetness loss that would otherwise be more pronounced in an all-aspartame system. This improvement is also a feature of the sweetener-sweetener salt.

The many advantages of aspartame and acesulfame K have fostered an enormous growth in their joint use, as well as a general acceptance that blended sweetener systems can often offer the consumer a better taste than single sweeteners alone, however "sugar-like" the latter are claimed to be. Yet mechanical blends of this pair of sweeteners are not without technological problems (Fry 1996; Hoek et al. 1999). There are issues of dissolution time, hygroscopicity, and the homogeneity of powder mixes, all of which bear on the ease of use of physical mixtures of these sweeteners and the quality of consumer products made with them. In the creation of the novel crystalline form of Twinsweet, where aspartame and acesulfame are combined at the molecular level, these issues have been largely resolved.

The way aspartame and acesulfame are combined in the salt gives rise to yet more advantages. For example, the molecular arrangement is such that, in the solid, access to the free amino group of the aspartyl moiety is hindered. The availability of this group is critical to the (in)stability of aspartame when used conventionally as a separate sweetener in certain low-moisture applications, such as sugar-free confectionery, especially chewing gum. Aspartame is widely and successfully used in chewing gum, where it is employed in both free and encapsulated forms. However, free aspartame can react with flavors high in aldehyde content such as cinnamon and cherry. This can shorten shelf life unacceptably, as there is simultaneous loss of both flavor and sweetness. The hindered structure of the solid aspartame-acesulfame salt, however, is less susceptible to aldehyde attack and the salt can be used successfully to create products of extended shelf life.

The advantages of aspartame-acesulfame are not confined to its physico-chemical properties as a solid. Because it provides two sweetening components in each molecule of the crystalline material, the salt represents a saving on the number of raw materials to be purchased, stored and handled. An additional benefit in Europe is that the salt carries its own unique E-number (E962). This single number can replace the two separate numbers for aspartame (E951) and acesulfame K (E950). This both shortens the ingredient list, which is seen as a marketing advantage, and reduces the number of E designations, to which some consumers are averse.

Another, separate advantage is that the salt is a more concentrated source of sweetness than the blend because it is free of potassium and lower in moisture content. As a result, the sweetener-sweetener salt provides 11% more sweetness on a weight-for-weight basis than the corresponding equimolar blend, which is a modest but real benefit to those handling large quantities of low-calorie sweeteners.

Finally, aspartame-acesulfame poses no new toxicological issues. The sweetener-sweetener salt dissociates immediately on dissolution in water. In doing so it releases the same sweetener molecules which would be present were a mere mixture of aspartame and acesulfame K to have been used. Consequently, the consumer is exposed only to known, permitted sweeteners.

For all the above reasons, aspartame-acesulfame was the sweetener-sweetener salt preferred for commercial development, and these and other attributes of Twinsweet are discussed further below.

Production

In the patented process (Fry and Van Soolingen 1997, 1998), aspartame-acesulfame is made by combining aspartame and acesulfame K in an aqueous acid solution. The trans-salification reaction is depicted in Figure 6.1. The sweetener-sweetener salt is subsequently crystallized, separated, washed and dried.

All the components used are commercially available and of food grade. No unusual forms of the sweeteners and no organic solvents are used. No additional purification of the salt is necessary. This is because, as is apparent from Figure 6.1, the preparation of the salt is, in effect, also a recrystallization of the starting materials. As might be expected of such a process, the resultant aspartame-acesulfame has a higher degree of purity than even the food grade raw materials. Moreover, the process introduces no new impurities, and so neatly combines synthesis with purification.

Physical and Other Data

Table 6.1 lists the main characteristics of aspartame-acesulfame.

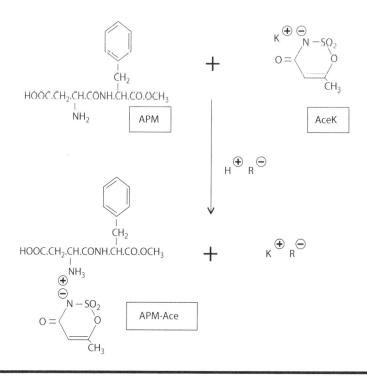

Figure 6.1 Reaction scheme for the preparation of aspartame-acesulfame (APM-Ace) from aspartame (APM) and acesulfame K (AceK). (Reprinted from Fry, J.C., Two in one—an innovation in sweeteners. *Low-Calorie Sweeteners. Openings in an Expanding European Market,* **Proceedings of 1996 International Sweeteners Association Conference, eds. S.G. Lisansky and A. Corti. Newbury: CPL Press. 58–68, 1996. With permission.)**

Relative Sweetness

As described above, each molecule of aspartame-acesulfame contains one molecule of aspartame and one of acesulfame which are released immediately as the salt is dissolved. From these observations, it is to be expected that the salt would exhibit the same sweetness as an equimolar blend of the two parent sweeteners, aspartame and acesulfame K. This is so in practice. Taste panel tests (Fry 1996, 1998), using solutions both in water and in a model soft drink base (a citrate buffer of pH value 3.2), show exactly the same sweetness for the salt and an appropriate equimolar blend of aspartame and acesulfame K, suitably adjusted for the weight due to the presence of extra potassium ions and moisture in the blend.

Figure 6.2 shows the concentration dependence of the sweetness of Twinsweet in these two solvents. The acid model system was included because, in practice, most high-potency sweeteners are used in acid foods and beverages and taste panel data obtained for such a system are more closely representative of the majority of real applications than tests conducted with water. Figure 6.2 indicates that Twinsweet is approximately 350 times as sweet as sucrose, in water, and 400 times as sweet as sucrose, in pH 3.2 citrate, at 4% sucrose equivalence. The higher relative sweetness in the citrate buffer is consistent with observations of the behavior of aspartame, which also tastes sweeter in acid solution than in water.

Table 6.1 Physical and Other Data of Aspartame-Acesulfame

Appearance	White, odorless, crystalline powder	
Taste	Clean sweet taste, with rapid onset and no lingering sweetness or off-taste	
Chemical formula	$C_{18}H_{23}O_9N_3S$	
Molecular weight	457.46	
Loss on drying	Not more than 0.5%	
Assay (on dried basis)	Not less than 63.0% and not more than 66.0% of aspartame, not less than 34.0% and not more than 37.0% of acesulfame Calculated as acid form	
Melting point	Decomposes before melting	
Solubility	Temperature (°C)	Solubility (% Weight in Water)
	10	1.82
	21	2.53
	40	4.64
	60	11.76
pH of solution	2–3 (0.3% by weight in water, room temperature)	
Shelf life	5 years	

Source: From The NutraSweet Company, Chicago, IL, USA.

Across the full range of concentration, the relative sweetness of Twinsweet is 11% higher than could be obtained from the same weight of an equimolar mixture of aspartame with acesulfame K. This is because the sweetener-sweetener salt contains only active sweeteners and no potassium. The latter accounts for 19.4% of acesulfame K by weight, but contributes no sweetness. Indeed, it is well known that potassium ions are associated with bitterness (Bartoshuk et al. 1988; Bravieri 1983; Breslin and Beauchamp 1995; Pangborn and Braddock 1989) and it is tempting to associate some of the bitter character of acesulfame K with this cation. However, at realistic use concentrations of acesulfame K, the potassium concentration is lower than that in saliva (Tenovuo 1989) and it is unlikely to contribute to the bitterness of acesulfame K.

As well as lacking the non-sweet potassium of acesulfame K, Twinsweet also has a very low moisture content. In contrast, the aspartame component of a physical blend can contain up to 4.5% moisture (strictly "loss on drying") while remaining within international specification limits. The overall saving on non-sweet components, namely potassium and moisture, means that the aspartame-acesulfame salt is a more effective sweetener on a weight-for-weight basis than a blend. This higher relative sweetness of Twinsweet is in addition to the synergy between aspartame and acesulfame, which is considered below.

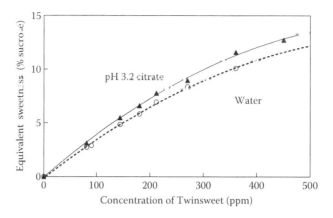

Figure 6.2 **Equivalent sweetness of aspartame-acesulfame (Twinsweet) in water and in pH 3.2 citrate buffer.**

Because the salt provides both aspartame and acesulfame, its relative sweetness is enhanced by the quantitative synergy between the two. That is, the salt is significantly sweeter than would have been predicted by a simple summation of the characteristics of the individual sweeteners tasted alone. This is illustrated in Figure 6.3, which shows how the sweetness of an aspartame:acesulfame K blend (400 ppm total sweeteners in pH 3.2 citrate) varies with the ratio of the two sweeteners.

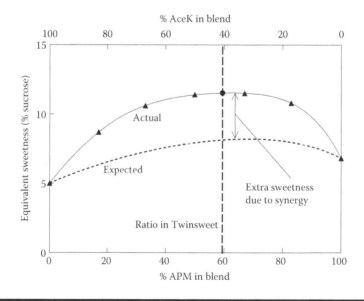

Figure 6.3 **Synergy in a blend of aspartame (APM) with acesulfame-K (AceK) as a function of blend ratio, also showing the fixed equimolar ratio of aspartame-acesulfame (Twinsweet). Solvent = pH 3.2 citrate buffer, blend concentration = 400 ppm at all ratios, aspartame-acesulfame concentration = 360 ppm. (Modified from Fry, J.C. Two in one—an innovation in sweeteners. In** *Proceedings of the 45th Annual Conference of the International Society of Beverage Technologists*, **Savannah, GA, 83–92, 1998.)**

Also shown in Figure 6.3 is the effective ratio of the sweeteners as provided by Twinsweet. Because the salt is an ionic compound, this ratio is fixed and dictated by the molecular weights of aspartame and acesulfame. The equimolar ratio of the two sweeteners combined in Twinsweet translates to a conventional blend ratio of approximately 60:40 aspartame:acesulfame K by weight. As can be seen, 60:40 is at or near the peak for quantitative synergy between aspartame and acesulfame K, and the salt thus provides the maximum quantitative synergy available. In the case of the system illustrated, this synergy boosts the "expected" sweetness by nearly 50%.

Technical Qualities

Twinsweet represents an advance over mechanical blends of aspartame with acesulfame K in three main areas. The aspartame-acesulfame salt dissolves more rapidly than the blend, is much less hygroscopic, and exhibits a higher stability than aspartame in certain aggressive environments. These are dealt with individually below.

Improved Dissolution Rate

Consumers take for granted that powder products, such as desserts, toppings and beverage mixes, can be reconstituted almost instantaneously, and that tabletop sweeteners dissolve immediately. Achieving such performance in sugar-free and diet products containing high-potency sweeteners poses difficulties for the formulation technologist. In particular, aspartame is relatively slow to dissolve, especially in cold systems such as might be involved in reconstituting a cold dessert or drink with refrigerated milk or cold water, or in sweetening iced tea. The speed of dissolution can be improved by reducing the particle size, as this exposes a greater surface area to the solvent. However, as explained further in the "Applications" section, it is generally desirable not to have very fine fractions in powder products as these contribute to dust and poor flow. Figure 6.4 contrasts the dissolution time of aspartame with aspartame-acesulfame in cold

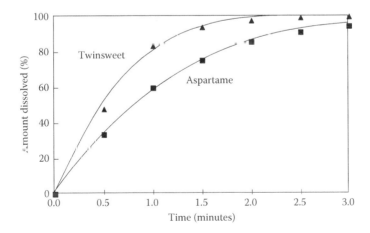

Figure 6.4 Dissolution profile of aspartame-acesulfame (Twinsweet) compared with aspartame. Solvent = stirred water at 10°C, particle size range of each sweetener = 100–250 μm.

water for powders of matched particle size. The more rapid dissolution of the salt is evident. In particular, the salt achieves the critical range of 80%–95% dissolved in approximately half the time of aspartame.

Absence of Hygroscopicity

Hygroscopicity (the tendency for materials to take up moisture from their surroundings) is an important property of food ingredients. Where powder mixes are concerned, hygroscopic materials can draw moisture from the atmosphere during manufacture and storage, or from other ingredients with which they are blended. The moisture taken up can change the powder flow characteristics. Together with other problems, this can lead to the self-agglomeration or clumping of one or more ingredients, thus causing further difficulties. Clumped ingredients, for example, may be hard for the mixer to disperse, or they may segregate during subsequent handling, to produce inhomogeneous mixtures. Agglomerates may even be visible and give rise to consumer complaint. Furthermore, on reconstitution, agglomerates may act as if they were single, large particles and dissolve only slowly. In addition, powders containing excess moisture are more likely to cake on storage, and may lose all ability to flow.

Twinsweet has virtually zero hygroscopicity and is remarkably immune to moisture uptake, even when exposed to very high relative humidities. This advantageous behavior of the aspartame-acesulfame salt is illustrated in Figure 6.5. The latter shows the moisture taken up by an equimolar mixture of aspartame with acesulfame K at various relative humidities (RH), a trend which rises steeply above 45% RH at room temperature. Twinsweet takes up little moisture, even in the region of 95% RH. Thus, in circumstances where moisture uptake is likely to create difficulty, the salt clearly out-performs the blend. This advantage is not confined to easing issues of powder mixing and flow but also simplifies packaging and storage requirements, both for the salt being handled as a bulk ingredient and for mixtures made with it.

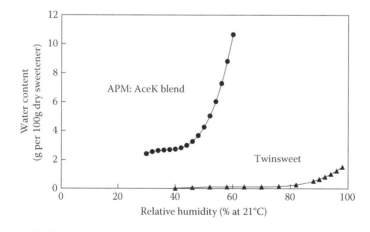

Figure 6.5 Hygroscopicity. Moisture uptake of aspartame-acesulfame (Twinsweet) is negligible compared with an equimolar physical mixture of aspartame with acesulfame K (APM: AceK blend).

Stability

Aspartame-acesulfame is stable as a dry solid. The salt shows no breakdown in either of the aspartame or acesulfame moieties on prolonged storage: periods as long as a year at abuse temperatures of 60°C have failed to show any change in composition. Similarly, there is no measurable breakdown after 60 minutes at 140°C. There is a suggestion that the salt might be more stable to abuse than aspartame. This may be due to the fact that, in the salt, the amino group of aspartame is blocked by the presence of acesulfame (Figure 6.1) and is thus hindered from taking part in the breakdown reactions that involve self-cyclization to diketopiperazine. This route of decomposition involves water also and, in practice, any blocking effect would be further enhanced by the non-hygroscopic nature of the salt previously mentioned.

Cyclization in the presence of water, however, is not the only hazard to aspartame stability. As a dipeptide, aspartame exhibits properties also found in other peptides and in proteins, including an ability to take part in Maillard-type reactions. In some low moisture applications, such as chewing gum, aspartame can be lost through reaction with the aldehyde groups of certain flavors. This reaction also depends on the accessibility of the free amino group of the sweetener and, it is surmised, this may be the reason why Twinsweet is so effective in extending the shelf life of such products. This is discussed further in the "Applications" section.

While aspartame-acesulfame exhibits advantageously high stability as a dry solid, it should be clear that the salt has no particular stability benefit once dissolved. At the moment of solution the salt releases only aspartame and acesulfame, which then behave in exactly the same way as if they had been released from a physical mixture of aspartame with acesulfame K. Consequently, in solution, the stability characteristics of aspartame contributed by Twinsweet are no different from those of aspartame from any other source. Use of the aspartame-acesulfame salt has no consequences, beneficial or deleterious, for the subsequent stability of the aspartame once the salt has been dissolved.

Applications

Aspartame-acesulfame can be used wherever both aspartame and acesulfame K are used jointly, as well as in most applications where these sweeteners might be used singly. Thus the salt is suitable for a wide range of products, including beverages, dairy products, tabletop sweeteners, confectionery and pharmaceutical preparations. The fixed composition of the salt means that it always delivers the equivalent of approximately a 60:40 weight ratio of aspartame to acesulfame K, but at 11% higher sweetness. It is the absence of the potassium ion which principally makes Twinsweet a more concentrated source of sweetness than a blend. This can be seen in Table 6.2, which shows guidelines for the amounts of Twinsweet for various applications. Generally, the concentration of Twinsweet required in any product is that which supplies the same number of molecules of aspartame and acesulfame as would be derived from a 60:40 mechanical blend of the two sweeteners.

However, although suited for use in any product where the parent sweeteners appear, the salt is especially beneficial for dry or low moisture materials, owing to its particular properties. Specifically, there are substantial advantages to be gained by using the salt in powder mix products such as instant beverages for cold reconstitution, mixes for hot cocoa-based drinks, instant desserts, toppings, tabletop sweeteners and sugar substitutes, as well as pharmaceutical powder preparations. In addition, aspartame-acesulfame has a number of benefits in sugar-free confectionery,

Table 6.2   Guidelines for Twinsweet Concentrations in Various Products

Product	Twinsweet Concentration Ready to Consume (ppm)
Beverages	
Carbonated lemon-lime	270
Hot cocoa mix	240
Cold chocolate mix	190
Instant lemon tea	200
Instant lemon drink	220
Desserts/Dairy	
Instant pudding mix	380
Gelatin mix	435
Confectionery	
Chewing gum	2700
Hard candy	1000
Chocolate	800
Tabletop Sweeteners	
Tablets (1 tablet = 1 teaspoon of sugar)	11 mg/tablet

Source: The NutraSweet Company, Chicago, IL, USA.

including chewing gum and hard candy, to which can be added medicated confectionery and chewable tablets. These are discussed in the following paragraphs.

Powder Mixes

These products are typically sold as a convenient package of pre-mixed powder suitable for "instant" reconstitution by the consumer. Consumption usually follows shortly after reconstitution and may be immediate, as in the case of certain pharmaceutical preparations which are stirred into water and swallowed directly. Another example is a sugar substitute added to a beverage, stirred briefly, and then taken. Key elements in successful powder mixes are rapid dissolution, essential to meet consumer expectations of an instant product, and the homogeneity of the mix, essential to deliver reproducible performance. In addition, there are factors which affect the manufacturer, such as ease of mixing of the ingredients and low dust content.

Unfortunately, aspartame and acesulfame K crystallize in different forms. Aspartame has needle-like crystals while those of acesulfame K are more cubic, which makes it technologically difficult to create and maintain mixtures of the sweeteners which are homogeneous and stay so throughout the manufacturing and retail chains. There is evidence that this difficulty influences the quality of powder products. Not only is there a tendency for the two conventional sweeteners to separate from each other, they can also redistribute themselves unevenly with respect to the other components of the mixture. For example, Hoek et al. (1999) have shown that, in segregation tests of aspartame and acesulfame K contained in a typical instant beverage powder, the sweeteners can separate to a degree which could be perceptible to consumers as differences in sweetness. In the aspartame-acesulfame salt, however, the two sweeteners are combined at the molecular level and cannot be separated from each other until the moment they are dissolved. Moreover, because of the rapid dissolution of the salt (see the "Technical Qualities" section), there is greater freedom to choose a particle size range which gives a mechanically stable, homogeneous mix. This is an advantage in comparison with aspartame, which dissolves relatively slowly, a factor which may drive the powder mix manufacturer to use very finely-ground material to increase dissolution speed. Such fine powders bring other difficulties, however, including a tendency to cohere, which leads finely-milled aspartame to behave like all fine powders; it does not flow easily and can clump. Perversely, such clumps can act as very large particles and take an extended time to dissolve. The tendency to form clumps is made worse by the uptake of moisture, a process encouraged by the hygroscopicity of the sweetener. In contrast, the complete absence of significant hygroscopicity in the salt has already been noted.

Comparative trials have been made of Twinsweet with an equisweet, equimolar physical blend of aspartame with acesulfame K in a typical instant beverage mix (Hoek et al. 1999). Under standardized mixing conditions, and in both tumbling and convective mixers, the sweetener-sweetener salt produced a more homogeneous mix. Coefficients of variation for the concentration of both sweetener components were half those found with the blend. The mixture based on the salt was also much more resistant to segregation. In tests, columns of powder were deliberately vibrated to induce segregation, and subsequently sectioned horizontally and analyzed. Following this treatment, the spread of sweetness with Twinsweet was much more narrowly distributed than that using the blend. More importantly, the variation in concentration of the simple blend was very likely to be perceptible to a proportion of consumers, whereas that of the aspartame-acesulfame salt was unlikely to be detected.

While Twinsweet may be used in a powder mix solely because of its advantageous powder flow properties, its rapid dissolution is also of obvious direct benefit. The salt dissolves in approximately half the time required for aspartame of matched particle size (see the "Technical Qualities" section), which means better products for impatient consumers, especially where dissolution is required in a cold solvent such as chilled milk or iced tea. In addition, when Twinsweet dissolves, it always releases aspartame and acesulfame in an exactly balanced ratio, and this ratio is constant throughout the dissolution process. This is not so when a physical blend of the high-potency sweeteners is used, because the individual sweeteners dissolve at different rates. This means that, until both sweeteners are fully dissolved, a process that can take several minutes, there is a mismatch in blend ratio causing the taste to differ from that which the product designers intended.

Chewing Gum

The aspartame-acesulfame salt has exciting advantages in chewing gum. Fry et al. (1998a) have demonstrated long-lasting sweetness, a noticeable boost to sweetness after some minutes of chewing, as well as improvements to stability and shelf life.

Long-lasting sweetness is a key feature of chewing gum, as sweetness and flavor perception are intimately related. A gum which is no longer sweet is also perceived as having reduced flavor. In general, ordinary sugar-free chewing gum gives an immediate sweetness which declines quite rapidly after the first 1 or 2 minutes. The immediate sweetness comes mainly from the bulk sweetener, a polyol, while the high potency sweeteners present take somewhat longer to chew out but also disappear quickly. Encapsulation of part of the high-potency sweetener can greatly lengthen this time, but increases cost. Twinsweet improves on this. In comparative trials, gum sweetened with aspartame-acesulfame was significantly sweeter at the end of 15 minutes chewing than gums made with other sweeteners, and was most preferred overall. This was achieved by direct incorporation of Twinsweet during gum manufacture. The salt was not encapsulated, although some of the sweeteners with which it was compared were coated to extend their sweetness release. Direct addition of the salt thus represents a useful simplification of the alternative, complex process of encapsulation and produces longer sweetness.

However, not only does aspartame-acesulfame extend gum sweetness overall, it also provides a remarkable boost to sweetness after some minutes chewing. This second peak of sweetness occurs following 5 to 8 minutes chewing. Again, the effect is achieved with Twinsweet as the sole sweetener and without the use of encapsulated material. The intensity and timing of the second peak depends on the type of gum base, sweetener concentration, product formulation and the gum manufacturing process. These offer the product developer considerable scope to tailor a sweetness release profile and use has been made of this flexibility to engineer some novel sweetness/time profiles in recent commercial gums.

As well as profound effects on the sweetness delivery of chewing gum, aspartame-acesulfame can be used to extend shelf life (Fry et al. 1998a). Gum is a concentrated, low-moisture system and is a good medium for unwanted reactions between flavor compounds and aspartame. Especially aldehyde-rich, cinnamon and cherry flavor gums host these reactions which reduce both flavor impact and sweetness. Twinsweet is much more resistant to attack by aldehydes than is aspartame (see the "Technical Qualities" section), and this is manifest in substantial improvements to storage stability and acceptability. Figure 6.6 shows a comparison of the sweetness during storage of two cinnamon-flavor gums which are identical except that one was made with Twinsweet while the other contained an equimolar blend of aspartame and acesulfame K. After 32 weeks storage, the gum sweetened with the blend was deemed barely acceptable by a taste panel after 2 minutes of chewing, and unpleasant after 15 minutes. The same gum with Twinsweet was found to be good at

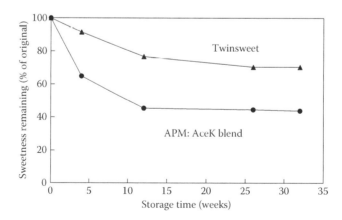

Figure 6.6 Shelf life of cinnamon-flavor chewing gum. Aspartame-acesulfame (Twinsweet) retains sweetness on storage longer than an equisweet, equimolar physical mixture of aspartame with acesulfame K (APM: AceK blend).

2 minutes, and still acceptable at 15 minutes, results which were maintained following storage for a further 20 weeks. The dramatic impact of the aspartame-acesulfame salt was underlined by the fact that, after 32 weeks storage, the gum with the salt released three times as much aspartame as that sweetened with the blend.

Hard Candy

Sugar-free candy, including medicated confectionery, is an area of increasing demand, and relies on the use of polyols as bulk substitutes for the sucrose and glucose syrup used in conventional high-boiled sweets. Often, the lower sweetness of the polyols needs a boost with a modest amount of a low-calorie sweetener. Such an addition is not always easy, particularly where aspartame is involved. While addition of aspartame to fruit-flavored candy is straightforward, non-acid flavors such as mints can create a problem, as sufficient aspartame cannot always be dispersed homogeneously throughout the hot candy mass (Fry et al. 1998b). This difficulty vanishes when Twinsweet is employed, as it disperses directly in the hot mass to give products with a fuller, more sugar-like sweetness and better flavor impact than achieved with an equisweet blend of sweeteners.

Chewable Tablets and Tabletop Sweetener Tablets

A number of the advantages of aspartame-acesulfame already cited above combine to make the salt well-suited to tabletting processes. The qualities of Twinsweet which contribute to mechanically stable, homogeneous powder mixes are directly relevant to the mixing of powdered ingredients prior to tabletting. The absence of hygroscopicity means that the salt flows reliably and will not change its flow characteristics on exposure to moist air. At the same time, the relative chemical stability of crystalline Twinsweet leads to a long shelf life for tablets by minimizing any degradative reactions with flavors or excipients, for example, the yellowing that can occur in aspartame-based tablets that use lactose as an excipient. This is another instance of an unwanted Maillard reaction; the lactose, although only weakly reducing, is capable of reacting with the free amino group of aspartame over time. Finally, rapid dissolution of the salt assists in giving an immediate release of sweetness when the tablet comes to be used.

In common with hard candy, the use of aspartame-acesulfame in a variety of low-moisture products, tablets, and so forth, is the subject of widespread patent applications (Fry et al. 1998b).

Other Applications

Aspartame-acesulfame can be used as a sole sweetener, but also in conjunction with any other permitted sweetener, whether caloric or high-potency. Thus, Twinsweet can be used to make multi-component blends in concert with saccharin, cyclamate, sucralose, neotame, the steviol glycosides and others. If required, the ratio of aspartame to acesulfame can be adjusted by adding one or other parent sweetener. Generally, this is best achieved by adding all of the aspartame in the form of the sweetener-sweetener salt and including additional acesulfame K as needed. This approach has the benefit of maintaining the rapid dissolution of the aspartame in the salt form.

The rapid dissolution of the salt has even raised interest among manufacturers of ready-to-drink beverages. Soft drinks were initially thought to be an area where the salt was not relevant because its advantageous properties depend on it being in the solid state. Surprisingly, such is the additional effort required to dissolve aspartame in bulk that beverage producers have considered the sweetener-sweetener salt as a viable ingredient, owing to savings in both process time and raw materials inventory control.

Toxicology

The toxicology of aspartame-acesulfame is straightforward. The synthesis of the salt is, in effect, a recrystallization and a purification of the raw material sweeteners. Those raw materials are already food grade, and the synthesis of Twinsweet simply results in further reduction of any trace impurities. No new impurities are introduced and, as a result, there are no toxicological issues associated with it's manufacture. Furthermore, the salt dissociates immediately on solution to release only known, widely-permitted sweetener molecules, namely aspartame and acesulfame. Once in solution then, aspartame-acesulfame salt behaves the same as a mixture of aspartame and acesulfame K from which the functionless potassium has been removed. Accordingly, human dietary exposure is not to Twinsweet itself, which does not exist in solution, but to the permitted sweeteners from which it is derived, and the toxicological fate of aspartame-acesulfame is the same as that of those existing, permitted sweeteners.

There remains the question of the amount of aspartame-acesulfame employed and whether the quantities likely to be used will affect dietary intakes of either aspartame or acesulfame. This is also easily resolved. The salt is used to provide the same amounts of aspartame and acesulfame as would have been present had these "parent" sweeteners been used separately. At the same time, no new uses have been proposed for the salt. It owes its existence to the unique way it overcomes technological problems and offers consumer benefits in existing applications. Accordingly, use of aspartame-acesulfame will not affect the human exposure data and predictions upon which regulatory approval for aspartame and acesulfame has been based. In short, from a toxicological point of view, there is no difference between Twinsweet and a physical blend of aspartame and acesulfame K, and the use of aspartame-acesulfame introduces no new toxicological issues.

Regulatory Status

It goes without saying that aspartame-acesulfame is only applicable to products where both aspartame and acesulfame K are permitted to be used jointly. There are countries where quantitative limits are applied to one or both of these sweeteners. In such locations, the use of the salt must conform to the limits in terms of the amounts of aspartame and/or acesulfame released by the salt when it dissolves. As the salt is a fixed, equimolar ratio, this is a simple calculation.

Both the US Food and Drug Administration (US FDA) and the Canadian Health Protection Branch regard Twinsweet as being covered by current regulations on aspartame and acesulfame K. Products containing the salt are required to declare this in their ingredients lists as "aspartame-acesulfame."

In June 2000, the Joint WHO/FAO Expert Committee on Food Additives (JECFA) concluded that the aspartame and acesulfame moieties in Twinsweet are covered by the acceptable daily intake (ADI) values established previously for aspartame and acesulfame K. The same year, in the EU, the Scientific Committee for Food (SCF) also concluded that the use of Twinsweet raised no additional safety concerns. As a result Directive 94/35/EC (the "Sweeteners Directive") has been amended to include aspartame-acesulfame as a permitted sweetener. Regulatory clearance has also been obtained in Australia, New Zealand, Mexico, Switzerland, China, and Russia. Clearance is being sought in South America and elsewhere. This is principally a matter of administrative process, necessitated because the wording of regulations in some countries does not encompass the concept of sweetener-sweetener salts.

Internationally-recognized specifications for aspartame-acesulfame are published in the form of a FAO Food and Nutrition Paper (Anonymous 2005) and a monograph in the *Food Chemicals Codex* (Anonymous 2010).

Conclusion

Aspartame-acesulfame is a sweetener-sweetener salt where the potassium ion of acesulfame K has, in effect, been replaced by aspartame. The result is a new high-potency sweetener molecule that can be produced commercially as a very pure and stable solid, possessing highly advantageous properties. Aspartame-acesulfame has a high relative sweetness, approximately 350 times as sweet as sucrose in water and 400 times as sweet in pH 3.2 citrate. This is because it comprises only synergistic, intensely-sweet molecules and contains no significant amounts of functionless potassium ions or moisture. It dissolves more rapidly than an equimolar mechanical mix of aspartame and acesulfame K, yet releases only the same sweetening molecules as this familiar and widely-accepted blend. Also, in contrast to a mixture of aspartame and acesulfame K, the salt is remarkably immune to moisture uptake. This, coupled with its rapid dissolution and excellent powder flow characteristics, makes it an ideal sweetener for use in powder mixes of all types. In addition, aspartame-acesulfame is very stable in low-moisture products, which can pose a particularly challenging environment to aspartame itself through potential reaction with aldehyde-rich flavors. Not only is the salt resistant to these reactions in products such as chewing gum, it is also responsible for a marked extension of the sweetness release of gum, both effects being achieved without the need to encapsulate the sweetener. Other low-moisture products such as confectionery, table-top sweeteners, sugar substitutes, pharmaceutical powders and tablets also benefit from the special attributes of aspartame-acesulfame. Naturally, there is no reason the salt cannot be used in any application permitted by regulation, including liquid beverages and dairy products.

The salt is made from existing, permitted high-potency sweeteners. Its manufacture even purifies further the food grade raw materials, and the process introduces no new impurities. The salt is stable on dry storage, including at elevated temperatures, and dissociates immediately when dissolved to provide an equimolar solution of aspartame and acesulfame. Thus, toxicologically, there is no difference between aspartame-acesulfame salt and an equimolar, mechanical mixture of aspartame and acesulfame K, and use of the salt introduces no new toxicological issues. The salt is regarded in the United States as being covered by current FDA regulations on aspartame and acesulfame K, and regulatory clearance has been granted in other key markets, including the EU.

Aspartame-acesulfame, its production and many applications are the subjects of international patents and patent applications now owned by The NutraSweet Company, which markets the sweetener-sweetener salt under their trademark Twinsweet™.

References

Anonymous. 2005. Aspartame-acesulfame salt. In *Compendium of Food Additive Specifications*, FAO Food and Nutrition Paper 52 Addendum 13. Rome: World Health Organization Food and Agriculture Organization of the United Nations.

Anonymous. 2010. Aspartame-acesulfame salt. In *Food Chemicals Codex*, 7th Ed., 75–76. Rockville: The United States Pharmacopeial Convention.

Bartoshuk, L.M., Rifkin, B., Marks, L.E., and Hooper, J.E. 1988. Bitterness of KCl and benzoate: Related to genetic status for sensitivity to PTC/PROP. *Chemical Senses* 13:517–528.

Bravieri, R.E. 1983. Techniques for sodium reduction and salt substitution in commercial processing. *Activities Report, Morton Salt Co.,* Chicago, Illinois 35(2):79–86.

Breslin, P.A.S., and Beauchamp, G.K. 1995. Suppression of bitterness by sodium: Variation among bitter taste stimuli. *Chemical Senses* 20:614–620.

Fry, J.C. 1996. Two in one—an innovation in sweeteners. In *Low-Calorie Sweeteners. Openings in an Expanding European Market*, Proceedings of 1996 International Sweeteners Association Conference, ed. S.G. Lisansky and A. Corti, 58–68. Newbury: CPL Press.

Fry, J.C. 1998. Two in one—an innovation in sweeteners. In *Proceedings of the 45th Annual Conference of the International Society of Beverage Technologists*, Savannah, GA, 83–92. Hartfield, VA, US: International Society of Beverage Technologists.

Fry, J.C., and Van Soolingen, J. 1997. Sweetener Salts. European Patent Application EP 0 768 041.

Fry, J.C., and Van Soolingen, J. 1998. Sweetener Salts. US Patent 5827562.

Fry, J.C., Hoek, A.C., and Kemper, A.E. 1998a. Chewing gums containing dipeptide sweetener with lengthened and improved flavour. International Patent Application WO-98/02048-A.

Fry, J.C., Hoek, A.C., and Vleugels, L.F. W. 1998b. Dry foodstuffs containing dipeptide sweetener. International Patent Application WO-98/02050-A.

Hoek, A.C., Vleugels, L.F.W., and Groeneveld, C. 1999. Improved powder mix quality with Twinsweet. In *Low-calorie Sweeteners: Present and Future*, World Review of Nutrition and Dietetics, Vol. 85, ed. A. Corti, 133–139. Basel: S. Karger AG.

Palomo Coll, D.A. 1986. Procedimiento para la preparación de nuevas sales fisiológicamente activas o aceptables de sabor dulce. Spanish Patent ES-A-8 604 766.

Pangborn, R.M., and Braddock, K.S. 1989. Ad libitum preferences for salt in chicken broth. *Food Quality and Preference* 1(2):47–52.

Tenovuo, J.O. 1989. *Human Saliva: Clinical Chemistry and Microbiology*. Boca Raton: CRC Press Inc.

Chapter 7

Cyclamate

Frances Hunt, Barbara A. Bopp, and Paul Price

Contents

Introduction

Cyclamate (Figure 7.1) was synthesized in 1937 by a University of Illinois graduate student, Michael Sveda, who accidentally discovered its sweet taste (Audrieth and Sveda 1944). The patent for cyclamate (US Patent 1942) eventually became the property of Abbott Laboratories, which

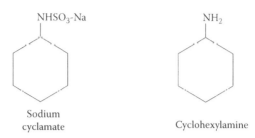

NHSO$_3$-Na

NH$_2$

Sodium
cyclamate

Cyclohexylamine

Figure 7.1 Structures of cyclamate and cyclohexylamine.

performed the necessary studies and submitted a New Drug Application for the sodium salt in 1950. Cyclamate was initially marketed as tablets that were recommended for use as a tabletop sweetener for diabetics and others who had to restrict their use of sugar. In 1958, after enactment of the Food Additive Amendment to the Food, Drug, and Cosmetic Act, the Food and Drug Administration (FDA) of the US classified cyclamate as a Generally Recognized as Safe (GRAS), sweetener. A mixture of cyclamate and saccharin, which had been found to have synergistic sweetening properties and an improved taste (Vincent et al. 1955; US Patent 1957), was subsequently marketed for use in special dietary foods. When soft drinks sweetened with the cyclamate-saccharin mixture became popular in the United States during the 1960s, the consumption of cyclamate increased dramatically. Prompted by the growing use of cyclamate, additional studies were initiated (Kojima and Ichibagase 1966). In 1969, a chronic toxicity study claimed to show a link between the consumption of cyclamate and bladder cancer in rats (Price et al. 1970). Cyclamate was subsequently removed from GRAS status (Federal Register 1969) and in 1970, it was banned in the United States (Federal Register 1970). Over the next few years, many additional toxicity and carcinogenicity studies were conducted with cyclamate but they failed to confirm these findings. On the basis of these additional studies, the industry filed a Food Additive Petition for the re-approval of cyclamate in 1973 (Food Additive Petition 1973, Federal Register 1980) and in 1982 (Food Additive Petition 1982). Despite the conclusion of the US FDA's Cancer Assessment Committee in 1984 that cyclamate is not carcinogenic (Cancer Assessment Committee 1984), the petition for the re-approval of cyclamate in the United States is still pending. Following extensive scientific studies showing that cyclamate is safe for human consumption, it has continued to be approved in more than 100 countries worldwide.

Production

The production of cyclamate, which is accomplished by the sulfonation of cyclohexylamine, was originally limited to Abbott Laboratories (Kasperson and Primack 1986). However, as cyclamate became more widely used in foods and beverages, other companies, including The Pillsbury Company, Pfizer Inc., Cyclamate Corporation of America, and Miles Laboratories, entered the market. By 1970, several foreign countries, including Japan, Taiwan, and Korea, also had production capabilities. When cyclamate was banned in the United States, all domestic producers, except Abbott Laboratories, ceased production. Production was also banned in Japan, and thus the Taiwanese and Koreans became the major producers. Today, the major producers and exporters of cyclamate are located in China, Indonesia, and Spain. They include Zhong Hua Fang Da (HK) which has the longest history of more than 50 years of continuous production.

Physical Characteristics

Cyclamic acid, or cyclohexylsulfamic acid ($C_6H_{13}NO_3S$; MW = 179.24), is a white crystalline powder, with a melting point of 169°C–170°C, good aqueous solubility (1 g/7.5 ml), and a lemon-sour sweetness. It is a strong acid, and the pH of a 10% aqueous solution is approximately 0.8–1.6. Sodium and calcium cyclamate are strong electrolytes, which are highly ionized in solution, fairly neutral in character, and have little buffering capacity. Both salts exist as white crystals or white crystalline powders. The molecular weight of the sodium salt ($C_6H_{12}NO_3S \cdot Na$) is 201.22 and that of the calcium salt ($C_{12}H_{24}N_2O_6S_2 \cdot Ca$) is 396.54 (432.58 as the dihydrate). They are freely soluble in water (1g/4–5 ml) at concentrations far in excess of those required for normal use, but have limited solubility in oils and nonpolar solvents. Cyclamate solutions are stable to heat, light, and air throughout a wide pH range.

Cyclohexylamine, which is the starting material in the synthesis of cyclamate and is also a metabolite of cyclamate, has distinctly different properties. Cyclohexylamine (MW = 99.17) is a base with a fishy odor and a bitter taste. It is a clear colorless liquid, which is miscible with water, alcohol, and non-polar solvents and has a boiling point of 134.5°C. The pH of a 0.01% aqueous solution is 10.5.

Relative Sweetness and Utility

In contrast to the sweet taste from sucrose, which appears quickly and has a sharp, clean cut-off, the sweetness from cyclamate builds to its maximal level more slowly and persists for a longer time (Beck 1980). Cyclamate is generally accepted as being 30 times sweeter than sucrose, but the relative sweetness of cyclamate tends to decrease at higher sweetness intensities. Cyclamate, for example, is approximately 40 times as sweet as a 2% sucrose solution but only 24 times as sweet as a 20% sucrose solution (Beck 1980). This trend may be at least partially due to the increasing levels of bitterness and aftertaste that characteristically appear at very high cyclamate concentrations (Vincent et al. 1955). This off-taste is, however, not a problem at concentrations normally used.

Calcium cyclamate is somewhat less sweet than sodium cyclamate, and the off-taste response starts at a lower concentration of the calcium salt than of the sodium salt. Vincent et al. (1955) suggested that the differences between the two salts might be related to differences in ionization because the sweet taste is due to the cyclamate ion, whereas the off-taste may be associated with the undissociated salt. The sweetening power of cyclamate also varies with the medium, and its actual relative sweetness should be determined in each different product. Most notably, the sweetening power of cyclamate is considerably enhanced in fruits. More detailed information on the relative sweetness and other properties of cyclamate can be found in articles by Beck (1957, 1980).

The primary use of cyclamate is as a non-caloric sweetener, generally in combination with other sweeteners, but cyclamate can also be used as a flavoring agent (i.e., to mask the unpalatable taste of drugs) (Kasperson and Primack 1986; Beck and Nelson 1963). Before being banned, cyclamate was used in the United States in tabletop sweeteners, in a variety of foods, in beverages, and in both liquid and tablet pharmaceuticals, and it is found in numerous products in more than 100 countries today (Table 7.1). Many of these same applications are proposed in the current food additive petition as the potential uses of cyclamate and include: Tabletop sweeteners, in tablet, powder, or liquid form, beverages, fruit juice drinks, beverage bases or mixes, processed fruits, chewing gum and confections, salad dressings and gelatin desserts, jellies, jams, and toppings.

Table 7.1   Regulatory Status of Cyclamate

Country	Food[a]	Beverage	Tabletop
Albania		+	
Angola			+
Antigua	+	+	
Argentina	+	+	+
Australia	+	+	+
Austria	+	+	+
Azerbaijan		+	
Bahamas	+	+	
Barbados		+	
Belarus		+	
Belgium	+	+	+
Bolivia		+	
Bosnia and Herzegovina		+	
Botswana		+	
Brazil	+	+	+
Bulgaria	+	+	+
Canada			+
Caribbean (Ind.)	+	+	+
Chile	+	+	+
China	+	+	
Comoros		+	
Costa Rica		+	
Croatia		+	
Cyprus	+	+	+
Czech Republic	+	+	+
Denmark	+	+	+
Djibouti		+	
Dominica	+	+	+
Dominican Republic		+	

Table 7.1 (Continued)   Regulatory Status of Cyclamate

Country	Food[a]	Beverage	Tabletop
Ecuador			+
El Salvador		+	
Estonia	+	+	+
Finland	+	+	+
France	+	+	+
French Guiana		+	
Germany	+	+	+
Georgia		+	
Greece	+	+	+
Grenada		+	
Guadeloupe		+	
Guatemala		+	+
Haiti	+	+	
Honduras		+	
Hong Kong	+	+	+
Hungary	+	+	+
Iceland	+	+	+
Indonesia	+	+	+
Iraq		+	
Ireland	+	+	+
Israel	+	+	I
Italy	+	+	+
Jamaica		+	
Kazakhstan		+	
Kenya		+	
Kuwait			+
Kyrgyzstan		+	
Latvia	+	+	+

(continued)

Table 7.1 (Continued) Regulatory Status of Cyclamate

Country	Food[a]	Beverage	Tabletop
Lesotho		+	
Lithuania	+	+	+
Luxembourg	+	+	+
Macau (Macao)		+	
Macedonia		+	
Madagascar		+	
Malawi		+	
Malta	+	+	+
Martinique	+	+	+
Mauritius		+	
Mayotte		+	
Mexico		+	
Moldova		+	
Montenegro		+	
Montserrat		+	
Morocco		+	
Mozambique		+	
Namibia		+	
Netherlands	+	+	+
New Zealand	+	+	+
Nicaragua	+	+	+
Norway	+	+	+
Oman			+
Pakistan	+		+
Panama		+	
Papua New Guinea	+	+	+
Paraguay	+	+	+
Peru	+	+	
Poland	+	+	+

Table 7.1 (Continued)   Regulatory Status of Cyclamate

Country	Food[a]	Beverage	Tabletop
Portugal	+	+	+
Romania	+	+	+
Russia	+	+	+
Rwanda		+	
Saint Helena		+	
Saint Kitts and Nevis		+	
Saint Lucia		+	
Saint Vincent		+	
Saudi Arabia	+		+
Serbia		+	
Seychelles		+	
Sierra Leone	+	+	+
Slovakia	+	+	+
Slovenia	+	+	+
South Africa	+	+	+
Spain	+	+	+
Sri Lanka			+
Suriname		+	
Swaziland		+	
Sweden	+	+	+
Switzerland	+	+	I
Taiwan	+	+	
Tajikistan		+	
Tanzania		+	
Thailand			+
Trinidad and Tobago	+	+	
Tunisia		+	
Turkey	+	+	+
Turkmenistan		+	

(*continued*)

Table 7.1 (Continued)   Regulatory Status of Cyclamate

Country	Food[a]	Beverage	Tabletop
Turks and Caicos Islands		+	
Ukraine		+	
United Arab Emirates			+
United Kingdom	+	+	+
Uruguay	+	+	+
Uzbekistan		+	
Venezuela		+	+
Vietnam	+		
West Bank and Gaza		+	
Zambia		+	
Zimbabwe	+		+

[a] May not apply to all food categories. + = Permitted; Information current as of 2010 but may be subject to change. (Worldwide Regulatory Status of Cyclamate 2010.)

Admixture Potential

Cyclamate has most frequently been used in combination with saccharin. In the 1950s, it was shown that cyclamate-saccharin mixtures are sweeter than would be expected from the known sweetness of either component alone and that any off-taste is minimized with the mixtures (Vincent et al. 1955; US Patent 1957). A synergism of approximately 10%–20% is observed when cyclamate and saccharin are used together. A combination of 5 mg saccharin and 50 mg cyclamate in a tabletop sweetener, for example, is as sweet as 125 mg cyclamate alone or 12.5 mg saccharin alone (Beck 1980). Although the ratio of cyclamate to saccharin can vary considerably from product to product, the 10:1 mixture is used most frequently. With this combination, each component contributes approximately equally to the sweetness of the mixture because saccharin is about 10 times sweeter than cyclamate.

More recently, numerous patents have described the use of cyclamate in combination with aspartame or aspartame and saccharin (e.g., British Patent 1971; Canadian Patent 1978). Applications have also been reported for cyclamate in combination with acesulfame K and other sweeteners (US Patent 1979). If and when cyclamate is again approved for use in the United States, it would undoubtedly be used primarily in combination with other more potent sweeteners because no one sweetener appears to meet all the technological requirements, and mixtures of sweeteners generally appear to offer improved taste.

Technical Qualities

Cyclamate has a number of technical qualities that make it a good alternative sweetener (Beck 1963; Beck 1980; Stein 1966). It is non-caloric and non-cariogenic. Although its relative sweetness is less than that of saccharin or aspartame, its sweetening power is adequate, especially when used in combination with other more intense sweeteners. Cyclamate has a favorable taste profile and does not leave an unpleasant aftertaste at normal use concentrations. It is better than sugar in masking bitterness, and it enhances fruit flavors. Cyclamate is compatible with most foods and food ingredients, as well as with natural and artificial flavoring agents, chemical preservatives, and other sweeteners. Its solubility is more than adequate for most uses, and at normal concentrations it does not change the viscosity or density of solutions. The stability of cyclamate is excellent, at both high and low temperatures, over a wide pH range, and in the presence of light, oxygen, and other food chemicals. Cyclamate is nonhygroscopic and will not support mold or bacterial growth. However, like most non-caloric sweeteners, cyclamate does not provide the bulk, texture, or body associated with sugar.

Utility

Although most of the product development work with cyclamate was done in the 1950s and 1960s, Beck has reviewed some of the applications and presented typical formulations sweetened with cyclamate or a cyclamate-saccharin mixture (Beck 1980). Perhaps the primary rule for the development of low-calorie foods and beverages is that cyclamate (or any other nonnutritive sweetener) cannot simply be substituted for the sugar; instead, the product must be reformulated (Beck 1980). The two most critical aspects for a successful product are its flavor and texture. Because a flavor may not taste the same in systems sweetened with sucrose and cyclamate (or mixtures of cyclamate and other nonnutritive sweeteners), the flavoring of a product frequently has to be modified. A balance between the taste effects of the acid and sweetener components of a product must also be achieved. If the lingering sweetness from non-caloric sweeteners results in a higher sweetness intensity than desired, for example, some compensation can be achieved by increasing the level of acidity. On the other hand, if an aftertaste develops at the level of sweetness desired, decreasing the acidity may permit a reduction in the sweetness level, hence minimizing the aftertaste. The other major problem in the development of low-calorie products with cyclamate is that of texture or body. This largely results from the elimination of sugar solids and can frequently be solved by the addition of a suitable hydrocolloid or bulking agent. An alternative approach that has been used in soft drinks involves achieving the proper balance between sweetness and tartness, the use of special flavors, and the adjustment of the carbonation level.

Cyclamate has always been particularly useful in fruit products because it enhances fruit flavors and, even at low concentrations, can mask the natural tartness of some citrus fruits (Beck 1963; Beck 1980; Stein 1966; Beck and Ziemba 1966; Salunkhe et al. 1963). The cyclamate solutions used for canned fruits have a lower specific gravity and osmotic pressure than sucrose syrups and hence do not draw water out of the fruit. Thus, fruits packed in cyclamate solutions tend to have a greater drained weight than those packed in sucrose. Cyclamate-sweetened gelatins are reasonably easy to formulate, requiring the use of high-bloom gelatins and crystalline sorbitol or mannitol as a bodying agent, dispersant, and filler (Beck 1966; Beck 1980). Thickening and consistency represent the major problems with jams, jellies, and puddings sweetened with cyclamate. Low methoxy pectin is usually used as a gelling agent in jams and jellies because it does not require sugar for gel formation (Beck 1966; Beck 1980; Stein 1966; Salunkhe et al. 1963). However, low-methoxy pectin needs more calcium than is normally present, and hence calcium cyclamate may be preferable to the sodium salt for this application. Because of the lower concentrations of osmotically active compounds, jams and jellies containing cyclamate may require a preservative to extend their shelf life. Body and thickening of puddings can be achieved with starches or a combination of non-nutritive gums and thickeners (Beck 1966; Beck 1980). Low-calorie salad dressings require the substitution of two basic ingredients: cyclamate (or a mixture of cyclamate and another sweetener) for the sugar and a hydrocolloid or thickener for the oil (Beck 1966, Beck 1980).

Baked goods are probably the most difficult foods to reformulate with non-caloric sweeteners (Beck 1966; Beck 1980; Salunkhe et al. 1963; Stone 1962). In addition to sweetness, sugar provides bulk and texture, has a tenderizing effect on gluten, and is important in the browning reaction. Cyclamate cannot furnish these properties, and hence the formulations must be modified to include bulking agents (e.g., modified starch or dextrins, carboxymethylcellulose) and a tenderizing agent (e.g., lecithin). Although proper browning is difficult to achieve, some success can be obtained by application of a caramel solution onto the surface. In yeast doughs, sugar also acts as an energy source for the fermentation reaction. However, satisfactory products containing cyclamate and only a small amount of sugar (1% or less) can be prepared if the salt content and the fermentation time are reduced (Stone 1962). As the amount of sugar in chemically aerated products is decreased, more liquid must be added to retain the proper consistency of the batter and the eggs must be used to their best advantage for structure and aeration. Good results were obtained with a slurry technique, in which the flour and other dry ingredients were mixed with water before being added to the whipped eggs (Stone 1962). However, preparation of high-quality, low-calorie baked goods still represents a major technical challenge.

In contrast to baked goods, the lack of a browning reaction with cyclamate can be an advantage in cured meats (Beck et al. 1958; Beck 1963; Beck 1966). When sugar-cured bacon or ham is fried, the sugar tends to caramelize, losing its sweet taste and giving the meats a darkened appearance. Because cyclamate has a higher melting point than sucrose, cyclamate-cured meats taste better, have an improved color, and do not scorch or stick in the frying pan.

Cyclamate has also found applications in pharmaceutical and oral hygiene products. It is particularly good at masking the bitterness and unpalatable taste of many drugs and hence is especially useful in syrups, other liquid formulations, and chewable tablets (Lynch and Gross 1960). Cyclamate imparts a high level of sweetness with a low solid content, thus providing suspensions that are more fluid and have fewer problems with caking (Lynch and Gross 1960) or tablets that disintegrate rapidly and have less bulk (Endicott and Gross 1959). Cyclamate is also useful as a sweetener in both film coating and compression coating of tablets, and the acid form can be used as an effervescent agent (Endicott and Gross 1959). Because cyclamate is non-cariogenic, it is suitable for use in toothpastes and mouthwashes.

Availability

Cyclamate is still available in the United States, and from foreign suppliers. The use of cyclamate as a food additive is currently prohibited in the United States, and therefore it may not be used in foods or beverages in this country at present. It is, however, legal to manufacture cyclamate-containing foods in the United States for export to foreign countries, where the use of cyclamate as a sweetener is permitted. This must, however, be done in compliance with Section 801(d) of the Federal Food, Drug, and Cosmetic Act 21 USC 381(d).

Shelf Life

Samples of tablets containing a cyclamate and saccharin mixture, which were manufactured in 1969 or earlier, showed no diminution in sweetening ability or any physical deterioration after at least seven years (Kasperson and Primack 1986). It would seem, therefore, that cyclamate in tablet form has an extremely long shelf life. Information about the possible shelf life of cyclamate in other applications, such as soft drinks or canned foods, was not available to the authors. When cyclamate was widely used in the United States, however, the shelf life was more than adequate to allow the products to be sold in the ordinary course of business. There is no known instance of a recall of products because of the degradation of the sweetening content from cyclamate. The cyclamate stability data indicate that an expiration date is not needed to ensure the identity, strength, quality, and purity of either the bulk food additive or foods and beverages containing cyclamate.

Transport

No known problem exists with the transport of the bulk material, and in the United States cyclamate is nonregulated with respect to transport.

General Cost/Economics

It is anticipated that should the US FDA once again allow the use of cyclamate as a food additive in the United States, the price would be approximately $3.00 per kilogram, estimated at today's production costs. The price from foreign producers has varied considerably from year to year, depending largely on product availability. As the price of sugar also fluctuates considerably, the only means of determining the economics of cyclamate use is to compare its cost at the time of production to the cost of equivalent sweetening from sugar at its prevailing price. Such comparisons generally indicate that cyclamate is one of the most economical non-caloric sweeteners.

Metabolism

Cyclamate is slowly and incompletely absorbed from the gastrointestinal tract (Bopp et al. 1986). In one study involving almost 200 subjects, the absorption of cyclamate averaged only 37% (Sonders 1967–1968). The volume of distribution for cyclamate in rats is approximately equal to the total body water content, hence cyclamate does not concentrate in most tissues (Bopp et al. 1986). Once absorbed, cyclamate is excreted unchanged in the urine by both glomerular filtration and active tubular secretion (Bopp et al. 1986).

Although early studies had indicated that cyclamate was not metabolized to any appreciable extent, in 1966 Kojima and Ichibagase (1966) detected cyclohexylamine in urine samples from humans and dogs receiving cyclamate. Subsequently, conversion was also demonstrated in mice, rats, guinea pigs, rabbits, monkeys, pigs, dogs and humans (Bopp et al. 1986). Cyclamate is, however, not metabolized to cyclohexylamine by mammalian tissues, but rather cyclohexylamine is formed by the action of microflora on the non-absorbed cyclamate remaining in the intestinal tract (Bopp et al. 1986).

Probably the most important feature of cyclamate metabolism is the extreme variability in cyclohexylamine formation. Not all individuals are able to convert cyclamate to cyclohexylamine, and even among converters, the extent of conversion varies greatly and often changes in the same individual over time. Retrospective analyses of studies that attempted to define the incidence of converters indicated that approximately 25% of the subjects were able to metabolize cyclamate to cyclohexylamine (Bopp et al. 1986; Sonders 1967–1968). The incidence of converters was slightly lower (~20%) among Europeans and North Americans who were given at least three daily doses of cyclamate but appeared to be higher among the Japanese (80% or more in studies involving approximately 60 subjects).

As noted previously, even among those individuals who can convert cyclamate to cyclohexylamine, there is substantial inter-subject and day-to-day variability in the extent of conversion, which can range from less than 0.1% to more than 60%. The frequency distribution curve for cyclamate conversion is, however, strongly skewed, and only a few individuals are able to form large amounts of cyclohexylamine. It has been estimated that approximately 3% of a population converts more than 20% of a cyclamate dose to cyclohexylamine, and 1% or less converts 60% or more (Food Additive Petition 1982; Renwick 1983; Collings 1989). The 60% level approaches, on the average, the maximal conversion possible because only the non-absorbed cyclamate can be metabolized, and the absorption of cyclamate averages about 40%.

The conversion of cyclamate to cyclohexylamine also appears to depend on continuous exposure to the sweetener (Bopp et al. 1986). A single dose of cyclamate will frequently not be metabolized, and daily ingestion of cyclamate is usually necessary to induce and maintain the converting ability at a high but still variable level. Furthermore, if cyclamate is withdrawn from the diet for even a few days, the ability to metabolize cyclamate is diminished and gradually lost. Hence, intermittent cyclamate use (which might be a typical pattern of use by some people) would tend to limit the converting ability of an individual.

Because the no-observed effect level (NOEL) in animal toxicity studies is based on cyclohexylamine, not cyclamate, determination of the extent of conversion is critical for establishing the safe level of use for the sweetener. However, the skewed nature of the cyclamate conversion curve makes this more difficult. It is not appropriate to use the mean conversion by an entire group of converters because most of these subjects would be forming only small amounts of cyclohexylamine. Instead, the best estimate can probably be derived from the average conversion by a subgroup of high converters. The available data suggest that among those subjects who converted at least 1% of the dose to cyclohexylamine, conversion averaged slightly less than 20% (Food Additive Petition 1982; Bopp et al. 1986). If the group was further restricted to only those subjects converting at least 5% of the dose, the average level increased only slightly (~25%) (Bopp et al. 1986; Renwick 1979). A few individuals would, of course, be converting at higher levels at least some of the time. It is probably not necessary, however, to use the maximal conversion rate in establishing the acceptable daily intake (ADI) because the large safety factor applied to food additives compensates for considerable inter-subject variability.

In contrast to cyclamate, cyclohexylamine is rapidly and completely absorbed from the gastrointestinal tract, even from the large intestine where it is formed (Bopp et al. 1986). The plasma half-life of cyclohexylamine in humans is dose-dependent, increasing from 3.5 hr with a 2.5 mg/kg oral dose to 4.8 hr with a 10 mg/kg dose (Eichelbaum et al. 1974). The apparent volume of distribution of cyclohexylamine is 2–3 L/kg, and tissue concentrations typically exceed those in plasma (Bopp et al. 1986; Eichelbaum et al. 1974).

Cyclohexylamine is primarily excreted unchanged in the urine by both glomerular filtration and a saturable transport mechanism (Eichelbaum et al. 1974; Roberts and Renwick 1989). Although cyclohexylamine is not extensively metabolized, its biotransformation shows some species differences (Bopp et al. 1986). The principal metabolic pathway in rats is ring hydroxylation, leading to the formation of the isomeric 3- or 4-aminocyclohexanols, which account for 5%–20% of the dose (Renwick and Williams 1972; Roberts and Renwick 1985; Roberts et al. 1989). Mice and humans, however, form negligible quantities of these metabolites (Renwick and Williams 1972;

Roberts and Renwick 1985; Roberts et al. 1989). Only the deamination products, cyclohexanol and *trans*-cyclohexane-l,2-diol, are found in humans given cyclohexylamine, and these metabolites represent only 1%–2% of the dose (Renwick and Williams 1972). Both ring hydroxylation and deamination occur in guinea pigs and rabbits (Renwick and Williams 1972).

The extent of metabolism of cyclamate to cyclohexylamine in humans is critical in the determination of the ADI for cyclamate. In the EU, the Scientific Committee for Food (SCF) evaluation of cyclamate in 1995 called for additional data on the variability in cyclamate metabolism in order to determine a fixed ADI. In 2004, Renwick et al. undertook an extensive clinical study in human volunteers, which provided clinical data on the variations in cyclamate metabolism during long-term administration in humans. The study concluded that the ADI of 0–11 mg/kg, as originally allocated by the Joint FAO/WHO Expert Committee on Food Additives (JECFA) in 1982 (WHO 1982), and which had been based on empirical considerations of the likely extent of human variability in gut flora metabolism of cyclamate to cyclohexylamine, is well supported by the results of this study.

Carcinogenicity

In 1969, a 2-year chronic toxicity study with a cyclamate-saccharin mixture was nearing completion at the Food and Drug Research Laboratories (FDRL) in Maspeth, New York. In the study, rats were fed diets containing a 10:1 mixture of sodium cyclamate and sodium saccharin to provide daily doses of 500, 1120, or 2500 mg/kg/day (Price et al. 1970; Oser et al. 1975). The ability of some of the rats from this study to convert cyclamate to cyclohexylamine had been documented (Oser et al. 1968), and after 78 weeks, the diets of half of the animals were supplemented with cyclohexylamine at levels corresponding to approximately 10% conversion of the cyclamate dose (i.e., 25, 56, and 125 mg/kg/day). At the conclusion of the study, tumors were found in the urinary bladder of eight high-dose rats, with four to eight of the tumors being classified as carcinomas (Price et al. 1970). In subsequent analyses, the number of tumors increased to 12 and all were considered carcinomas (Oser et al. 1975). Calcification of the urinary tract and bladder parasites were observed in some of these rats and could possibly have affected the results of this study as both of those factors are known to contribute to the development of bladder tumors in rats. Furthermore, six of the tumors occurred in consecutively numbered and presumably consecutively housed male rats, suggesting that some extraneous environmental factor could also have been involved in the development of the tumors. Nevertheless, the results of the study were interpreted as implicating cyclamate as a bladder carcinogen in rats (Price et al. 1970) and led to its removal from GRAS status, and finally to its being banned for use in foods and beverages in the United States (Federal Register 1969, 1970).

In the years following this observation, the carcinogenic potential of cyclamate and cyclohexylamine was reevaluated in a group of well-designed and well-controlled bioassays, which were performed by independent investigators throughout the world. Cyclamate, including both the sodium and calcium salts, was tested in at least four separate studies with rats given doses up to 2.5 g/kg/day (Schmähl 1973; Schmähl and Habs 1984; Taylor et al. 1980; Furuya et al. 1975; Ikeda et al. 1973; Homburger 1973), in three separate studies with mice given doses up to 7–9 g/kg/day (Homburger 1973; Brantom et al. 1973; Kroes et al. 1977), and in one study with hamsters given doses up to 3 g/kg/day (Althoff et al. 1975). The 10:1 cyclamate-saccharin mixture that had been used in the original FDRL study has been tested twice in rats (Schmähl 1973; Schmähl and Habs 1984; Furuya et al. 1975; Ikeda et al. 1973) and once in mice (Kroes et al. 1977). Two studies, one in rats (Taylor et al. 1980) and one in mice (Kroes et al. 1977), even included in utero exposure of the animals to cyclamate. In addition to these conventional rodent bioassays, two studies have been conducted in monkeys given cyclamate, with treatment lasting eight years in one (200 mg/kg/day) (Coulston et al. 1977) and 20+ years in the other (100 or 500 mg/kg/day) (Sieber and Adamson 1979; Takayama et al. 2000). Cyclohexylamine, as the hydrochloride or sulfate salt, was tested in three studies with rats given maximal doses of 150–300 mg/kg/day (Schmähl 1973; Gaunt et al. 1976; Oser et al. 1976) and in two studies with mice given maximal doses of 400–600 mg/kg/day (Kroes et al. 1977; Hardy et al. 1976). A 9-year study was also performed in dogs (Industrial Bio-Test Laboratories 1981). None of these studies confirmed the original findings that implicated cyclamate as a bladder carcinogen. Instead, when the results of these studies are evaluated in accordance with recognized toxicological and statistical procedures, they provide strong evidence that neither cyclamate nor cyclohexylamine is carcinogenic in animals.

In 1976, the Temporary Committee of the National Cancer Institute that was charged with reviewing the cyclamate data concluded "the present evidence does not establish the carcinogenicity of cyclamate or its principal metabolite, cyclohexylamine, in experimental animals" (Temporary Committee for the Review of Data on Carcinogenicity of Cyclamate 1976, 48). On receipt of this report, the Bureau of Foods of the US FDA apparently concurred and stated in a memorandum to the Commissioner "consider the carcinogenicity issue resolved." (Roberts Actions Memorandum, undated, 13). In 1977, the World Health Organization's JECFA reached a similar conclusion, stating that it is now possible to conclude that cyclamate has been demonstrated to be non-carcinogenic in a variety of species (WHO 1978). Nevertheless, carcinogenicity was a major issue in the 1980 decision of the US FDA commissioner, who stated that "cyclamate has not been shown not to cause cancer" and thus denied cyclamate food additive status (Federal Register 1980, 61474). However, some toxicological and statistical principles used in this decision were subsequently challenged by the American Statistical Association (1981) and the Task Force of the Past Presidents of the Society of Toxicology (1982).

After submission of the second Food Additive Petition for cyclamate in 1982, the Cancer Assessment Committee (CAC) of the US FDA's Center for Food Safety and Applied Nutrition reviewed all the cyclamate data and concluded that "there is very little credible data to implicate cyclamate as a carcinogen at any organ/tissue site to either sex of any animal species tested," that "the collective weight of the many experiments... indicates that cyclamate is not carcinogenic," and further that "no newly discovered toxic effects of cyclamate are likely to be revealed if additional standardized studies were performed" (Cancer Assessment Committee 1984, 62). In 1985, the National Academy of Sciences (NAS) reaffirmed the CAC conclusion stating that "... the totality of the evidence from studies in animals does not indicate that cyclamate or its major metabolite cyclohexylamine is carcinogenic by itself" (National Academy of Sciences 1985, 3). The NAS report, however, raised some concern about a possible role of cyclamate as a promoter. This was largely predicated on two studies involving direct exposure of the urinary bladder to cyclamate. One study involved implantation of a pellet containing cyclamate in the bladder of mice (Byran and Erturk 1970) and the other involved intravesicular instillation of *N*-methyl-*N*-nitrosourea into the urinary bladder of female rats, which were then given cyclamate in either their food or water (Hicks et al. 1975). At the request of the US FDA, the Mitre Corporation evaluated these models and concluded that "both types of direct bladder exposure studies, pellet implantation and intravesicular catheterization, are considered unsuitable for predicting human carcinogenic risk" (DeSesso et al. 1987, 6-14). It appears that the carcinogenicity issue has been settled, at least from a scientific point of view, and it can be concluded that cyclamate and cyclohexylamine are not carcinogenic in animals.

The possible association between cancer, particularly bladder cancer, and the consumption of non-caloric sweeteners by humans has been extensively studied during the past 30 years. Since the widespread use of cyclamate in foods and beverages was restricted to a relatively short time span in many countries, these studies are more applicable to the assessment of the possible carcinogenicity of saccharin than cyclamate. The cyclamate issue was specifically addressed, however, in some of the studies. The epidemiology studies with the non-caloric sweeteners have been reviewed by others and it has generally been agreed that there is no conclusive evidence of an increased risk of bladder cancer associated with the use of these sweeteners (Morgan and Wang 1985; Jensen 1983; Arnold et al. 1983).

Mutagenicity

In the 1960s and 1970s, the mutagenic potential of cyclamate and cyclohexylamine was evaluated in many different test systems, including the Ames test and other microbial gene mutation assays, studies in *Drosophila*, *in vitro* cytogenetic studies, *in vivo* cytogenetic studies with both somatic and germ cells, dominant lethal tests, and others. Although there were instances of conflicting or discordant results, the preponderance of evidence provided by this battery of tests suggested that neither cyclamate nor cyclohexylamine represented a significant mutagenic hazard (Bopp et al. 1986; Cattanach 1976; Cooper 1977). Nevertheless, in the 1980 decision the Commissioner of the US FDA still concluded that "cyclamate has not been shown not to cause heritable genetic damage" (Federal Register 1980, 61474).

In 1985 the NAS Committee evaluated the mutagenicity tests with cyclamate and cyclohexylamine and found little evidence to conclude that either compound was a DNA-reactive carcinogen (National Academy of Sciences 1985). It was recommended, however, that additional tests for mammalian cell DNA damage, mammalian cell gene mutation tests, and more definitive cytogenetic studies be conducted to complete and strengthen the database for the two compounds. After consultation with the US FDA, both calcium cyclamate and cyclohexylamine were tested in the Chinese hamster ovary HGPRT forward mutation assay, for unscheduled DNA synthesis in rat hepatocytes, and in the *Drosophila* sex-linked recessive lethal assay (Brusick et al. 1989). The results of these three tests gave no evidence of any intrinsic genotoxicity from either compound. In addition, dominant lethal and heritable translocation tests were conducted in male mice given the maximal tolerated dose of calcium cyclamate for six weeks as the sodium salt had been used in almost all the previous studies with mammalian germ cells (Cain et al. 1988). The results of these two tests were also negative, indicating that the calcium salt did not induce any transmissible chromosomal aberrations in the germ cells of mice. Thus, these studies, performed according to the currently accepted standards, confirm the previous conclusions that cyclamate and cyclohexylamine are not mutagenic and do not cause heritable genetic damage.

Other Toxicity

Cyclamate

Although a great many toxicity studies have been conducted with cyclamate and the cyclamate-saccharin mixture, they have revealed very few pathophysiological effects attributable to the sweetener (Bopp et al. 1986). Perhaps the most frequently reported effects in animals and humans given cyclamate are softening of the stools and diarrhea. These effects, however, only occur at excessively high doses and are due to the osmotic activity of the non-absorbed cyclamate remaining in the gastrointestinal tract (Hwang 1966).

Cyclohexylamine

Cyclohexylamine is considerably more toxic than cyclamate, and its toxicity limits the use of the sweetener. Because the toxicity studies with cyclohexylamine have thoroughly been reviewed (Bopp et al. 1986), only the two areas of major concern, testicular atrophy and cardiovascular effects, will be discussed herein.

Testicular Effects of Cyclohexylamine

Numerous toxicological studies have shown that the testes of rats are the organs most sensitive to adverse effects from cyclohexylamine, and this effect has been used by JECFA and others as the basis for establishing the ADI intake for cyclamate. The results of three 90-day studies in rats given cyclohexylamine in the diet indicated that the testes were not adversely affected at concentrations of 0.2% (approximately 100 mg/kg/day) and below, but at concentrations of 0.6% (approximately 300 mg/kg/day) and above, testicular atrophy, characterized by decreased organ weight, decreased spermatogenesis, and/or degeneration of the tubular epithelium, was observed (Collings and Kirby 1974; Gaunt et al. 1974; Crampton 1975; Mason and Thompson 1977). Because the concentration of cyclohexylamine in the food of these animals was held constant, the milligram per kilogram per day dose ingested by the animals declined over the 3-month study period as the food consumption of the rats relative to their body weight progressively decreased. To circumvent this problem and to more precisely define the no-adverse effect dose, an additional study was conducted by Dr. Horst Brune (Brune et al. 1978) in Germany. Groups of 100 young male rats were fed diets containing cyclohexylamine to provide constant daily doses of 50, 100, 200, or 300 mg/kg/day for three months. Ad libitum and pair-fed control groups were also included in the experimental design of this study. The results clearly demonstrated that 100 mg/kg/day was a NOEL. Slight, but statistically significant, histopathological changes were noted in the testes of the rats receiving the 200 mg/kg/day dose, and marked effects were seen at 300 mg/kg/day. Analysis of all the studies involving an effect of cyclohexylamine on the rat testes showed that the data were quite consistent and confirmed the steep nature of the dose-response curve, as demonstrated by the Brune study (Bopp et al. 1986). The lack of significant histopathological changes in the testes of rats given 150 mg/kg/day cyclohexylamine for two years (Oser et al. 1976) and the presence of only mild, sporadic effects in rats given ~175 mg/kg/day for 90 days (Collings and Kirby 1974) are consistent with the steep dose-response curve and suggest that the no-effect level might be as high as 150 mg/kg/day (Food Additive Petition 1982). However, 100 mg/kg, which was clearly shown to be a NOEL in the study of Brune, was adopted as the NOEL by JECFA (1982).

Early studies had indicated that mice were less sensitive to the effects of cyclohexylamine than rats as no testicular changes had been seen in mice given doses of ~300 mg/kg/day (0.3% in the diet) (Hardy et al. 1976). Renwick and Roberts suggested that this species difference in testicular toxicity might be related to differences in metabolism because the ring-hydroxylated metabolites of cyclohexylamine (i.e., 3- or 4-aminocyclohexanols) account for approximately 15%–20% of the dose in male rats but are formed in negligible amounts in mice (Roberts and Renwick 1985; Roberts et al. 1989). To test this hypothesis, the metabolism and testicular effects of 400 mg/kg/day cyclohexylamine were compared in mice, Wistar rats, and DA rats, a strain deficit in hydroxylating ability (Roberts et al. 1989). Consistent with the previous findings, the mice formed only small amounts of the ring-hydroxylated metabolites and did not develop testicular toxicity. Although the Wistar rats had higher concentrations of the aminocyclohexanol metabolites in their plasma and testes than the DA rats, the DA strain appeared to be more sensitive to the cyclohexylamine-induced testicular effects than the Wistar rats. Thus, the development of testicular atrophy could not be attributed to the ring-hydroxylated metabolites. Instead, the species differences in toxicity were related to differences in the pharmacokinetics of cyclohexylamine (Roberts and Renwick 1989). The concentrations of cyclohexylamine in the plasma and testes of rats showed a nonlinear relationship to dietary intake, with substantially elevated levels occurring at intakes greater than 200 mg/kg/day. This nonlinear accumulation of cyclohexylamine in the rat testes correlated well with the previously noted steep dose-response curve for the testicular atrophy. In contrast, both the plasma and testicular cyclohexylamine concentrations in mice increased linearly with intake, even at doses far in excess of 200 mg/kg/day. Furthermore, the dose-normalized cyclohexylamine concentrations at levels greater than 200 mg/kg/day were considerably higher in rats than in mice, suggesting that the greater exposure of the rat testes to cyclohexylamine probably contributed to the greater toxicity in that species. Because these studies clearly demonstrated that the testicular toxicity of cyclohexylamine was attributable to the parent compound per se and that the rat was more sensitive to this effect than the mouse, it is appropriate to use the NOEL in rats for determining the ADI of cyclamate in humans.

Cardiovascular Effects of Cyclohexylamine

The other major question about the safety of cyclohexylamine involves its possible effects on blood pressure. Cyclohexylamine is an indirectly acting sympathomimetic agent, similar to tyramine but more than 100 times less potent than tyramine (Bopp et al. 1986). Intravenous administration of cyclohexylamine to anesthetized animals causes vasoconstriction and increases both the blood pressure and heart rate. Despite these acute effects, a rise in blood pressure was not seen in sub-chronic or chronic toxicity studies with orally administered cyclohexylamine in rats, even at high doses (0.4%–1% in the diet or approximately 200–400 mg/kg/day) (Schmähl 1973; Collings and Kirby 1974).

The cardiovascular effects of single oral doses of cyclohexylamine in humans have been thoroughly characterized by Eichelbaum et al. (1974). They found that a 10 mg/kg bolus oral dose of cyclohexylamine caused a 30-mm rise in the mean arterial blood pressure of healthy human volunteers. A smaller increase was seen after a 5 mg/kg dose, but no significant change in blood pressure occurred with a 2.5 mg/kg dose. A reflex-mediated, slight decrease in heart rate accompanied the vasopressor effects of the two higher doses. The cyclohexylamine levels in the plasma of these subjects were closely correlated with the increases in blood pressure, and it was estimated that the lowest cyclohexylamine concentration to cause a significant hypertensive effect was about 0.7–0.8 µg/ml. The blood pressure in these subjects, however, rapidly returned toward normal despite the presence of cyclohexylamine concentrations that were associated with a pressor effect during the absorptive phase. These observations suggested the rapid development of tolerance or tachyphylaxis to the hypertensive effects of cyclohexylamine.

Despite the potential of cyclohexylamine to increase blood pressure, there is no evidence to suggest that such effects actually occur after the administration of cyclamate. Periodic cardiovascular monitoring in some subjects participating in cyclamate metabolism studies revealed no changes in blood pressure or heart rate, even in those individuals who received high doses of cyclamate and were forming large amounts of cyclohexylamine (Sonders 1967–1968; Collings

1989; Litchfield and Swan 1971; Collings 1969). Moreover, the cyclohexylamine blood levels in some of these subjects approached the concentrations associated with a rise in blood pressure after a bolus oral dose of cyclohexylamine, yet blood pressure was apparently not affected (Collings 1989).

Two human studies conducted in Germany during the 1970s also addressed the questions of cyclamate conversion, cyclohexylamine blood levels, and possible cardiovascular effects. In one study, the high converters among a group of 44 regular users of cyclamate were initially identified on the basis of their urinary cyclohexylamine levels. After ingestion of 1.2–1.7 g of sodium cyclamate, the cyclohexylamine blood levels in three of the high converters ranged from 0.03–0.42 μg/ml and were correlated with the urinary cyclohexylamine levels (Schmidt 1987). In the other study, the blood pressure and heart rate of 20 subjects were monitored three times a day during a 10-day dosing period with sodium cyclamate (Dengler et al. 1987). Half of these subjects did not convert cyclamate to cyclohexylamine, five were low converters, and the other five were high converters. No significant changes in blood pressure or heart rate were seen, however, even in the high converters. Furthermore, noninvasive hemodynamic studies, which were conducted in six subjects known to be good converters, failed to reveal any changes that were attributable to cyclamate or cyclohexylamine, even though the cyclohexylamine levels in serum ranged from approximately 0.2–0.6 μg/ml in most subjects and reached 1 μg/ml in one subject (Dengler et al. 1987).

Buss et al. (1992) conducted a controlled clinical study investigating the conversion of cyclamate to cyclohexylamine and its possible cardiovascular consequences in 194 diabetic patients who were given calcium cyclamate (1 g/day as cyclamic acid equivalents) for 7 days. The incidence and extent of cyclohexylamine formation in this group of subjects were similar to those reported in the earlier studies; 78% of subjects excreted less than 0.1% of the daily dose as cyclohexylamine in urine, but 4% (eight subjects) excreted more than 20% of the daily dose as cyclohexylamine. Similar inter-subject variability was observed in the cyclohexylamine concentrations in plasma, with concentrations of 0–0.01 μg/ml in 168 subjects, 0.01–0.3 μg/ml in 18 subjects, 0.3–1.0 μg/ml in six subjects, and more than 1.0 μg/ml in two subjects. A significant correlation was found between the cyclohexylamine in plasma and urine. The changes in mean arterial blood pressure and heart rate in the eight subjects with plasma cyclohexylamine concentrations between 0.3 and 1.9 μg/ml were similar to those found in 150 subjects with very low (less than 0.01 μg/ml) cyclohexylamine concentrations. In a second study conducted 10–24 months later, 20 of the subjects were given calcium cyclamate for 2 weeks at a dose of 2 g cyclamic acid equivalents/day (0.66 g tid). Cyclohexylamine concentrations in plasma, blood pressure, and heart rate were measured every 30 minutes during the final 8-hr dosing interval. Twelve of the subjects had plasma cyclohexylamine concentrations of 0.09–2.0 μg/ml at the start of the dose interval. Transient increases or decreases in the cyclohexylamine concentrations were not observed during the 8-hr dosing interval nor were there any significant changes in blood pressure or heart rate measurements. Collectively, these results indicate that the metabolism of cyclamate (2 g/day) to cyclohexylamine does not significantly affect blood pressure or heart rate, even in those few individuals who are high converters and have relatively high concentrations of cyclohexylamine in plasma.

The lack of cardiovascular effects in the subjects who had plasma cyclohexylamine concentrations of 0.3–1.9 μg/ml in the study by Buss et al. (1992) contrasts with the observation by Eichelbaum et al. (1974) that cyclohexylamine concentrations of 0.7–0.8 μg/ml were sufficient to increase the blood pressure of subjects given a bolus oral dose of cyclohexylamine. The difference in the cyclohexylamine plasma concentration-time profiles from the two routes of delivery may offer an explanation for this discrepancy. A bolus oral dose of cyclohexylamine leads to a rapid rise and subsequent fall in the plasma concentrations. In contrast, during periods of cyclamate ingestion, the cyclohexylamine concentrations would increase gradually as metabolizing activity develops and then remain relatively constant. To further explore this possibility, Buss and Renwick (1992) investigated the relationship between cyclohexylamine concentrations and blood pressure changes in rats given intravenous infusions of cyclohexylamine. The blood pressure increases in the rats were inversely related to the duration of the infusion, despite the presence of similar plasma concentrations at the end of the infusion. In addition, the plasma concentration-effect relationship showed a clockwise hysteresis, indicative of tachyphylaxis and similar to that seen in humans in the study by Eichelbaum et al. (1974) Thus, the hypertensive effects of cyclohexylamine primarily occur when the plasma concentrations are rapidly increasing. Because the kinetics of cyclamate metabolism would not lead to a rapid increase in the cyclohexylamine concentrations, hypertensive effects would not be expected to occur in those individuals who are ingesting large amounts of cyclamate and are high converters.

Regulatory Status

Cyclamate was given a positive safety assessment by the JECFA in 1982 (WHO 1982) and by the SCF of the European Commission, now the European Food Safety Authority (EFSA), in 2000.

JECFA has set an ADI for cyclamate at 0–11mg/kg, expressed as cyclamic acid (WHO 1982). Following a review of the ADI by the SCF (now EFSA) in Europe in 2000, the temporary ADI of 0–11mg/kg was replaced by a full ADI of 0–7mg/kg (SCF 2000).

In 2010, based on JECFA's dietary exposure assessment for cyclamate in beverages (WHO 2009), the Codex Alimentarius Commission (2010) agreed to adopt a level of 350mg/kg for the use of cyclamate in beverages, thereby completing the list of permissions for the use of cyclamate in food and beverages in the Codex General Standard for Food Additives (GSFA).

Permissions for cyclamate were also recently reviewed by Food Standards Australia New Zealand (FSANZ) in 2007. FSANZ confirmed that the established ADI of 0–11mg/kg is adequately protective of consumers (FSANZ 2007).

Cyclamate is now approved in more than 100 countries worldwide. As summarized in Table 7.1 (Regulatory Status of Cyclamate), some countries allow the use of cyclamate in foods, beverages or both, whereas others only allow tabletop use. Pharmaceutical use is also permitted in some countries.

In the EU, cyclamate is authorised under the Sweetener Directive 94/35/EC. Currently, cyclamate is not approved in the United States. A petition for the re-approval of cyclamate is currently under review by the US FDA.

References

Arnold, D.L., Krewski, D., and Munro, I.C. 1983. Saccharin: A toxicological and historical perspective. *Toxicol* 27:179–256.
Althoff, J., Cardesa, A., Pour, P., and Shubik, P.A. 1975. A chronic study of artificial sweeteners in Syrian golden hamsters. *Cancer Lett* 1:21–24.
American Statistical Association. April 7, 1981. Letter to the Commissioner of the Food and Drug Administration.

Audrieth, L.F., and Sveda, M. 1944. Preparation and properties of some N-substituted sulfamic acids. *J Org Chem* 9:89–101.

Beck, K.M. 1957. Properties of the synthetic sweetening agent, cyclamate. *Food Technol* 11:156–158.

Beck, K.M. 1980. Nonnutritive sweeteners: Saccharin and cyclamate. In *CRC handbook of food additives,* ed. T.E. Furia, Vol. II, 2nd edition. Boca Raton, FL: CRC Press.

Beck, K.M., Jones, R., and Murphy, L.W. 1958. New sweetener for cured meats. *Food Eng* 30:114.

Beck, K.M., and Nelson, A.S. 1963. Latest uses of synthetic sweeteners. *Food Eng* 35:96–97.

Beck, K.M., and Ziemba, J.V. 1966. Are you using sweeteners correctly? *Food Eng* 38:71–73.

Bopp, B.A., Sonders, R.C., and Kesterson, J.W. 1986. Toxicological aspects of cyclamate and cyclohexylamine. *CRC Crit Rev Toxicol* 16:213–306.

Brantom, P.G., Gaunt, I.F., and Grasso, P. 1973. Long-term toxicity of sodium cyclamate in mice. *Food Cosmet Toxicol* 11:735–746.

British Patent 1,256,995. 1971. Saccharin-dipeptide sweetening compositions. Scott, D. (GD Searle).

Brune, H., Mohr, U., and Deutsch-Wenzel, R.P. 1978. Establishment of the no-effect dosage of cyclohexylamine hydrochloride in male Sprague-Dawley rats with respect to growth and testicular atrophy. Unpublished report.

Brusick, D., Cifone, M., Young, R., and Benson, S. 1989. Assessment of the genotoxicity of calcium cyclamate and cyclohexylamine. *Environ Mol Mutagen* 14:188–199.

Buss, N.E., and Renwick, A.G. 1992. Blood pressure changes and sympathetic function in rats given cyclohexylamine by intravenous infusion. *Toxicol Appl Pharmacol* 115:211–215.

Buss, N.E., Renwick, A.G., Donaldson, K.M., and George, C.F. 1992. The metabolism of cyclamate to cyclohexylamine and its cardiovascular consequences in human volunteers. *Toxicol Appl Pharmacol* 115:199–210.

Byran, G.T., and Erturk, E. 1970. Production of mouse urinary bladder carcinomas by sodium cyclamate. *Science* 167:996–998.

Cain, K.T., Cornett, C.V., Cacheiro, N.L.A., Hughes, L.A, Owens, J.G., and Generoso, W.M. 1988. No evidence found for induction of dominant lethal mutations and heritable translocations in male mice by calcium cyclamate. *Environ Mol Mutagen* 11:207–213.

Canadian Patent 1,043,158. 1978. Sweetening compositions containing APM and other adjuncts Finucane TP (General Foods).

Cancer Assessment Committee (CAC). 1984. Center for Food Safety and Applied Nutrition, Food and Drug Administration. Scientific review of the long-term carcinogen bioassays performed on the artificial sweetener. cyclamate, April.

Cattanach, B.M. 1976. The mutagenicity of cyclamates and their metabolites. *Mutat Res* 39:1–28.

Codex Alimentarius Commission (CAC). July 2010. Report of the Codex Alimentarius Commission. ALINORM 10/33/REP.

Codex General Standard for Food Additives (GSFA). GSFA Online: http://www.codexalimentarius.net/gsfaonline.

Collings, A.J. 1969. Effect of dietary level and the role of intestinal flora on the conversion of cyclamate to cyclohexylamine. Unilever Research Laboratory. Unpublished report.

Collings, A.J. 1989. Metabolism of cyclamate and its conversion to cyclohexylamine. *Diabetes Care* 12(suppl. 1):50–55.

Collings, A.J., and Kirkby, W.W. 1974. The toxicity of cyclohexylamine hydrochloride in the rat, 90-day feeding study, Unilever Research Laboratory. Unpublished report.

Cooper, P. 1977. Resolving the cyclamate question. *Food Cosmet Toxicol* 15:69–70.

Coulston, F., McChesney, E.W., and Benitz, K-F. 1977. Eight-year study of cyclamate in Rhesus monkeys. *Toxicol Appl Pharmacol* 41:164–165.

Crampton, R.F. 1975. British Industrial Biological Research Association, correspondence to H. Blumenthal, Food and Drug Administration. December 19.

Dengler, H.J., Hengstmann, J.H., and Lydtin, H. 1987. Extent and frequency of conversion of cyclamate into cyclohexylamine in a German patient group, Supplement to Food Additive Petition 2A3672, December 14.

DeSesso, J.M., Kelley, J.M., and Fuller, B.B. 1987. Relevance of direct bladder exposure studies to human health concerns. The MITRE Corporation. Unpublished report.

Eichelbaum, M., Hengstmann, J.H., Rost, H.D., Brecht, T., and Dengler, H.J. 1974. Pharmacokinetics, cardiovascular and metabolic actions of cyclohexylamine in man. *Arch Toxikol* 31:243–263.

Endicott, C.J., and Gross, H.M. 1959. Artificial sweetening of tablets. *Drug Cos Ind* 85:176–177, 254–256.

Federal Register. 34(202):17063 (October 21, 1969).

Federal Register. 35(167):13644 (August 27, 1970).

Federal Register. 45:61474 (September 16, 1980).

Food Additive Petition 4A2975 (November 15, 1973). Cyclamate. Abbott Laboratories.

Food Additive Petition 2A3672 (September 22, 1982). Cyclamate. The Calorie Control Council and Abbott Laboratories.

Food Standards Australia New Zealand (FSANZ). 12th December 2007. Final Assessment Report, Proposal P287, Review of cyclamate permissions.

Furuya, T., Kawamata, K., Kaneko, T., Uchida, O., Horiuchi, S., and Ikeda, Y. 1975. Long-term toxicity study of sodium cyclamate and sodium saccharin in rats. *Japan J Pharmacol* 25(suppl. 1):55P.

Gaunt, I.F., Hardy, J., Grasso, P., Gangolli, S.D., and Butterworth, K.R. 1976. Long-term toxicity of cyclohexylamine hydrochloride in the rat. *Food Cosmet Toxicol* 14:255–267.

Gaunt, I.F., Sharratt, M., Grasso, P., Lansdown, A.B.G., and Gangolli, S.D. 1974. Short-term toxicity of cyclohexylamine hydrochloride in the rat. *Food Cosmet Toxicol* 12:609–624.

Hardy, J., Gaunt, I.F., Hooson, J., Hendy, R.J., and Butterworth, K.R. 1976. Long-term toxicity of cyclohexylamine hydrochloride in mice. *Food Cosmet Toxicol* 14:269–276.

Hicks, R.M., Wakefield, J.St.J., and Chowaniec, J. 1975. Evaluation of a new model to detect bladder carcinogens or co-carcinogens: Results obtained with saccharin, cyclamate and cyclophosphamide. *Chem-Biol Interactions* 11:225–233.

Hombuger, F. 1973. Studies on saccharin and cyclamate. Bio-Research Consultants, Inc., unpublished report.

Hwang, K. 1966. Mechanism of the laxative effect of sodium sulfate, sodium cyclamate and calcium cyclamate. *Arch Int Pharmacodyn Ther* 163:302–340.

Ikeda, Y., Horiuchi S., Furuya, T., Kawamata, K., Kaneko, T., and Uchida, O. 1973. Long-term toxicity study of sodium cyclamate and sodium saccharin in rats. National Institute of Hygienic Sciences, Tokyo, Japan, unpublished report.

Industrial Bio-Test Laboratories. April 21, 1981. Chronic oral toxicity study with cyclohexylamine sulfate in beagle dogs. Unpublished report.

Jensen, O.M. October 1983. Artificial sweeteners and bladder cancer: Epidemiological evidence. Third European Toxicology Forum. Geneva, Switzerland.

Kasperson, R.W., and Primack, N. 1986. Cyclamate. In *Alternative sweeteners*, eds. L. O'Brien-Nabors and R.C. Gelardi, 71–87. New York: Marcel Dekker.

Kojima, S., and Ichibagase, H. 1966. Studies on synthetic sweetening agents. VIII. Cyclohexylamine, a metabolite of sodium cyclamate. *Chem Pharm Bull* 14:971–974.

Kroes, R., Peters, P.W.J., Berkvens, J.M., Verschuuren, H.G., deVries, T., and Van Esch, G.J. 1977. Long-term toxicity and reproduction study (including a teratogenicity study) with cyclamate, saccharin, and cyclohexylamine. *Toxicol* 8:285–300.

Litchfield, M.H., and Swan, A.A.B. 1971. Cyclohexylamine production and physiological measurements in subjects ingesting sodium cyclamate. *Toxicol Appl Pharmacol* 18:535–541.

Lynch, M.J., and Gross, H.M. 1960. Artificial sweetening of liquid pharmaceuticals. *Drug Cos Ind* 87:324–326, 412–413.

Mason, P.L., and Thompson, G.R. 1977. Testicular effects of cyclohexylamine hydrochloride in the rat. *Toxicol* 8:143–156.

Morgan, R.W., and Wang, O. 1985. A review of epidemiological studies of artificial sweeteners and bladder cancer. *Food Chem Toxicol* 23:529–533.

National Academy of Sciences. 1985. National Research Council, Committee on the Evaluation of Cyclamate for Carcinogenicity. Evaluation of cyclamate for carcinogenicity. Washington, DC: National Academy Press.

Oser, B.L., Carson, S., Cox, G.E., Vogin, E.E., and Sternberg, S.S. 1975. Chronic toxicity study of cyclamate:saccharin (10:1) in rats. *Toxicol* 4:315–330.

Oser, B.L., Carson, S., Cox, G.E., Vogin, E.E., and Sternberg, S.S. 1976. Long-term and multigeneration toxicity studies with cyclohexylamine hydrochloride. *Toxicol* 6:47–65.

Oser, B.L., Carson, S., Vogin, E.E., and Sonders, R.C. 1968. Conversion of cyclamate to cyclohexylamine in rats. *Nature* 220:178–179.

Price, J.M., Biava, C.G., Oser, B.L., Vogin, E.E., Steinfeld, J., and Ley, H.C. 1970. Bladder tumors in rats fed cyclohexylamine or high doses of a mixture of cyclamate and saccharin. *Science* 167:1131–1132.

Renwick, A.G. 1979. The metabolism, distribution and elimination of non-nutritive sweeteners. In *Health and sugar substitutes*, ed. B. Guggenheim, 41–47. Basel: S. Karger.

Renwick, A.G. 1983. The fate of cyclamate in man and rat. Third European Toxicology Forum, Geneva, Switzerland.

Renwick, A.G., Thompson, J.P., O'Shaughnessy, M., and Walter, E.J. April 2004. The metabolism of cyclamate to cyclohexylamine in humans during long-term administration. *Toxicol Appl Pharmacol* 196:367–380.

Renwick, A.G., and Williams, R.T. 1972. The metabolites of cyclohexylamine in man and certain animals. *Biochem J* 129:857–867.

Roberts, A., and Renwick, A.G. 1985. The metabolism of ^{14}C-cyclohexylamine in mice and two strains of rat. *Xenobiotica* 15:477–483.

Roberts, A., and Renwick, A.G. 1989. The pharmacokinetics and tissue concentrations of cyclohexylamine in rats and mice. *Toxicol Appl Pharmacol* 98:230–242.

Roberts, A., Renwick, A.G., Ford, G., Creasy, D.M., and Gaunt, I. 1989. The metabolism and testicular toxicity of cyclohexylamine in rats and mice during chronic dietary administration. *Toxicol Appl Pharmacol* 98:216–229.

Roberts H. Actions Memorandum. Undated. Food Additive Petition for Cyclamate (FAP 4A2975). Unsigned to Commissioner Schmidt.

Salunkhe, D.K., McLaughlin, R.L., Day, S.L., and Merkley, M.B. 1963. Preparation and quality evaluation of processed fruits and fruit products with sucrose and synthetic sweeteners. *Food Technol* 17:85–91.

Schmähl, D. 1973. Absence of carcinogenic activity of cyclamate, cyclohexylamine, and saccharin in rats. *Arzneim-Forsch/Drug Research* 23:1466–1470.

Schmähl, D., and Habs, M. 1984. Investigations on the carcinogenicity of the artificial sweeteners sodium cyclamate and sodium saccharin in rats in a two-generation experiment. *Arzneim-Forsch/Drug Research* 34:604–606.

Scientific Committee for Food (SCF). 2000. Revised opinion on cyclamic acid and its sodium and calcium salts, SCF/CS/EDUL/192 final. March 9.

Schmidt, U. 1987. Cyclamate conversion study in humans. Supplement to Food Additive Petition 2A3672, December 14.

Sieber, S.M., and Adamson, R.H. 1979. Long-term studies on the potential carcinogenicity of artificial sweeteners in non-human primates. In *Health and sugar substitutes,* ed. B. Guggenheim, 266–271. Basel: S. Karger.

Sonders, R.C. 1967–1968. Unpublished studies, Abbott Laboratories.

Stein, J.A. 1966. Technical aspects of non-nutritive sweeteners in dietary products. *Food Technol.* March.

Stone, C.D. 1962. Sucaryl. Calorie reduction in baked products. *Bakers Digest* 36:53–56, 59–64.

Takayama, S., Renwick, A.G., Johansson, S.L., Thorgeirsson, U.P., Tsutsumi, M., Dalgard, D.W., and Sieber, S.M. 2000. Long-term toxicity and carcinogenicity study of cyclamate in nonhuman primates. *Toxicol Sci* 53:33–39.

Taylor, J.M., Weinbuger, M.A., and Friedman, L. 1980. Chronic toxicity and carcinogenicity to the urinary bladder of sodium saccharin in the in utero exposed rat. *Toxicol Appl Pharmacol* 54:57–75.

Task Force of Past Presidents of the Society of Toxicology. 1982. Animal data in hazard evaluation: Paths and pitfalls. *Fundam Appl Toxicol* 2:101–107.

Temporary Committee for the Review of Data on Carcinogenicity of Cyclamate. 1976. Report. DHEW Publication No. (NIH)77-1437, Department of Health, Education and Welfare. February.

US Patent No. 2,275,125 (March 3, 1942). Audrieth, L.F., Sveda, M.

US Patent No. 2,803,551 (August 20, 1957).

US Patent No. 4,158,068 (June 12, 1979). Hoechst AG.

Vincent, H.C., Lynch, M.J., Pohley, F.M., Helgren, F.J., and Kirchmeyer, F.J. 1955. A taste panel study of cyclamate–saccharin mixture and of its components. *J Am Pharm Assoc* 44:442–446.

World Health Organization (WHO). 1977. 21st Report of the Joint FAO/WHO Expert Committee on Food Additives (JECFA). April 18–27, 1977. Technical Report Series 617, 1978.

World Health Organization (WHO). 1982. 26th Report of the Joint FAO/WHO Expert Committee on Food Additives (JECFA). April 19–28, 1982. Technical Report Series 683, 1982.

World Health Organization (WHO). 2009. 71st Report of the Joint FAO/WHO Expert Committee on Food Additives (JECFA). June 16–24, 2009. Technical Report Series 956, 2009.

Worldwide Regulatory Status of Cyclamate. 2010. Calorie Control Council. Updated by International Sweeteners Association, 2010. Unpublished material.

Chapter 8

Neohesperidin Dihydrochalcone

Francisco Borrego

Contents

Introduction

The sweet taste of neohesperidin dihydrochalcone (neohesperidine DC) was discovered by Horowitz and Gentili in 1963 while studying the relationships between structure and bitter taste in citrus phenolic glycosides. To their surprise, the product resulting of the hydrogenation of the bitter flavanone neohesperidin yielded an intensely sweet substance: neohesperidine DC (Horowitz 1964; Horowitz and Gentili 1961, 1963a, 1963b, 1969). Since then, a very large number of variants of the original sweetener have been synthesized. Some of these represent only small variations on the theme while others are more radical departures. All were made to gain a better understanding of taste-structure relations, to improve taste quality or raise solubility. From a practical standpoint, it appears that even the best of the new derivatives was not significantly better than the original compound. For this reason we shall focus on neohesperidine DC, which is the only sweet dihydrochalcone currently in use.

Currently, neohesperidine DC is receiving renewed interest due to its approval both as a sweetener and as a flavor ingredient. Today, there is an increasing trend to explore the use of sweetener blends in foods in order to (a) produce a sweetness profile closer to that of sucrose, (b) mask after-taste, (c) improve stability, (d) meet cost restraints, and (e) reduce the daily intake of any particular sweetener. Recent developments have shown that, if neohesperidine DC is used at low levels and in combination with other intense or bulk sweeteners, there can be significant technological advantages in terms of sweetness synergy, positive impact on taste quality, and flavor-enhancing and flavor-modifying properties.

Origin and Preparation

Neohesperidine DC is a glycosidic flavonoid. Flavonoids are a family of natural substances ubiquitous in the plant kingdom, as they occur in all higher plants. Depending upon their chemical structure, they are classified in different groups including flavanones, flavones, chalcones, dihydrochalcones, and anthocyanins. Dihydrochalcones are open-ring derivatives of flavanones and are defined by the presence of two C_6 rings joined by a C_3 bridge. There are several flavanone glycosides unique to *Citrus* (Horowitz and Gentili 1977). The peels of oranges and grapefruit contain hesperidin and naringin as main flavanone glycosides, which have different applications as a pharmaceutical raw material and as a bittering agent for the food industry, respectively.

Starting material for the commercial production of neohesperidine DC is either neohesperidin, which can be extracted from bitter orange (*Citrus aurantium*), or naringin, which is obtained from grapefruit (*Citrus paradisii*). Synthesis from extracted neohesperidin involves hydrogenation in the presence of a catalyst under alkaline conditions. Synthesis from extracted naringin is based in the conversion to phloroacetophenone-4'-β-neohesperidoside, which can be condensed with isovanillin (3-hydroxy-4-methoxybenzaldehyde) to yield neohesperidin. The reactions involved have been described in detail (Horowitz and Gentili 1968, 1991; Krbechek et al. 1968; Robertson et al. 1974; Veda and Odawara 1975) and are summarized in Figure 8.1.

Attempts to produce sweet dihydrochalcones by microbiological conversion of flavanone glycosides have also been reported in the literature (Linke et al. 1973). A novel biotechnological approach to the synthesis of neohesperidine DC is being explored by Jonathan Gressel and coworkers at the Weizmann Institute in Israel (Lewinsohn et al. 1988; 1989a, 1989b, 1989c; Bar-Peled et al. 1991, 1993). They have characterized a rhamnosyl transferase from *Citrus* that is able to catalyze the transfer of rhamnose from UDP-rhamnose to the C-2 hydroxyl group of glucose that is attached via C-7-O to naringenin or hesperetin. A "proof-of-concept" for production of neohesperidin (and consequently neohesperidine DC) from hesperetin-7-β-D-glucoside by biotransformation using metabolically engineered plant cell suspension cultures has recently been published (Frydman et al. 2005). Hesperetin 7-β-D-glucoside would be obtained by partially hydrolyzing hesperidin (a readily available by-product of the orange processing industry) to remove the rhamnose attached at position 6 of the glucose.

Although neohesperidine DC has not yet been found in nature, several structurally related dihydrochalcones have been identified so far in about 20 different families of plants (Bohm 1988). Some of them have been consumed for a long time as natural sweetening agents. Sweet dihydrochalcones have been detected, for example, in the leaves of *Lithocarpus litseifolius*, a variety of sweet tea from China (Rui-Lin et al. 1982). Other sweet dihydrochalcones were detected in the fruit of *Iryanthera laevis*, which is used in the preparation of a Colombian sweet food (Garzón et al. 1987). As a recent example of occurrence in processed foods, dihydrochalcone glycosides have been detected and quantified in different apple juices and jams, at levels up to 15 ppm in an apple compote (Tomás-Barberán et al. 1993).

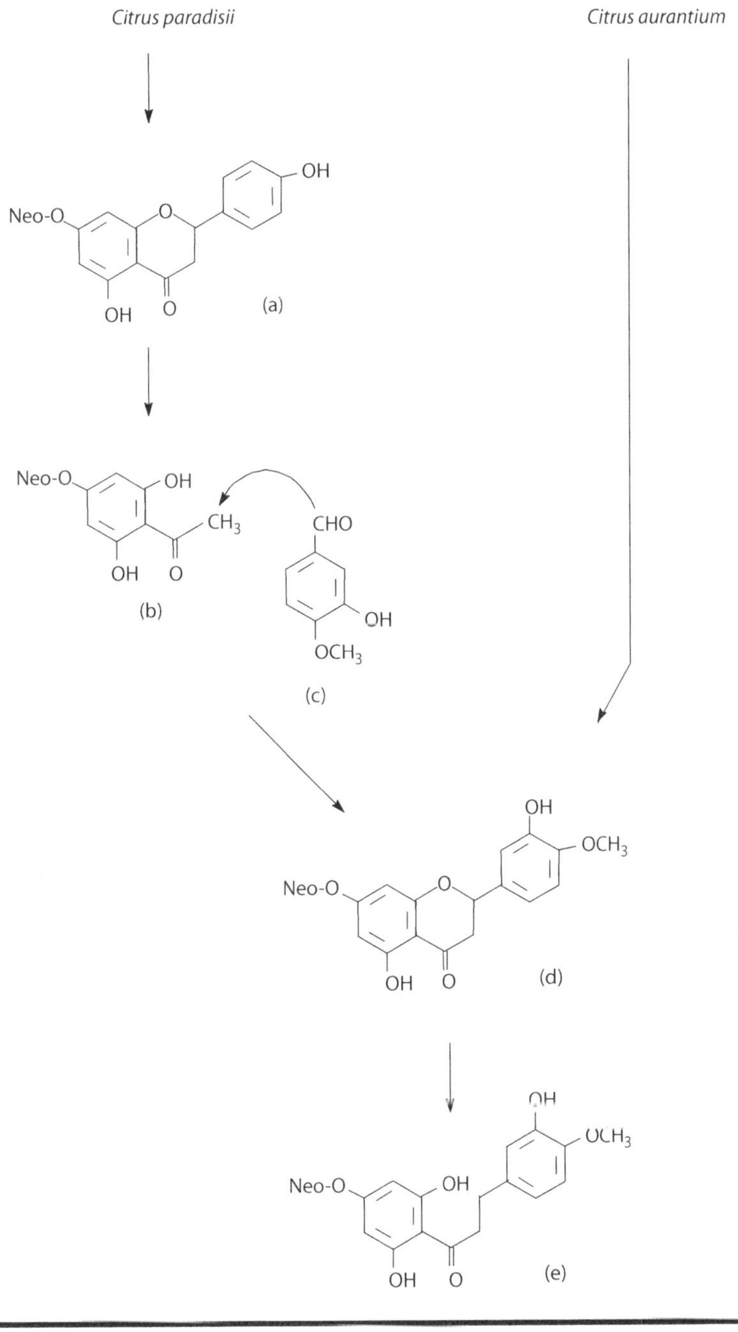

Figure 8.1 The process of producing neohesperidine DC from flavanones. (a) Naringin; (b) Phloroacetophenone-4′-β-neohesperidoside; (c) Isovanillin; (d) Neohesperidin; (e) Neohesperidine DC (Neo) Neohesperidoside.

Physical Properties and Stability

Neohesperidine DC is an off-white crystalline powder. The crystals are monoclinic, space group P2$_1$. X-ray diffraction demonstrated that the molecule is more or less 'J'-shaped, with the β-neohesperidosyl residue forming the curved part of the 'J' and the hesperetin dihydrochalcone aglycone forming the linear segment (Wong and Horowitz 1986; Shin et al. 1995).

The solubility of neohesperidine DC in distilled water at room temperature is low (0.4–0.5 g/l), being freely soluble both in hot (80°C) water and aqueous alkali. As with other citrus flavonoids (hesperidin, naringin) it exhibits higher solubility in an alcohol-water mixture than in water or ethanol alone (Bär et al. 1990). Several methods for solubility enhancement have been described in the patent literature, such as preparation of the sodium and calcium salts (Westall and Messing 1977) and combinations with water soluble polyols such as sorbitol (Dwivedi and Sampathkumar 1981a, 1981b). Solubility in water may also be enhanced when administered in propylene glycol or glycerol solutions (Borrego et al. 1995). In any case, as it is used at very low concentrations, the solubility in cold water is not a limitation for application in food and beverages (Lindley et al. 1991).

Stability studies on neohesperidine DC have shown that the product can be stored over three years at room temperature without any sign of decomposition. It is slightly hygroscopic taking up water in a saturated environment up to a maximum of 15% (unpublished observations).

In liquid media, and under certain conditions of high temperature and low pH, the glycosidic bonds are hydrolyzed forming the aglycone hesperetin dihydrochalcone, glucose and rhamnose (Figure 8.2).

The stability of neohesperidine dihydrochalcone in aqueous model systems at various pHs and temperatures has been studied (Crosby and Furia 1980; Inglett et al. 1969; Canales et al. 1993). Aqueous solutions at room temperature are quite resistant to hydrolysis to the free sugars and the aglycone as long as the pH does not fall below 2. Even if hydrolysis were to occur, there would not be complete loss of sweetness since the aglycone, hesperetin dihydrochalcone, is itself sweet, though not very soluble. The degradation of neohesperidine DC was studied at pH values from 1 to 7 and temperatures ranging from 30°C to 60°C for up to 140 days (Canales et al. 1993). The hydrolysis of neohesperidine DC could be represented as a first order reaction across the range of temperatures and pH tested. Optimum pH was 4, although half-life values indicate that no stability problem would be expected in the pH range 2–6 (Table 8.1). The Arrhenius plot could be used to predict stability under temperature conditions other than those checked empirically. Using this calculation model, it can be estimated that after 12 months at 25°C, the percentage of neohesperidine DC remaining unchanged would be 94.8% at pH 3.

It is widely accepted that buffered aqueous solutions are simplified systems and the data obtained indicate only general trends as they do not consider potential interactions between neohesperidine DC and typical food constituents. Thus, the stability of neohesperidine DC has also been evaluated under processing and storage conditions in a number of complex foods.

The stability of neohesperidine DC has been studied during pasteurization of fruit-based soft-drinks under different temperature and acidity conditions. No significant hydrolysis was detected in orange, lemon, apple, and pineapple even under severe conditions not used at industrial scale (1 hr at 90°C). Only at the lowest pH value tested (2) was there a significant loss of neohesperidine DC (8%) after 1 hr at 90°C (Montijano and Borrego 1996).

With regard to long-term stability, no loss of neohesperidine DC was found after a one year storage period in a lemonade system, either in the dark or when exposed to light. Similarly, on storage of samples at 40°C for up to 3 months, there was no significant change in neohesperidine DC concentration (Montijano et al. 1997a).

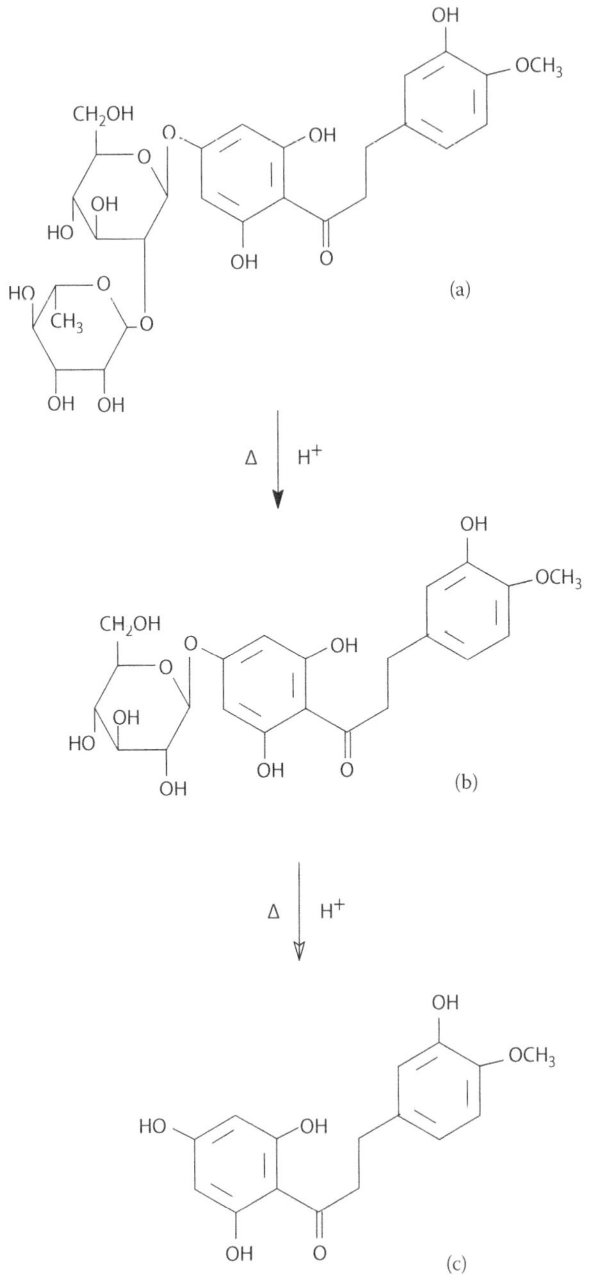

Figure 8.2 Neohesperidine DC degradation under strong acid conditions and high temperature. (a) Neohesperidine DC; (b) Hesperetin dihydrochalcone 4′-β-glucoside; (c) Hesperetin dihydrochalcone.

Table 8.1   Half-Lives of Neohesperidin Dihydrochalcone in Aqueous
Buffered Solutions at pH 1 to 7 and Different Temperatures

pH	Half Life (Days)		
	20°C	40°C	60°C
1	1360	55	3
2	2344	248	33
3	8268	1100	172
4	9168	1925	475
5	3074	976	357
6	2962	624	103
7	368	50	10

Source: Adapted from Canales et al., *J Food Sci* 57:589–591, 643, 1993. Data at 20°C
are extrapolated values using the Arrhenius plot.

No neohesperidine DC decomposition was noted under the temperature conditions prevailing during a fruit jam manufacturing process (102°C–106°C during 35–40 minutes), as judged by the identical values of nominal concentration included in the formulation and the neohesperidine DC concentration in jam, after processing, at storage time zero. A statistically non-significant degradation of 11% of the initial neohesperidine DC was observed after 18 months storage at room temperature (Tomás-Barberán et al. 1995).

Neohesperidine DC has been shown to remain stable after pasteurization and fermentation of milk, and cold storage of yogurt. These stability properties, together with the lack of sweetener-related effects in the course of acidification, suggest that neohesperidine DC could be added, where appropriate, at an early stage of the yogurt manufacturing process (Montijano et al. 1995a).

The degradation of neohesperidine DC at high temperature and pH 6–7 has been studied in order to estimate losses during thermal processing of non-fermented milk-based products. An increase in the pH from 6 to 7 produced an increase in rate constants by a factor of 5. Loss of neohesperidine DC was below 0.5% for the pasteurization and UHT sterilization conditions tested, and about 9%–10% for in-container sterilization at 120°C for 10 min at pH 7 (Montijano et al. 1995b).

Sensory Properties and Applications

Several studies of the sweetness intensity of neohesperidine DC have been published. At or near threshold, neohesperidine DC is about 1800 times sweeter than sucrose on a weight basis (Guadagni et al. 1974). As concentration increases, the sweetness of neohesperidine DC decreases relative to that of sucrose, so that at the 5% sucrose level it is about 250 times sweeter. However, in other studies, higher levels of sweetness of 1000 and 600 times that of sucrose were reported at sucrose concentrations of 5 and 8.5%, respectively (Bär et al. 1990). From a practical standpoint, the maximum sweetness contribution to the vast majority of food products from neohesperidine DC will not exceed that delivered by 3% sucrose. At this sucrose equivalent concentration, neohesperidine DC is described to be 1500 times sweeter than sucrose (Schiffman et al. 1996).

Temporal characteristics of neohesperidine DC as the sole sweetener contrast with those of sugar and other sweeteners such as aspartame, due to its long onset and persistence times, also eliciting a lingering liquorice-like cooling aftertaste (Montijano et al. 1996). An explanation for the slow taste onset of neohesperidine DC is that some modifications of the molecule must occur within the oral cavity before the active glucophore is produced. Explanations for the lingering taste would involve a strong and slowly reversible binding to the sweet receptor site with the neohesperidine DC molecule adopting a 'bent' active conformation in elicitation of sweet taste (DuBois et al. 1977a; Crosby et al. 1979; Hodge and Inglett 1974). X-ray studies contradict these earlier suggestions (Shin et al. 1995) as it is proposed than the partially extended form of the dihydrochalcone is the sweet conformer.

Many attempts have been described to overcome these limitations by both chemical derivatization and combination with other substances (Crosby and DuBois 1976, 1977; DuBois et al. 1977a, 1977b, 1980, 1981a, 1981b; DuBois and Crosby 1977; DuBois and Stephenson 1981; Horowitz and Gentili 1974, 1975; Rizzi 1973). Thus, for example, it has been reported that complexation with β-cyclodextrin significantly reduces both aftertaste and sweet taste of neohesperidine DC (Caccia et al. 1998). However, from a practical point of view, the most adequate way to overcome the negative slow onset of sweetness and lingering aftertaste is to blend it at low concentrations with other low calorie sweeteners. With appropriate choice of blend concentrations, this approach proves to be a fully successful method of taking advantages of its synergy and flavor-modifying properties.

Substantial enhancement of sweetness in blends of neohesperidine DC with the relevant intense and bulk sweeteners have should be has reported with acesulfame K (Lipinsky and Lück 1979), sucralose (Jenner 1992) and according to other studies (Chung 1997; Schiffman et al. 1996) with the full range of low-calorie and bulk sweeteners. Therefore, it plays an important role as a minor component of sweetener blends, in which its contribution to the total sweetness would be not more than 10%, for any low calorie foods in which intense sweeteners are normally used.

It is relatively common for sweeteners to modify and enhance flavor, while also eliciting sweetness. In many cases, their flavor-modifying characteristics are perceived at concentrations above the threshold and could therefore be considered a consequence of sweetness, rather than a structure-induced effect. On the contrary, neohesperidine DC has been consistently shown to modify flavor and enhance mouthfeel, not only at supra-, but also at sub-threshold concentrations, this being independent of sweetness induction (Foguet et al. 1994; Lindley 1996; Lindley et al. 1993). Functionality of neohesperidine DC as a flavor is closely parallel to that of other flavoring substances widely used in flavoring preparations, maltol and ethyl maltol. In addition, remarkable synergistic effects have been described between neohesperidine DC and maltol and/or ethyl maltol (Engels and Stagnitti 1996).

These flavoring effects of incorporating neohesperidine DC at levels below the sweetness threshold have been studied in detail in a range of sweet and savory products by using Quantitative Descriptive Analysis (QDA) sensory profiles. In all products tested, most of the statistically significant changes were considered beneficial for overall product quality. In general terms, the flavor-modifying effects of neohesperidine DC may be described as increases in mouthfeel perception and a general smoothing or blending of the individual elements of the product flavor profile (Lindley et al. 1993).

It has been proposed that these flavor effects may be a consequence of the ability of neohesperidine DC to reduce the perception of bitterness. The bitterness-reducing effects of neohesperidine DC were originally reported on limonin and naringin (Montijano and Borrego 1996), the compounds responsible for the bitter taste of grapefruit juice. Thus, for example, at 1% sucrose equivalence neohesperidine DC increased the threshold of limonin 1.4-fold, being much more effective than sucrose to mask the bitterness of limonin. Today, bitterness amelioration has been described in other bitter tasting substances, such as paracetamol, dextromethorphan and other pharmaceuticals (Borrego and Montijano 1995), or the formulation of special foods such as energy-boosting drinks (Bruna et al. 1996).

The ability of neohesperidine DC to mask the unpleasant taste of many pharmaceuticals and to improve the organoleptic properties of other sweeteners led us to think about the potential use of this substance in medicated feedstuffs. The positive effects of neohesperidine DC in feedstuffs for farm animals have been demonstrated, in particular when used as a flavor modifier in blends with saccharin, due to the ability of neohesperidine DC not only to reduce saccharin aftertaste, but also to mask the bitter taste of some components of feed. Surprisingly, neohesperidine DC has also been shown to act as an attractant for certain fish such as sea-bass and rainbow trout, which suggests its use as an ingredient in new feeds for cultured fish (Foguet et al. 1993).

Toxicology and Metabolism

The safety assessment of neohesperidine DC can be based on data from a number of toxicological studies conducted at the Western Regional Research Laboratory, Albany, California. These studies include subacute feeding trials in rats, a three-generation reproduction and teratogenicity study in rats, a 2-year chronic carcinogenicity/toxicity in rats, and a 2-year feeding study in dogs (Gumbmann et al. 1978). The data showed no evidence of any increased incidence of tumors that could be associated with the ingestion of neohesperidine DC, as well as no adverse effect of toxicological significance even in the high dose group.

Neohesperidine DC has been checked for mutagenicity in the Ames test; it is nonmutagenic, regardless of which of the various *Salmonella* tester strains is used (Batzinger et al. 1977; Brown et al. 1977; Brown and Dietrich 1979; MacGregor and Jurd 1978). In mice, the compound causes no increase in the normal frequency of micronucleated polychromatic erythrocytes in bone marrow (MacGregor 1979).

In 1985–1986, a detailed study on the "subchronic (13 wk) oral toxicity of neohesperidine DC in rats" was carried out at the TNO-CIVO Toxicology and Nutrition Institute in the Netherlands (Lina et al. 1990). Neohesperidine DC was fed to groups of 20 male and 20 female Wistar rats at dietary levels of 0%, 0.2%, 1.0%, and 5.0% for 91 days. Only at the 5% level were there any treatment-related effects, i.e., marginal changes in body weight and food consumption, cecal enlargement, and slight changes in some of the clinical chemistry variables. These phenomena were judged to be of little, if any, toxicological significance. Neither the low- nor intermediate-dose groups showed any compound-related effects, and none of the groups showed any ophthalmoscopical, hematological, or histopathological changes. It was concluded that the intermediate dose, which translates to about 750 mg/kg/day of neohesperidine DC, was the no-effect level. Based on these and the earlier results, the Scientific Committee for Food of the European Community recognized neohesperidine DC as "toxicologically acceptable" in 1987 and assigned it an acceptable daily intake (ADI) of 0–5 mg/kg (SCF 1987). This ADI is adequate for a wide range of uses.

More recently, the embryotoxicity/teratogenicity of neohesperidine DC was examined in Wistar rats (Waalkens-Berendsen et al. 2004). Neohesperidine DC was fed at dietary concentrations of 0%, 1.25%, 2.5% or 5% to groups of 28 mated female rats from day 0 to 21 of gestation. No adverse effects were observed at neohesperidine DC levels of up to 5% of the diet, the highest dose level tested, at which the rats consumed about 3.3 g/kg/day. The observed cecal enlargement is a well known physiological adaptive response to the ingestion of high doses of a low digestible substance and is generally accepted to lack toxicological relevance.

The metabolism of neohesperidine DC and that of flavonoid glycosides ingested in substantial amounts with ordinary foods (Kuhnau 1976) share many features and result partly in the formation of the same end-products (Horowitz and Gentili 1991). As with other flavonoid glycosides, it appears that metabolism is carried out largely by the action of intestinal microflora. After formation of the aglycone, hesperetin dihydrochalcone is split by bacterial glycosidases into phloroglucinol and dihydroisoferulic acid (representing ring A and B of the parent molecule). The latter compound is subsequently converted to a spectrum of metabolites that, like phloroglucinol, result also from the metabolism of certain naturally occurring flavonoids (DeEds 1971; Braune et al. 2005).

Excretion studies using [^{14}C] neohesperidine DC showed that, when oral doses of up to 100 mg/kg were administered to rats, more than 90% of the radioactivity was excreted in the first 24 hours, primarily in the urine. After 24 hours, only traces of radioactivity could be detected in various tissues.

The caloric value of neohesperidine DC has been estimated to be not more than 2 cal/g, based on the assumption that the sugar residues are hydrolytically split and metabolized and that the aglycone is not extensively metabolized. Because of its high potency, neohesperidine DC would probably afford not more than 1/1000 as many calories as an equivalent amount of sucrose.

Neohesperidine DC has been proposed as a non-cariogenic, non-fermentable sweetening agent, based on the finding that it is relatively inert to the action of cariogenic bacteria (Berry and Henry 1983).

Regulatory Status

With the adoption and publication of the EU Sweeteners Directives (Directives 94/35/EC and 96/83/EC) (EC 1994, 1997) and after their implementation in national food regulations, the use of neohesperidine DC as a sweetener is authorized in a wide range of foodstuffs. Other countries such as Switzerland, the Czech Republic and Turkey have adopted EU legislation.

As a flavor-modifying substance, neohesperidine DC has also been included in the Directive 95/2/EC on Food Additives other than Colours and Sweeteners (EC 1995), for use in an additional number of foodstuffs most of which (i.e., margarine, meat products, and vegetable protein products) are clearly not sweet.

During recent years the functionality of neohesperidine DC as a flavor and flavor modifier has steadily gained recognition in regulatory circles. Thus, neohesperidine DC has been recognized as GRAS by the Expert Panel of FEMA for use as a flavor ingredient in many food categories (Waddell et al. 2007) at the levels of use specified in Table 8.2. Other countries, such as Japan, Australia and New Zealand, have authorized the use of neohesperidine DC as a flavoring without any use limitation.

Analysis in Foods

HPLC is the most effective analytical method for precise and accurate analysis of neohesperidine DC, both as a raw material (Castellar et al. 1997) and in foodstuffs (Borrego et al. 1995; Hausch 1996). Other reported methods such as TLC (thin-layer chromatography" and "UV" (ultraviolet)

Table 8.2   GRAS Use Levels for Neohesperidin Dihydrochalcone as a Flavor Ingredient

Food Category	Usual Use (ppm)	Maximum Use (ppm)
Baked goods	4	4
Beverages (alcoholic and non alcoholic)	5	10
Breakfast cereal	3	3
Cheese	3	4
Chewing gum	200	200
Condiments and Relishes	2	3
Confectionery	3	3
Egg products	2	3
Fats/oils	4	4
Fish products	2	3
Frozen dairy	2	3
Fruit ices	1	2
Gelatins/puddings	2	3
Gravies	3	4
Hard candy	5	15
Imitation dairy	3	4
Instant coffee/tea	2	3
Jams/jellies	2	3
Meat products	2	3
Milk products	3	6
Nut products	3	4
Other grains	3	4
Poultry	2	3
Processed fruit	2	3
Processed vegetables	2	3
Reconstituted vegetables	2	3
Seasonings/flavors	2	3
Snack food	3	3
Soft candy	2	3
Soups	5	10
Sugar substitutes	4	4
Sweet sauces	2	3

or UV spectrophotometry with and without chemical derivatization, lack adequate selectivity and are only useful for quality control of the pure material. Neohesperidine DC has a different regulatory status, and hence different conditions of use, depending upon the area. It is therefore important that it can be detected in products at the proposed level of use, and that it be possible to analytically differentiate between flavoring and sweetening use levels.

HPLC methods to quantitate neohesperidine DC in foodstuffs have been reported (Schwarzenbach 1976; Fisher 1976a, 1976b; Montijano et al. 1997b). While these previous methods are useful starting points, they were limited to one single product or studied food and beverages with unrealistically high sweetener concentrations. An analytical method to detect and quantitate neohesperidine DC in foodstuffs has been developed and validated, yielding adequate results in terms of precision, accuracy, selectivity and ruggedness to quantitate neohesperidine DC both at flavoring and sweetening use levels in soft-drinks (Montijano et al. 1997b). The method has also successfully assayed complex foods such as dairy products, confectionery and fat-based foods which require selective extraction of the sweetener with appropriate solvents (dimethyl sulfoxide, methanol and their blends with water). Acceptable recoveries (more than 90%) were found in all tested samples both at flavoring and sweetening use levels. This method provides sufficient separation between the neohesperidine DC peak and the corresponding hydrolysis acid products.

Neohesperidine DC is used as a minor component of sweetener blends and therefore at very low concentrations. However, HPLC techniques allow detection and quantitation of neohesperidine DC at levels below those which are normally used for sweetening and flavoring purposes. Thus detection and quantitation limits for neohesperidine DC, determined by the method based on the standard deviation of the response and the slope, are 0.2 and 0.7 mg/l, respectively. These values are below the minimum concentration which shows a technological function in the final food.

Usually, extraction in dimethyl sulfoxide or alcohols are sufficient for selective extraction of neohesperidine DC, however adsorption of neohesperidine DC onto Amberlite XAD and subsequent fractionation on Sephadex were judged to be essential steps for successful quantitation of neohesperidine DC in blackcurrant jam, due to the fact that anthocyanin-related compounds may interfere with neohesperidine DC in crude extracts (Tomás-Barberán et al. 1995)

Availability and Patent Situation

Neohesperidine DC is produced and marketed worldwide by Ferrer HealthTech under the trade name of Citrosa.

The original patents which covered the manufacture of neohesperidine DC are expired by now, although neohesperidine DC has a patent portfolio that covers its use and applications (Foguet et al. 1993, 1994, 1995).

References

Bär, A., Borrego, F., Benavente, O., Castillo, J., and del Río, J.A. 1990. Neohesperidin dihydrochalcone: Properties and applications. *Lebens Wiss u Technol* 23:371–376.

Bar-Peled, M., Lewinsohn, E., Fluhr, R., and Gressel, J. 1991. UDP-rhamnose flavanone-7-O-glucoside-2'-O-rhamnosyl transferase, purification and characterization of an enzyme catalyzing the production of bitter compounds in Citrus. *J Biol Chem* 266:20953–20959.

Bar-Peled, M., Fluhr, R., and Gressel, J. 1993. Juvenile-specific localization and accumulation of a rhamnosyl transferase and its bitter flavonoids in foliage, flowers and young fruits. *Plant Physiol* 103:1377–1384

Batzinger, R.P., Ou, S.Y.L., and Bueding, E. 1977. Saccharin and other sweeteners: Mutagenic properties. *Science* 198:944–946.

Berry, C.W., and Henry, C.A. 1983. Baylor College of Dentistry, Meeting of the American Association for Dental Research, as reported in Food Chemical News, March 21, pp. 25.

Bohm, B.A. 1988. The minor flavonoids. In *The flavonoids, advances in research since 1980*, ed. T.H. Harborne, 329–348. New York: Chapman & Hall.

Borrego, F., and Montijano, H. 1995. Anwendungsmöglichkeiten des Süssstoffs Neohesperidin Dihydrochalkon in Arneimitteln. *Pharm Ind* 57:880–882.

Borrego, F., Canales, I., and Lindley, M.G. 1995. Neohesperidin dihydrochalcone: State of knowledge review. *Z Lebensm Unters Forsch* 200:32–37.

Braune, A., Engst, W., and Blaut, M. 2005. Degradation of neohesperidin dihydrochalcone by human intestinal bacteria. *J Agric Food Chem* 53:1782–1790.

Brown, J.P., Dietrich, P.S, and Brown, R.J. 1977. Frameshift mutagenicity of certain naturally occurring phenolic compounds in the *Salmonella*/microsome test: Activation of anthraquinone and flavonol glycosides by gut bacterial enzymes. *Biochem Soc Trans* 5:1489–1492.

Brown, J.P., and Dietrich, P.S. 1979. Mutagenicity of plant flavonols in the *Salmonella*/mammalian microsome test: Activation of flavonol glycosides by mixed glycosidases from rat cecal bacteria and other sources. *Mutation Res* 66:223–240.

Bruna, M.A., Miralles, R.J., Poch, J.A., and Redondo, F.L. 1996. Compositions for alkalinizing and energy-boosting drinks. WO Patent No. 96/03059.

Caccia, F., Dispenza, R., Fonza, G., Fuganti, C., Malpezzi, L., and Mele, A. 1998. Structure of neohesperidin dihydrochalcone/β-cyclodextrin inclusion complex: NMR, MS, and X-ray spectroscopy investigation. *J Agric Food Chem* 46:1500–1505.

Canales, I., Borrego, F., and Lindley, M.G. 1993. Neohesperidin dihydrochalcone stability in aqueous buffer solutions. *J Food Sci* 57:589–591, 643.

Castellar, M.R., Iborra, J.L., and Canales, I. 1997. Analysis of commercial neohesperidin dihydrochalcone by high performance liquid chromatography. *J Liq Chrom & Rel Technol* 20:2063–2073.

Chung, H.J. 1997. Measurement of synergistic effects of binary sweetener mixtures. *J Food Sci Nutr* 2:291–295.

Crosby, G.A., and DuBois, G.E. 1976. Sweetener derivatives. US Patent No. 3,976,687.

Crosby, G. A., and DuBois, G.E. 1977. Sweetener derivatives. US Patent No. 4,055,678.

Crosby, G.A., and Furia, T.E. 1980. New sweeteners. In *CRC handbook of food additives,* ed. T.E. Furia, Vol. II, 2nd edition, 204. Boca Raton, FL: CRC Press.

Crosby, G.A., DuBois, G.E., and Wingard, R.E. 1979. The design of synthetic sweeteners. In *Drug design*, ed. E.J. Ariéns, Vol. 8, 215–310. New York: Academic Press.

DeEds, F. 1971. Flavonoids metabolism. *Compr Biochem* 125:417–423.

DuBois, G.E., and Crosby, G.A. 1977. Dihydrochalcone oligomers. US Patent No. 4,064,167.

DuBois, G.E., and Stephenson, R.A. 1981. Sulfamo dihydrochalcone sweeteners. US Patent No. 4,283,434.

DuBois, G.E., Crosby, G.A., and Saffron, P. 1977a. Nonnutritive sweeteners: Taste-structure relationships for some new simple dihydrochalcones. *Science* 195:397–399.

DuBois, G.E., Crosby, G.A., Stephenson, R.A., and Wingard, Jr., R.E. 1977b. Dihydrochalcone sweeteners: Synthesis and sensory evaluation of sulfonate derivatives. *J Agric Food Chem* 25:763–771.

DuBois, G.E., Stephenson, R.A., and Crosby, G.A. 1980. Alpha amino acid dihydrochalcones. US Patent No. 4,226,804.

DuBois, G.E., Crosby, G.A., Lee, J.F., Stephenson, R.A., and Wang, P.C. 1981a. Dihydrochalcone sweeteners. Synthesis and sensory evaluation of a homoserine dihydrochalcone conjugate with low aftertaste sucrose-like organoleptic properties. *J Agric Food Chem* 29:1269–1276.

DuBois, G.E., Crosby, G.A., and Stephenson, R.A. 1981b. Dihydrochalcone sweeteners. A study of the atypical temporal phenomena. *J Med Chem* 24:408–428.

Dwivedi, B.K., and Sampathkumar, P.S. 1981a. Artificial sweetener composition and process for preparing and using same. US Patent No. 4,304,794.

Dwivedi, B.K., and Sampathkumar, P.S. 1981b. Preparing neohesperidin dihydrochalcone sweetener composition. US Patent No. 4,254,155.

EC. 1994. European Parliament and Council Directive 94/35/EC of 30 June 1994 on sweeteners for use in foodstuffs. Off J Eur Comm (10.09.94) No. L237/3.

EC. 1995. European Parliament and Council Directive 95/2/EC of 20 February 1995 on food additives other than colours and sweeteners. Off J Eur Comm (18.03.95) No. L61/1.

EC. 1997. Directive 96/83/EC of the European Parliament and of the Council of 19 December 1996. Off J Eur Comm (19.02.97) No. L 48/16.

Engels, L., and Stagnitti, G. 1996. Flavor modifying composition. WO Patent No. 96/17527.

Fisher, J.F. 1976a. A high-pressure liquid chromatographic method for the quantitation of neohesperidin dihydrochalcone. *J Agric Food Chem* 25:682–683.

Fisher, J.F. 1976b. Review of quantitative analyses for limonin, naringin, naringenin rutinoside, hesperidin and neohesperidin dihydrochalcone in citrus juice by high performance liquid chromatography. *Proc Int Soc Citric* 3:813–816.

Foguet, R., Borrego, F., Campos, J., Madrid, J.A., and Zamora, S. 1993. New feed for cultured fish. Eur Patent Application No. 0655 203 A1.

Foguet, R., Cisteró, A., and Borrego, F. 1994. Body and mouthfeel potentiated foods and beverages containing neohesperidin dihydrochalcone. US Patent No. 5,300,309.

Foguet, R., Cisteró, A., and Borrego, F. 1995. Use of neohesperidin dihydrochalcone for potentiating body and mouthfeel of foods and beverages. Eur Patent Specification No. 0 500 977 B1.

Frydman, A., Weisshaus, O., Huhman, D.V., Sumner, L.W., Bar-Peled, M., Lewinsohn, E., Fluhr, R., Gressel, J., and Eyal, Y. 2005. Metabolic engineering of plant cells for biotransformation of hesperidin into neohesperidin, a substrate for the production of the low-calorie sweetener and flavor enhancer NHDC. *J Agric Food Chem* 53:9708–9712.

Garzón, N.L., Cuca, S.L.E., Martínez, V.J.C., Yoshida, M., and Gottlieb, O.R. 1987. The chemistry of Colombian *Myristicaceae*. Part 5. Flavonolignoid from the fruit of Iryanthera laevis. *Phytochemistry* 26:2835–2837.

Guadagni, D.G., Maier, V.P., and Turnbaugh, J.H. 1974. Some factors affecting sensory thresholds and relative bitterness of limonin and naringin. *J Sci Food Agric* 25:1199–1205.

Gumbmann, M.R., Gould, D.H., Robbins, D.J., and Booth, A.N. 1978. Toxicity studies of neohesperidin dihydrochalcone. In *Proceedings, sweeteners and dental caries*, eds. J.H. Shaw and G.G. Roussos, 301–310. Washington, DC: Information Retrieval Inc.

Hausch, M. 1996. Simultaneous determination of neohesperidin dihydrochalcone and other sweetening agents by HPLC. *Lebensmittelchemie* 50:31–32.

Hodge, J.E., and Inglett, G.E. 1974. Structural aspects of glycosidic sweeteners containing (1-2)-linked disaccharides. In *Symposium: Sweeteners*, ed. G.E. Inglett, 216–234. Westport, CT: Avi Publishing Co.

Horowitz, R.M. 1964. Relations between the taste and structure of some phenolic glycosides. In *Biochemistry of phenolic compounds*, ed. J.B. Harborne, 545–571. New York: Academic Press.

Horowitz, R.M., and Gentili, B. 1961. Phenolic glycosides of grapefruit: A relation between bitterness and structure. *Arch Biochem Biophys* 92:191–192.

Horowitz, R.M., and Gentili, B. 1963a. Dihydrochalcone derivatives and their use as sweetening agents. US Patent No. 3,087,821.

Horowitz, R.M., and Gentili, B. 1963b. Flavonoids of citrus VI. The structure of neohesperidose. *Tetrahedron* 19:773–782.

Horowitz, R.M., and Gentili, B. 1968. Conversion of naringin to neohesperidin and neohesperidin dihydrochalcone. US Patent No. 3,375,242.

Horowitz, R.M., and Gentili, B. 1969. Taste and structure in phenolic glycosides. *J Agric Food Chem* 17:696–700.

Horowitz, R.M., and Gentili, B. 1974. Dihydrochalcone xylosides and their use as sweetening agents. US Patent No. 3,826,856.

Horowitz, R.M., and Gentili, B. 1975. Dihydrochalcone galactosides and their use as sweetening agents. US Patent No. 3,890,296.

Horowitz, R.M., and Gentili, B. 1977. Flavonoid constituents of citrus. In *Citrus science and technology*, eds. S. Nagy, P.E. Shaw, and M.K Veldhuis, Vol. 1, 397 426. Westport, CT. The AVI Publishing Company.

Horowitz, R.M., and Gentili, B. 1991. Dihydrochalcone sweeteners from citrus flavanones. In *Alternative sweeteners*, eds. L. O'Brien and R.C. Gelardi, 2nd edition, 97–115. New York: Marcel Dekker.

Inglett, G.E., Krbechek, L., Dowling, B., and Wagner, R. 1969. Dihydrochalcone sweeteners-sensory and stability evaluation. *J Food Sci* 34:101–103.

Jenner, M.R. 1992. Sweetening agents. Eur Patent Application No. 0 507 598 A1.

Krbechek, L., Inglett, G.E., Holik, M., Dowling, B., Wagner, R., and Riter, R. 1968. Dihydrochalcones: Synthesis of potential sweetening agents. *J Agric Food Chem* 16:108–112.

Kuhnau, J. 1976. The flavonoids. A class of semiessential food components: Their role in human nutrition. *World Rev Nutr Diet* 24:117–191

Lewinsohn, E., Britsch, L., Mazur, Y., and Gressel, J. 1988. Flavanone glycoside biosynthesis in Citrus: Chalcone synthase, UDP-glucose-flavanone-7-O-glucosyl transferase and rhamnosyl transferase activities in cell-free extracts. *Plant Physiol* 91:1323–1328.

Lewinsohn, E., Gavish, H., Berman, E., Fluhr, R., Mazur, Y., and Gressel, J. 1989a. Towards biotechnologically synthesized neohesperidin for the production of its dihydrochalcone sweetener. International Conference on Sweeteners. Los Angeles, CA: American Chemical Society, September 22–25, abstract No. 39.

Lewinsohn, E., Berman, E., Mazur, Y., and Gressel, J. 1989b. 7-glycosylation and (1-6) rhamnosylation of exogenous flavanones by undifferentiated Citrus cell cultures. *Plant Science* 61:23–28.

Lewinsohn, E., Britsch, L., Mazur, Y., and Gressel, J. 1989c. Flavanone glycoside biosynthesis in Citrus. *Plant Physiol* 91:1323–1328.

Lina, B.A.R., Dreef-van der Meulen, H.C., and Leegwater, D.C. 1990. Subchronic (13 wk) oral toxicity of neohesperidin dihydrochalcone in rats. *Food Chem Toxicol* 28:507–513.

Lindley, M.G. 1996. Neohesperidin dihydrochalcone: Recent findings and technical advances. In *Advances in sweeteners*, ed. T.H. Grenby, 240–252. Glasgow, UK: Chapman & Hall.

Lindley, M.G., Canales, I., and Borrego, F. 1991. A technical re-appraisal of the intense sweetener neohesperidin dihydrochalcone. In *FIE Conference Proceedings*, 48–51. Maarsen, The Netherlands: Expoconsult Publishers.

Lindley, M.G., Beyts, P.K., Canales, I., and Borrego, F. 1993. Flavor modifying characteristics of the intense sweetener neohesperidin dihydrochalcone. *J Food Sci* 58:592–594+666.

Linke, A.H., Natarajan, S., and Eveleigh, D.E. 1973. A new screening agent to detect the formation of dihydrochalcone sweeteners by microorganisms. The Annual meeting of the American Society of Microbiology, abstract No. 184, pp. 171.

Lipinsky, R.G.W., and Lück, E. 1979. Sweetener mixture. US Patent No. 4,158,068.

MacGregor, J.T. 1979. Mutagenicity studies of flavonoids in vivo and in vitro. *Toxicol Appl Pharmacol* 48:A47.

MacGregor, J.T., and Jurd, L. 1978. Mutagenicity of plant flavonoids: Structural requirements for mutagenic activity in Salmonella typhimurium. *Mutation Res* 54:297–309.

Montijano, H.F., Tomás-Barberán, A., and Borrego, F. 1995a. Stability of the intense sweetener neohesperidine DC during yogurt manufacture and storage. *Z Lebensm Unters Forsch* 201:541–543.

Montijano, H.F. Tomás-Barberán, A., and Borrego, F. 1995b. Accelerated kinetic study of neohesperidine DC hydrolysis under conditions relevant for high temperature processed dairy products. *Z Lebensm Unters Forsch* 204:180–182.

Montijano, H., and Borrego, F. 1996. Neohesperidine DC. In *LFRA ingredients handbook (Sweeteners)*, ed. J.M. Dalzell, 181–200. Surrey, UK: Leatherhead Food Research Association, Leatherhead.

Montijano, H., Coll, M.D., and Borrego, F. 1996. Assessment of neohesperidine DC stability during pasteurization of juice-based drinks. *Int J Food Sci Technol* 31:397–401.

Montijano, H., Borrego, F., Tomás-Barberán, F.A., and Lindley, M.G. 1997a. Stability of neohesperidine DC in a lemonade system. *Food Chem* 58:13–15.

Montijano, H., Borrego, F., Canales, I., and Tomás-Barberán, F.A. 1997b. Validated high-performance liquid chromatographic method for quantitation of neohesperidin dihydrochalcone in foods. *J Chromatogr A* 758:163–166.

Rizzi, G.P. 1973. Dihydrochalcone sweetening agents. US Patent No. 3,739,064.

Robertson, G.H., Clark, J.P., and Lundin, R. 1974. Dihydrochalcone sweeteners: Preparation of neohesperidin dihydrochalcone. *Ind Eng Chem Prod Res Devel* 13:125–129.

Rui-Lin, N., Tanaka, T., Zhow, J., and Tanaka, O. 1982. Phlorizin and trilobatin, sweet dihydrochalcone glucosides from leaves of Lithocarpus litseifolius (Hance) Rehd. (Fagaceae). *Agric Biol Chem* 46:1933–1934.

SCF. 1987. Report of the Scientific Committee for Food on Sweeteners. Commission of the European Communities (CS/EDUL/-Final).

Schiffman, S.S., Booth, B.J., Carr, B.T., Loose, M.L., Sattely-Miller, E.A., and Graham, B.G. 1996. Investigation of synergism in binary mixtures of sweeteners. *Brain Res Bull* 50:72–81.

Schwarzenbach, R. 1976. Liquid chromatography of neohesperidin dihydrochalcone. *J Chromatogr* 129:31–39.

Shin, W., Kim, S.J., Shin, M., and Kim, S.H. 1995. Structure-taste correlations in sweet dihydrochalcone, sweet dihydroisocoumarin and bitter flavone compounds. *J Med Chem* 38:4325–4335.

Tomás-Barberán, F.A., García Viguera, C., Nieto, J.L., Ferreres, F., and Tomás-Lorente, F. 1993. Dihydrochalcone from apple juices and jams. *Food Chem* 46:33–36.

Tomás-Barberán, F.A., Borrego, V., Ferreres, V., and Lindley, M.G. 1995. Stability of the intense sweetener neohesperidin dihydrochalcone in blackcurrant jams. *Food Chem* 52:263–265.

Veda, K., and Odawara, V. 1975. Neohesperidin from naringin. *Jpn Kokai* 75:154, 261.

Waalkens-Berendsen, D.H., Kuilman-Wahls, M.E., and Bär, A. 2004. Embritoxicity and teratogenicity study with neohesperidin dihydrochalcone in rats. *Regul Toxicol Pharmacol* 40:74–79.

Waddell, W.J., Cohen, S.M., Feron, V.J., Goodman, J.L., Marnett, L.J., Portoghese, P.S., Rietjens, I.M.C.M., et al. 2007. GRAS flavoring substances 23. *Food Technol* 61:22–61.

Westall, E.B., and Messing, A.W. 1977. Salts of dihydrochalcone derivatives and their use as sweeteners. US Patents Nos. 3,984,394 (1976) and 4,031,260.

Wong, R.Y., and Horowitz, R.M. 1986. The X-ray crystal and molecular structure of neohesperidin dihydrochalcone sweetener. *J Chem Soc Perkin Trans* I:843–848.

Chapter 9

Neotame

Dale A. Mayhew, Brad I. Meyers, W. Wayne Stargel,
C. Phil Comer, Sue E. Andress, and Harriett H. Butchko

Contents

Introduction

Neotame (N-(*N*-(3,3-dimethylbutyl)-L-α-aspartyl)-L-phenylalanine 1-methyl ester) is a high potency sweetener and flavor enhancer that is currently approved for use in 69 countries worldwide. Neotame is 7,000 to 13,000 times sweeter than sucrose. This zero calorie sweetener has a clean sweet taste with no undesirable taste characteristics; it is functional and stable in a wide array of beverages and foods and requires no special labeling for phenylketonuria (PKU). Given its high potency and resulting low required usage in food and beverage applications, neotame has proven to be one of the most cost-effective sweeteners on the market.

Neotame resulted from a long-term research program designed to discover high potency sweeteners with optimized performance characteristics. The French scientists Claude Nofre and Jean-Marie Tinti invented neotame from a simple N-alkylation of aspartame (Nofre and Tinti 1991; Glaser et al. 1995). The NutraSweet Company holds the rights to a wide range of patents related to neotame.

Neotame Taste Profile

Neotame has a clean, sweet taste similar to sugar with no significant bitter, metallic or other off-tastes. Moreover, this taste profile is maintained over the range of concentrations required in applications. Taste testing has shown that the sweetness of neotame increases with concentration. However, taste attributes other than sweetness remain below threshold levels even as neotame concentrations increase.

Sweetness Potency

Sweetness Related to Sucrose

"Sucrose equivalence" or "% SE" is the standardized sweetness intensity scale established in sweetener research using sucrose as the reference. An x% SE is equivalent in sweetness to an x% sucrose water solution. Neotame is functional as a sweetener in food applications at typical sweetness levels ranging from 3% SE (e.g., tabletop) to 10% SE (e.g., carbonated soft drinks). The concentration-response (C-R) curve for neotame in water was established by using a trained sensory panel to evaluate the sweetness intensity of five solutions of neotame at increasing concentrations. The results are presented in Figure 9.1.

The sweetness potency of neotame is at least orders of magnitude greater than most other high potency sweeteners. In water, at concentrations equivalent to about 3% to 10% sucrose, neotame is 30 times sweeter than saccharin, about 60 to 100 times sweeter than acesulfame potassium (acesulfame K) and 300 to 400 times sweeter than cyclamate (DuBois et al. 1991). Due to its remarkable sweetness potency, neotame is used in food at considerably lower concentrations than other high potency sweeteners.

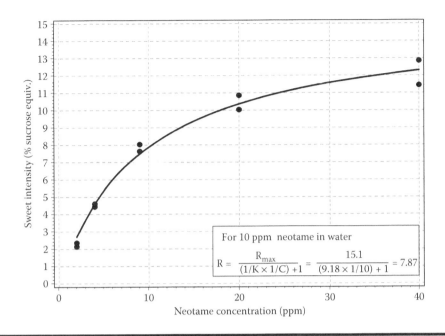

Figure 9.1 Neotame concentration-response curve in water. R Observed response; R$_{max}$ Maximum observed response; C Sweetener concentration; 1/K Concentration which yields half-maximal response.

Neotame reaches a maximum sweetness intensity of 15.1% SE in water and 13.4% SE in a cola drink. In contrast, sweeteners such as acesulfame K, cyclamate and saccharin attain their maximum sweetness intensity (plateau) at, respectively, 11.6% SE, 11.3% SE and 9% SE in water (DuBois et al. 1991). Although neotame can be used as a stand alone sweetener in a broader range of applications than can acesulfame K, cyclamate, and saccharin, it is typically applied as part of a multi-sweetener blend in food products.

Sweetness Potency of Neotame

All high potency sweeteners exhibit a potency range depending upon the application and desired intensity of sweetness. Neotame is 7,000 to 13,000 times sweeter than sucrose or 30 to 60 times sweeter than aspartame. For example, sensory evaluation of neotame formulated in cola-flavored carbonated soft drinks indicates a sweetness potency 31 times that of aspartame in similar products.

Sensory Profile of Neotame versus Sucrose

Neotame exhibits a clean taste profile at a number of usage levels relevant to product applications and functions effectively as a sweetener and flavor enhancer in foods and beverages. A trained descriptive panel evaluated a number of different sweeteners, including neotame and sucrose, at comparable sweetness levels in water. The taste profile of neotame is very similar to that of sucrose, with the predominant sensory characteristic of neotame being a very clean sweet taste. Other taste attributes such as metallic flavors, bitterness, and sour are all below threshold levels. The flavor

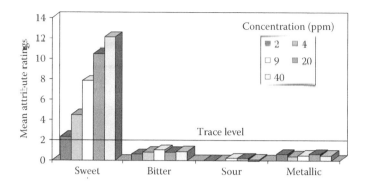

Figure 9.2   Taste profile of neotame at various concentrations in water.

profile of neotame in water at varying concentrations is shown as 10 mg/L compared to an 8% sucrose solution is shown in Figure 9.2.

The sweetness temporal profile is a key to the functionality of a sweetener and is complementary to its taste profile. It demonstrates the changes in the perception of sweetness over time. Every sweetener exhibits a unique characteristic onset time of response and subsequent extinction time. As shown in Figure 9.3, the sweetness temporal profile of neotame in water is close to that of aspartame, with a slightly slower onset but without significantly longer linger.

Like other high potency sweeteners, the temporal profile for neotame is application specific. In confectionery applications such as sugar-free chewing gum, neotame positively extends both

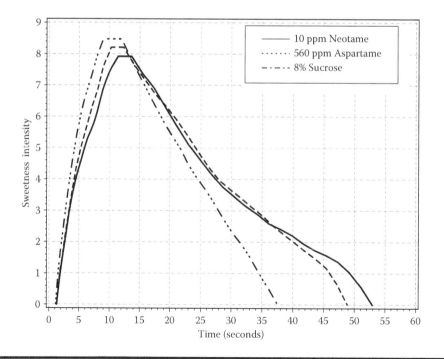

Figure 9.3   Comparative temporal profile of neotame vs. sugar and aspartame at isosweet concentrations in water.

sweetness and flavor. In other applications such as powdered soft drinks, the temporal parameters for neotame are similar to those of sucrose.

Neotame can be used as a sweetener either alone or blended with other sweeteners in foods and beverages. The functionality of neotame in carbonated soft drinks, powdered soft drinks, cakes, yogurt, hot packed lemon tea, chewing gum, and as a tabletop sweetener in hot coffee and iced tea has been demonstrated.

Flavor-Enhancing Property

Flavor enhancers are substances which can be added to foods to supplement, enhance, or modify the original taste and/or aroma of a food without imparting a characteristic taste or aroma of its own (US FDA 1999a). In addition to being a sweetener, neotame enhances certain flavors when used at non-sweetening levels. Sensory testing demonstrated flavor enhancement for neotame at very low sub-sweetening concentrations.

Flavor enhancing properties combined with potent sweetness provide unique functionality to neotame as a food additive. The flavor synergy of neotame in fruit-based drinks reduces the amount of juice required to maintain the mouthfeel of a higher juice level. Neotame also maintains tartness, which can allow for a reduction in acid use. Moreover, neotame can allow a significant reduction in mint flavoring required for chewing gum with no decrease in flavor intensity. Further, neotame masks bitter tastes of other food ingredients (i.e., caffeine) and sweeteners (i.e., saccharin). This unique combination of properties allows for both improved taste of products and a reduction in requirements for other added flavors.

Utility in Food Applications and Beverage Systems

Solubility and Dissolution

The solubility of neotame in water, ethyl acetate, and ethanol at a range of temperatures illustrates how neotame behaves in various food matrices. Neotame is very soluble in ethanol at all temperatures tested. The solubility of neotame increases in both water and ethyl acetate with increasing temperature. This solubility may create opportunities for food and beverage manufacturers to use neotame in a wide range of liquid delivery systems. The solubility of neotame in water, ethanol, and ethyl acetate is shown in Table 9.1.

Table 9.1   Neotame Solubility in Water, Ethanol, and Ethyl Acetate

Solubility of Neotame (grams/100 grams solvent)			
Temperature	Water	Ethyl Acetate	Ethanol
15°C	1.06	4.36	>100
25°C	1.26	7.70	>100
40°C	1.80	23.8	>100
50°C	2.52	87.2	>100
60°C	4.75	>100	>100

Neotame dissolves rapidly in aqueous solutions due to the very low usage levels required for sweetening. For example, the dissolution rate of neotame was determined in water; at 2 minutes 93.0% of the total neotame had dissolved and within 5 minutes 98.6% of the total amount of neotame had dissolved.

Admixture Potential

Blending sweeteners to achieve synergy is a common industry practice. Synergy can be classified as either quantitative or qualitative. Quantitative synergy is defined as occurring when the sweetness of a mixture of sweeteners is greater than the sum of their respective, individual sweetness intensities. Qualitative synergy is defined as using two or more sweeteners with complimentary temporal profiles to achieve a closer approximation to that of sucrose. Blends or combinations of two or more sweeteners are being used more frequently to achieve both the desired level of sweetness and temporal profile in food and beverage products. Neotame is very compatible in admixture with other sweeteners, including sugar. Blending neotame with sugar allows for significant calorie reduction in products compared to sugar alone without compromising taste. Moreover, blending neotame with other high potency sweeteners will contribute to a more sugar-like sweetness profile while potentially mitigating off tastes.

The taste profile of sweetened foods and beverages can be tailored to specific taste requirements by blending. Sweetener blends offer taste advantages such as lower off-flavors and more sucrose-like temporal profiles (quicker onset and shorter sweet linger) than individual sweeteners. In the same way, blends of neotame with other sweeteners will provide opportunities for food manufacturers to formulate better tasting products. Furthermore, because of the low cost-in-use of neotame, a sweetener blend with neotame can offer a significant cost advantage to the manufacturer. The flavor enhancement quality of neotame, which masks bitter flavors even at sub-sweetening levels, makes neotame in admixture blends a desirable choice with other sweeteners which possess potentially undesirable or more complex taste profiles. Tables 9.2 through 9.5 detail suggested use levels and blend ratios for neotame in a variety of beverage and non-beverage applications.

Table 9.2 Representative Example Formulations for Blends of Neotame and Nutritive Sweeteners in Carbonated, Still, and Powdered Beverages

Carbonated Soft Drinks	Sugar-Containing Ingredients Mono- and Di-Saccharides (% as Formulated)	Sugar Sweetness Contribution (% of Total Sweetness in Blend)	Sugar: Neotame: Acesulfame K Blend (% as Formulated)		
			Sugar	Neotame	Acesulfame K
Cola	9–12	80	7.2–9.6	0.00018–0.0003	0.000–0.002
		75	6.75–9.0	0.00023–0.00039	0.000–0.0035
		67	6.0–9.0	0.00021–0.00038	0.001–0.005
		50	4.5–6.0	0.00043–0.00055	0.0025–0.011
Orange	10–14	80	8.0–11.2	0.00024–0.00029	—
		75	7.5–10.5	0.00032–0.00039	—
		67	6.7–9.4	0.00032–0.00004	0.0035–0.0065
		50	5.0–7.0	0.00045–0.00005	0.0090–0.0130

Table 9.2 (Continued)   Representative Example Formulations for Blends of Neotame and Nutritive Sweeteners in Carbonated, Still, and Powdered Beverages

Carbonated Soft Drinks	Sugar-Containing Ingredients Mono- and Di-Saccharides (% as Formulated)	Sugar Sweetness Contribution (% of Total Sweetness in Blend)	Sugar: Neotame: Acesulfame K Blend (% as Formulated)		
			Sugar	Neotame	Acesulfame K
Lemon Lime	9–11	80	7.2–8.8	0.00019–0.00023	—
		75	6.75–8.25	0.00023–0.00028	—
		67	6.0–7.3	0.00025–0.00035	0.003–0.006
		50	4.5–5.5	0.00033–0.00041	0.0070–0.0115

Still/RTD Beverages	Sugar-Containing Ingredients Mono- and Di-Saccharides (% as Formulated)	Sugar Sweetness Contribution (% of Total Sweetness in Blend)	Sugar: Neotame: Acesulfame K Blend (% as Formulated)		
			Sugar	Neotame	Acesulfame K
Fruit Punch	10–12	80	8.0–9.6	0.00023–0.00029	—
		75	7.5–9.0	0.00023–0.00034	—
		67	6.7–9.0	0.00024–0.00031	0.0025–0.0045
		50	5.0–6.0	0.00041–0.00054	0.003–0.0075
Orange/ Citrus	10–12	80	8.0–9.6	0.00021–0.00031	—
		75	7.5–9.0	0.00023–0.00033	—
		67	6.7–9.0	0.00023–0.000034	0.0025–0.004
		50	5.0–6.0	0.00041–0.000049	0.0050–0.090

Powdered Soft Drinks on Dry Mix Basis	Sugar-Containing Ingredients Mono- and Di-Saccharides (% as Formulated)	Sugar Sweetness Contribution (% of Total Sweetness in Blend)	Sugar: Neotame: Acesulfame K Blend (% as Formulated)		
			Sugar	Neotame	Acesulfame K
Fruit Punch	90–95	80	87–94	0.0024–0.0045	—
		75	87–93	0.0040–0.0075	0.000–0.20
		67	85–93	0.00035–0.0085	0.055–0.105
		50	82–90	0.0050–0.0115	0.1075–0.285
Lemonade	90–95	80	87–94	0.0015–0.0045	—
		75	87–93	0.0030–0.0065	0.00–0.17
		67	85–93	0.0035–0.0070	0.04–0.09
		50	82–90	0.0045–0.0095	0.125–0.285

Table 9.3 Representative Example Formulations for Sugar-Free Blends of Aspartame, Neotame, and Acesulfame K in Carbonated, Still, and Powdered Beverages

Product Type	Sweetness level (as % Sugar Equivalent)	Neotame: Aspartame: Acesulfame K Blend (% of total sweetness) **(Assume 25%–33% APM:Ace K Synergy)**		
		Neotame	Aspartame	Acesulfame K
Carbonated	9–14	20	75	5
		23	70	7
		30	50	20
		32	56	12
Still/RTD	8–13	21	67	12
		22	46	32
		28	55	17
		34	45	21
Powdered (As Finished Beverage)	8–12	20	60	20
		22	49	29
		29	22	49
		31	48	21

Table 9.4 Representative Example Formulations for Sugar-Free Blends of Sucralose, Neotame, and Acesulfame K in Carbonated, Still, and Powdered Beverages

Product Type	Sweetness Level (as % Sugar Equivalent)	Neotame: Sucralose: Acesulfame K Blend (% of Total Sweetness)		
		Neotame	Sucralose	Acesulfame K
Carbonated	9–14	20	73	7
		29	50	21
		34	41	25
		46	32	22
Still/RTD	8–13	13	80	7
		23	56	21
		30	40	30
		37	45	18
Powdered (As Finished Beverage)	8–12	22	50	28
		34	31	35
		35	40	25
		39	21	40

Table 9.5   Representative Example Formulations for Blends of Neotame and Nutritive Sweeteners in a Variety of Non-Beverage Applications

Product *Many of these products would require bulking agents. The sweetness contribution of the bulking agents has been factored into this table.*	*Sugar-Containing Ingredients Mono- and Di-Saccharides (% as Formulated)*	*Sugar: Neotame Blend 75:25 Sweetness Contribution (% as Formulated)*	
		Sugar	*Neotame*
Frozen Deserts	14–22	10.5–16.5	0.0003–0.001
Ice Cream	14–22	10.5–16.5	0.0003–0.0006
Yogurt, plain	3–5	2.25–3.75	0.00005–0.0001
Yogurt, fruited/flavored	6–12	4.5–9	0.00012–0.0003
Yogurt Fruit Preparations	45–63	34–47	0.0015–0.003
Chewing Gum, Bubble Gum	50–60	37.5–45	0.001–0.0075
Confections	54–98	40.5–73.5	0.0004–0.002
Jams, Jellies, Fruit Spreads, Toppings	80–90	60–67.5	0.001–0.003
OTC Tablets	60–80	45–60	0.001–0.008
Fiber Laxatives on dry mix basis	45–55	33.75–41.25	0.0016–0.003
Cough Lozenges	97–98	72.75–73.5	0.001–0.002
Powdered Dessert Mixes on dry mix basis	50–90	43–87	0.005–0.011
Refrigerated Chocolate Milk	5–7	3.75–5.25	0.0001–0.00015
Baked Goods	10–25	7.5–18.75	0.0002–0.0008
Salad Dressings	3–25	2.25–18.75	0.00006–0.0008
Sauces	5–33	3.75–24.75	0.0001–0.0004

Physical Characteristics and Chemistry

Structure and Synthesis

Neotame is manufactured from aspartame and 3,3-dimethylbutyraldehyde via reductive alkylation followed by purification, drying, and milling. The N-alkyl component of neotame, 3,3-dimethylbutanoic acid, occurs naturally in goat cheese (Pierre et al. 1998). The chemical structures of aspartame and neotame are compared in Figure 9.4.

Dry Stability

Dry bulk chemical stability of neotame has been demonstrated for five years. Dry neotame is extremely stable with the major degradation product being de-esterified neotame. De-esterified

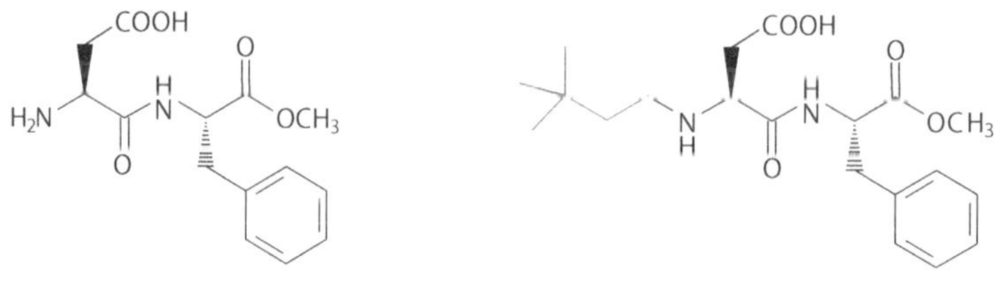

Aspartame

L-aspartyl-L-phenylalanine

1-methyl ester

Neotame

N-[*N*-(3,3-dimethylbutyl)-L-α-aspartyl]-

L-phenylalanine 1-methyl ester

Figure 9.4   Comparison of chemical structures of neotame and aspartame.

neotame is formed at extremely low levels by the simple hydrolysis of the methyl ester group from neotame. It is also the major aqueous degradant product and the major *in vivo* metabolite of neotame in humans and animals. More strenuous conditions involving higher temperatures and humidity result in the formation of increased amounts of de-esterified neotame without significant amounts of other degradation products; thus neotame maintains an excellent material balance even under adverse conditions. Fluorescent lighting and polyethylene packaging have no effect on the stability of neotame. Stability studies have confirmed that products containing dextrose, maltodextrin, and neotame are stable when stored for extended periods of time at relevant storage conditions of ambient temperature and humidity.

Stability in Food Applications

Neotame is stable under conditions of intended use as a sweetener across a wide range of food and beverage applications. The stability of neotame has been assessed through a systematic approach and demonstrated to be similar to that of aspartame, with the exception that neotame has greater stability in heat-processed and dairy goods. Unlike aspartame, a diketopiperazine derivative is not formed.

Neotame maintains stability and functionality over a range of pHs, temperatures, and storage times representing relevant conditions of use. Traditionally, functionality and stability have been demonstrated for high potency sweeteners by developing data for a large number of categories. This approach generated extensive redundant data. Aspartame stability work, for example, was carried out for over thirty different food categories even though many of the categories have similar processing and storage conditions. A systematic, chemistry-based approach to stability testing was therefore established for neotame as determined by the key chemical and physical parameters that impact the stability and functionality of a food additive (Pariza et al. 1998).

Matrix System for Stability Determinations

Functionality and chemical stability studies establish that neotame is functional as a sweetener and flavor enhancer in a variety of food applications when used in accordance with good manufacturing practices. The functionality of neotame was demonstrated using a three-dimensional food matrix representing the intended conditions of use in foods. Based on experience with aspartame

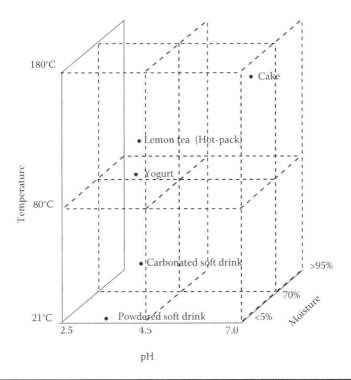

Figure 9.5   Matrix model for neotame applications.

and knowledge of the chemistry of neotame, the three key properties of this food matrix are temperature, pH, and moisture; these determine the stability of neotame and thus its functionality under intended conditions of use. Studies on the functionality of aspartame have validated this matrix approach as a predictor of its stability and functionality in foods within the defined matrix. Similarly, if neotame is functional in foods at the edges of the matrix, it will be functional for all food applications that are within the limits of the matrix. The food applications tested in the matrix under commercially relevant processing and storage conditions are: carbonated soft drinks, powdered soft drinks, baked goods (cake), yogurt, and hot packed beverages. These foods represent the ranges of temperature, pH, and moisture relevant to neotame applications. As predicted by its chemistry, neotame functionality and stability are similar to those of aspartame, with the exception of greater stability in heat-processed goods. The representative positions of these categories within the matrix are shown in Figure 9.5.

Carbonated soft drinks (CSD), powdered soft drinks (PSD), cake, yogurt, and hot packed still beverages comprise the majority of commercial applications for high potency sweeteners. Taken together these five representative products, along with tabletop products, account for greater than 90% of all uses for high potency sweeteners. Thus, these formulations were investigated for stability over the expected shelf life of the products. Additional applications selected to assess functionality and stability of neotame included powdered tabletop preparations and chewing gum.

The greatest stability of neotame is in low moisture products such as tabletop preparations and PSD mixes; in these formulations neotame shows little or no degradation following significant storage times. Neotame is also stable following normal commercial processing and storage times for products represented in the matrix such as hot packed still beverages, cake, yogurt and CSD.

Results for the product matrix model demonstrate that neotame is functional and stable as a sweetener in a wide range of product applications. In all products assessed in the matrix model, the stability of neotame was demonstrated to be comparable or improved relative to aspartame.

Degradation Products

Neotame Is Not Subject to Diketopiperazine (DKP) Formation

Degradation studies demonstrate excellent mass balance for neotame. Unlike aspartame, no DKP is formed from the intra-molecular cyclization of the aspartame moiety of neotame due to the presence of the N-alkyl substitution on the aspartyl amino group. This results in excellent stability of neotame in heat-processed applications. The possible formation of Maillard reaction products or *in vitro* nitrosation of neotame was also assessed to be negligible. Stability studies have confirmed that products containing dextrose, maltodextrin, and neotame are stable when stored for extended periods of time at relevant storage conditions of ambient temperature and humidity. Neotame is similarly inert to a number of food components such as flavoring agents and reducing sugars, including fructose.

De-esterified Neotame Is the Major Degradant

The major route of degradation is the hydrolysis of the methyl ester moiety of neotame to form de-esterified neotame.

De-esterified neotame is the only degradant formed to any extent and is also the major metabolite of neotame found in humans and animals. Under relevant conditions of use (pH 3.2 and 20°C), approximately 89% of neotame remained in mock beverage formulations after eight weeks of storage. Based on product survey data, 90% of diet carbonated beverages are purchased and consumed within eight weeks of production.

Carbonated soft drinks represent the largest use category for high potency sweeteners; for example, approximately 80% of all aspartame produced is used in carbonated soft drinks. The relative pattern for neotame usage is not expected to deviate significantly from that of aspartame. Stability studies done with neotame in carbonated soft drinks at anticipated use levels did not result in detectable levels of degradants other than de-esterified neotame.

Safety

Projected Consumption/Exposure

The mean consumption of neotame for all users is estimated to be 0.04 mg/kg/day. Consumption for 90th percentile users of neotame is estimated to be 0.10 mg/kg/day. These anticipated consumption levels of neotame provide extremely wide margins of safety when evaluated in the light of results from animal safety studies. Anticipated levels of neotame consumption are based upon known 14-day estimates of aspartame consumption (Butchko and Kotsonis 1991, 1996; Butchko et al. 1994), a conservative sweetening potency ratio for neotame of 31 times that of aspartame, and the potential for neotame to replace aspartame in the market.

Safety Evaluation of Neotame

Neotame has been subjected to extensive investigations in safety studies to establish its safe use in foods (US FDA 1998, 1999b). Numerous studies including subchronic, chronic, and special studies have been done in rats, mice, dogs, and rabbits. Long-term feeding studies were carried out in rodents to evaluate carcinogenic potential; developmental (teratogenicity) toxicity studies were done in rats and rabbits to determine the potential for effects on the fetus; a two-generation study was carried out in rats to determine the potential for reproductive effects, and for any effects on pregnancy and development/growth of newborn rats. Studies including bacterial and mammalian cell test systems evaluated the mutagenic potential of both neotame and its major degradant/metabolite (de-esterified neotame). Chronic and carcinogenicity studies in the rat included in utero exposure to neotame insuring that animals were dosed throughout the duration of the studies from conception to study termination. Safety studies were done according to the general principles in current US Food and Drug Administration (US FDA) guidelines (US FDA 1993) and other international guidelines.

Neotame was generally administered in the diet, as this route of exposure is the most relevant to human consumption. Diets were formulated and concentrations adjusted regularly to provide dosages in milligram per kilogram per day rather than dosing at fixed percentages of neotame in the diet. Microscopic examinations were performed on tissues from all animals at all doses in all key toxicity studies. In addition, standard and supplemental parameters were evaluated for evidence of immunotoxicity or neurotoxicity.

General Toxicology and Carcinogenicity

The safety of a wide range of dose levels of neotame was established in dietary studies in rats, mice, and dogs. Evaluations included clinical observations, body weight, body weight gain, food and water consumption, hematology, clinical chemistry, urinalysis parameters, ophthalmologic examinations, electrocardiograms (in dogs), and assessments of gross necropsy findings and microscopic evaluations.

Neotame was well-tolerated in all subchronic (90 day) studies and was without adverse effects even at very high doses (3000 mg/kg/day in the rat, 8000 mg/kg/day in the mouse, and 1200 mg/kg/day in the dog). There was no mortality, and there were no changes in appearance/behavior or other parameters during these subchronic studies at dietary doses 12,000 to 80,000 times the anticipated 90th percentile level of human consumption.

Food refusal or spillage was observed when neotame was present at higher concentrations in the diet for mice, rats and dogs. Diets containing extremely high levels of neotame (up to 5% in diet) were immediately disliked and partially rejected. This immediate decrease in food consumption was generally followed by accommodation to near control values and suggests poor palatability of the diet. Furthermore, the effect of neotame on food consumption was a function of the concentration of neotame in food and not the dosage of neotame consumed. Special dietary preference studies in rats confirmed that certain concentrations of neotame reduced the palatability of diet. Poor palatability of neotame-containing diet was observed, particularly in rats and mice, as small but consistent reductions in food consumption as animals approached the body weight plateau of adulthood.

One-year dietary studies were carried out in rats and dogs and two-year carcinogenicity studies were done in rats and mice. In rats, the studies included parental exposure prior to mating and in utero exposures for offspring; thus, animals used in safety studies were exposed to neotame from conception, during prenatal and postnatal development (childhood to adult), and throughout the duration of the studies. The highest doses in the one year studies were 1000 mg/kg/day in rats and 800 mg/kg/day in dogs. There was no evidence of toxicity in chronic studies in either the rat or dog. In addition, there was no evidence of target organ toxicity or carcinogenicity when neotame was administered for two years to rats with in utero exposure at dosages up to 1000 mg/kg/day, or to mice at dosages up to 4000 mg/kg/day. All non-neoplastic findings were consistent with those expected of aging rodents.

There was no toxicity at any doses tested in definitive toxicology studies in rats, mice, and dogs. Given that consumer exposure to neotame even at the 90th percentile is so small, margins of safety based on each of these species are tens of thousands times greater than projected human exposure, ranging from 8,000- to 40,000-fold.

Reproduction and Developmental (Teratology) Studies

Sprague-Dawley rats were fed doses of neotame up to 1000 mg/kg for 10 weeks (males) and 4 weeks (females) prior to pairing. Dosing continued throughout pairing, gestation and lactation to weaning of the F_1 offspring at Day 21 of age. At approximately four weeks of age, animals were selected for the F_1 generation and were exposed for 10 more weeks before being bred to produce the F_2 generation. The F_2 generation was raised to Day 21. Consumption of high doses of dietary neotame by animals through two successive generations did not result in any effects on reproductive performance, pregnancy, postnatal development (growth to adults), or on development of the embryo or fetus.

In a developmental (teratology) study, female Sprague-Dawley rats were fed neotame-containing diet for 4 weeks before pairing and throughout gestation until Day 20 after mating. There were no effects on parameters evaluated or any evidence of effects upon the dams or upon their litters. Examination of the fetuses for external, visceral, and skeletal abnormalities revealed no fetotoxic or developmental effects.

No developmental effects occurred in female rabbits dosed by oral gavage with neotame between Days 6 and 19 after mating. There were no effects on parameters evaluated or postmortem effects in dams or in litters. There were no developmental effects of neotame upon fetal examination. The rat and rabbit reproductive and developmental studies demonstrated that neotame, even at high doses, had no effect on reproduction, pregnancy, the development of the embryo or fetus or development/growth to adults.

Genetic Toxicology

Neotame was not mutagenic in the Ames assay in six bacterial strains when tested with or without metabolic activation and there was no evidence of mutagenic activity in the mouse lymphoma cell gene mutation assay with and without metabolic activation. No chromosomal aberrations were detected in Chinese hamster ovary cells after exposure to neotame with or without metabolic activation.

Neotame administered by oral gavage to mice in a bone marrow micronucleus assay did not induce any changes in the ratio of polychromatic erythrocytes to total erythrocytes or in the frequency of micronucleated polychromatic erythrocytes in this assay system.

In addition to the mutagenicity evaluations with neotame, the major *in vivo* metabolite that is also the major *in vitro* degradant, de-esterified neotame, was not mutagenic in the Ames assay and the xanthine-guanine phosphoribosyl transferase mutation assay in Chinese hamster ovary cells.

Based on the results of *in vivo* and *in vitro* genotoxicity studies, neither neotame nor de-esterified neotame, the major metabolite and degradant, has the potential for mutagenicity or clastogenicity.

Metabolism

Data from both humans and animals show that neotame has an excellent metabolic and pharmacokinetic profile. All metabolites identified in humans were also present in species used in safety studies. Neotame is rapidly but incompletely absorbed in all species and radiolabeled doses of neotame are completely eliminated. The major route of metabolism is to de-esterified neotame. Methanol is produced in equimolar quantities during the formation of de-esterified neotame whether through metabolism or degradation. The maximum amount of methanol exposure at projected 90th percentile consumption of neotame is negligible compared to levels of methanol considered to be safe (US FDA 1996). Neotame does not induce liver microsomal enzymes.

Neotame and de-esterified neotame have short plasma half-lives with rapid and complete elimination. Neither neotame nor de-esterified neotame accumulate following repeated dosing in humans or toxicology species. Absorbed neotame is rapidly excreted in the urine and feces. In radiolabeled studies, the majority of total radioactivity was excreted as the de-esterified neotame in the feces in all species and likely reflects largely unabsorbed neotame in the gastrointestinal tract.

A comparison of metabolism and pharmacokinetic data in animals and humans demonstrates that the species used in the toxicology studies are relevant for predicting human safety.

Overview of Clinical Studies

The results from clinical studies demonstrate that neotame is safe and well-tolerated in humans, even in amounts well above projected chronic consumption levels. These clinical studies included single doses ranging from 0.1 to 0.5 mg/kg, repeated doses of 0.25 mg/kg hourly for eight hours totaling 2 mg/kg/day, and daily doses including 1.5 mg/kg/day for up to 13 weeks in healthy populations including male and female subjects and doses of 1.5 mg/kg/day in a population of male and female individuals with non-insulin dependent diabetes mellitus (Type 2). Specifically, there were no clinically significant or neotame-related changes in any vital signs, electrocardiogram results, or ophthalmologic examinations. There were no neotame-related observations of clinical significance in any biochemical, hematological, physiological, or subjective findings. There were no statistically significant differences between neotame and placebo treatments in reported adverse experiences. Furthermore, neotame had no significant effect on plasma glucose or insulin concentrations or glycemic control in the population with non-insulin dependent diabetes mellitus.

The 0.25 mg/kg hourly dose in the repeated dose study is equivalent to consuming about one liter of beverage sweetened with 100% neotame every hour for an eight hour period. The high dose of 1.5 mg/kg/day in the 13-week study is equivalent to consuming about six liters of beverage sweetened with neotame daily for 13 weeks. Thus, the results of the clinical studies with neotame using doses up to 20 times the projected 90th percentile consumption clearly confirm the safety of neotame for use by the general population.

Patents

Neotame was invented by the French scientists Claude Nofre and Jean-Marie Tinti and patented in 1992 (Nofre and Tinti 1991; Glaser et al. 1995). The NutraSweet Company holds the rights to a wide range of patents related to neotame.

Table 9.6 Neotame Countries of Approval

Algeria	Japan
Argentina	Kazakhstan
Australia	Mexico
Bangladesh	New Zealand
Belarus	Nigeria
Brazil	Norway
Canada	Pakistan
Chile	Peru
China	Philippines
Colombia	Puerto Rico
Costa Rica	Russia
Ecuador	South Africa
European Union[a]	Sri Lanka
Fiji	Taiwan
Georgia	Thailand
Guatemala	Trinidad and Tobago
Hong Kong	Turkey
India	United Arab Emirates
Indonesia	United States
Iran	Venezuela
Israel	Vietnam

[a] Austria, Belgium, Bulgaria, Cyprus, Czech Republic, Denmark, Estonia, Finland, France, Germany, Greece, Hungary, Ireland, Italy, Latvia, Lithuania, Luxembourg, Malta, Netherlands, Poland, Portugal, Romania, Slovakia, Slovenia, Spain, Sweden, the United Kingdom.

Regulatory Status

Neotame is currently approved for use in 69 countries as detailed in Table 9.6. Additional approvals are pending in a number of countries.

In the US FDA approval of neotame (US FDA 2008, 86), it is stated as follows: "the food additive neotame may be safely used as a sweetening agent and flavor enhancer in foods generally, …in accordance with good manufacturing practice, in an amount not to exceed that reasonably required to accomplish the intended technical effect." Neotame is approved in a similar broad manner in Mexico, China and other countries without use levels specified. The Joint WHO/FAO

Expert Committee on Food Additives (JECFA), the EU, Japan and Australia/New Zealand and other countries determined an acceptable daily intake (ADI) for neotame of 0–2 mg/kg/day. In the JECFA Assessment of Intake they stated the following: "Conservative calculations based on its (neotame) lowest sweetness potency (7,000 times that of sugar) suggest that intakes of 2 mg neotame/kg of body weight per day would correspond to the replacement of 840 kg of sugar in the diet of a 60 kg adult" (WHO 2003, 35). Therefore even a total replacement of sugar with neotame would not lead to the ADI being exceeded.

Conclusion

Neotame is a non-caloric, high potency sweetener and flavor enhancer that is 7,000 to 13,000 times sweeter than sucrose. Neotame is structurally related to and chemically derived from aspartame, but is 30 to 60 times sweeter and exhibits greater heat-stability. Neotame does not require special labeling for phenylketonuria and does not degrade to a diketopiperazine.

Neotame can favorably modify and enhance flavors and taste at or below sweetening levels. It provides a clean sweetness with no bitter, metallic, or sour aftertaste. The sweet quality of neotame will contribute to better tasting food products.

On a sweetness equivalency basis versus existing sweetener alternatives, neotame offers the potential to deliver improved cost structure due to its high sweetness potency and low levels of usage. The small amounts of neotame required for sweetening will also reduce the cost of delivering and handling sweeteners.

The way in which goods are manufactured in terms of their impact on the environment is coming under increasing scrutiny by consumers and environmentalists. Factors such as resource consumption, energy consumption, waterborne and airborne emissions, and solid waste have been analyzed to attempt to quantify the environmental impact of neotame manufacture. Due to the high potency and ease of manufacture of neotame, the analysis indicates that it will outperform other high potency sweeteners in terms of its environmental impact.

Consumer exposure to neotame in foods is estimated to be 0.10 mg/kg/day at the 90th percentile level of consumption. Clinical studies have demonstrated neotame to be safe and well-tolerated in amounts up to 20 times the projected 90th percentile of use. In addition, neotame does not alter glycemic control in subjects with non-insulin dependent diabetes mellitus. Margins of safety for neotame in the various species employed for safety testing are tens of thousands-fold greater than the estimated human exposure levels. Thus, the very large safety margins in animals and the demonstrated safety in humans at multiples many times 90th percentile consumption levels establish the safety of neotame for its intended use as a sweetener and flavor enhancer.

The technical and functional qualities of neotame make this new high potency sweetener and flavor enhancer desirable in a wide variety of food and beverage preparations. Neotame offers product developers a tool to formulate better tasting products, reduce calories, and lower costs.

References

Butchko, H.H., and Kotsonis, F.N. 1991. Acceptable daily intake versus actual intake: The aspartame example. *J Am Coll Nutr* 10(3):258–266.

Butchko, H.H., and Kotsonis, F.N. 1996. Acceptable daily intake and estimation of consumption. In *The clinical evaluation of a food additive: Assessment of aspartame*, eds. C. Tschanz, H.H. Butchko, W.W. Stargel, and F.N. Kotsonis, 43–53. Boca Raton, FL: CRC Press.

Butchko, H.H., Tschanz, C., and Kotsonis, F.N. 1994. Postmarketing surveillance of food additives. *Reg Toxicol Pharmacol* 20:105–118.

DuBois, G.E., Walters, D.E., Schiffman, S.S., Warwick, Z.S., Booth, B.J., Pecore, S.D., Gibes, K., Carr, B.T., and Brands, L.M. 1991. Concentration-response relationships of sweeteners. In *Sweeteners discovery, molecular design, and chemoreception*, eds. D.E. Walters, F.T. Orthoefer, and G.E. DuBois. Washington, DC: American Chemical Society.

Glaser, D., Tinti, J., and Nofre, C. 1995. Evolution of the sweetness receptor in primates. I. Why does alitame taste sweet in all prosimians and simians, and aspartame only in Old World simians? *Chem Senses* 20(5):573–584.

Nofre, C., and Tinti, J.M. 1991. In quest of hyperpotent sweeteners. In *Sweet-taste chemoreception*, eds. M. Mathlouthi, J.A. Kanters, and G.G. Birch, 205–236. London: Elsevier.

Pariza, M.W., Ponakala, S.V., Gerlat, P.A., and Andress, S. 1998. Predicting the functionality of direct food additives. *Food Technol* 52(11):56–60.

Pierre, A., Le Quere, J.L., Famelart, M.H., Riaublanc, A., and Rousseau, F. 1998. Composition, yield, texture and aroma compounds of goat cheeses as related to the A and O variants of alpha s1 casein in milk. *Lait* 78:291–301.

US FDA. 1993. Toxicological principles for the safety assessment of direct food additives and color additives used in food: "Redbook II." Washington, DC: Food and Drug Administration.

US FDA. 1996. Food additives permitted for direct addition to food for human consumption; dimethyl dicarbonate. *Fed Regist* 61:26786–26788.

US FDA. 1998. Monsanto Co.: Filing a food additive petition. *Fed Regist* 63(27):6762.

US FDA. 1999a.Flavor enhancers. 21 CFR 170.3(o)(11).

US FDA. 1999b. Monsanto Co.: Filing a food additive petition. *Fed Regist* 64(25):6100.

US FDA. 2008. Multipurpose Additives. 21 CFR 172.829.

WHO. 2003. Evaluation of certain food additives and contaminants. Joint FAO/WHO expert committee on food additives:35.

Chapter 10

Saccharin

Abraham I. Bakal and Lyn O'Brien Nabors

Contents

Introduction

Saccharin, chemically known as o-sulfabenzamide, 2,3-dihydro-3-oxobenzisosulfonazole, is an intense sweetener that, according to Dubois (1991), is 200–800 times sweeter than sugar depending on its concentration in aqueous solutions. It was discovered in 1878 by Ira Remsen and Constantine Fahlberg working at John Hopkins University in Baltimore, Maryland. The surreptitious discovery of the intense sweetness of this compound led Fahlberg to aggressively pursue commercializing this invention. Fahlberg coined the name saccharin and proceeded to patent the compound without credit to Remsen (Pearson 2001).

The history of the early production of saccharin is described by Pearson (2001). Saccharin was first produced by Fahlberg in a pilot plant facility in New York. The production was then moved to Germany. The manufacturing of saccharin in the United States began in 1901 by a newly formed company, the Monsanto Chemical Company. Szmrecsayi and Alvarez (1998) provide an excellent review of saccharin and other sweeteners including a thorough discussion of the economics involved in selecting a sweetener.

Saccharin was not viewed favorably when it was introduced in the United States. Some criticized saccharin because it was fake, others because it had "no food value or no calories" and others attacked saccharin as "unhealthy." Saccharin, however, remained in the marketplace in the United States because of the support of President Theodore Roosevelt. The President was quoted in 1906 as saying that "anyone who says saccharin is injurious to health is an idiot" (Whelan 1977, 54) In a letter dated July 7, 1911, President Roosevelt wrote "I always completely disagreed about saccharin both as to the label and as to being deleterious … I have used it myself for many years as a substitute for sugar in tea and coffee without feeling the slightest bad effects. I am continuing to use it now" (Pearson 2001, 149).

Sugar cost and sugar rationing during World War I and World War II increased the use of saccharin and by 1917 saccharin was widely used as a tabletop sweetener in the United States and Europe (Smith 1979).

Production Processes

There are two basic processes for the manufacture of saccharin; the Remsen and Fahlberg process and the Maumee process. Both are described at length by Pearson (2001).

The Remsen and Fahlberg process starts with the treatment of toluene with chlorosulfonic acid resulting in ortho- and para-toluene sulfonyl chloride. Treatment with ammonia results in the formation of the corresponding toluenesulfonamides. The ortho-toluenesulfonamide is separated from the para-toluenesulfonamide and is oxidized to form saccharin (Warner 2008). It is possible to start saccharin production with ortho-toluenesulfonamide as the starting material which is oxidized to form saccharin acid. Saccharin acid is then converted into sodium or calcium saccharin through several dissolution, concentration and crystallization steps. A simplified schematic presentation of the various steps involved in this process for making sodium saccharin is given in Figure 10.1.

Saccharin acid, sodium saccharin and calcium saccharin are commercially produced using this basic process in South Korea, India and other countries.

The Maumee process was developed in the early 1950s by the Maumee Chemical Company and production using this novel process began in 1954. This process has been modified and improved since that time. The starting material is purified methyl anthranilate (MA), which is a substance naturally occurring in grape flavor (Bedoukian 1967). The Maumee process today is a continuous one whereby MA is diazotized to form 2-carbomethoxybenzene-diazonium chloride. Sulfonation followed by oxidation results in 2-carbomethoxybenzenesulfonyl chloride. This intermediate is amidated, followed by acidification to form insoluble saccharin acid. Sodium and calcium saccharin are produced by the addition of sodium hydroxide and calcium hydroxide, respectively (Pearson 2001). A simplified schematic presentation of the Maumee process for producing sodium saccharin is given in Figure 10.2.

Until recently, PMC Specialties Group was the sole producer of saccharin in the United States using the Maumee process. Currently this process and modifications thereof are widely practiced in China and India as well as in other countries.

Properties

As mentioned previously, there are three major forms of saccharin that are in wide use; sodium saccharin also known as soluble saccharin, calcium saccharin, and saccharin acid. Other forms of saccharin have been synthesized but are not commercially available.

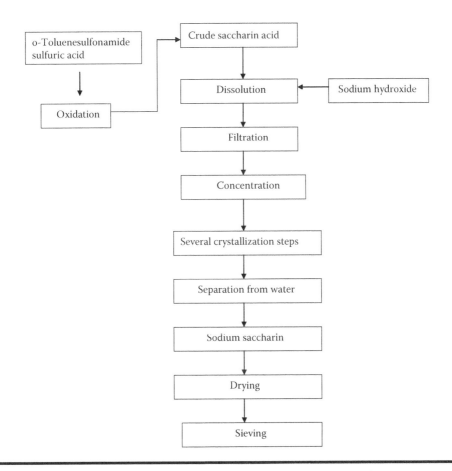

Figure 10.1   Schematic presentation of the Remsen and Fahlberg saccharin process.

Saccharin acid is an odorless, white crystalline powder which is only slightly soluble in water. It is used commercially in various applications including toothpastes, mouthwash, cosmetics, pharmaceuticals, tobacco and so forth (JMC 2008).

Sodium saccharin is the most widely used salt because of its high solubility and ease of production. Calcium saccharin is also used in a variety of food applications. The solubility of these three forms of saccharin is given in Table 10.1.

It is interesting to note that the form of saccharin has no effect on its sweetness intensity.

Sweetening Potency and Admixture

Most publications characterize saccharin as 300 times sweeter than sugar. Paul (1921) (cited by Pearson 2001) discovered that the sweetness potency of aqueous solutions of saccharin is inversely related to its concentration. This phenomenon is not specific to saccharin and is a characteristic of most sweeteners (Amerine et al. 1965).

DuBois (1991) reports data on the potency of aqueous solutions of saccharin. DuBois, for example, reports that at 1% sucrose concentration, saccharin is over 800 times sweeter than sugar, at 3% sucrose concentration saccharin is about 500 times sweeter than sugar and at 7% sucrose concentration it is over 300 times sweeter than sucrose.

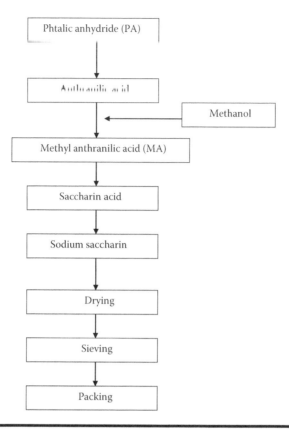

Figure 10.2   Schematic presentation of the Maumee process for saccharin.

Paul (1921) found that when blends of saccharin are used, the total sweetness in some cases is additive, that is, the total perceived sweetness is the sum of the sweetness of each sweetener at the concentration used. Subsequent studies found that in some cases, the total sweetness is higher than the additive effect, in other words a synergy is observed. This synergistic phenomenon is observed in blends of saccharin with most of the other intense sweeteners. The synergistic effect

Table 10.1   Solubility (g/100g water) of Saccharin Salts in Water as a Function of Temperature

	Acid Saccharin	*Sodium Saccharin*	*Calcium Saccharin*
20°C	0.2	100	37
35°C	0.4	143	82
50°C	0.7	187	127
75°C	1.3	254	202
90°C	—	297	247

Source: From Salant, A. *Handbook of Food Additives*, 2nd edition, Furia, T.F. (Ed.), p. 523. Cleveland, OH: CRC Press, 1972.

was commercially utilized in blends of saccharin and cyclamates before the ban of cyclamate in the United States in 1969. In a US patent, Helgren (1957) describes the use of blends of these two sweeteners. Beck (1975) suggested that the taste profile of 10:1 cyclamate:saccharin is closely similar to sucrose. This blend is commonly used in numerous applications including tabletop applications in countries where cyclamate is approved. As discussed by Bakal in Chapter 26 ("Mixed Sweetener Functionality") of this volume, blends of other sweeteners with saccharin also exhibit synergistic effects. Blends of saccharin with aspartame, sucralose, alitame, stevia, neotame, sucrose and fructose exhibit synergy. No synergy was reported between saccharin and acesulfame potassium (acesulfame K).

Blends of saccharin and other sweeteners have significantly less aftertaste than saccharin alone. This is probably, to a large extent, related to the previously mentioned phenomenon that saccharin aftertaste increases with increased saccharin concentration. Lavia and Hill (1972) disclose in a French patent that the addition of a very small amount of aspartame to saccharin sweetened products will significantly reduce the saccharin aftertaste and mask its bitter aftertaste. This was verified in experiments conducted in our laboratory. Similar observations were made in our laboratory when small amounts of neotame were added to saccharin sweetened products.

Another approach to overcoming the taste limitations of saccharin has been the search for masking agents or sweetness enhancers. Cumberland Packing Corp., for example, was granted a US patent for the use of cream of tartar and other ingredients in saccharin based tabletop sweetener (Eisenstadt 1971). The inventor claims that the use of cream of tartar reduces the perceived saccharin aftertaste. Bakal (1983) as well as Daniels (1973) describe some of the materials claimed to improve the taste of saccharin sweetened products. PMC Specialties (Pearson 2001) reports that calcium saccharin has a lower level of aftertaste than sodium saccharin. This has not been reported by other researchers.

Stability

In the dry form, saccharin and its salts are very stable. They are also very stable in aqueous solutions over a wide range of pH. Saccharin is also stable over a wide temperature range normally experienced in the food industry. Pearson (2001) provides a detailed discussion of the stability and shelf life of saccharin.

Uses

Saccharin has been widely used in beverages, foods, pharmaceuticals and cosmetics. The approval and introduction of new intense sweeteners has reduced uses of saccharin in some food applications. It is still widely used in several non food applications such as toothpaste, mouthwash, pharmaceuticals and cosmetics.

Pearson (2001) lists some non food applications of saccharin including:

- Nickel electroplating
- Agricultural intermediates
- Animal feed as a sweetener
- Pharmaceutical intermediates
- Biocide intermediate

■ Anaerobic adhesive accelerator
■ Personal care products

Metabolism

Saccharin is not metabolized by humans and is excreted in the urine and to a small extent, in the feces, unchanged. Metabolism studies have shown no metabolites of saccharin, clearly confirming it is not metabolized. Renwick (1986) carried out extensive studies on the metabolism of saccharin confirming these conclusions.

Toxicological Evaluation

Saccharin is one of the most studied ingredients in the food supply. It has been the subject of extensive scientific studies and generated numerous debates. The totality of available research clearly shows that saccharin is safe for human consumption. However, there has been controversy over its safety. The controversy rests primarily on the findings that consumption of large daily doses of sodium saccharin by male rats is associated with increased incidence in bladder tumors (Tisdel et al. 1974; Taylor et al. 1980; Arnold et al. 1980).

Recent research has concentrated on understanding the mechanism of the association of saccharin consumption and increase in tumor formation in the male rat. These studies have shown that this phenomenon is the result of a sequence of events that begins with the administration of high doses of sodium saccharin. This leads to the formation of a urinary milieu conducive to the formation of a calcium phosphate-containing precipitate that is cytoxic to the rat urothelium. This, in turn, leads to a regenerative hyperplasia, which persists over the lifetime of the animal if the administration of sodium saccharin continues, resulting in the formation of bladder tumors. The no effect level for the hyperplastic response to sodium saccharin is 1% of the diet, and this is identical to the no-effect level for the hyperplastic response to sodium saccharin, tumor promotion and to the tumors observed in the two-generation bioassays (Cohen 1997). Ellwein and Cohen (1990) and Cohen et al. (1998) have shown that administration of other sodium salts (e.g., sodium ascorbate) produces effects similar to those of high doses of sodium saccharin.

Extensive research in human populations has shown no increased risk of bladder cancer in humans. More than 30 human studies have been completed, supporting the safety of saccharin at human levels of consumption (Elcock and Morgan 1993).

In 1997, a special International Agency for Research on Cancer (IARC) panel determined that the increased incidence in bladder tumors in male rats associated with the ingestion of sodium saccharin is not relevant to man. In 1998, the IARC downgraded saccharin from a Group 2B substance, possibly carcinogenic to humans, to a Group 3 substance. Substances may be placed in Group 3 when there is strong evidence that the mechanism of carcinogenicity in experimental animals does not operate in humans. This was the first time the IARC had considered mechanistic data (Rice et al. 1999).

In 1996, the Calorie Control Council submitted a petition to the US National Toxicology Program (NTP) requesting that, under the program's new criteria allowing the consideration of mechanistic data, saccharin be delisted as a possible carcinogen from the NTP's Ninth Report on Carcinogens. In May 2000, the NTP released the 9th edition and announced that saccharin has been delisted (NTP 2000).

Regulatory

Saccharin is the most widely known sweetener and is used in more than 100 countries. In the United States, it is approved for use under an interim food additive regulation, permitting use for special dietary and certain technological purposes (US FDA 2010).

In 1977, the US Food and Drug Administration (US FDA) proposed a ban on saccharin as a result of studies reporting bladder tumors in some male rats fed high doses of sodium saccharin. The US Congress passed a moratorium preventing the proposed ban. The moratorium was extended seven times. In 1991, the FDA formally withdrew its 1977 proposal to ban saccharin. On December 21, 2000, President Bill Clinton signed legislation removing the saccharin notice requirement for products sweetened with saccharin (H.R. 4577).

In 1993, the Joint Food and Agriculture/World Health Organization Expert Committee on Food Additives (JECFA) reviewed the safety of saccharin and doubled its acceptable daily intake (ADI). The JECFA noted that the animal data which earlier raised questions about saccharin, are not considered relevant to man (JECFA 1993).

The Codex Committee of Food Additives (CCFA 2010) has reviewed saccharin and its salts and adopted provisions for levels of use of saccharin in numerous food categories, which are now included in the Codex Alimentarius General Standard for Food Additives (GSFA 2010).

Conclusion

Saccharin is an extremely stable and versatile sweetener, suitable for numerous food and beverage applications and one of the most extensively evaluated substances. It has had a long and controversial history since its discovery in 1879 and continuing until recent years. The issue of bladder tumors was ultimately shown to be a high dose phenomenon that occurred only in rats. The methodology and issues utilized in the research of saccharin have greatly extended the sciences of toxicology, epidemiology and risk assessment. The fact that saccharin does not pose a cancer risk to humans is now accepted (Cohen et al. 2008). Saccharin is a success story of modern science.

References

Amerine, M.A., Pangborn, R.M., and Roessler, E.B. 1965. *Food Science and Technology*, p. 86. New York: Academic Press.

Arnold, D.L., Moodie, C.A., Grice, H.C., Charbonneau, S.M., Stavric, B., Collins, B.T., Mcguire, P.F., Zawidzka, Z.Z., and Munro, I.C. 1980. Long term toxicity study of orthotoluene sulfonamide and sodium saccharin in the rat. *Toxicol Appl Parmacol* 52:113–152.

Bakal, A.I. 1983. *Functionality of Combined Sweeteners in Several Food Applications*, p. 700. Chemistry and Industry, Society of Chemical Industry, London. September 19.

Bock, K.M. 1975. Practical considerations of synthetic sweeteners. *Food Prod Develop* 9:47–54.

Bedoukian, P. 1967. *Perfumery and Flavoring Synthetics*, p. 41. New York: Elsevier.

Codex Committee on Food Additives (CCFA). 2010. Report of Forty-second Session. Beijing. March 15–19. http://www.codexalimentarius.net/web/archives.jsp?year=10. Accessed April 21, 2011.

Cohen, S.M. 1997. Saccharin safety update. In *Low-Calorie Sweeteners: Proceedings of a seminar in New Delhi, India, sponsored by the International Sweeteners Association and the Confederation of Indian Food Trade and Industry*. February 5–6. Newbury: CPL Press.

Cohen, S.M., Anderson, T.A., De Oliveira, L.M., and Arnold, L.L. 1998. Tumorgenicity of sodium ascorbate in male rats. *Cancer Res* 58:2557–2561.

Cohen, S.M., Arnold, L.L., and Emerson, J.L. 2008. AgroFOOD industry. *Hi-Tech* 19(6):24–26.

Daniels, R. 1973. *Sugar Substitutes and Enhancers*, pp. 80–103, Park Ridge, NJ: Noyes Data Corporation.

DuBois, G.E. 1991. In *Sweeteners: Discovery, Molecular Design and Chemoreception*, Walters, D.E., Orthefer, F.T. and DuBois, G.E. (Eds.), p. 274. Washington, D.C.: American Chemical Society.

Eisenstadt, M.E. 1971. Cyclamate free, calorie free sweetener. US Patent 3,625,711.

Elcock, M., and Morgan, R.W. 1993. Update of artificial sweetener and bladder cancer. *Reg Toxicol Pharmacol* 17:35–43.

Ellwein, L.B., and Cohen, S.M. 1990. The health risks of saccharin revisited. *Crit Rev Toxicol* 20:311–326.

General Standard for Food Additives. 2010. Saccharins (INS 954). Codex Alimentarius. Updated up to the 33rd Session of the Codex Alimentarius Commission. http://www.codexalimentarius.net/gsfaonline/groups/details.html?id=98. Accessed October 28, 2010.

Helgren, F.J. 1957. Sweetening compositions and method of production the same. US Patent 2,803,551.

H.R. 4577. Making appropriations for the Department of Labor, Health and Human Services and Education and related agencies for the fiscal year ending September 30, 2001 and for other purposes. Signed by the President (effective date), December 21, 2000.

JMC Corporation. 2008. Company Brochure.

Joint FAO/WHO Expert Committee on Food Additives (JECFA). 1993. Evaluation of Certain Food Additives and Contaminants. Forty-first report. WHO Technical Report Series. World Health Organization, Geneva, Switzerland.

Lavia, A.F., and Hill, J.A. 1972. Sweeteners with masked aftertaste. Fr. Patent 2,087,843.

National Toxicology Program (NTP). 2000. 9th Report on Carcinogens. US Government Printing Office.

Paul, T. 1921. Physical chemistry of food stuffs. V. degree of sweetness of sugar. *Chem Zeit* 45:38–39.

Pearson, R.L. 2001. Saccharin. In *Alternative Sweeteners*, Nabors, L.O. (Ed.), p. 147. NY: Marcel Dekker.

PMC Specialties Group, Inc. Cincinnati, OH. Unpublished data. Cited by R.L. Pearson. In Saccharin. Alternative Sweeteners, Nabors, L.O. (Ed.), p. 147. NY: Marcel Dekker.

Renwick, A.G. 1986. The metabolism of intense sweeteners. *Xenobiotica* 16(10/11):1057–1071.

Rice, J.M., Braan, R.A., Blettner, M., Genevois-Charmeau, C., Grosses, Y., McGreggor, D.B., and Partensky, C. 1999. Rodent tumors of urinary bladder, renal cortex and thyroid gland in IARC monographs evaluations of carcinogenic risk to humans. *Toxicol Sciences* 49:166–171.

Salant, A. 1972. Nonnutritive sweeteners. In *Handbook of Food Additives*, 2nd edition, Furia, T.F. (Ed.), p.523. Cleveland, OH: CRC Press.

Smith, T. 1979. Saccharin. Washington, DC: American Council on Science and Health.

Szmrecsayi, T., and Alvarez, V.M P. 1998. The search for a perfect substitute: Technological and economic trajectories of synthetic sweeteners, from saccharin to aspartame (c. 1880–1980). Presented at the 12th International Economic History Congress, Session C-36 on Sugar and Alternative Sweeteners in Historical Perspectives, Madrid, Spain.

Taylor, J.M., Weinberger, M.A., and Friedman, L. 1980. Chronic toxicity and carcinogenicity of the urinary bladder of sodium saccharin in the utero-exposed rat. *Toxicol Appl Pharmacol* 54:57–75.

Tisdel, M.O., Nees, P.O., Harris, D.L., and Derse, P.H. 1974. Long term feeding of saccharin in rats. In *Symposium: Sweeteners* by Inglett, G. E., (Ed.), 145–158. Westport, CN: Avi Publishing Co.

US Food and Drug Administration (US FDA). 2010. Code of Federal Regulations, Title 21, Section 180.37, Saccharin, ammonium saccharin, calcium saccharin, and sodium saccharin. http://www.access data.fda.gov/scripts/cdrh/cfdocs/cfcfr/CFRSearch.cfm?fr=180.37&SearchTerm=saccharin. Accessed October 28, 2010.

Whelan, E.M. 1977. The Conference Board Review. 16:54. The Conference Board, New York.

Warner, D. 2008. Ira Remsen, saccharin, and the linear model. *Ambix* 55(1):50–61.

Chapter 11

Steviol Glycosides

Michael Carakostas, Indra Prakash, A. Douglas Kinghorn,
Christine D. Wu, and Djaja Djendoel Soejarto

Contents

Introduction

The leaves of *Stevia rebaudiana* (Bertoni) Bertoni (Asteraceae) (henceforth referred to as "stevia"), a South American plant, have been known for over a century in the border regions of Paraguay and Brazil for their sweet taste, but their specific use as a sweetener in western diets is relatively new. Crude stevia extracts were first commercialized as sweeteners in Japan in the early 1970s. Products extracted from *S. rebaudiana* leaves have remained popular in Japan, and their use has now spread to South Korea, the People's Republic of China, and other countries beyond South America and Asia. Stevia leaves and relatively crude extracts have been available as dietary supplements in the United States and a few other countries since the mid-1990s, but their sale or use as a tabletop sweetener or an ingredient for foods and beverages was prohibited in most major markets until very recently (Kinghorn 2002). High-purity, zero-calorie stevia extracts are of great interest to the global food industry because their natural source appeals to many consumers.

Ten sweet-tasting steviol glycosides have been identified in stevia (Bakal and Nabors 1986; Kinghorn and Soejarto 1991; Kinghorn et al. 2001; Prakash et al. 2008). All are glycosides of a common aglycone, steviol (*ent*-13-hydroxykaur-16-en-19-oic acid) (Figure 11.1). Steviol glycosides differ in the number and the type of sugars attached, as shown in Table 11.1. The earliest commercially produced stevia extracts contained predominantly stevioside, which was also the focus of most early safety research. More recently, highly purified rebaudioside A (purity greater than 95%, also called rebiana) and high purity steviol glycoside mixtures (purity greater than 95% of rebaudiosides A, B, C, D and F, stevioside, steviolbioside, dulcoside A and rubusoside) have become available commercially (Cargill Incorporated 2008; McNeil Nutritionals, LLC 2008). There has been some interest in producing high-purity rebaudioside D extracts, but so far this steviol glycoside is not available commercially.

Much new information about steviol glycosides has been published in the scientific literature since the last edition of this book. Over the past decade the food industry has also migrated away from extracts containing primarily stevioside to extracts of predominantly, or nearly entirely, rebaudioside A due to its superior taste characteristics. A flurry of research and

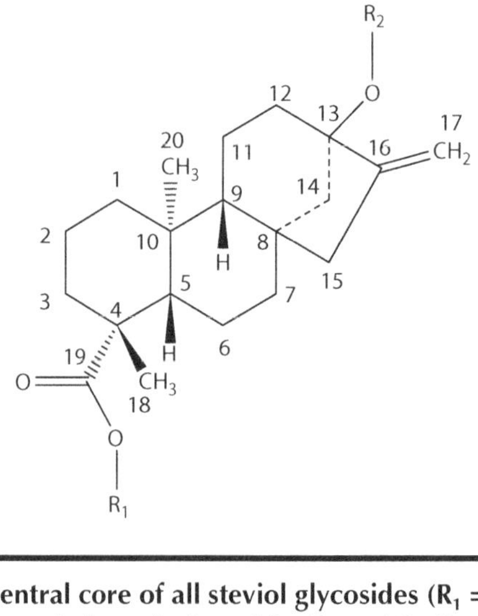

Figure 11.1 Steviol is the central core of all steviol glycosides (R_1 = OH, R_2 = H). Steviol glycosides will have various sugar moieties substituted for R_1 and R_2 (See Table 11.1).

Table 11.1 Comparison of Steviol Glycosides Structures and Sweetness Potency (Sucrose = 1)

| Compound | R-Groups in Steviol Backbone | | Sweetness Potency |
	R_1	R_2	
Dulcoside A	β-glc-	α-rha-β-glc-	30
Rebaudioside A	β-glc-	(β-glc)$_2$-β-glc-	200–300
Rebaudioside B	H	(β-glc)$_2$-β-glc-	150
Rebaudioside C	β-glc-	(β-glc, α-rha-)-β-glc-	30
Rebaudioside D	β-glc-β-glc-	(β-glc)$_2$-β-glc-	221
Rebaudioside E	β-glc-β-glc-	β-glc-β-glc-	174
Rebaudioside F	β-glc-	(β-glc, β-xyl)-β-glc-	200
Rubusoside	β-glc-	β-glc-	114
Steviolbioside	H	β-glc-β-glc-	90
Stevioside	β-glc-	β-glc-β-glc-	150–250

development work preceded the establishment of acceptable specifications, the resolution of safety concerns and the submission of regulatory dossiers leading to the appearance of commercial food and beverage products containing steviol glycosides in a number of major markets world-wide. Many attempts were made to secure the approval of international and national food safety authorities to market steviol glycosides as a food ingredient beyond just Japan, South Korea and a few other countries. This chapter will focus on the information that was used to overcome the barriers to steviol glycoside development as a legal food ingredient in a much larger number of countries since 2008. The reader is referred to earlier editions of this series for additional information of historical interest (Bakal and Nabors 1986; Kinghorn and Soejarto 1991; Kinghorn et al. 2001).

Production

Stevia rebaudiana is a member of a New World genus of 150–300 species that belong to the tribe Eupatorieae of the family Asteraceace (sunflower family). The plant may reach a height of 80 cm when fully grown and is native to Paraguay in the Department of Amambay on the border of Paraguay with Brazil. Comprehensive reviews have appeared in the literature on both the botany and the ethnobotany of the *S. rebaudiana* and the genus *Stevia* (Soejarto 2002a, 2002b). Stevia plants require long daylight periods and water and can be grown all around the world, although most commercial stevia agriculture is done in China. Typically, stevia leaves contain 7%–15% steviol glycosides (4.0%–8.5% stevioside, 1.5%–5% rebaudioside A, 0.1%–2.5% rebaudioside C, 0.1%–1% dulcoside A), but improved plant breeding has resulted in a higher steviol glycoside content (13%–20%) and a higher percentage of rebaudioside A (5%–12%) in new varieties (Kinghorn 2002).

Work to elucidate the chemical structures of *S. rebaudiana* sweeteners began in the early twentieth century, but proceeded slowly. The structure of the parent compound, stevioside (Figure 11.2) was not fully determined until 1963 (Kinghorn 2002). During the 1970s, additional sweet components, including rebaudiosides A–E, were isolated from *S. rebaudiana* leaves and characterized by Osamu Tanaka and co-workers at Hiroshima University in Japan (Kinghorn et al. 2010). However, some evidence exists that rebaudioside B and steviolbioside are not native *S. rebaudiana* constituents, but are formed by partial hydrolysis during the extraction process (Hanson and De Oliviera 1993; Tanaka 1997).

Stevioside and rebaudioside A are obtained from *S. rebaudiana* leaves in two stages. First, steviol glycosides are extracted from the leaves with water and the extract dried. In the second stage, further purification via alcohol/water crystallization yields very pure stevioside and/or rebaudioside A. While adhering to this general approach, several process variations have been described (Prakash et al. 2008). In modern processes, hot-water extraction of stevia leaves gives a "primary extract" from which other plant components are then removed by flocculation. The cleared solution is passed through adsorption resins to concentrate the steviol glycosides, which are then eluted with alcohol. The dried eluate, comprising mixed steviol glycosides, may be stored and transported in this form before final purification. In this last step, the dried eluate is re-dissolved in a lower alcohol (pure or an aqueous solution) and re-crystallized. Refluxing the dried solid in a methanol solution followed by cooling enables the isolation of stevioside (Payzant et al. 1999). Rebaudioside A is obtained either from the filtrate after stevioside is removed or by directly crystallizing the solid with alcohol or aqueous alcohol. Traditionally, methanol has been used as the solvent, but ethanol has the advantage of selectively increasing the yield of rebaudioside A. Finally, the crystallized product is filtered and dried (Prakash et al. 2008).

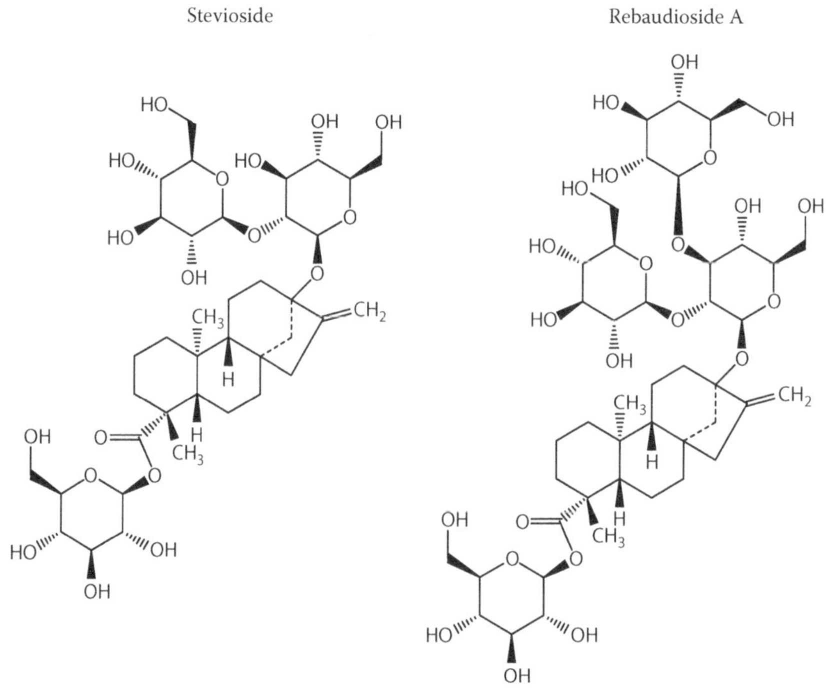

Figure 11.2 Chemical structures of stevioside and rebaudioside A.

Physical and Chemical Properties

Steviol glycosides, when crystalline, have relatively high melting points and can exist in various polymorphic crystalline forms. They can also be isolated in amorphous forms (Table 11.2).

Amorphous and anhydrous forms of rebaudioside A and stevioside readily form supersaturated solutions (more than 20 g/100 g at 25°C in 5 minutes) in water on simple stirring. Methanolate or ethanolate forms of rebaudioside A and stevioside can be obtained on crystallization of amorphous or anhydrous forms from pure methanol or ethanol. The solubilities of rebaudioside A and stevioside are only 0.8% (w/v) and 0.13% (w/v) in water at 25°C, respectively. If the hydrate of rebaudioside A is isolated by filtration and drying, however, it exhibits a low rate of dissolution in water (less than 0.2 g/100 g at 25°C in 5 minutes) (Prakash et al. 2008). Rebaudioside A and stevioside both show minimal dissolution in ethanol.

The viscosities of 0.01% and 0.5% aqueous solutions of rebaudioside A were determined using a laboratory viscometer at a temperature of 22°C and were found to be 1.67 and 4.79 cP (centipose), respectively. Viscosities of the aqueous rebaudioside A solutions increase as the concentration increases. The surface tensions of 0.01% and 0.5% aqueous solutions of rebaudioside A determined at 22°C were found to be 64.21 and 49.37 mN/m (milli-Newton/meter), respectively. The surface tension of an aqueous solution of rebaudioside A decreases as the concentration of this compound is increased. The refractive indices of 0.01%, 0.1%, and 0.5% aqueous solutions of rebaudioside A were evaluated at 24°C versus a water blank using a laboratory refractometer and were determined as 1.3401, 1.3405, and 1.3410, respectively.

Stability of Stevioside and Rebaudioside A

As a dry powder, stevioside is stable for at least two years and rebaudioside A is stable for at least three years at ambient temperature and under controlled humidity conditions. In solution,

Table 11.2 Physical and Solubility Data of the Steviol Glycosides

Sweetener	Melting Point (°C)	Specific Rotation	Solubility in Water at 25°C (%)
Dulcoside A	193–195	−52.22	0.58
Rebaudioside A	242–244	−30.7	0.80
Rebaudioside B	245–250	−26.44	0.11
Rebaudioside C	225–227	−38.11	0.21
Rebaudioside D	245–247	−28.91	0.10
Rebaudioside E	205–207	−34.2	1.7
Rebaudioside F	204–206	−30.96	0.60
Rubusoside	180–182	−57.4	0.08
Steviolbioside	195–200	−39.78	0.03
Stevioside	237–242	−43.52	0.13

rebaudioside A is most stable between pH values 4–8 and noticeably less stable below pH 2. As expected, stability decreases with increasing temperature (Prakash et al. 2008).

In aqueous solutions (pH 2–8), the major reaction pathways leading to a loss of rebaudioside A are as shown in Figure 11.3.

1. Isomerization of the C-16 olefin to form the C-15 isomer (III)
2. Hydration of the C-16 olefin to yield alcohol (IV)
3. Hydrolysis of the glycosyl ester at C-19 to form rebaudioside B (V).

Some rebaudioside A may degrade to stevioside first, which then follows the same degradation pathways listed above. The C-15 isomer (III), alcohol (IV) and rebaudioside B (V) are also sweet (Prakash et al. 2008).

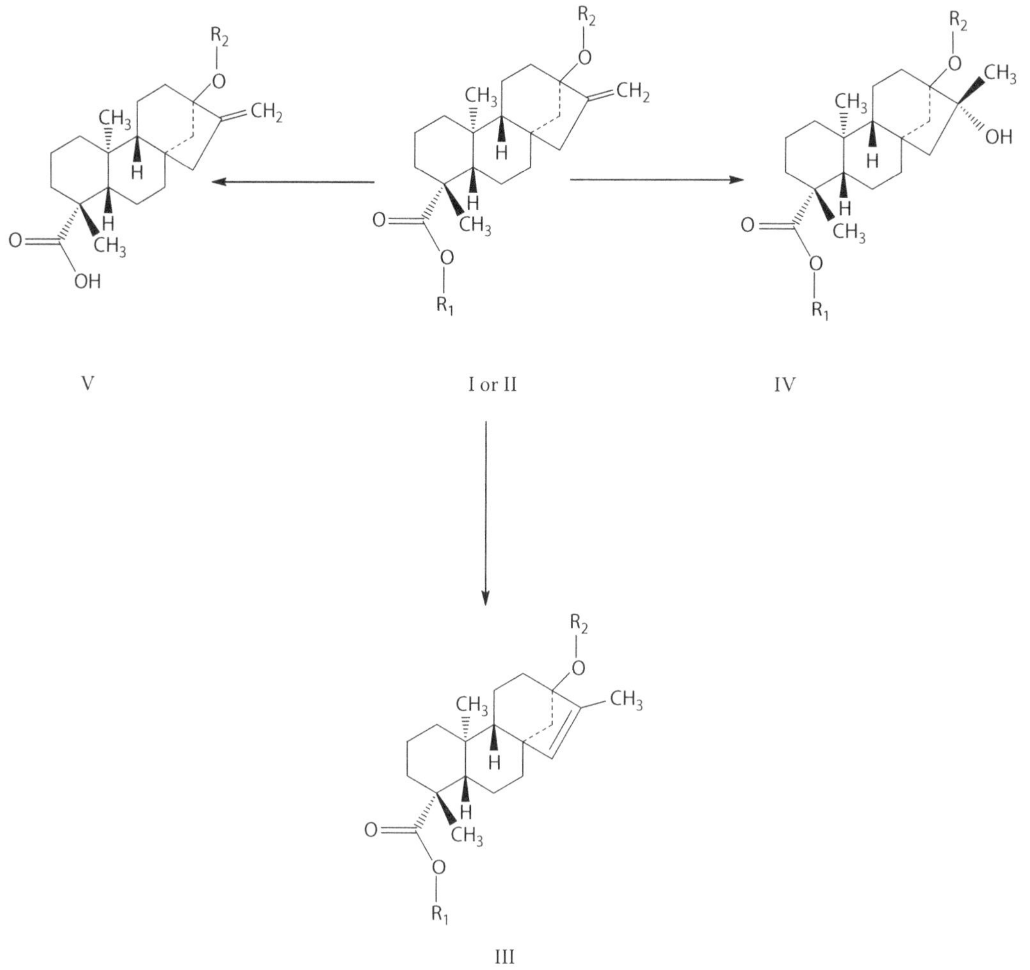

Figure 11.3 **Major pathways of degradation of stevioside (I) or rebaudioside A (II) under hydrolytic conditions.**

Stevioside and rebaudioside A are stable following exposure to sunlight (Clos et al. 2008). Rebaudioside A is more stable than aspartame in both low and high pH applications. In heat-processed beverages, such as flavored iced-tea, juices, sport drinks, flavored milk, drinking yogurt and non-acidified teas, the sweetener shows good stability during High Temperature-Short Time heat processing and on subsequent product storage. Rebaudioside A is also stable in yogurt with live cultures as well as baked good applications such as cake (Prakash et al. 2008).

Organoleptic Properties

Sweetness Potency

In the literature, stevioside and rebaudioside A are often reported as 150–250 and 200–300 times sweeter than sucrose, respectively. However, sweetness potency is strongly dependent on sucrose equivalency (SE) level for all high potency sweeteners (HPS). Therefore, it is important to state the SE level at which sweetness potency has been determined. Sweetness potency is also system dependent, therefore, it is important to also define the medium (e.g., water or phosphoric acid at pH 2.5). For comparison of different HPS, the most common medium is water (in the scientific literature where the medium is not specified, it is assumed to be water). While the choice of test medium is important, SE level is the major determinant of sweetness potency. There is, however, no industry-wide agreement on a common SE at which to report sweetness potencies.

Typical use-levels of HPS in beverages are generally in the range of 4% to 8% SE. Consequently, 6% SE represents a reasonable average value at which to compare the potencies of HPS in a plain water vehicle. At 6% SE, the sweetness potency of rebaudioside A is 200 times that of sucrose (Prakash et al. 2008). This is similar to the sweetness potency of aspartame, a HPS that is widely marketed and approved for addition to numerous foods and beverages in many countries. In comparison to sucrose, aspartame is 180 times as potent at 6% SE (DuBois et al. 1991). Although the concentration-response functions for rebaudioside A ($R = 8.2C/(194 + C)$) and aspartame ($R = 25.5C/(1160 + C)$) differ slightly in water, the two sweeteners are similar in sweetness over the range of SE laevels where the two ingredients are commonly used (also see Figure 8 in Prakash et al. 2008).

Flavor Profile

As with most high-potency sweeteners, stevioside and rebaudioside A exhibit clean sweetness at low SE levels, but have other negative taste attributes (e.g., bitterness and black licorice) at higher SE levels. Stevioside exhibits much more bitterness than rebaudioside A, so most new steviol glycoside sweetener products now contain high levels of rebaudioside A or are nearly pure rebaudioside A. Rebaudioside D also exhibits a clean sweet taste without any bitter aftertaste, but is present at relatively low levels in currently available stevia plants.

A trained descriptive panel evaluated rebaudioside A in water (Prakash et al. 2008). At low to medium SE levels (i.e., SE 6 or less), bitter and black licorice attributes were low/negligible and at higher SE levels (i. e., SE more than 6), they become notable. No sour, salty, savory, metallic or other negative taste attributes, however, were detected (also see Figure 9 in Prakash et al. 2008).

Sweetness Temporal Profile

Sweetness temporal profiles demonstrate changes in perception of sweetness over time. This property is key to the utility of a sweetener in foods and beverages, and is complementary to its flavor profile. Every sweetener exhibits a characteristic Appearance Time (AT) and Extinction Time (ET) (Dubois and Lee 1983). All high potency sweeteners, in contrast to carbohydrate sweeteners, display prolonged ETs. This can be beneficial in some products such as chewing gum, where prolonged sweetness is desirable.

The sweetness temporal profiles of aspartame at 531 mg/L, rebaudioside A at 529 mg/L, and sucrose at 8% in water at room temperature were compared over a period of 3 minutes (samples were swallowed at 5 seconds). The AT maximum was shortest for sucrose, slightly longer for aspartame and longest for rebaudioside A. The ET was longest for rebaudioside A, followed by aspartame and then sucrose, with the ET of rebaudioside A being significantly longer than sucrose (Prakash et al. 2008).

Uses and Admixtures

Food Applications

The functionality and stability of rebaudioside A was demonstrated with a three-dimensional food matrix model representing the intended conditions of food use (Pariza et al. 1998). Based on known results from aspartame and neotame, the key factors likely to affect rebaudioside A stability were product moisture, process temperature, and product pH. Accordingly, these were selected to represent the three dimensions of the matrix. Products comprised carbonated and still non-alcoholic beverages, tabletop sweetener formulations, chewing gum, yogurt and cake. These were packed, stored (mostly at 25°C and 60% relative humidity (RH)) and evaluated at intervals using both chemical (HPLC) and sensory analyses. Sweetness was assessed using panels consisting of 35 to 50 persons. Samples were evaluated using a five-point scale of categories ranging from 5 (much too sweet) to 1 (not at all sweet). Samples were considered satisfactory if at least 80% of the panelists rated the sweetness in category 3 (just about right) or above. Key findings were (Prakash et al. 2008):

1. Soft drinks (100–600 ppm)[*]: Rebaudioside A remained acceptably sweet throughout 26 weeks storage (cola and lemon-lime). For comparison, most soft drinks are consumed within 16 weeks of production.
2. Tabletop sweetener (200–2000 ppm): Rebaudioside A remained stable for at least 52 weeks in all formulations tested.
3. Chewing gum (300–10000 ppm): Rebaudioside A was considered stable and functional in chewing gum for 26 weeks.
4. Plain yogurt (100–1000 ppm): No significant loss was measured during pasteurization (190°F for 5 minutes) and fermentation. Rebaudioside A sweetness was stable throughout a 6-week storage period (40°F).
5. White cake (200–1000 ppm): No significant loss of sweetness was measured during the baking process (350°F for 20–25 minutes) or during 5 days subsequent storage (25°C and 60% RH).
6. Other products (100–1000 ppm): Rebaudioside A has been successfully formulated into cereals and cereal-based foods, dietary supplements, pharmaceuticals, edible gels, and confectionery products.

---

[*] Typical use ranges for rebaudioside A in the food category.

Blending

Blending of certain high-potency sweeteners is often found to result in sweetness synergy. Blends also benefit from improved flavor and temporal taste profiles, cost reductions and often improvements in stability (Paul 1921; Scott 1971; Lavia and Hill 1972; Verdi and Hood 1993; Walters 1993; Schiffman et al. 1995).

Due to low maximal response, sweetness linger and "off" taste (i.e., bitter and black licorice) at higher concentration levels (i.e., greater than 6% SE levels), rebaudioside A is unlikely to be used as a sole sweetener in several important food and beverage categories (e.g., zero-calorie soft drinks). However, this limitation of rebaudioside A is easily addressed by blending with any of a number of sweeteners. A number of non-caloric and caloric sweeteners are suitable partners for rebaudioside A. Examples of the former include erythritol, Luo Han Guo (an extract of *Siraitia grosvernorii* and its major sweet constituent mogroside V), aspartame, acesulfame potassium, brazzein, cyclamate, monatin, saccharin salts and sucralose (Prakash et al. 2007). Caloric sweeteners that may be blended advantageously with rebaudioside A include polyols (e.g., sorbitol and xylitol) and carbohydrates such as glycerol, glucose, fructose, sucrose, and high fructose corn syrup (HFCS). For example, a blend with sucrose where rebaudioside A contributes 15%–50% of the sweetness exhibits flavor and temporal profiles very close to those of sugar (short AT) with fewer calories. At the same time, bitter and licorice notes are imperceptible. Sweet-tasting amino acids such as glycine, alanine, and serine also improve the taste of rebaudioside A (Prakash et al. 2007). These too permit the formulations of good-tasting, natural, blended sweetener systems with lower energy content than with carbohydrate sweeteners alone.

Metabolism

Considerable progress in the understanding of steviol glycoside metabolism has occurred since the previous edition of this book was published. Stevioside and rebaudioside A are hydrolyzed to a common metabolite, the aglycone steviol (Figure 11.1), by colonic and/or cecal bacteria, although, some will pass through the intestinal tract completely or partially intact. There is virtually no absorption of intact steviol glycosides in the gastrointestinal tract. Glucose moieties, released by the microbial metabolism of steviol glycosides, are presumably used for energy by colonic bacteria, as there is no evidence glucose is absorbed. Steviol is absorbed from the colon and eliminated via the feces through enterohepatic re-circulation, or via urine as a glucuronide. There is no evidence that steviol accumulates in the body from successive ingestions of steviol glycosides (Roberts and Renwick 2008).

Enzymes and acids found in the human mouth, stomach or small intestine do not hydrolyze steviol glycosides (Hutapea et al. 1997; Koyama et al. 2003a). *In-vivo* and *in-vitro* studies have shown that stevioside and rebaudioside A are hydrolyzed to the aglycone steviol by bacteria in the colon of humans and cecum and colon of rats (reviewed in Renwick and Tarka 2008). Steviol glycosides undergo the successive removal of glucose moieties in the presence of intestinal microflora. This appears to explain the somewhat slower hydrolysis of rebaudioside A compared to stevioside due to the presence of an additional glucose in the latter compound. The conversion of steviol glycosides to steviol in the lower intestinal tract occurs almost entirely due to the action of *Bacteroides* sp.

Radioactivity from labeled stevioside was found primarily in the feces and bile of Wistar rats and it was inferred from the results that significant enterohepatic circulation of steviol occurred

(Nakayama et al. 1986; Koyama et al. 2003a). A metabolism study comparing stevioside and rebaudioside A in Sprague-Dawley rats revealed similar results and confirmed the presence of such enterohepatic circulation (Roberts and Renwick 2008). Radioactivity from orally administered steviol glycosides slowly increased in plasma over a period of hours, and was excreted primarily via the feces within 48 hours of oral dosing. While the half life of plasma radioactivity was 5 hours in male rats and 10 hours in female rats, other kinetic parameters were similar in males and females. Both steviol and steviol glucuronide were identified in plasma, but steviol plasma concentrations were often below the detectable limit. Peak plasma levels of radioactivity were slightly lower for rebaudioside A compared to stevioside. As indicated above, this is not unexpected given the additional time required to remove a fourth glucose moiety present in rebaudioside A. Less than 2% of the radioactivity was found in the urine and virtually no residual radioactivity was observed in any organ 96 hours after dosing. The predominant compound observed in bile from cannulated rats was steviol glucuronide, while steviol was the predominant compound found in rat feces. Radioactivity in the feces accounted for 97%–98% of the administered dose of both stevioside and rebaudioside A demonstrating that excretion of steviol glucuronide via bile is the major excretory route of steviol in the rat.

Metabolism of steviol glycosides in humans and rats is the same, but the pattern of metabolite excretion is different (Wheeler et al. 2008). In humans, as in rats, both stevioside and rebaudioside A are metabolized by bacteria in the lower gut to steviol, which is absorbed into the portal blood system and transported to the liver where it is glucuronidated. In humans, however, most steviol glucuronide appears in the plasma instead of the bile. Peak plasma concentrations of steviol glucuronide in humans occur approximately 8 and 12 hours post-dosing for stevioside and rebaudioside A, respectively. Like the rat, peak plasma metabolite concentrations were lower after rebaudioside A ingestion than after stevioside ingestion. The half-life of steviol glucuronide in human plasma is approximately 14 hours. Steviol glucuronide is the major excretion form of absorbed steviol in humans, and excretion occurs primarily via the urine rather than the feces. Only a small amount of steviol excretion occurs via the feces in humans.

Comparative metabolism studies are important to ensure that high-dose studies performed in rodents are appropriate models for human risk assessment. Metabolism of steviol glycosides in rats and humans to a common core metabolite, steviol, is important for demonstrating that safety studies performed on one steviol glycoside are applicable for risk assessment across the entire group of steviol glycosides. The difference in the major route of steviol excretion between rats and humans is presumably due to a species difference in the molecular weight threshold for excretion of compounds via the bile (Carakostas et al. 2008). Neither humans nor rats have much systemic exposure to free steviol. Steviol glucuronide is formed in both species and the larger human systemic exposure to steviol glucuronide is considered to be without toxicological significance. Glucuronidation is a common metabolic process in humans. Glucuronides are common and stable detoxification products that are quickly eliminated via the urine in humans.

The metabolism of orally ingested steviol is interesting, but not relevant for the understanding of steviol glycoside metabolism or safety. Humans do not ingest steviol and orally ingested steviol kinetics are quite different compared to those of steviol glycosides. Steviol is rapidly absorbed from the upper intestinal tract following ingestion. Peak steviol concentrations in plasma appear within about 15 minutes of steviol ingestion in rats (Roberts and Renwick 2008). Peak steviol concentrations in plasma following stevioside or rebaudioside A ingestion were not observed for up to 8 hours following ingestion (Koyama et al. 2003b; Roberts and Renwick 2008). This kinetic pattern fits well with the evidence that ingested steviol glycosides transit the intestinal tract to the colon where they are converted to steviol by bacteria and only then is steviol slowly absorbed. Orally ingested steviol administered experimentally, on the other hand, appears to be quickly absorbed in the upper gastrointestinal tract.

Safety Studies

Since all steviol glycosides are metabolized to steviol, exposure to steviol is an important consideration in the evaluation of steviol glycoside safety. For the purposes of comparing exposure to steviol from the ingestion of different steviol glycosides with different structures and molecular weights, steviol glycoside amounts are converted to steviol equivalents—the amount of steviol present in a specific amount of a steviol glycoside. To convert to steviol equivalents, stevioside quantities are multiplied by 0.4 and rebaudioside A quantities by 0.33 due to their different molecular weights. This conversion allowed regulatory and food safety authorities to establish a single safe dietary exposure limit for all steviol glycosides based on the relative exposure to steviol. The steviol glycoside exposures described below are the amounts tested without conversion to steviol equivalents. However, consumption amounts and regulatory levels in later sections of this chapter are stated in steviol equivalents (and noted as such) when it is appropriate to do so.

General Pre-Clinical Toxicity Studies

A number of acute, subchronic and chronic toxicity studies have been reported on various extracts of *S. rebaudiana* (Kinghorn and Soejarto 1991; Kinghorn et al. 2001). Some early studies were performed with crude or poorly defined stevia extracts that made interpretation of findings difficult. The Joint FAO/WHO Expert Committee on Food Additives (JECFA) established temporary specifications for steviol glycosides at their 63rd meeting in 2004 (Wallin 2004). Stevia extracts meeting current JECFA specifications can consist of nine steviol glycosides in any combination, but total steviol glycoside content must be 95% or more. For practical purposes, most extracts contain primarily rebaudioside A and/or stevioside with the remainder consisting of seven minor "named" steviol glycosides (rebaudiosides B, C, D, and F, steviolbioside, dulcoside A and rubusoside). Some studies performed before the establishment of JECFA specifications used test material that met present steviol glycoside specifications for purity and so were useful in several recently conducted safety evaluations. Many early studies, however, were conducted with crude extracts or simply ground stevia leaves with little or no chemical characterization available. Effects reported from these studies are difficult to interpret given the uncertainty about the composition of the material tested.

The acute toxicity of stevioside and of the aglycone steviol has been investigated in the mouse, rat and hamster. Stevioside was not lethal in any test animal at doses up to 15 g/kg (Toskulkao et al. 1997). The oral LD_{50} for steviol in hamsters was 5.20 and 6.10 g/kg in males and females, respectively. Death was attributed to acute renal failure and proximal renal tubular degeneration was observed histopathologically. The oral LD_{50} results for steviol in mice and rats were higher than the result observed in hamsters. Both mice and rats had LD_{50} results above 15 g/kg.

Aze et al. (1991) and Xili et al. (1992) reported virtually no toxicity from daily oral exposure to stevioside lasting approximately 13 weeks at dietary concentrations up to 2500 mg/kg/day in male and female rats. Nikiforov and Eapen (2008) and Curry and Roberts (2008) conducted similar 13-week toxicity studies on rebaudioside A, except that the latter study used higher dietary exposures. Curry and Roberts (2008) reported the highest no-observable-adverse-effect level (NOAEL) of any reported 13-week study, 4161 mg/kg/day for male rats and 4645 mg/kg/day for female rats. Reduced mean body weights and reduced body weight gains were observed in several of the high dose groups, but this was due to taste aversion and not toxicity (see discussion in the "Carcinogenicity" section). These authors also reported a NOAEL of approximately 10000 mg/kg/day in rats from a 4-week oral toxicity study. Importantly, at these very high exposure levels in both the 13-week and 4-week studies, male and female reproductive organs and kidneys were unaffected, answering questions raised by earlier toxicity studies that used crude or poorly characterized stevia-extracts (Carakostas et al. 2008).

Genetic Toxicity

The genetic toxicity of steviol glycosides has received considerable scrutiny over the years (Kinghorn and Soejarto 1991; Kinghorn et al. 2001; Brusick 2008). After several reviews, the JECFA concluded in 2004 that stevioside and rebaudioside A are not genotoxic (JECFA 2005). The JECFA also noted that steviol has been found to be genotoxic in several specific *in-vitro* assays, but was not genotoxic in *in-vivo* assays at high doses.

Rebaudioside A has been evaluated for genotoxicity in numerous *in-vivo* and *in-vitro* assays, which were uniformly negative (Pezzuto et al. 1985; Nakajima 2000a, 2000b; Sekihashi et al. 2002; Williams and Burdock 2009). Stevioside had positive results in two assays, a single Ames assay using *Salmonella typhimurium* TA98 and an *in-vivo* Comet assay (Suttajit et al. 1993; Nunes et al. 2007). However, there are many other published negative genotoxicity test results with stevioside, including four other negative Ames assays (one with TA98) (Brusick 2008). The single positive Ames assay using *Salmonella typhimurium* strain TA98 has recently been attributed to the presence of a low-level contaminant in the test material based on an analysis of the published results (Brusick 2008).

A report by Nunes et al. (2007) on DNA damage in a Comet assay in rats dosed with stevioside has been the subject of several critical evaluations on this study's technical conduct and interpretation of results, raising concerns about its validity (Geuns 2007; Williams 2007). The study did not use a positive control and DNA damage was not observed until week 5 of the study. Brusick (2008) pointed out in his review of steviol glycoside genotoxicity, that the Comet assay is designed to detect short-lived genetic damage that disappears quickly due to normal repair processes. Two stevia extracts containing mostly stevioside were negative in Comet assays conducted in mice (Sasaki et al. 2002; Sekihashi et al. 2002). The JECFA and other national food safety authorities have evaluated the entire spectrum of genotoxicity results for stevioside as part of the overall evaluation of steviol glycosides and have concluded that steviol glycosides in food are not genotoxic.

Steviol has also been evaluated on many occasions in both *in-vitro* and *in-vivo* assays considered standard genetic toxicity tests, with predominantly negative results. A few positive results, however, have received quite a bit of attention, both in the scientific and popular literature on stevia. In 1996 positive results in mammalian cell chromosomal aberration and gene mutation assays were reported (Matsui et al. 1996). A review of this study and others concluded that the positive results were caused by cytotoxicity due to the use of high steviol concentrations in the cell culture (Brusick 2008).

The frequently cited positive forward mutation test for steviol appears to be a real phenomenon, but does not indicate that steviol is likely to cause genetic damage in humans. Steviol is negative in standard Ames assays, but a positive result in a forward mutation assay using *S. typhimurium* TM677 has been observed (Pezzuto et al. 1985). This unique bacterial strain is repair deficient and contains both the plasmid pKM101 and rfa mutations and is not used in standardized Ames assays used for human risk assessment (Brusick 2008). Steviol is only mutagenic in this strain if it has been converted to a reactive intermediate by an S9 mix from rats exposed to polychlorinated biphenyls. This mutagenic effect of steviol appears only under highly specific experimental conditions in a single unique bacterial strain when induced hepatic microsomal enzymes modify steviol, so this result appears to have little biological significance for human risk assessment. Steviol is not mutagenic in standardized forward mutation assays using bacterial strains without the rfa mutation and that are not repair deficient (Brusick 2008). *In-vivo* genetic toxicity studies of steviol, including Comet assays in mice and micronucleus tests in mice, rats, and hamsters, have all been negative at doses of up to 2000 mg/kg/day for the Comet assay and at least 4000 mg/kg/day for the micronucleus test.

Carcinogenicity

Steviol glycosides have been evaluated for carcinogenicity primarily through studies on stevioside. Stevioside exposure at 5% in the diet did not cause bladder cancer in F344 rats and did not enhance the development of bladder tumors or pre-neoplastic lesions caused by *N*-nitrosobutyl-*N*-(4-hydroxybutyl) amine, a known bladder carcinogen (Hagiwara et al. 1984).

An early attempt to evaluate carcinogenicity potential was performed using a relatively low-purity stevioside (85%) that did not meet current JECFA specifications of 95% purity (Xili at al. 1992). The two-year dietary exposure study was also performed with a low exposure concentration of only 600 mg/kg/day. No tumors or other adverse effects, however, were reported.

A group from Japan reported a 13-week pre-oncogenicity study and a two-year carcinogenicity study on stevioside using Fischer 344 rats (Aze et al. 1991; Toyoda et al. 1997). The purity of the test material was 95.6%. No adverse effects were observed in the 13-week study, so dose-levels for the carcinogenicity study were set at 0%, 2.5% and 5% of the diet. Toyoda and associates reported no evidence of treatment-related increases in neoplastic or non-neoplastic lesions in the two-year study. The incidence of interstitial cell tumors of the testes and large granular cell leukemia were high in all groups, including controls, as is typical of this rat strain. The authors of this report considered the NOAEL to be 2.5% dietary exposure (equivalent to 970 mg/kg/day) based on reduced mean body weights observed in the high-dose group males and females, and a reduced survival rate in high-dose males. The reduced survival rate in high-dose males appeared to be caused by the strain-related interstitial cell tumors. The reduced body weight results appeared to be due to a taste-aversion induced reduction in diet intake, particularly during the early part of the study.

Reduced body weights and reduced body weight gains were reported in several subchronic toxicity studies performed on rebaudioside A, as well as the Toyoda et al. carcinogenicity study with stevioside. Taste aversion, with resultant decreases in food intake, is a common finding in high-potency sweetener safety studies and can cause reduced body weight in adults or reduced body weight gain in growing test animals (Flamm et al. 2003; Mayhew et al. 2003). The taste aversion-induced body weight reduction has no relevance for human exposure at typical ingestion levels and is not caused by systemic toxicity in the test animals, but rather is an artifact of extreme exposure levels required in toxicity studies to demonstrate a large margin of safety. Reduced body weights caused by taste aversion are usually considered non-adverse in determining a NOAEL in a toxicity study.

The decision by Toyoda et al. (1997) to set the NOAEL at the intermediate dose level based on the body weight and survival observations in the high-dose (5%) group appears to be very conservative, as a good argument can be made that both the body weight and reduced survival effects observed in the 5% group are not treatment-related. This NOAEL decision became very important because results from the two-year study by Toyoda et al. were used by the JECFA to establish the acceptable daily intake (ADI) for steviol glycosides. The permanent JECFA ADI established in 2008 is 0–4 mg steviol equivalents/kg/day (JECFA 2009; ADI calculation reviewed in Carakostas et al. 2008).

Reproduction Safety

Concerns about adverse reproductive system effects caused by stevia consumption circulated on the Internet for many years and was one of the regulatory issues preventing the approval of steviol glycoside use in food and beverages in many countries. Some of these Internet stories were based on documented reports that Paraguayan Matto Grasso Indians used leaves and stems from *S. rebaudiana* for their contraceptive and anti-fertility effects. Several early rodent studies also reported adverse effects on reproductive function. In a study reported in 1968, reduced fertility rates were observed in female rats given a crude aqueous extract of stevia at a dose of 10 ml/kg (Mazzei-Planas and Kuc 1968). Oliveira-Filho et al. (1989) and Melis (1999) reported reduced male reproductive organ weights in separate studies using semi-purified stevia extracts. However, in the recent subchronic and chronic toxicity studies discussed in the "General Pre-Clinical Toxicity Studies" section, in which highly pure rebaudioside A or stevioside were used, reproductive organ weight changes or other evidence of toxicity were not observed even at extremely high exposure levels. There are also three additional studies reporting no adverse reproductive system effects, one in hamsters and two in rats, in which stevioside of known purity was used. In a three-generation reproductive safety study in hamsters using 90% stevioside, no adverse reproductive outcomes were observed at doses up to 2500 mg/kg/day (the highest dose) (Yodyingyuad and Bunyawong 1991). Two teratology studies in rats, with 96.5% or 95.6% stevioside, showed no adverse developmental effects at doses up to 1000 mg/kg/day in one study and 3000 mg/kg/day in the other (Mori et al. 1981; Usami et al. 1995).

In their 51st and 63rd meetings, the JECFA stated that based on the total body of evidence, steviol glycosides meeting their specifications were not a reproductive hazard to humans. However, other food safety officials continued to raise concerns about reproductive safety (Carakostas et al. 2008). As a consequence of these stated concerns, a two-generation reproduction safety study using high-purity rebaudioside A was conducted as part of an overall project to gain much wider food-use approval for steviol glycosides, particularly in the United States and Europe (Curry et al. 2008). In this study, Wistar rats were exposed to 97% rebaudioside A via the diet at concentrations up to 25000 ppm. The study was conducted using current internationally accepted guidelines for two-generation reproductive safety studies and included a wide array of endpoints. No adverse reproductive effects related to rebaudioside A consumption were observed in either the F0 or F1 generations, including mating performance, fertility, gestation length and estrus cycles. Sperm motility, concentration, and morphology were also unaffected by treatment. Offspring in the F1 and F2 generations were not adversely affected by rebaudioside A exposure. The NOAEL was considered to be the highest dietary exposure level, 25000 ppm.

Previous reports about potential reproductive toxicity from exposure to high levels of steviol in hamsters do not appear to be relevant for human risk assessment (Wasuntarawat et al. 1998). Humans are not exposed to steviol through direct oral consumption, but rather at very low levels in the portal system through the slow process of microbial metabolism in the colon. While the Wasuntarawat et al. (1998) study using hamsters reported significant toxicity in both dams and fetuses at steviol doses as low as 500 mg/kg/day, other studies in hamsters and rats with steviol glycosides at higher steviol-equivalent doses have not demonstrated similar toxicity. The authors speculated that steviol might be more toxic to hamsters than to other species. The rapid uptake and high initial exposures to steviol from absorption in the upper gastrointestinal tract, however, likely played a major role in the toxicity of orally ingested steviol that is not observed following steviol glycoside ingestion.

Dihydroisosteviol was reported in an earlier study to exhibit antiandrogenic effects in a chick-comb bioassay, but was not found to affect testosterone activity in rats (Bakal and O'Brien Nabors 1986; Hanson and De Oliveira 1993; Melis 1999). Subsequent analytical work on purified stevia extracts containing more than 95% rebaudioside A failed to find dihydroisosteviol as either a contaminant or degradation product (Cargill Incorporated 2008; Prakash et al. 2008). Given the current consensus that steviol glycosides do not present a reproductive hazard to humans, this early finding on dihydroisosteviol does not appear to be relevant for steviol glycoside risk assessment.

Clinical Studies

Two clinical trials on the effects of stevioside conducted in hypertensive Chinese subjects have been reported. In the first study, hypertensive subjects were taken off medication and given either 750 mg/day of stevioside (purity unknown) or a placebo for 12 months (Chan et al. 2000). The second study was a two-year follow-up of the first study that used patients newly diagnosed with hypertension who were given either 1500 mg/day of stevioside or a placebo (Hsieh et al. 2003). Patients given stevioside in both studies had a significant reduction in systolic and diastolic blood pressure compared to baseline. However, other studies have failed to demonstrate an effect of steviol glycosides on blood pressure in both normotensive and mildly hypertensive subjects (Ferri et al. 2006; Geuns et al. 2007),

A number of studies conducted in animals given stevioside demonstrated an antidiabetic effect through an improved glucose response to insulin. Such an effect would only be relevant in type-2 diabetics. This led to several clinical studies being conducted. One study reported a mild reduction in the postprandial glycemic response following a meal containing 1000 mg of stevioside (Gregersen et al. 2004). However, a later more comprehensive study failed to confirm this effect. Jeppesen and associates (2006) demonstrated that no reduction in glycosylated hemoglobin occurred in subjects with type-2 diabetes given 1500 mg of stevioside daily for three months.

Studies reporting therapeutic effects on human subjects usually create barriers for the development of that substance as a food ingredient. Food ingredients are typically consumed over a wide range of intakes and without any medical supervision. Food safety authorities, therefore, are often reluctant to allow the unrestricted use of substances with a medicinal effect, even if the effect appears generally beneficial. Studies reporting potential antihypertensive and glucose-lowering effects of stevioside created a number of questions from JECFA and food safety officials in both the United States and Europe during early reviews (JECFA 2005). Specific clinical studies were requested by the JECFA at their 2004 meeting to answer questions about whether steviol glycosides have any effects on subjects with low blood pressure or subjects with type-2 diabetes. Specific concerns of the JECFA were for subjects with asymptomatic hypotension, and for well-controlled type-2 diabetics who might be adversely affected by steviol glycoside effects on glucose homeostasis.

Barriocanal et al. (2008) approached this request with a single study combining the issues of hypotension and type-2 diabetes. Subjects with normal glucose homeostasis and type-1 or type-2 diabetes were given 750 mg of steviol glycoside mixture daily for three months. No effects on glucose homeostasis or blood pressure were observed. The study, however, lacked sufficient statistical power to conclusively evaluate these effects and the test material did not meet JECFA specifications.

Maki et al. (2008a) took a different approach and conducted separate studies to evaluate potential effects on blood pressure and diabetes. High purity rebaudioside A meeting JECFA specifications was provided to 50 subjects with normal and low-normal blood pressure at a dose of 1000 mg/day for four weeks. Fifty additional subjects with normal and low-normal blood pressure were given a placebo. Rebaudioside A had no effect on resting seated diastolic or systolic blood pressure, heart rate, mean arterial pressure or 24-hour ambulatory blood pressure.

Sixty subjects with type-2 diabetes were given 1000 mg of high purity rebaudioside A daily over a 16-week period to allow time for detectable changes in hemoglobin A1c to occur (Maki et al. 2008b). An equal number of diabetic control subjects were dosed with a placebo. No changes in hemoglobin A1c, fasting glucose, insulin, and C-peptide were observed during the treatment period. Blood pressure was also measured in this study and no treatment-related effects were observed.

The potential effects of steviol glycosides on colon microflora in humans were evaluated using fecal samples from 11 healthy human volunteers (Gardana et al. 2003). Fecal suspensions were incubated under anaerobic conditions with 40 mg of stevioside or 40 mg of rebaudioside A for 72 hours. Minor changes in the fecal bacteria cultures were observed, but were considered clinically insignificant. Stevioside caused a small reduction in the growth of anaerobic bacteria and rebaudioside A caused a small reduction in the growth of aerobic bacteria. The possibility that steviol glycoside metabolism could affect human colonic microbiotia has been reviewed by Renwick and Tarka (2008) who concluded that the likelihood of such an effect was remote.

Cariogenicity

Pure stevioside and rebaudioside A were tested for cariogenicity in an albino rat model. Sixty Sprague-Dawley rats were colonized with *Streptococcus sobrinus* and divided into groups fed diets containing either 0.5% stevioside, 0.5% rebaudioside A, 30% sucrose or the basal diet without an added sweetener. After five weeks of treatment, the rats were killed and viable *S. sobrinus* were counted and caries evaluated according to Keyes' technique. Neither stevioside nor rebaudioside A were cariogenic under the conditions of the study (Das et al. 1992). In an *in-vitro* study, eight steviol glycosides (stevioside, rebaudiosides A-E, dulcoside A and steviolbioside) and two hydrolytic products of stevioside (steviol and isosteviol) were tested against a panel of cariogenic and periodontopathic oral bacteria. The test compounds were evaluated for their ability to inhibit sucrose-induced adherence, glucan binding, and glucosyltransferase (GTF) activity. None of the test compounds suppressed the growth or acid production of cariogenic organism, *Streptococcus mutans*, affected sucrose-induced adherence or altered GTF activity. Rebaudiosides B, C and E, steviol and isosteviol, however, inhibited the glucan-induced aggregation of the *S. mutans* to some extent, which could provide some oral health benefits by interfering with cell surface functions of cariogenic bacteria (Wu et al. 1998).

Recently, a clinical trial was conducted evaluating the *in-vivo* dental plaque pH in human volunteers (Goodson et al. 2010). Twenty-four subjects were evaluated in a double-blind cross-over trial in which a solution of rebiana (rebaudioside A) isosweet with 4.7% sucrose was compared to water, 4.7% sucrose, and isosweet sucralose. Dental plaque pH measurements were taken 12 times during a 60-minute period post-exposure from six teeth using a hand-held touch electrode on the mesiobuccal tooth surface. The study showed that a rebaudioside A solution isosweet with a 4.7%

sucrose solution was no more acidogenic than a water rinse and was significantly less acidogenic than a sucrose solution. The authors concluded that rebiana was non-acidogenic and meets the criteria set by the US Food and Drug Administration (US FDA) for a non-cariogenic sweetener.

Safety and Estimated Consumption Levels

The ADI is defined as an amount of a food additive that can be ingested daily over a lifetime without an appreciable health risk. Individuals consuming an amount higher than the ADI over short periods of time are not necessarily exposed to a hazardous level of the food ingredient. The ADI is an estimate of safety for an entire population over a lifetime of exposure and should not be interpreted to be an upper safe level of intake by an individual on any single day or series of days.

The ADI for steviol glycosides was tentatively established by the JECFA in 2004 based on the NOAEL from the two-year rat chronic toxicity/carcinogenicity study discussed above (970 mg/kg/day) and applying a 200-fold safety factor. The typical safety factor used to determine an ADI when a two-year safety study is available is 100, but the JECFA applied an additional 2× safety factor until clinical studies were completed. When those studies were made available in 2008, the ADI was increased to 0–4 mg/kg/day on a steviol equivalent basis, which converted for rebaudioside A is 0–12 mg/kg/day.

The evaluation of food additives by regulatory authorities requires, among many other items, a comparison of the estimated intake of the ingredient by average and high-intake consumers with the ADI. Steviol equivalent consumption estimates of 0.9–3.5 mg/kg/day for Europe, Latin America, Africa and Asia were published by the JECFA in its 63rd report using the international Global Environment Monitoring System (GEMS)/Food database (JECFA 2005). These estimates were based on the assumption that steviol glycosides would replace all caloric and non-caloric sweeteners in the diet. Upper level (90th percentile) intakes for Japan and the United States were also developed, but were updated at a later meeting. At its 2008 meeting, the JECFA updated its intake estimates for the United States and Japan resulting in intake estimates of 5.8 mg/kg/day and 3.0 mg/kg/day, respectively (in steviol equivalents). The JECFA considered these steviol glycoside intake estimates to be excessively conservative because no low-calorie sweetener would be expected to replace the consumption of all sugar and other caloric sweeteners. Actual intakes were considered to be only 20%–30% of the maximum intake estimates established in 2004 and 2008. For the United States, which had the highest estimated steviol glycoside intake of any country based on the total sweetener replacement model, the adjustment brought the JECFA estimates in line with other steviol glycoside intake estimates at approximately 1.7 mg steviol equivalents/kg/day for the highest consumption group, non-diabetic children (JECFA 2009).

Renwick published a rebaudioside A intake assessment based on his previously reported technique of using existing low-calorie sweetener consumption surveys from the United States and Europe (Renwick 2008). He reported average expected intakes of rebaudioside A of 1.3 and 2.1 mg/kg/day in the general population and children, respectively. High-intake adults and children were estimated to consume 3.4 and 5.0 mg/kg/day, respectively. High-intake diabetic children were estimated to consume 4.5 mg/kg/day. Converted to steviol equivalents, Renwick's rebaudioside A consumption results for non-diabetic children become approximately 1.7 mg/kg/day.

Safety for population-wide consumption is established if the estimated intake of steviol glycosides is less than the ADI. The highest 90th percentile intake of steviol glycosides from non-diabetic children (approximately 1.7 mg steviol equivalents/kg/day) is well within the ADI of 0–4 mg/kg/day.

Regulatory Status

Leaves of *S. rebaudiana* and/or extracts containing primarily stevioside have been approved food additives in Argentina, Brazil, Japan, South Korea, the People's Republic of China, Peru, and Russia for some years. However, a number of the individual country regulations were confusing because the word "stevioside" was sometimes used in a manner that could be interpreted to mean all steviol glycosides and specifications were frequently not stated. Thus, it has been difficult to know what steviol glycoside forms were actually permitted. Publication of final JECFA specifications in 2007 alleviated much of this confusion, as most countries recognize these specifications outright, or used them as the basis for changes or updates for their own regulations. In 2008, establishment of a permanent JECFA ADI further facilitated the approval of steviol glycosides meeting JECFA specifications in a number of countries. Some countries, such as Australia, New Zealand and Mexico, conducted their own separate evaluations prior to approving steviol glycosides, but final JECFA actions clearly paved the way for many individual country approvals since 2008.

Stevia held a particularly confusing status in the United States for many years as a dietary supplement. Although certain segments of the population clamored for widespread use of stevia as a sweetener in popular foods and beverages, most did not understand that such products were illegal under existing US regulations. Dietary supplements cannot be marketed as food ingredients; hence the sale of a food or beverage containing stevia in any form was illegal until recently. Advertising claims and widespread information on the Internet that stevia supplements could be used as a sweetener added to consumer confusion about the regulatory status of stevia in the United States.

In mid-2008, the US FDA was officially "notified" by two companies intending to market a steviol glycoside ingredient as a sweetener using the Generally Recognized As Safe (GRAS) process that is unique to the US food regulatory system. In late 2008, the US FDA responded that they had no questions about these GRAS Notifications. Steviol glycoside sweeteners made of primarily rebaudioside A began appearing in the US market shortly thereafter. Subsequently, at least a half-dozen other GRAS Notifications for steviol glycoside products meeting JECFA specifications have been filed with the US FDA.

In Canada and the European Union (EU), large complex regulatory dossiers must be filed with the authorities and review times are long. At the present time, steviol glycosides are not yet approved in either Canada or the EU. The European Food Safety Authority stated in early 2010 that steviol glycosides were safe, possibly paving the way for approval sometime in late 2011 or 2012.

In Canada, stevia leaves can be sold to consumers for their personal use in food provided no health claims are made, but foods containing stevia in any form may not be sold. Health Canada has approved the sale of more than 100 specific "natural health products" containing stevia leaves or stevia extracts. However, general sale or use of stevia in any form as a food additive or sweetener is currently prohibited.

References

Aze, Y., Toyoda, K., Imaida, K., Hayashi, S., Imazawa, T., Hayashi, Y., and Takahashi, M. 1991. Subchronic oral toxicity study of stevioside in F344 rats. *Eisei Shikenjo Hokoku* 109:48–54.

Bakal, A.I., and Nabors, L.O. 1986. Stevioside. In *Alternative sweeteners*, eds. L. O'Brien Nabors, and R.C. Gelardi, 295–307. New York: Marcel Dekker.

Barriocanal, L., Palacios, M., Benitez, G., Benitez, S., Jimenez, J.T., Jimenez, N., and Rojas, V. 2008. Apparent lack of pharmacological effect of steviol glycosides used as a sweetener in humans. A pilot study of repeated exposures in some normotensive and hypotensive individuals and type 1 and type 2 diabetics. *Regul Toxicol Pharmacol* 51:37–41.

Brusick, D.J. 2008. A critical review of the genetic toxicity of steviol and steviol glycosides. *Food Chem Toxicol* 46(Suppl. 7S):S83–S91.

Carakostas, M.C., Curry, L.L., Boileau, A.C., and Brusick, D.J. 2008. Overview: The history, technical function and safety of rebaudioside A, a naturally occurring steviol glycoside, for use in food and beverages. *Food Chem Toxicol* 46(Suppl. 7S):S1–S10.

Cargill Incorporated. 2008. GRAS exemption claim for rebiana (rebaudioside A). GRAS 253. http://www.accessdata.fda.gov/scripts/fcn/fcnNavigation.cfm?rpt=grasListing.

Chan, P., Tomlinson, B., Chen, Y., Liu, J., Hsieh, M., and Cheng, J. 2000. A double-blind placebo-controlled study of the effectiveness and tolerability of oral stevioside in human hypertension. *Br J Clin Pharmacol* 50:215–220.

Clos, J.F., DuBois, G.E., and Prakash, I. 2008. Photostability of rebaudioside A and stevioside in beverages. *J Agric Food Chem* 56:8507–8513.

Curry, L.L., and Roberts, A. 2008. Subchronic toxicity of rebaudioside A. *Food Chem Toxicol* 46(Suppl. 7S):S11–S20.

Curry, L.L., Roberts, A., and Brown, N. 2008. Rebaudioside A: Two-generation reproductive toxicity study in rats. *Food Chem Toxicol* 46(Suppl. 7S):S21–S30.

Das, S., Das, A.K., Murphy, R.A., Punwani, I.C., Nasution, M.P., and Kinghorn, A.D. 1992. Evaluation of the cariogenic potential of the intense natural sweeteners stevioside and rebaudioside A. *Caries Res* 26:363–366.

DuBois, G.E., Walters, D.E., Schiffman, S.S., Warwick, Z.S., Booth, B.J., Pecore, S.D., Gibes, K., Carr, B.T., and Brands, L.M. 1991. Concentration-response relationships of sweeteners: A systematic study. In *Sweeteners: Discovery, molecular design, and chemoreception*, eds. D.E. Walters, F.T. Orthoefer, and G.E. DuBois, 261–276. ACS Symposium Series, no. 450. Washington, DC: American Chemical Society Books.

DuBois, G.E., and Lee, J.F. 1983. A simple technique for the evaluation of temporal taste properties. *Chem Senses* 7:237–247.

Ferri, L.A.F., Wilson, A.D.P., Yamada, S.S., Gazola, S., Batista, M.R., and Bazotte, R.B. 2006. Investigation of the antihypertensive effect of oral crude stevioside in patients with mild essential hypertension. *Phytother Res* 20:732–736.

Flamm, W.G., Blackburn, G.L., Comer, C.P., Mayhew, D.A., and Stargel, W.W. 2003. Long-term food consumption and body weight changes in neotame safety studies are consistent with the allometric relationship observed for other sweeteners and during dietary restriction. *Regul Toxicol Pharmacol* 38:144–156.

Gardana, C., Simonetti, P., Canzi, E., Zanchi, R., and Pietta, P. 2003. Metabolism of stevioside and rebaudioside A from *Stevia rebaudiana* by human microflora. *J Agric Food Chem* 51:6618–6622.

Geuns, J.M.C. 2007. Letter to the editor. *Food Chem Toxicol* 45:2601–2602.

Geuns, J.M.C., Buyse, J., Vankeirsbilck, A., and Temme, E.H.M. 2007. Metabolism of stevioside by healthy subjects. *Exp Biol Med* 232:164–173.

Goodson, J.M., Cugini, M., Floros, C., Roberts, C., Boileau, A., and Bell, M. 2010. Effect of a Truvia™ rebiana on the pH of dental plaque. General Session and Exhibition of the International Association for Dental Research, Barcelona, Spain, July 14–17.

Gregersen, S., Jeppesen, P.B., Chen, P.B., Holst, J.J., and Hermansen, K. 2004. Antihyperglycemic effects of stevioside in type 2 diabetic subjects. *Metabol* 53:73–76.

Hagiwara, A., Fukushima, S., Kitaori, M., Shibata, M., and Ito, N. 1984. Effects of three sweeteners on rat urinary bladder carcinogenesis initiated by N-butyl-N-(4-hydroxybutyl)-nitrosamine. *Gann* 75:763–768.

Hanson, J.R., and De Oliveira, B.H. 1993. Stevioside and related sweet diterpenoid glycosides. *Nat Prod Rep* 10:301–309.

Hsieh, M.H., Chan, P., Sue, Y.M., Liu, J.C., Liang, T.H., Huang, T.Y., Tomlinson, B., Chow, M.S., Kao, P.F., and Chen, Y.J. 2003. Efficacy and tolerability of oral stevioside in patients with mild essential hypertension: A two-year, randomized, placebo-controlled study. *Clin Ther* 25:2797–2808.

Hutapea, A.M., Toskulkao, C., Buddhasukh, D., Wilairat, P., and Glinsukon, T. 1997. Digestion of stevioside, a natural sweetener, by various digestive enzymes. *J Clin Biochem Nutr* 23:177–186.

JECFA. 2005. Steviol glycosides. In *Safety evaluation of certain food additives*, 117–144. WHO Food Additives Series 54, World Health Organization: Geneva.

JECFA. 2009. Evaluation of certain food additives. *Sixty-ninth report of the joint FAO/WHO expert committee on food additives,* 50–55. WHO Technical Report Series 952. Geneva: World Health Organization.

Jeppesen, P.B., Barriocanal, L., Meyer, M.T., Alacios, M., Canete, F., Benitez, S., Logwin, S., Schupmann, T., and Benitez, G. 2006. Efficacy and tolerability of oral stevioside in patients with type 2 diabetes: A long-term randomized, double-blinded, placebo-controlled study. *Diabetologia* 49(Suppl. 1):511–512.

Kinghorn, A.D., ed. 2002. *Stevia: The genus* Stevia. London: Taylor & Francis.

Kinghorn, A.D., and Soejarto, D.D. 1991. Stevioside. In *Alternative sweeteners,* 2nd edition (revised and expanded), eds. L. O'Brien Nabors and R.C. Gelardi, 157–171. New York: Marcel Dekker.

Kinghorn, A.D., Wu, C.D., and Soejarto, D.D. 2001. Stevioside. In *Alternative sweeteners,* 3rd edition (revised and expanded), eds. L. O'Brien Nabors and R.C. Gelardi, 166–183. New York: Marcel Dekker.

Kinghorn, A.D., Chin, Y.-W., Pan, L., and Jia, Z. 2010. Natural products as sweeteners and sweetness modifiers. In *Comprehensive natural product chemistry—II, vol. 3,* ed. R. Verpoorte, 269–315. Oxford, UK: Elsevier.

Koyama, E., Kitazawa, K., Ohori, Y., Izawa, O., Kakegawa, K., Fujino, A., and Ui, M. 2003a. In vitro metabolism of the glycosidic sweeteners, stevia mixture and enzymatically modified stevia in the human intestinal microflora. *Food Chem Toxicol* 41:359–374.

Koyama, E., Sakai, N.Y. Ohori, K., Kakegawa, K., Izawa, O., Fujino, A., and Ui, M. 2003b. Absorption and metabolism of the glycosidic sweeteners stevia mixture and their aglycone, steviol, in rats and humans. *Food Chem Toxicol* 41:875–883.

Lavia, A., and Hill, J. 1972. Sweeteners with masked saccharin aftertaste. French patent 2,087,843.

Maki, L.C., Curry, L.L., Carakostas, M.C., Tarka, S.M., Reeves, M.S., Farmer, M.V., McKenney, J.M., et al. 2008a. The hemodynamic effects of rebaudioside A in healthy adults with normal and low-normal blood pressure. *Food Chem Toxicol* 46(Suppl. 7S):S40–S46.

Maki, K.C., Curry, L.L., Reeves, M.S., Toth, P.D., McKenney, J.M., Farmer, M.V., Schwartz, S.L., et al. 2008b. Chronic consumption of rebaudioside A, a steviol glycoside, in men and women with type 2 diabetes mellitus. *Food Chem Toxicol* 46(Suppl. 7S):S54–S60.

Matsui, M., Matsui, K., Kawasaki, Y., Oda, Y., Nogushi, T., Kitagawa, Y., and Sawada, M. 1996. Evaluation of the genotoxicity of stevioside and steviol using six *in-vitro* and one *in-vivo* mutagenicity assays. *Mutagenesis* 11:573–579.

Mayhew, D.A., Comer, C.P., and Stargel, W.W. 2003. Food consumption and body weight changes with neotame, a new sweetener with intense taste: Differentiating effects of palatability from toxicity in dietary safety studies. *Regul Toxicol Pharmacol* 38:124–143.

Mazzei-Planas, G., and Kuc, J. 1968. Contraceptive properties of *Stevia rebaudiana. Science* 162:1007.

McNeil Nutritionals, LLC. 2008. Generally Recognized As Safe (GRAS) notification for the use of *Stevia Rebaudiana* (Bertoni) Bertoni. GRAS 275. http://www.accessdata.fda.gov/scripts/fcn/fcnNavigation.cfm?rpt=grasListing.

Melis, M.S. 1999. Effects of chronic administration of *Stevia rebaudiana* on fertility in rats. *J Ethnopharmacol* 67:157–161.

Mori, N., Sakanoue, M., Takeuchi, M., Simpo, K., and Tanabe, T. 1981. Effect of stevioside on fertility in rats. *Shokuhin Eiseigaku Zasshi* 22:409–414.

Nakajima, M. 2000a. Chromosome aberration assay of rebaudioside A in cultured mammalian cells. Test Number 5001, Biosafety Research Center, Japan. (Unpublished report cited in reference; JECFA 2005).

Nakajima, M. 2000b. Micronucleus test of rebaudioside A in mice. Test Number 5002, Biosafety Research Center, Japan. (Unpublished reported cited in reference; JECFA 2005).

Nakayama, K., Kasahara, D., and Yamamoto, F. 1986. Absorption, distribution, metabolism and excretion of stevioside in rats. *Shokuhin Eiseigaku Zasshi* 27:1–8.

Nikiforov, A.I., and Eapen, A.K. 2008. A 90-day oral (dietary) toxicity study of rebaudioside A in Sprague-Dawley rats. *Int J Toxicol* 27:65–80.

Nunes, A.P.M., Ferreira-Machado, S.C., Nunes, R.M., Dantas, F.J.S., De Mattos, J.C.P., and Caldeira-de-Araújo, A. 2007. Analysis of genotoxicity potential of stevioside by comet assay. *Food Chem Toxicol* 45:662–666.

Oliveira-Filho, R.M., Uehara, O.A., Minetti, C.A.S.A., and Valle, L.B. 1989. Chronic administration of aqueous extract of *Stevia rebaudiana* (Bert.) Bertoni in rats: Endocrine effects. *Gen Pharmacol* 20:187–191.

Pariza, M., Ponakala, S., Gerlat, P., and Andress, S. 1998. Predicting the functionality of direct food additives. *Food Technol* 52:56–60.

Paul, T. 1921. Physical chemistry of foodstuffs. V. Degree of sweetness of sugars. *Chem Z* 45:38–39.

Payzant, J.D., Laidler, J.K., and Ippolito, R.M. 1999. Method of extracting selected sweet glycosides from *Stevia rebaudiana* plant. US patent 5,962,678.

Pezzuto, J.M., Compadre, C.M., Swanson, S.M., Nanayakkara, N.P.D., and Kinghorn, A.D. 1985. Metabolically-activated steviol, the aglycone of stevioside, is mutagenic. *Proc Natl Acad Sci USA* 82:2478–2482.

Prakash, I., DuBois, G.E., Jella, P., King, G.A., San Miguel, R.I., Sepcic, K.H., Weerasinghe, D.K., and White, N.R. 2007. Natural high-potency sweetener compositions with improved temporal profile and/or flavor profile, methods for their formulation, and uses. US patent application 20070128311.

Prakash, I., DuBois, G.E., Clos, J.F., Wilkens, K.L., and Fosdick, L.E. 2008. Development of rebiana, a natural, non-caloric sweetener. *Food Chem Toxicol* 46:S75–S82.

Renwick, A.G. 2008. The use of a sweetener substitution method to predict dietary exposures for the intense sweetener rebaudioside A. *Food Chem Toxicol* 46(Suppl. 7S):S61– S69.

Renwick, A.G., and Tarka, S.M. 2008. Microbial hydrolysis of steviol glycosides. *Food Chem Toxicol* 46(Suppl. 7S):S70–S74.

Roberts, A., and Renwick, A.G. 2008. Comparative toxicokinetics and metabolism of rebaudioside A, stevioside and steviol in rats. *Food Chem Toxicol* 46(Suppl. 7S):S31–S39.

Sasaki, Y.F., Kawaguchi, S., Kamaya, A., Ohshita, M., Kabasawa, K., Iwama, K., Taniguchi, K., and Tsuda, S. 2002. The comet assay with 8 mouse organs: Results with 39 currently used food additives. *Mutat Res* 519:103–119.

Schiffman, S., Booth, B., Carr, B., Losee, M., Sattely-Miller, E., and Graham, B. 1995. Investigation of synergism in binary mixtures of sweeteners. *Brain Res Bull* 38:105–120.

Scott, D. 1971. Saccharin-dipeptide sweetening compositions, British patent 1,256,995.

Sekihashi, K., Saitoh, H., and Sasaki, Y.F. 2002. Genotoxicity studies of stevia extract and steviol by the comet assay. *J Toxicol Sci* 27:1–8.

Soejarto, D.D. 2002a. Botany of *Stevia* and *Stevia rebaudiana*. In *Stevia: The genus* Stevia, ed. A.D. Kinghorn, 18–39. London: Taylor & Francis.

Soejarto, D.D. 2002b. Ethnobotany of *Stevia* and *Stevia rebaudiana*. In *Stevia: The genus* Stevia, ed. A.D. Kinghorn, 40–67. London: Taylor & Francis.

Suttajit, M., Vinitketkaummuen, U., Meevatee, U., and Buddhasukh, D. 1993. Mutagenicity and human chromosomal effect of stevioside, as sweetener from *Stevia rebaudiana* Bertoni. *Environ Health Perspect* (Suppl.) 101:53–56.

Tanaka, O. 1997. Improvement of taste of natural sweeteners. *Pure Appl Chem* 69:675–683.

Toskulkao, C., Chaturat, L., Temcharoen, P., and Glinsukon, T. 1997. Acute toxicity of stevioside, a natural sweetener, and its metabolite, steviol, in several animal models. *Drug Chem Toxicol* 20:31–44.

Toyoda, K., Matsui, H., Shoda, T., Uneyama, C., Takada, O., and Takahashi, M. 1997. Assessment of the carcinogenicity of stevioside in F344 rats. *Food Chem Toxicol* 35:597–603.

Usami, M., Sakemo, K., Kawashima, K., Tsuda, M., and Ohno, Y. 1995. Teratology study of stevioside in rats. *Eisei Shikenjo Hokoku* 113:31–35.

Verdi, R.J., and Hood, L.L. 1993. Advantages of alternative sweetener blends. *Food Technol* 47:94–102.

Wallin, H. 2004. Steviol glycosides, chemical and technical assessment. UN Food and Agricultural Organization, Rome.

Walters, E. 1993. High intensity sweetener blends. *Food Prod Design* 3:86–92.

Wasuntarawat, C., Temcharoen, P., Toskulkao, C., Munkornkarn, P., Suttajit, M., and Glinsukon, T. 1998. Developmental toxicity of steviol, a metabolite of stevioside, in the hamster. *Drug Chem Toxicol* 21:207–212.

Wheeler, A., Boileau, A.C., Winkler, P.C., Compton, J.C., Prakash, I., Jiang, I., and Mandarino, D.A. 2008. Pharmacokinetics of rebaudioside A and stevioside after single oral doses in healthy men. *Food Chem Toxicol* 46(Suppl. 7S):S54–S60.

Williams, G. 2007. Letter to the editor. *Food Chem Toxicol* 45:2597–2598.

Williams, L.D., and Burdock, G.A. 2009. Genotoxicity studies on high-purity rebaudioside A preparation. *Food Chem Toxicol* 47:1831–1836.

Wu, C.D., Johnson, S.A., Srikantha, R., and Kinghorn, A.D. 1998. Intense natural sweeteners and their effects on cariogenic bacteria. Annual Meeting of the American Association for Dental Research, Minneapolis, MN, March 4–7.

Xili, L., Chomgjiany, B., Eryi, X., Reiming, S., Yuengming, W., Haodong, S., and Zhiyian, H. 1992. Chronic oral toxicity and carcinogenicity study of stevioside in rats. *Food Chem Toxicol* 30:957–965.

Yodyingyuad, V., and Bunyawong, S. 1991. Effect of stevioside on growth and reproduction. *Hum Reprod* 6:158–165.

Chapter 12

Sucralose

V. Lee Grotz, Samuel Molinary, Robert C. Peterson,
Mary E. Quinlan, and Richard Reo

Contents

Introduction

The high quality sweetness of sucralose (SPLENDA® Brand Sweetener), was discovered as a consequence of a research program conducted at Queen Elizabeth College at the University of London during the 1970s (Hough and Khan 1978). Hough and his colleagues, with the support of Tate & Lyle PLC, showed that the selective chlorination of sugar could result in intensely sweet compounds (Jenner 1989). This discovery led to a series of studies, which ultimately identified sucralose (1,6-dichloro-1,6-dideoxy-β-D-fructofuranosyl-4-chloro-4-deoxy-α-D-galactopyranoside) (Figure 12.1) as a most promising ideal sweetener.

Sucralose is made from sucrose (common table sugar) by the selective replacement of three hydroxyl groups with chlorine, a process that occurs with inversion of configuration at the 4 position of the galacto-analog. The result is a sweetener that is remarkably different from sucrose in its sweetness intensity and stability, although it has a comparable taste quality. On a weight-for-weight basis, sucralose is about 600 times sweeter than sugar. Although sucralose is not a sugar, it has a pleasant, sugar-like, sweet taste with no unpleasant aftertaste. Like sugar, sucralose is a white, crystalline, nonhygroscopic, free-flowing powder. It is highly soluble in water, ethanol and methanol, and has a negligible effect on the pH of solutions. Sucralose exerts negligible lowering of surface tension.

The glycosidic linkage of sucralose is significantly more resistant to acid and enzymatic hydrolysis than that of the parent molecule, sucrose. The resistance of the glycosidic bond is responsible for the inability of mammalian species to digest the molecule and metabolize it as an energy source. Therefore, sucralose is noncaloric. This resistant bond is also the reason that oral microorganisms responsible for dental plaque formation cannot utilize sucralose for energy, consistent with the fact that sucralose is noncariogenic.

The safety of sucralose has been established through one of the most thorough testing programs ever performed. Scientists and regulators around the world have evaluated the significant breadth of data that comprises this program, which includes more than 100 scientific studies, representing over twenty years of research. The studies conducted were designed to determine the biological fate of sucralose and to evaluate its potential to have any adverse effects. Sucralose is safe for all individuals. It is permitted for use in more than 100 countries and is consumed daily by people around the world. Sucralose provisions for use have been adopted by the Codex Alimentarius Commission and an acceptable daily intake (ADI) of 15 mg/kg/day was established by the Joint FAO/WHO Expert Committee on Food Additives (JECFA).

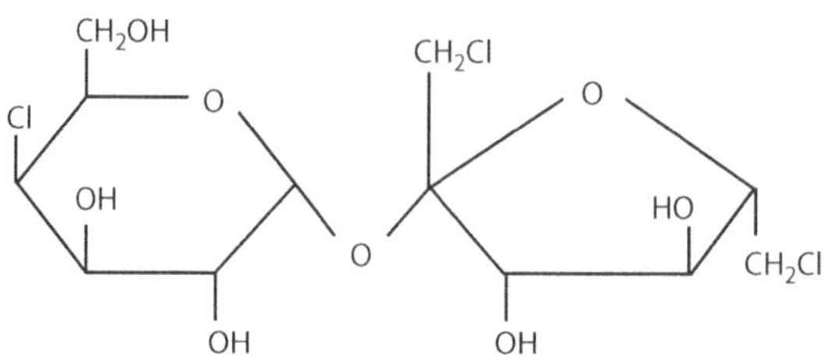

Figure 12.1   Molecular structure of sucralose.

Sensory Characteristics of Sucralose

Sweetness Potency

The potency of sucralose is in the order of 600 times that of sucrose but, as with other high potency sweeteners, this factor varies depending on the level of sucralose being used and the influence of other ingredients. Figure 12.2 shows the dose response curve for sucralose in an unflavored acidified system at pH 3.0.

This demonstrates that sucralose is approximately 750 times sweeter than sucrose at a concentration equisweet to a 3% sucrose solution. At 6% sugar equivalence, sucralose is around 600 times sweeter than sucrose and at 9%–10% sugar equivalence the potency is about 480 times that of sucrose. As a general guideline, sucralose is regarded as being 600 times sweeter than sucrose.

Due to the intense sweetness of sucralose, an extremely small amount is necessary to achieve the desired sucrose iso-sweetness level. As an example, approximately 200 mg of sucralose per liter (200 ppm) is needed to provide the sweetness for beverages that would typically contain 9% or 10% sugar. It is important to note that the actual use levels of sucralose are influenced by characteristics such as the pH, temperature, other ingredients and the viscosity of the food or beverage system in which sucralose is incorporated. Consequently, the amount of sucralose used should be optimized for each product.

Temporal Qualities

Temporal properties of sweeteners, such as time of sweetness onset, time to maximum sweetness intensity, and rate of sweetness decay, are of interest to food and beverage formulators. Time intensity measurements have demonstrated that the temporal sweetness profile of sucralose is similar to that of sugar (Figure 12.3). At 5% sugar equivalence, typical of sweetened tea and coffee, sucralose displays a rapid onset of sweetness and similar sweetness duration to sucrose.

Taste Quality

The taste quality of sucralose is demonstrated by the myriad of products around the world that are sweetened with it. Additionally, sucralose is available globally in consumer tabletop forms

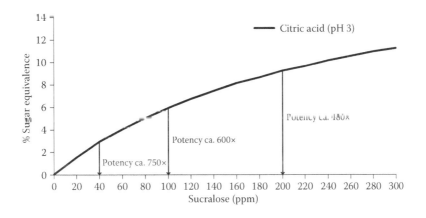

Figure 12.2　Sucralose sweetness potency as a function of concentration.

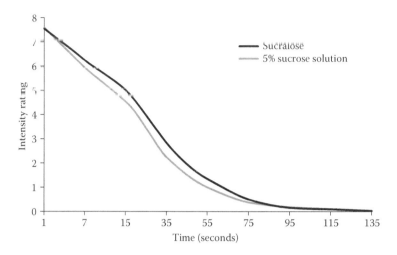

Figure 12.3 Time intensity profile at 5% sugar equivalence.

(e.g., granular, tablet, packet, etc.) as a sugar replacement. Scientifically, the flavor attributes of isosweet concentrations of aqueous solutions of sucralose and sucrose, at a 9% sucrose equivalence, were identified and quantified by trained sensory panelists using a sip and spit procedure. Both sweet and nonsweet aftertastes were evaluated. The experimental series was repeated twice and the taste scores averaged across replicates. The flavor profiles of sucrose and sucralose are presented in Figure 12.4. This comparison demonstrates that sucralose is a high quality sweetener with no bitter or metallic aftertastes, although the sucralose sweetness displays a slightly longer duration than sucrose at this sweetness level. However, the food or beverage system in which sucralose is used and the dose level, impact these flavor attributes, and adaptations to the formulation can be made to optimize the flavor characteristics (Wiet and Beyts 1988).

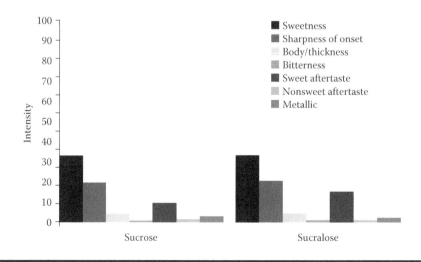

Figure 12.4 Flavor profiles of sucrose and sucralose at 9% sugar equivalence.

Safety and Regulatory Status of Sucralose

Sucralose is one of the most tested food ingredients available today. It has been found safe for its intended use by health and food safety experts from around the world (Grotz and Munro 2009). Sucralose is permitted for use in more than 100 countries. It is used in thousands of food and beverage products worldwide and is safe for use over an entire lifetime.

More than 100 scientific studies conducted to describe the safety of sucralose, represent a methodical, intentional and broad-range research program, as required by prominent health and food safety authorities. In the United States, for example, the standard set for approval of any new food ingredient, quite different from the standards set for drug products, is "reasonable *certainty* of no harm" (Rulis and Levitt 2009, 22; emphasis added). As such, research studies conducted in describing the safety of a new food ingredient must be rigorous and comprehensive.

The studies conducted to assess the safety of sucralose investigated possible effects with short-term exposures and long-term, essentially lifetime, exposures, from conception to advanced adulthood (Grotz and Munro 2009). Many of the sucralose research studies utilized very high daily doses of sucralose, doses far greater than what could be expected to be consumed, to understand margins for safe use. Use of such high daily doses was particularly employed in the core sucralose research studies, in accordance with international standards for studies designed to determine potential risk (Rulis and Levitt 2009).

Key Laboratory Studies

Metabolism studies show that the biological fate of sucralose is wholly consistent with its physico-chemical properties (Grotz and Munro 2009). Sucralose is a small (MW less than 400), poly-hydroxylated, highly-water soluble, poorly lipid-soluble, and relatively non-reactive molecule. While it is a substituted disaccharide, mammalian and yeast disaccharidase studies show that it is not susceptible to cleavage at the glycosidic bond to yield its constituent monosaccharide-like moieties. In the body, sucralose is not able to be digested for energy. Further, sucralose is relatively poorly absorbed from the gastrointestinal tract (GIT) (Grice and Goldsmith 2000; John et al. 2000; Roberts et al. 2000; Sims et al. 2000; Wood et al. 2000). The GIT is replete with enzymes that facilitate rapid uptake of monosaccharides, however, there is no digestion of sucralose at the level of the GIT (or elsewhere in the body). Thus, these enzymes have no impact on sucralose absorption, and specific studies show no active uptake of sucralose. About 85% of consumed sucralose is unabsorbed from the GIT and excreted unchanged in stool. A small amount (approximately 15%) of consumed sucralose is absorbed via passive diffusion across the GIT lumen. Consistent with sucralose being a highly water soluble, small and relatively inert substance, absorbed sucralose is distributed to essentially all tissues, and readily excreted in urine. There is no active transport into milk, transplacentally, or across the blood-brain barrier into the central nervous system (Grice and Goldsmith 2000; Sims et al. 2000). Sucralose is not used as a source of energy. It is not dechlorinated, and there is no evidence of degradation to any smaller chlorinated compounds (Grice and Goldsmith 2000; John et al. 2000; Roberts et al. 2000; Sims et al. 2000; Wood et al. 2000). Sucralose does not bind to proteins in the body, consistent with its relatively non-reactive nature. Most absorbed sucralose, and all unabsorbed sucralose, is excreted unchanged within 24 hours. About 2%–3% of consumed sucralose undergoes common phase II biotransformation, leading to the formation of highly water-soluble, non-toxicologically significant glucuronide conjugates (Roberts et al. 2000). Studies show no evidence of cytochrome P450 (CYP450) enzyme induction (Hawkins et al. 1987; Brusick et al. 2009). Estimated half-life of sucralose (including the minor glucuronide conjugates) in humans is approximately 13 hours (Roberts et al. 2000). Consistent with its low lipid-solubility, sucralose does not bioaccumulate (McLean Baird et al. 2000; Sims et al. 2000).

The safety of sucralose may be anticipated based on its physicochemical characteristics and metabolic fate, which shows that sucralose is poorly absorbed, non-catabolized, readily excreted, and essentially non-reactive or inert (Grotz and Munro 2009).

Acute toxicity studies in mice and rats showed no adverse effects at the highest dose tested. Intakes of up to 10000 to 16000 mg/kg, respectively, were without adverse effects (Goldsmith 2000). These amounts are the sweetness-equivalent of 990 to 1584 pounds of sugar for a 165 pound (75 kg) person.

Similarly, the results of subchronic and chronic toxicity studies in relevant laboratory species showed no toxicity with an average intake of, minimally, 500 mg/kg. For a 75 kg adult, this is equivalent in sweetness to an intake of at least 40 pounds of sugar, consumed every day over the lifetime (Mann et al. 2000a, 2000b).

As with other studies of non-nutritive sweeteners and/or other poorly absorbed substances, there were some differences noted between test and control animals, which relate to, respectively, issues of decreased diet palatability and physiological (but not toxicological) effects that are consequent to unabsorbed material in the GIT (Grice and Goldsmith 2000; Mann et al. 2000b; Flamm et al. 2003; Mayhew et al. 2003). Diet containing extremely high sucralose concentrations, for example, like those used in the core sucralose toxicity studies, is less preferable to rats than is control diet. Not surprisingly, when rats are fed these treated diets, they eat less total diet than rats fed control diet. The diets with sucralose are also less nutritionally dense, given the fact that sucralose is non-nutritive. In all, treated rats had less energy intake than control rats and not surprisingly had a slightly lower, but healthy, body weight. Lower food intake and body weight was also correlated with slight decreases in certain organ weights and some related changes, however, no toxicological changes were found. Importantly, these changes are not seen when rats are fed high daily doses of sucralose by gavage, a route of administration that avoids potential diet palatability effects. The absence, in gavage studies, of effects like those seen in the dietary studies confirms that the dietary study effects are palatability-related.

In addition to the numerous acute, subchronic and chronic sucralose toxicity studies conducted, a battery of other studies were conducted to further describe the potential for sucralose to have adverse effects (Grice and Goldsmith 2000; Grotz and Munro 2009). Genotoxicity studies and two-year carcinogenicity studies in two species demonstrated that sucralose is neither genotoxic nor carcinogenic (Grice and Goldsmith 2000; Mann et al. 2000a, 2000b; Brusick et al. 2010). Two-generation reproduction studies evaluated potential effects on mating, reproduction, fertility, gestation, conception, and neonatal growth and development (Kille et al. 2000; Mann et al. 2000b). As with the carcinogenicity studies, there were no adverse effects with extremely high, repeated sucralose intakes, which were equivalent in sweetness to over 100 pounds of sugar per day for a 75 kg person. Teratology studies were also negative, confirming that sucralose does not cause birth defects (Kille et al. 2000). Sucralose was also found not to be neurotoxic or immunotoxic, in studies specifically designed to investigate the potential for such effects (Grice and Goldsmith 2000). It has also been found to be non-allergenic, which may be predicted by its physicochemical characteristics, that is, it is a small, essentially unreactive molecule that is neither a protein nor protein-based (Grotz 2008). All of these studies support the conclusion that sucralose is safe for its intended use.

Additionally, a complete battery of environmental studies was carried out as required by the US Food and Drug Administration (US FDA) (Sucralose Food Additive Petition 1987) and the Organization for Economic Co-operation and Development (OECD) Environment and Safety Programme. In addition to assessing the physical nature of events surrounding sucralose in the environment, biological studies were conducted in relevant aquatic species including fish, algae, and crustaceans. This environmental assessment was submitted as part of the food additive petition and demonstrated that trace amounts of sucralose present in receiving waters would neither bioaccumulate nor be toxic to any of the tested species. Subsequent environmental studies in plants and animals conducted by several government and academic laboratories have confirmed these results. It is concluded that the presence of sucralose has no negative impact on the environment.

Key Clinical Studies

Clinical trials are an important part of the data that describe the safety of sucralose. Studies in humans include pharmacokinetic and metabolism studies, short-term and longer-term tolerance studies, and studies in persons with diabetes and in children (Mezitis et al. 1996; Roberts et al. 2000; McLean-Baird et al. 2000; Grotz et al. 2003; Reyna et al. 2003; Rodearmel et al. 2007). Comparison of human and laboratory animal metabolic and pharmacokinetic responses to sucralose shows similarity between the species, supporting the fact that the laboratory animals used in the core sucralose toxicity studies are good surrogates for humans. Human tolerance studies show that sucralose is as well-tolerated as placebo at intakes 3–7 times the maximum estimated average daily intake, consumed daily for up to six months, by normoglycemic and diabetic subjects (Sucralose Food Additive Petition 1996; McLean Baird et al. 2000; Grotz et al. 2003). No adverse effects are attributed to sucralose consumption. Although sucralose is poorly absorbed from the GIT, unlike sugar alcohols, which are also poorly absorbed from the GIT, studies do not support GIT side effects with sucralose consumption. The absence of such an effect can be anticipated because there is no GIT fermentation of sucralose, to yield potentially disturbing gas production in the lower intestine (Grotz and Munro 2009). Additionally, total human intake is so low (estimated maximum average daily intake is less than 3 mg/kg/day) that it precludes the likelihood of GIT-disturbing (or any) osmotic effects. The absence of GIT-related effects associated with sucralose use by humans is supported by both the clinical trials conducted and the high-dose essentially lifetime rodent feeding studies (Grice and Goldsmith 2000; Mann et al. 2000a, 2000b). The collective data indicate that there are no adverse gastrointestinal effects with the use of sucralose. The potential for sucralose to have effects related to carbohydrate metabolism and/or diabetes was specifically studied in light of the disaccharide origin of the sucralose molecule. Sucralose was found to have no effect on blood glucose, insulin or hemoglobin A1c (HbA1c) levels or overall glucose control (Mezitis et al. 1996; Sucralose Food Additive Petition 1996; McLean Baird et al. 2000; Grotz et al. 2003). Significant research shows that sucralose is not recognized as a carbohydrate. Similarly, sucralose is not recognized as a sugar by blood or urine glucose monitoring devices and it has no impact on HbA1c measurement or monitoring. Thus, sucralose will have no impact on measurement of these outcomes that are important for diabetes management.

While a multitude of studies show that no-calorie sweeteners can be helpful for calorie control in programs designed to help decrease or maintain weight, the helpful utility of foods and beverages containing sucralose was also demonstrated in a lifestyle change program for families with overweight children (Rodearmel et al. 2007). This study showed that such a program can help reduce weight gain in excess of that attributable to normal growth in overweight children.

Several clinical trials show that sucralose has no potential to induce dental caries (Meyerowitz et al. 1996; Steinberg et al. 1995,1996). This is consistent with results from both dental bacteria and rodent dental caries studies, which show that sucralose does not support the growth of bacteria involved in dental caries and does not promote or induce caries formation (Mandel and Grotz 2002). The sum of the data show that sucralose is noncariogenic.

Sucralose Hydrolysis Products

As previously discussed, sucralose has excellent stability in foods and beverages. No loss of sweetness is detected in any product tested. Several prototype baking studies, utilizing radiolabeled sucralose, show no hydrolysis of sucralose to its constituent monosaccharide-like moieties or to other materials, and no loss of chlorine from sucralose is known to occur in food systems. A very small amount of hydrolysis of sucralose can occur in certain food or beverage products, depending on pH, time and temperature. The potential intake is exceedingly small. The safety of the sucralose hydrolysis products was rigorously investigated in a manner similar to the investigation of sucralose (Grice and Goldsmith 2000). No safety concerns are associated with these products potentially resulting from sucralose use (US FDA 1998, 1999).

Chemical and Stability Characteristics of Sucralose

Physicochemical Characteristics of Sucralose

Physiochemical properties, such as solubility, provide insight into how a food ingredient can be used in the food manufacturing environment. Table 12.1 provides a summary of some of the key characteristics for sucralose.

Sucralose is easy to use in a manufacturing environment. It is freely soluble in water, even at low temperatures, and exhibits negligible lowering of surface tension for dilute solutions. This means that it is unlikely to cause excessive foaming when being pumped or mixed. In addition, the

Table 12.1 Physicochemical Properties of Sucralose

Physical Form	*White Powder*
Odor	Practically odorless
Caloric content	Zero
Solubility	Freely soluble in water (28.2 g/100ml at 20°C) Insoluble in corn oil (less than 0.1 g/100 g at 20°C)
Specific optical rotation	α_{20}^{D} + 85.8 degrees
Octanol/water partition co-efficient	0.32 (20°)
Surface tension of aqueous solutions	71.8 mN/m (20°C, 0.1 g/100ml)
Melting (decomposition) point	125°C (when heated from 115°C at 5°C/min
Specific gravity (10% aqueous solution)	1.04 (20°)
Viscosity	Follows Newtonian behavior
Refractive index	Linear with respect to concentration. Degrees brix readings (20°C) give a direct measurement of sucralose concentrations in aqueous solutions containing at least 3% sucralose
pH	Ca. pH 6
Molecular weight	397.64

comparatively low viscosities observed for sucralose solutions, along with a Newtonian behavior, means that no viscosity issues are created when mixing or dispersing solutions.

Sucralose is also readily soluble in ethanol and propylene glycol, both of which are often used in food manufacturing. Sucralose is, however, insoluble in corn oil, a result consistent with its hydrophilic nature. Consequently, sucralose will always partition with the aqueous portion of a food system comprising aqueous and lipid phases (Jenner and Smithson 1989).

Although valid and precise high-performance liquid chromatography (HPLC) and gas chromatography (GC) analytical methods for the determination of sucralose are available, more rapid and simpler techniques may be useful for food manufacturers, for quality assurance and quality control purposes. The use of refractive index provides a simple means of determining the concentration of sucralose in aqueous solutions. The refractive index of sucralose solutions is linear with respect to concentration over the range 3% w/w to 50% w/w and the degrees brix (i.e., percent solids, calibrated for pure sugar in water) gives a direct measurement of the sucralose concentration. However, the presence of other soluble solids will obviously affect the reading and correction factors have to be applied if other components are present.

Chemical Interactions

In the process of chlorination of the sucrose molecule to synthesize sucralose, the primary reaction sites on the molecule are occupied, and consequently, sucralose is essentially an inert ingredient.

The reactivity of sucralose was evaluated in a model system study where 1% solutions of sucralose were prepared with 0.1% of one of the following:

Bases (niacinamide, monosodium glutamate)
Oxidizing and reducing agents (hydrogen peroxide, sodium metabisulfite)
Aldehydes and ketones (acetaldehyde, ethyl acetoacetate)
Metal salt (ferric chloride)

The compounds evaluated are typical food, beverage and pharmaceutical system components, and the classes of compounds evaluated represent the broad range of compounds typically found in food and beverage formulations. The solutions were held at 40°C for 7 days, at pH 3.0, 4.0, 5.0, or 7.0, and analyzed for sucralose content by HPLC. Table 12.2 shows the sucralose retention in each of the solutions. As can be seen from the data, there are no chemical interactions of concern. These data demonstrate that sucralose is effectively chemically inert, and therefore chemical interactions of sucralose with ingredients in formulations are not likely to occur.

Stability of the Ingredient

Due to its physicochemical characteristics and inherent stability, sucralose can be offered as an ingredient in dry or liquid form. The stability of pure dry sucralose and aqueous solutions of sucralose have been evaluated, to determine shelf life and proper storage and distribution conditions.

Stability of Pure Dry Sucralose

The shelf life of pure dry sucralose is at least two years when stored at 25°C or below. Dry sucralose is, however, sensitive to elevated temperatures. Storage at high temperatures for extended periods can result in color formation. The shelf life can also be affected by packaging materials and container head space, so care should be taken to adhere to the recommended packaging and storage conditions.

Table 12.2 Sucralose Interaction Study

| Sample | Sucralose Retention (%) After 7 Days at 40°C | | | |
	pH 3.0	pH 4.0	pH 5.0	pH 7.0
Control	100	99.8	98.9	99.3
Sucralose and hydrogen peroxide	100	—	99.7	—
Sucralose and sodium metabisulfite	99.9	—	99.9	—
Sucralose and acetaldehyde	100	—	100	—
Sucralose and ethyl acetoacetate	98.8	—	100	—
Sucralose and ferric chloride	95.9	—	98.0	—
Sucralose and niacinamide	—	100	—	100
Sucralose and monosodium glutamate	—	99.8	—	100

Stability of Sucralose Liquid Concentrate

The solubility and inherent aqueous stability of sucralose allows for a sucralose liquid concentrate for industrial use, which can provide operational efficiencies in manufacturing.

A liquid concentrate product is commercially available as a 25% (w/w) sucralose solution with a dual preservative system of 0.1% sodium benzoate and 0.1% potassium sorbate, buffered (0.02 M citrate) to maintain a pH of 4.4. No chemical or physical changes have been found in sucralose liquid concentrate product stored for five years at 68°F (20°C). Triple challenge microbiological studies on fresh and aged liquid sucralose concentrate have clearly demonstrated that the liquid ingredient will not be a vector for microbiological contamination when good manufacturing practices are observed.

This product form provides the end user with an ingredient system that is extremely stable, chemically and microbiologically, and compatible with most food unit operations.

Utilization of Sucralose in Food Applications

Stability of Sucralose in Aqueous Model Systems

To assess the potential stability of sucralose in food systems, the aqueous stability of sucralose was determined over various pHs, temperatures, and times in model systems studies.

Figure 12.5 demonstrates the stability of sucralose at 20°C over a range of pH conditions and confirms that there is no significant loss of sucralose.

Similarly, Figure 12.6 shows the stability of sucralose solutions stored at 30°C for up to one year at pH 3.0, 4.0, 6.0, and 7.5. Under these conditions there would be negligible loss at pHs 4.0 and pH 6.0, about 1% loss at pH 7.5 and less than 3% loss of sucralose at pH 3.0. This level of stability provides food manufacturers and consumers with food products that have a consistent level of sweetness throughout their shelf life since a loss of 10%–15% of a sweetener is necessary before a loss of sweetness can be detected sensorally.

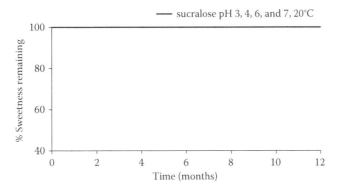

Figure 12.5   pH Stability of model aqueous systems at 20°C.

Use in Beverages

Sucralose is widely used in a variety of beverage products worldwide. Table 12.3 shows some typical use levels for different beverage products.

Having confirmed the stability of sucralose in model systems, further studies were undertaken to demonstrate its stability in formulated products. Table 12.4 gives the results for two cola products stored at 20°C while Figure 12.7 compares the stability of sucralose, acesulfame potassium (acesulfame K) and aspartame in a cola drink stored at 35°C. In both cases the sweetener levels were measured by HPLC analysis.

Sucralose-sweetened beverages maintain a consistent sweetness throughout the processing and shelf life of the product due to the remarkable stability of sucralose.

Stability to Food Manufacturing Processes

Before commercialization, the stability of sucralose to a range of typical manufacturing processes was assessed during a series of processing trials. Table 12.5 summarizes the results of these trials and again demonstrates the stability of sucralose under a variety of conditions.

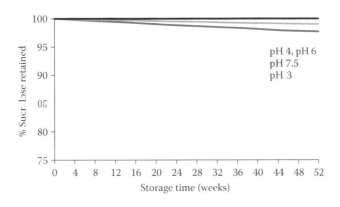

Figure 12.6   Aqueous stability of Sucralose—effect of pH at 30°C.

Table 12.3 Typical Use Levels for Sucralose in Beverage Products

Carbonated drinks	9%–11%	0.018%–0.0225%
Still drinks	7%–9%	0.012%–0.018%
Dry mix (Ready to drink)	9.5%	0.019%
Strawberry milk	4%	0.0058%

In addition, to confirm the stability of sucralose to baking, a radiolabeled stability study was performed. The objective of this study was to demonstrate the stability of sucralose in a variety of common baked goods prepared under typical baking conditions. For this study, ^{14}C-labeled sucralose was used to minimize the difficulties associated with the recovery and detection of low levels of sucralose in the presence of carbohydrates found in baked goods. Cakes, cookies, and graham crackers were selected for this study, since they represent a cross-section of common bakery ingredients and typical baking conditions encountered in the baking industry.

The baked goods were prepared under simulated baking time and temperature conditions. The sucralose was recovered by aqueous extraction, and the extracts were "cleaned up" by a methanol precipitation, prior to TLC analysis.

The total radioactivity expected for 100% recovery of the ^{14}C-labeled material from each baked product was calculated and compared to the actual level of radioactivity recovered in the aqueous extract. This comparison demonstrated that essentially 100% of the sucralose radioactivity that was added to each formulation prior to baking was recovered by the extraction of the baked goods.

With the demonstration of complete recovery of the radiolabled material, the extracts were subjected to thin-layer chromatography using two solvent systems independently as eluents. The solvent systems were selected because of their ability to separate sucralose from its potential breakdown products. The amount of radioactivity was quantitatively determined. The quantity and distribution of radioactivity contained in each baked product extract was compared with the quantity and distribution of a sucralose standard solution.

The most significant findings of this study were that no products other than sucralose were found under the conditions of this experiment, and the TLC distribution of radioactivity from the baked goods extracts corresponded almost exactly with that of the sucralose standard. These findings demonstrate that sucralose is suitable for use as a sweetener in baked goods and that it is

Table 12.4 Shelf Stability of Sucralose in Cola at 20°C

	Sucralose	
	Cola at pH 3	*Cola at pH 2.7*
Zero time	191 ppm	184 ppm
5 weeks	198 ppm	182 ppm
10 weeks	192 ppm	183 ppm
17 weeks	194 ppm	183 ppm
25 weeks	194 ppm	184 ppm

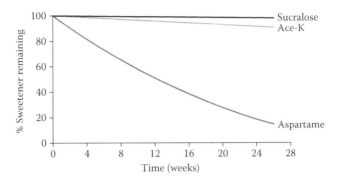

Figure 12.7 Sweetener stability in cola (pH 3.2) at 35°C.

stable in baking, with no evidence of any sucralose breakdown or reaction with other ingredients (Barndt and Jackson 1990).

The stability data in model systems was used to predict the efficacy of using sucralose in actual food products. Since commercialization, sucralose has been used to formulate a broad variety of food products that maintain their sweetness throughout processing, distribution and shelf life.

Use in Food Products

The taste, stability and physicochemical properties of sucralose mean that it is a very versatile sweetener suitable for use in a wide variety of food products. This is borne out by the number and diversity of products in which it is used around the world. Sucralose can be used to replace the sweetness provided by nutritive and non-nutritive sweeteners in virtually all food products. However, due to its high potency, only low levels are required, as shown in Table 12.6. Consequently, sucralose does not give the bulk or textural characteristics that sugar provides in

Table 12.5 Sucralose Stability to Food Manufacturing Processes

	pH	Process Conditions	Sucralose Post Processing
Pasteurization			
Tropical Beverage	2.8	93°C for 24 seconds	100%
Tomato Ketchup	3.8	93°C for 51 minutes	100%
Canned Fruit—Pears	3.3	100°C for 12 minutes	100%
Sterilization			
Beans in Sauce	5.6	121°C for 80 minutes	100%
UHT			
Dairy Dessert	6.7	140°C for 15 seconds	100%
Vanilla Milk	6.5	141°C for 3.5 seconds	100%

Table 12.6   Typical Use Levels for
Sucralose in Food Products

Category	Sucralose Use Level (%)
Yogurts	0.010–0.015
Ice Cream	0.010–0.014
Canned Fruit	0.012–0.025
Boiled Sweets	0.020–0.030
Chewing Gum	0.100–0.300
Biscuits	0.010–0.035

products such as confectionery, baked goods or ice cream. In these applications, sucralose can be used in combination with common bulking agents such as polydextrose, soluble fibers and polyols.

Extensive research in support of commercialization of sucralose demonstrates the utility of sucralose in sweetening high quality food and beverage products. The incorporation of sucralose into thousands of products in countries around the world substantiates the data developed in those studies. The optimal sucralose sweetness level for any product will vary depending upon the other ingredients in the formulation, including flavors.

Analytical Methodology

Significant efforts have been made to develop analytical methods for the recovery and analysis of sucralose from food products. HPLC has been shown to be the most effective analytical method for precision and accuracy in the analysis of sucralose.

As with much of food analysis, extraction methodology is the key to acceptable recoveries. The solubility of sucralose in water and in alcohol facilitates the selective extraction of sucralose from food products. When required, additional sample clean up and concentration of sucralose is achieved using solid phase extraction (SPE) techniques. Nonpolar C-18 or Alumina SPE cartridges have been most effective in a majority of the applications evaluated. An Alumina SPE cartridge has been most effective for dairy products.

Reversed-phase HPLC using a C-18 column, with a mobile phase of 70% water and 30% methanol, is used in most analyses. Occasionally, acetonitrile is used in place of methanol to achieve a specific separation. Injection volumes typically in the range of 100–150 μl are used, but they vary with the concentration of sucralose in the sample being evaluated.

Although there are several options regarding detection, refractive index is typically the most appropriate. Refractive index provides a detection system that has fewer problems with interference than ultraviolet detection and can be used effectively when temperature is controlled (Mulligan et al. 1988).

Conclusion

Sucralose is a high-quality, high-potency, noncaloric sweetener. While not a sugar, sucralose is derived by the selective chlorination of sucrose and has demonstrated sugar-like taste. The resulting ingredient is extraordinarily stable, with physicochemical characteristics that permit its use in a wide variety of applications. Prominent health and food safety authorities confirm that sucralose is safe for its intended use. The availability of sucralose provides the food, beverage, and pharmaceutical industry with a unique opportunity to improve existing products and develop totally new products that will meet the ever-growing consumer demand for good-tasting, lower-sugar, high-quality products.

References

Barndt, R.L., and Jackson, G. 1990. Stability of sucralose in baked goods. *Food Technology* 44:62–66.

Brusick, D., Grotz, V.L., Slesinski, R., Kruger, C.L., and Hayes, A.W. 2010. The absence of genotoxicity of sucralose. *Food and Chemical Toxicology* 48(2010):3067–3072.

Brusick, D., Williams, G., Kille, J., Gallo, M., Hayes, A.W., Pi-Sunyer, F.X., Burks, W., Williams, C., and Borzelleca, J.F., 2009. Expert panel report on a study of Splenda in male rats. *Regulatory Toxicology and Pharmacology* 55:612.

Flamm, W.G., Blackburn, G.L., Comer, C.P., Mayhew, D.A., and Stargel, W.W. 2003. Long-term food consumption and bodyweight changes in neotame safety studies are consistent with the allometric relationship observed for other sweeteners and during dietary restrictions. *Regulatory Toxicology and Pharmacology* 38:144–156.

Goldsmith, L.A. 2000. Acute and subchronic toxicity of sucralose. *Food and Chemical Toxicology* 38(Supplement 2):S53–S69.

Grice, H.C., and Goldsmith, L.A. 2000. Sucralose—an overview of the toxicity data. *Food and Chemical Toxicology* 38(Supplement 2):S1–S6.

Grotz, V.L. 2008. Sucralose and migraine. *Headache* 48:164–165.

Grotz, V.L., Henry, R.R., McGill, J.B., Prince, M.J., Shamoon, H., Trout, R., and Pi-Sunyer, X.F. 2003. Lack of effect of sucralose on glucose homeostasis in subjects with type 2 diabetes. *Journal of the American Dietetic Association* 103:1607–1612.

Grotz, V.L., and Munro, I.C. 2009. An overview of the safety of sucralose. *Regulatory Toxicology and Pharmacology* 55:1–5.

Hawkins, D.R., Wood, S.W., Waller, A.R., and Jordan, M.C. 1987. Enzyme induction studies of TGS and TGS-HP in the rat. Unpublished report from Huntingdon Research Centre, Huntingdon, UK (1987). Submitted to the World Health Organization by Tate & Lyle, cited in JECFA.

Hough, L., and Khan, R.A. 1978. Intensification of sweetness. *Trends in Biochemical Sciences* 3:61–63.

Jenner, M.R. 1989. Sucralose: Unveiling its properties and applications. In *Progress in sweeteners*, ed. T. Grenby,121–141(Chapter 5). London: Elsevier.

Jenner, M.R., and Smithson, A. 1989. Physicochemical properties of the sweetener sucralose. *Journal of Food Science* 54:1646–1649.

John, B.A., Wood, S.G., and Hawkins, D.R. 2000. The pharmacokinetics and metabolism of sucralose in the mouse. *Food and Chemical Toxicology* 38(Supplement 2):S107–S110.

Kille, J.W., Tesh, J.M., McAnulty, P.A., Ross, F.W., Willoughby, C.R., Bailey, G.P., Wilby, O.K., and Tesh, S.A. 2000. Sucralose: Assessment of teratogenic potential in the rat and rabbit. *Food and Chemical Toxicology* 38(Supplement 2):S43–S52.

Mandel, I.D., and Grotz, V.L. 2002. Dental considerations in sucralose use. *Journal of Clinical Dentistry* 13:116–118.

Mann, S.W., Yuschak, M.M., Amyes, S.J., Aughton, P., and Finn, J.P. 2000a. A carcinogenicity study of sucralose in the CD-1 mouse. *Food and Chemical Toxicology* 38(Supplement 2):S91–S98.

Mann, S.W., Yuschak, M.M., Amyes, S.J., Aughton, P., and Finn, J.P. 2000b. A combined chronic toxicity/carcinogenicity study of sucralose in Sprague–Dawley rats. *Food and Chemical Toxicology* 38(Supplement 2):S71–S89.

Mayhew, D.A., Comer, C.P., and Stargel, W.W. 2003. Food consumption and bodyweight changes with neotame, a new sweetener with intense taste: Differentiating effects of palatability from toxicity in dietary safety studies. *Regulatory Toxicology and Pharmacology* 38:124–143.

McLean Baird, I., Shephard, N.W., and Merritt, R.J. 2000. Repeated dose study of sucralose tolerance in human subjects. *Food and Chemical Toxicology* 38(Supplement 2):S123–S130.

Mezitis, N.H., Maggio, C.A., Koch, P., Quddoos, A., Allison, D.B., and Pi-Sunyer, F.X. 1996. Glycemic effect of a single high oral dose of the novel sweetener sucralose in patients with diabetes. *Diabetes Care* 19:1004–1005.

Mulligan, V., Quinlan, M.F., and Jenner, M.R. 1988. General Procedures for the Analytical Determination of the High Intensity Sweetener Sucralose in Food Products. 49th Annual Meeting of the Institute of Food Technologists, New Orleans, LA.

Meyerowitz, C., Syrrakou, E.P., and Raubertas, R.F. 1996. Effect of sucralose – alone or bulked with malto-dextrin and/or dextrose—on plaque pH in humans. *Caries Research* 30:439–444.

Reyna, N.Y., Cano, C., Bermudez, V.J., Medina M.T., Souki, A.J., Ambard M., Nunez, M., Ferrer M.A., and Inglett, G.E. 2003. Sweeteners and beta glucans improve metabolic and anthropometrics variables in well controlled type 2 diabetic patients. *American Journal of Therapeutics* 10:438–443.

Roberts, A., Renwick, A.G., Sims, J., and Snodin, D.J. 2000. Sucralose metabolism and pharmacokinetics in man. *Food and Chemical Toxicology* (Supplement 2):S31–S41.

Rodearmel, S.J., Wyatt, H.R., Stroebele, N., Smith, S.M., Ogden, L.G., and Hill, J.O. 2007. Small changes in dietary sugar and physical activity as an approach to preventing excessive weight gain: The America on the move family study. *Pediatrics* 120:e869–e879.

Rulis, A.M., and Levitt, J.A. 2009. FDA's food ingredient approval process. *Regulatory Toxicology and Pharmacology* 53:20–31.

Sims, J., Roberts, A., Daniel, J.W., and Renwick, A.G. 2000. The metabolic fate of sucralose in rats. *Food and Chemical Toxicology* 38(Supplement 2):S115–S121.

Steinberg, L.M., Odusola, F., Yip, J., and Mandel, I.D. 1995. Effect of aqueous solutions of sucralose on plaque pH. *American Journal of Dentistry* 8:209–211.

Steinberg, L.M., Odusola, F., and Mandel, I.D. 1996. Effect of sucralose in coffee on plaque pH in human subjects. *Caries Research* 30:138–142.

Sucralose Food Additive Petition (FAP 7A3987), February 9, 1987.

Sucralose Food Additive Petition (FAP 7A3987), August 16, 1996. pp. 1–357. A 12-week study of the effect of sucralose on glucose homeostasis and HbA1c in normal healthy volunteers. On file at: Center for Food Safety and Applied Nutrition, US FDA.

US FDA (US Food and Drug Administration). 1998. Food additives permitted for direct addition to food for human consumption: Sucralose [21CFR Part 172; Docket No. 87F-0086]. Federal Register US 63:16417–16433.

US FDA (US Food and Drug Administration). 1999. Food additives permitted for direct addition to food for human consumption: sucralose [21CFR Part 172; Docket No. 99F-0001]. Federal Register US 64:43908-43909.

Wiet, S.G., and Beyts, P.K. 1988. Psychosensory Characteristics of the Sweetener Sucralose, 49th Annual Meeting of the Institute of Food Technologists, New Orleans, LA.

Wood, S.G., John, B.A., and Hawkins, D.R., 2000. The pharmacokinetics and metabolism of sucralose in the dog. *Food and Chemical Toxicology* 38(Supplement 2):S99–S106.

Chapter 13

Tagatose (D-tagatose)

Christian M. Vastenavond, Hans Bertelsen,
Søren Juhl Hansen, Rene Soegaard Laursen,
James Saunders, and Kristian Eriknauer

Contents

Introduction

D-tagatose is a low calorie bulk sweetener with the following properties:

- It has 92% the sweetness of sucrose.
- It has a reduced caloric value.
- It is non-cariogenic.
- It is a prebiotic.
- It is a flavor enhancer.

D-tagatose, or tagatose, is a ketohexose in which the fourth carbon is chiral and is a mirror image of the respective carbon atom of the common D-sugar, fructose. The CAS number for D-tagatose is 87-81-0 and the empirical formula is $C_6H_{12}O_6$. The molecular weight of D-tagatose is 180.16. The structural formula for D-tagatose is depicted in Figure 13.1, along with that of D-fructose. Tagatose is a naturally occurring low-calorie bulk sweetener. It occurs in Sterculia

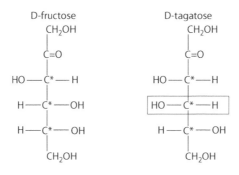

Figure 13.1   The fourth chiral carbon is a mirror image of fructose.

setigera gum, a partially acetylated acidic polysaccharide (Hirst et al. 1949). D-tagatose is also found in heated cows milk, produced from lactose (Troyano et al. 1992), and occurs in various other dairy products as well.

As shown in Table 13.1, D-tagatose occurs naturally in several food products, but not in sufficient quantities for commercialization (Petersen 2006).

Based on a similarity in sweetness and physical bulk to sucrose, D-tagatose is intended to be used as a reduced-calorie bulk sweetener and sugar replacer in ready-to-eat cereals, diet soft drinks, frozen yogurt/nonfat ice cream, soft confectionery, chocolate confectionery, hard confectionery, bakery products, frosting, and chewing gum.

Biospherics Inc. of Beltsville, Maryland, patented in 1989 the production of D-tagatose and in 1991 the use of D-tagatose in food products (Beadle et al. 1989). In 1996, MD Foods Ingredients

Table 13.1   Natural Occurence of Tagatose in Several Food Products

Food	mg/kg	Source
Apples	3500	Eurofins, 2005
Pineapples	1800	Eurofins, 2005
Oranges	1500	Eurofins, 2005
Cranberry concentrate	800	Eurofins, 2004
Raisins	700	Eurofins, 2004
Dates	700	Eurofins, 2004
Whole wheat	100	Eurofins, 2005
Dried white beans	100	Eurofins, 2005
Tropical date tree	30% of sugar	Biospherix Incorporated
Sterilized cows milk	2000–3000	Troyana ef aJ.
Powdered cows milk	800	Richards and Chandrasekhara
Hot cocoa with milk	1 40–1000	Biospherix Incorporated
Ultra high temperature milk	5	Biospherix Incorporated
Enfamil® infant formula	23	Biospherix Incorporated
BA® nature yoghurt	29	Biospherix Incorporated
Roquefort cheese	20	Biospherix Incorporated
Feta cheese	17	Biospherix Incorporated
Gjetost cheese	15	Biospherix Incorporated
Parmesan cheese	10	Biospherix Incorporated
Lactobacillus and *Streptococcus* metabolite	Variable range to be determined	Biospherix Incorporated

Source: From Petersen, S.U. Tagatose. Sweeteners and Sugar Alternatives in Food Technology, 262–294, 2006.

amba of Denmark bought the exclusive license to all patents and know-how pertaining to tagatose in foods and beverages, assuming responsibility for the production and commercialization of the sweetener. MD Foods Ingredients amba, set up a joint production company (50/50) with the German sugar group Nordzücker in Nordstemmen under the name of SweetGredients. SweetGredients Gmbh and Co. KG is a joint venture company of Arla Foods Ingredients amba and Nordzucker AG.

Notwithstanding the progress made on the introduction of D-tagatose in the world market, the costs of the chemical manufacturing process for D-tagatose crystals were prohibitive. In 2006, SweetGredients decided to close down the manufacturing of D-tagatose in Nordstemmen, Germany.

In June 2006, Nutrilab NV., the ingredients manufacturing arm of the Belgian Damhert group, entered negotiations with SweetGredients/Arla to buy the D-tagatose stock of SweetGredients Gmbh and Co KG., pending the approval of its enzymatic D-tagatose production process. In 2007, Nutrilab NV. purchased the remaining stock of D-tagatose. In this way, Damhert and Nutrilab NV. were able to perform preliminary experiments showing the feasibility to develop functional food products with tagatose as a sweetener, bringing it to the market and compiling market information on its acceptability and impact on production, taste, and distribution costs (trademark Tagatesse®).

On June 30, 2010, the High Health Council authorized the enzymatic production process filed by Nutrilab NV. in Brussels. The amendment on the EU novel food approval on D-tagatose was published December 14, 2005. Full scale industrial production is expected in mid 2011.

Process

Chemical Pathway

The chemical production of D-tagatose occurs in a stepwise manner, starting from the raw material lactose, which is then altered with the use of enzymes and various fractionating, isomerization and purification techniques. The food-grade lactose is hydrolyzed to galactose and glucose by passing the solution through an immobilized lactase column.

The sugar mixture from the enzyme hydrolysis is fractionated by chromatography. The chromatographic separation of glucose and galactose is essential and is similar to the normal industrial separation of glucose and fructose, using the same type of United States Food and Drug Administration (US FDA)-approved calcium-based cationic resins. The galactose fraction from chromatography is converted to D-tagatose under alkaline conditions by adding a suspension of technical-grade $Ca(OH)_2$ and optionally, a technical-grade catalyst, $CaCl_2$. The reaction is stopped by adding technical-grade sulfuric acid (H_2SO_4). A limited number of side reactions is observed as part of the isomerization process of D-galactose. The reactions are well known and are typical of those that occur between all common hexose monosaccharides (like D-fructose and D-glucose) and any source of hydroxide ion. D-tagatose is stable under the conditions of the isomerization process.

Upon removal of the gypsum formed, the resulting filtrate is further purified by means of demineralization and chromatography. The purified D-tagatose solution is then concentrated and crystallized to give a white crystalline product (greater than 99% pure) (Figure 13.2).

Enzymatic Pathway

Production of Tagatose from D-Galactose

D-tagatose is produced through enzymatic isomerization of D-galactose. Galactose is found in the highest concentrations in lactose. Lactose is an O-β-D-galactopyranosyl-1,4-D-glucopyranose

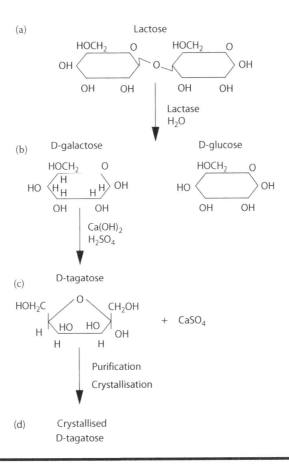

(a) Lactose

(b) D-galactose   D-glucose

Lactase
H_2O

$Ca(OH)_2$
H_2SO_4

(c) D-tagatose

$+ \; CaSO_4$

Purification

Crystallisation

(d)   Crystallised
D-tagatose

Figure 13.2 The two step tagatose production process.

which shows mutarotation and is present in a concentration of 4%–6% in milk. During hydrolysis, equimolar amounts of D-glucose and D-galactose are formed.

Lactose hydrolysis can be carried out chemically or enzymatically. The chemical hydrolysis of lactose is characterized by very severe pH and temperature conditions (pH 1–2 and temperature of 100°C–150°C). A high degree of hydrolysis can be obtained, but chemical conversion also shows some disadvantages. At first, the method cannot be applied to the hydrolysis of milk and protein-containing lactose solutions, because of denaturation of proteins occurring at high temperature and low pH. Therefore, use of whey permeate is required. Secondly, the presence of salts in whey causes deactivation of the acid. Therefore, a prior demineralization step is required. In addition, undesirable by-products are released and stringent requirements are necessary to resist chemical aggressive process conditions as far as plant construction is concerned (Gekas and Lopez-Leiva 1985).

The enzymatic hydrolysis can be accomplished under milder circumstances (pH 3.5–8.0 and temperature of 5°C–60°C). A β-galactosidase catalyzes hydrolysis of β-1,4-D-galactosidic bonds (Figure 13.3). Lactase is the common name for the β-galactosidase enzyme or more formerly called β–D-galactoside galactohydrolase. In practice, it is also possible that, depending on the circumstances, oligosaccharides are formed.

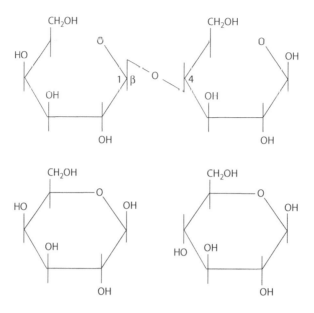

Figure 13.3   Structure of respectively lactose, D-galactose and D-glucose.

Possible sources of the enzyme are: plants, animal organs, bacteria, yeasts (intracellular enzyme), fungi or molds (extracellular enzyme). The enzyme properties depend on the source. Table 13.2 provides an overview of possible sources of lactase preparations (Gekas and Lopez-Leiva 1985). The optimal temperature and pH differ according to the source and even according to the particular commercial preparation. Enzyme immobilization, method of immobilization, and type of carrier can also influence those optima. In general, fungal lactases have pH optima in the acid range (2.5–4.5) and yeast and bacterial lactases in the almost neutral range (6–7 and 6.5–7.5, respectively). These specific optima make each lactase suitable for a specific application. Thus, fungal lactases are used for acid whey hydrolysis while yeast and bacterial lactases are suited for hydrolysis of milk (pH 6.6) and sweet whey (pH 6.1). The properties of several lactases are given in Table 13.3.

Product inhibition (namely inhibition by galactose) is another property which also depends on the lactase source. The enzyme from *Aspergillus Niger (A. niger)* is more strongly inhibited by galactose than that from *A. oryzae*. An enzyme preparation from a *Bacillus sp.* showed less inhibition than that from *Kluyveromyces fragilis (K. fragilis)* (Gekas and Lopez-Leiva 1985).

From Table 13.3, it can also be noticed that the optimal temperature for these lactases are in a temperature range of 35°C–80°C. The optimal β-galactosidase for treatment of whey permeate is preferably active at a pH of 6-7 and a temperature of 4°C–25°C.

The use of cold-active lactases from psychrophilic organisms, therefore, are preferred for hydrolysis of whey permeate. Psychrophilic organisms are able to live at low temperatures and are found in permanent cold places on earth. They produce cold-active enzymes which compensate the reduction in chemical reaction rate by low temperatures. The adaptation to cold is usually accomplished by a reduction in activation energy, which is caused by a higher flexibility of certain structures (D'Amico et al. 2002).

Nutrilab NV. developed, with the assistance of the University of Liege (ULg, Belgium) and the Engineering College ST Lieven from Gent (Belgium), a β-galactosidase from the Antarctic

Table 13.2 Sources of Lactases (Gekas and Lopez-Leiva 1985)

Plants	Animal Organs	Yeast	Bacteria	Fungi
Peach	Intestine	*Kluyveromyces*	*Escherichia coli*	*Neurospora crassa*
Apricot	Brain and skin tissue	*(Saccharomyces) lactis*	*Bacillus megaterium*	*Aspergillus foetidus*
Almond		*Kluyveromyces*	*Thermus aquaticus*	*Aspergillus Niger*
Kefir grains		*(Saccharomyces) fragilis*	*Streptococcus lactis*	*Aspergillus flavus*
Tips of wild roses		*Candida pseudotropicalis*	*Lactobacillus bulgaricus*	*Aspergillus oryzae*
Alfalfa seed		*Brettanomyces anomalus*	*Lactobacillus helveticus*	*Aspergillus phoenicis*
Coffee berries		*Wingea roberstsii*	*Bacillus sp.*	*Mucor pucillus*
			Bacillus circulans	*Mucor miehei*
			Bacillus stearothermophilus	*Scopuloriopsis*
			Lactobacillus sporogenes	*Alternaria palmi*
				Curvularia inaegualis
				Fusarium moniliforme
				Alternaria alternara

psychrophilic organism *Pseudoalteromonas haloplanktis*, for the needed lactose hydrolysis in whey permeate. Hydrolysis is executed by Nurilab at optimal temperatures between 5°C and 10°C over 24 hours, resulting in 100% hydrolysis.

After this enzymatic hydrolysis, an equimolar amount of glucose and galactose is obtained. Hence, the process conditions must be defined for separation of glucose and galactose.

Nutrilab NV. opted for a selective fermentation of glucose to ethanol because ethanol is easier to separate from galactose and optimized the enzymatic isomerization process

Enzyme

A number of bacterial L-arabinose isomerases (araA) were evaluated for their ability to convert D-galactose to D-tagatose. The most efficient enzyme was produced heterologously in *Escherichia coli* and characterized. The arabinose isomerase from *Geobacillus stearothermophilus* was selected as the most appropriate enzyme. Amino acid sequence comparisons indicated that the enzyme is only distantly related to the group of previously known araA sequences in which the sequence

Table 13.3 Properties of Lactases

Sources	pH Optimum	pH Change on Immoblization	Temperature Optimum (°C)	Molecular Weight (kD)	Activating Jobs. Other Remarks
1. A. niger	3.0–4.0	(0)–(−1½)	55–60	124	
2. A. Oryzae	5.0		50–55	90	
3. K. fragilis	6.6	(0)–(−0.5)	37	201	Mn^{2+}, K^+
4. K. lactis	6.3–7.3	(0)–(−0.6)	35	135	Mn^{2+}, Na^+
5. E. coli	7.2		40	540	Na^+, K^+
6. L. thermophilus	6.2–7.1		55–57	530	
7. C. Inaegualis	3.4–4.3		30–55		
8. B. circulans	6.0		60–65		High activity for skim milk
9. Bacillus sp	6.8		65		Less inhibition by galactose
10. Fusarium moniliorme	3.8–5.0		50–60		
11. L. bulgaricus	7.0	(−0.2)–(−0.5)	42–45		
12. Leuconostoc citrovorum	6.5		60		
13. Scopulariopsis	3.6–5		50–65		
14. B. stearothermophilus	6.0–6.4		65	215	
15. Streptococcus thermophilus	6.5–7.5		55	500–600	
16. Mucor pucillus	4.5–6		60		
17. Alternaria alternara	4.5–5.5		50–70		
18. Thermus aquaticus	4.5–5.5		80	570	

Source: From Gekas, V., and Lopez-Leiva, M. Hydrolysis of lactose: A literature review. *Process Biochem* 20:2–11, 1985.

similarity is evident. The substrate specificity and the Michaelis-Menten constants of the enzyme determined with L-arabinose, D-galactose and D-fructose also indicated that this enzyme is an unusual, versatile L-arabinose isomerase which is able to isomerize structurally related sugars. The enzyme was immobilized and used for initial lab-scale production of D-tagatose at 60°C. Starting from a 20% solution of D-galactose, L-arabinose isomerases fall within the general class of intra-molecular oxidoreductases and more specifically, the group of aldose isomerases, which are capable of interconverting aldoses in their corresponding ketoses. The idea of producing tagatose commercially, through galactose isomerization, is based on commercial production of high fructose corn sirup (HFCS) from glucose. L-arabinose isomerases in nature catalyze the isomerization of L-arabinose to L-ribulose. D-Galactose to D-tagatose conversion is considered to be a side activity of most arabinose isomerase enzymes (Figure 13.4). Kinetic studies indicated that the relative D-galactose activity of L-arabinose isomerase increases at elevated temperature and equilibrium studies of the enzyme showed a shift towards the ketose at elevated temperature. L-arabinose isomerases operating at high temperature, therefore, are particularly useful for D-tagatose production (Jørgensen et al. 2004).

The stability and specificity of the L-arabinose isomerase from *Geobacillus stearothermophilus* are of primary importance for the economical feasibility of tagatose production from galactose. Changing the specificity towards galactose was needed to increase tagatose production. The specificity had to be changed by point mutations.

In silico stability studies, therefore, were carried out by Nutrilab NV. Through computer-aided modeling and analysis of the 3D structure of the L-arabinose isomerase from *Geobacillus*

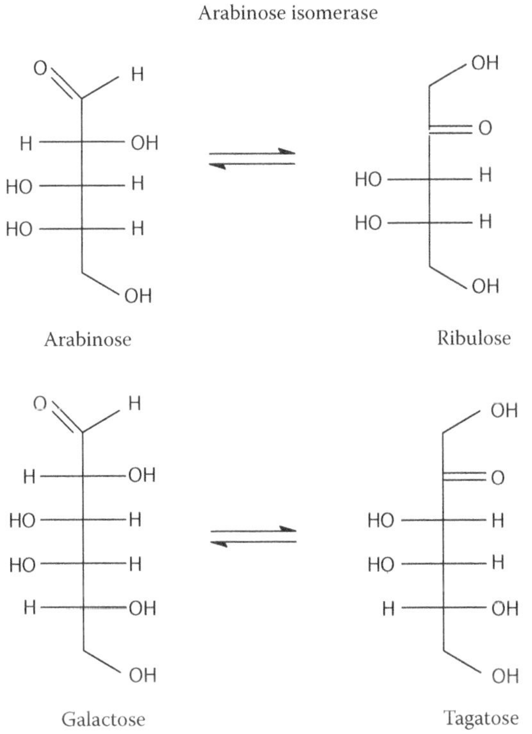

Figure 13.4 Enzymatic isomerization of galactose to tagatose by means of an L-arabinose isomerase.

stearothermophilus, as well as data available in the literature, the structure of the enzyme has been redesigned and optimized to a maximal economically productive level of 48% conversion.

In order to obtain crystalline tagatose, tagatose is separated from the remaining galactose fraction for the selective fermentation of galactose and for further downstream processing.

Toxicology and Regulatory Aspects

World Health Organization's Joint Expert Committee on Food Additives (JECFA) evaluated the safety of D-tagatose in June 2004 and stated there is no need for a limited acceptable daily intake (ADI) of D-tagatose. The JECFA has, therefore, established an ADI "not specified," the safest category in which the JECFA can place a food ingredient.

MD Foods Ingredients amba attained GRAS (Generally Recognized As Safe) status in 2001 under US FDA regulations, thereby permitting its use as a sweetener in foods and beverages. All relevant studies to document the safety of D-tagatose have been performed at US FDA approved research institutes. The key articles documenting the safety of D-tagatose appeared in the April 1999 issue of *Regulatory Toxicology and Pharmacology.*

The US FDA issued on October 25, 1999, a letter to MD Foods amba, recognizing the caloric value of D-tagatose as 1.5 kcal/g for the purpose of nutrition labeling. D-tagatose is approved in the EU as a novel food ingredient since December 14, 2005. In May 2010, the European parliament amended the definition of "sugars," defining sugars as all monosaccharides and disaccharides present in food, excluding polyols, isomaltulose and D-tagatose (European Parliament 2010). The regulation is under discussion in the permanent EU representative office. Energy content and new labeling requirements are under evaluation with the European Food Safety Authority (EFSA). An opinion is expected in early 2011. D-tagatose is accepted for use in Hong Kong, Australia and New Zealand.

Gastrointestinal Fate of D-tagatose

The rather small difference in chemical structure of D-tagatose compared to fructose has big implications on the overall metabolism of the sugars. The fructose carrier-mediated transport in the small intestine has no affinity for D-tagatose, and only approximately 20% of ingested D-tagatose is absorbed in the small intestine. The absorbed part is metabolized in the liver by the same pathway as fructose. The major part of ingested D-tagatose is fermented in the colon by indigenous microflora, resulting in the production of short-chain fatty acids (SCFAs).

Absorption in the Small Intestine

Among the different dietary monosaccharides, glucose and galactose are absorbed by an active energy-consuming transport mechanism. Fructose and xylose are absorbed by carrier-mediated, so-called "facilitated" diffusion. All other sugars are absorbed by passive absorption. Although D-tagatose and D-fructose are similar in terms of molecular structure, they are not taken up in the small intestine by the same carrier mechanism. In vitro experiments showed that D-tagatose does not inhibit the absorption of fructose or glucose (at concentrations up to 100 times higher than those of fructose) (Crouzoulon 1978; Sigrist-Nelson and Hopfer 1974). Binding of D-tagatose to the glucose and fructose carriers followed by transport across the mucosa can, therefore, be excluded.

The estimated absorption of radiolabelled D-tagatose in rats is approximately 20% (Saunders et al. 1999). Similarly, less than 26% of ingested D-tagatose was absorbed in pigs according to measurements of the disappearance of D-tagatose from the digesta (Lærke and Jensen 1999). Absorption rates of this magnitude are typical for monosaccharides and polyols of similar size which are absorbed by passive diffusion (e.g., mannitol (Bär 1990) or L-fructose and L-gulose (Levin et al. 1995)). Since the absorption of D-tagatose is likely to be even lower than that of L-rhamnose (because L-rhamnose has a slightly lower molecular weight and is slightly more lipophilic), the fractional absorption of D-tagatose in humans probably does not exceed 20%.

Excretion of D-tagatose in Urine

A metabolic study with D-[U-^{14}C]-tagatose in rats demonstrated that about 4% of ingested D-tagatose or 20% of absorbed D-tagatose was excreted with the urine (Saunders et al. 1999). The same study showed that 43% of an intravenous dose was excreted in the urine of unadapted rats over 48 hours, with most (more than 90%) being cleared by the kidneys within 6 hours of dosing.

In a pig study using 20% D-tagatose in the diet, 5% of the ingested D-tagatose was excreted in the urine and this was independent of adaptation to D-tagatose (Jensen and Laue 1998). A human study with 30 g of D-tagatose showed 0.7%–5.3% of ingested D-tagatose excreted in the urine (Buemann et al. 1998a).

Metabolism

The steps involved in the metabolism of fructose and D-tagatose are identical (Rognstad 1982). However, an important difference occurs in the rate of metabolism at the reaction step where fructose 1-P and D-tagatose 1-P are split by aldolase to glyceraldehyde and dihydroxyacetone phosphate. Although aldolase acts on both fructose and D-tagatose, evidence suggests that the rate at which aldolase cleaves D-tagatose 1-P is approximately 10% that for fructose 1-P (Lardy 1951) (see Figure 13.5).

Despite the low absorption of D-tagatose, one would expect a small transient increase in tagatose-1-P and concomitant decreases in Pi and ATP. This is because of the low affinity of aldolase b for tagatose-1-P. An increase of D-tagatose-1-P stimulates hexokinase activity, inhibits glycogen phosphorylase activity, and stimulates glycogen synthetase activity. The consequence of these effects on liver enzymes is an increase in the net deposition of glycogen in the liver. This has in fact been observed in the rat (Bar et al. 1999).

Fermentation

Unabsorbed D-tagatose is fermented by intestinal microorganisms to SCFAs. No D-tagatose was found in feces of pigs ingesting a 10% D-tagatose diet (Lærke and Jensen 1999). Similarly, no D-tagatose was found in human feces after a 30 g intake of D-tagatose (Buemann et al. 1998a). In adapted rats fed a diet with 10% D-tagatose, about 2% of the ingested dose of ^{14}C-D-tagatose was recovered in feces (Saunders et al. 1999). In unadapted rats, the recovery of D-tagatose in feces was much higher, approximately 25% of the ingested dose. Adaptation of the microflora in pigs, which results in increased in vitro fermentation of D-tagatose, is indicated by the finding of increased numbers of D-tagatose degrading bacteria and a disappearance of watery stools after a few days of D-tagatose ingestion (Saunders et al. 1999; Lærke and Jensen 1999).

Bertelsen et al.

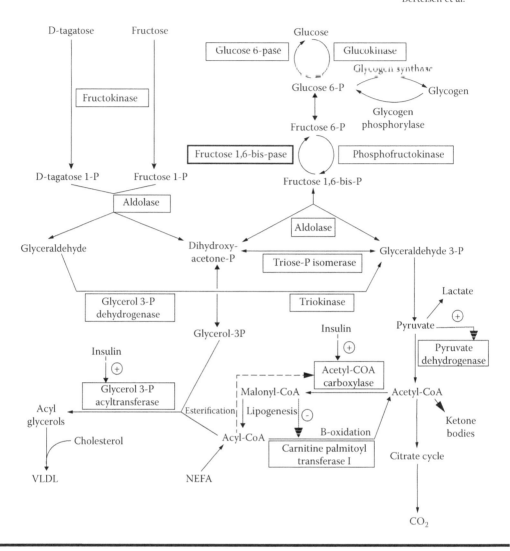

Figure 13.5   Fructose and D-tagatose utilization in the liver.

Use of D-tagatose Based on Biological Properties

D-tagatose is a sweet-tasting monosaccharide with interesting nutritional and physiological properties. It is only partly absorbed in the small intestine, and the major part is fermented in the colon, where it is converted into biomass, SCFAs, CO_2 and H_2. D-tagatose can be considered as a prebiotic based on the promotion of beneficial bacteria and an increase in the generation of SCFAs, specifically an increased level of butyrate. Furthermore, the reduced absorption and special fermentation mean that the caloric value of D-tagatose is a maximum of 1.5 kcal/g. D-tagatose consumption does not induce an increase of blood glucose or insulin levels, and even blunts the glucose level when D-tagatose is taken before glucose or sucrose. This makes D-tagatose a desirable sugar substitute for diabetics.

D-tagatose is so slowly converted to organic acids by tooth plaque bacteria that it does not cause dental caries. It has satisfied the Swiss regulation as safe for teeth. The US FDA has approved the use of a dental health claim for products containing D-tagatose as long as the retail product complies with all the requirements for a tooth-friendly product. The approved dental health claim may state, "Tagatose sugar does not promote tooth decay" or "Tagatose sugar may reduce the risk of tooth decay" (US FDA 2010).

Caloric Value of D-tagatose

Ingested D-tagatose is incompletely absorbed from the small intestine of animals and man. The fractional absorption is about 20% in rats and 25% in pigs. In humans, the fractional absorption is estimated at not more than 20% based on data of a structurally related carbohydrate (L-rhamnose). The absorbed fraction of D-tagatose is readily metabolized through the glycolytic pathway yielding 3.75 kcal/g. The unabsorbed fraction of D-tagatose reaches the large intestine where it is completely fermented by the intestinal microflora. The formed SCFAs are absorbed almost completely, and are metabolized. The metabolic fate of D-tagatose resembles, therefore, that of other incompletely digested carbohydrates (e.g., polyols).

The energy value of D-tagatose was evaluated in two studies in rats and one study in pigs. A net metabolizable energy value of −0.12 kcal/g was obtained for D-tagatose in one of the rat studies. This may be explained by an inhibition of absorption of sucrose which was present in the basal diet in very high concentrations and/or to the relatively low amount of fermentable fiber in the basal diet. The second rat study suggested a metabolizable energy value of about 1.2 kcal/g. Most relevance was attached to the pig study since the digestive tracts of pigs and humans show many similarities. The pig study resulted in a net metabolizable energy value of 1.4 kcal/g for D-tagatose.

Estimation of the net metabolizable energy value of D-tagatose by the factorial method gave a range of 1.1–1.4 kcal/g. In this method, the energy contributed by each metabolic step is evaluated separately, taking into account data from all pertinent experiments (in vitro, in vivo, in humans, in experimental animals). The factorial approach also takes into account losses of energy which are caused indirectly by the fermentation of D-tagatose (e.g., increased fecal excretion of biomass and nonbacterial mass). These studies formed the basis of the request to the US FDA for approval at 1.5 kcal/g.

D-tagatose and Diabetes

No Rise in the Glycemic Index

A clinical study at the University of Maryland School of Medicine published by Donner et al. (1996) showed that oral intake of 75 g of tagatose gave no increase in plasma glucose or serum insulin in either normal persons or Type 2 diabetic persons.

The above was confirmed in an eight person (normal subjects) study at the Research Department of Human Nutrition, Copenhagen, where oral intake of 30 g of tagatose similarly did not lead to changes in plasma glucose and serum insulin (Buemann et al. 2000).

Glucose Blunting

Oral tagatose ($t = -30$ min) blunts the rise in plasma glucose and serum insulin seen after oral glucose or sucrose ($t = 0$) in normal and diabetic persons (Donner et al. 1996).

In the Copenhagen study the 8 persons consumed 30 g of tagatose in water at $t = 0$ and a sucrose-rich lunch after 4 hours. The 30 g tagatose blunted the rise in plasma glucose and serum insulin after lunch in comparison with the same persons on either water or 30 g fructose at $t = 0$.

Tooth-Friendly Properties of D-tagatose

D-tagatose has been demonstrated to be non-cariogenic in two studies using human volunteers. The evaluation consisted of two phases: (a) an evaluation of the cariogenicity of D-tagatose using telemetric techniques, and (b) an evaluation of the cariogenic potential involving subjects' adaptation to D-tagatose.

During the first phase of testing, a 10% aqueous solution of D-tagatose was tested for cariogenicity in human volunteers using intraoral plaque-pH-telemetry (University of Zurich 1996). Telemetrically recorded plaque-pH-values after the ingestion of a substance at or above the pH limit of 5.7 can be regarded as a criterion of a low cariogenic potential of the tested food.

For the studies, six persons in generally good health served as test subjects. The subjects were asked not to alter their eating habits, and with the single exception of water rinses, they were instructed to refrain from all oral hygiene measures. Daily recordings of changes in pH were measured after subjects rinsed for two minutes with D-tagatose. Changes in pH were also measured after rinsing with 0.3 mol/L (10%) sucrose, which served as the positive control. No critical decreases (i.e., below 5.7) in the pH of interdental plaque due to bacterial fermentation of D-tagatose were noted during either the 2-minute D-tagatose rinsing periods, or during the 30-minute periods following rinsing with D-tagatose. The pH decreases that occurred subsequently to the 0.3 mol/L (10%) sucrose rinses provided adequate evidence of plaque metabolism and that the pH-telemetric equipment had functioned properly (Figure 13.6).

In phase II, researchers tested whether D-tagatose would be fermented by plaque bacteria either before adaptation or after bacteria had a chance to adapt to D-tagatose during a period of frequent exposure. Any fermentation would cause more acidification in the plaque layers upon exposure to D-tagatose. The same six subjects who participated in the acute D-tagatose telemetric evaluation also served as test subjects in this study. The subjects were asked not to alter their eating habits, and with the single exception of water rinses, they were instructed to refrain from all oral hygiene measures. During the individual periods of plaque accumulation and adaptation, the subjects rinsed five times a day. Each of the five rinsing applications consisted of two sequential 2-minute rinses with 15 ml of 10% aqueous solutions of D-tagatose. Two-minute sucrose rinses at 0.3 mol/L

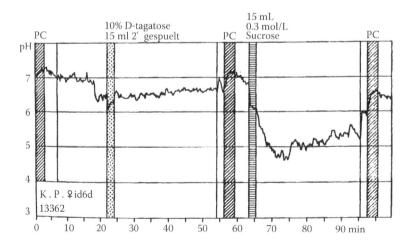

Figure 13.6   Measurement of oral pH in un-adapted dental trial.

(10%) served as the positive control. Periods of adaptation to D-tagatose rinsing (from 3–7 days in duration) were arranged to determine whether adaptation influenced fermentation in the mouth.

The results showed no critical decrease in the pH (i.e., below pH of 5.7) of interdental plaque, either during the rinsing periods or the 30-minute periods following rinsing with D-tagatose. This contrasts with plaque pH after rinsing with sucrose, which always fell below the critical pH-value of 5.7, due to the glycolytic production of bacterial acids. A comparison of the pH values of interdental plaque occurring in subjects who had been adapted to D-tagatose from 3–7 days, shows that these values are similar to the pH values of interdental plaque in the same subjects when they were unadapted to D-tagatose. It can therefore be concluded that plaque layers having grown under the constant exposure to D-tagatose, were not more acidified by a 10% D-tagatose rinse than unadapted plaque layers in the same volunteers (Figure 13.7). D-tagatose has thus been demonstrated to be non-cariogenic in both studies.

Prebiotic Properties of D-tagatose

The mucosal surfaces of the intestinal tract are one of the main sites of cell replication in the human body. In the colon, the epithelial cells are exposed not only to the circulation and to the endogenous secretions of other mucosal cells, but also to the contents of the colonic lumen, which is rich in food residues, and to the metabolic products of the microflora (Johnson 1995). Epidemiological and animal studies suggest that dietary fat and protein may promote carcino-genesis in the colon, whereas increased fiber and complex carbohydrates in the diet may protect against colon cancer. Colonic luminal butyrate concentrations are postulated to be the key protec-tive component of high-fiber diets against colon cancer (Velázquez et al. 1996).

Butyrate is one of the SCFAs that are the C2-C5 organic acids. These compounds are formed in the gastrointestinal tract of mammals as a result of anaerobic bacterial fermentation of undi-gested dietary components, and are readily absorbed by the colonic epithelium. Dietary fiber is the principal substrate for the fermentation of SCFA in humans, however intake of fibers is often low in a typical Western diet. Other undigested components, like starch and proteins, contribute to the production of SCFAs, but also the low molecular weight oligosaccharides, sugars and poly-ols which escape digestion and absorption in the small intestine contribute to the production of

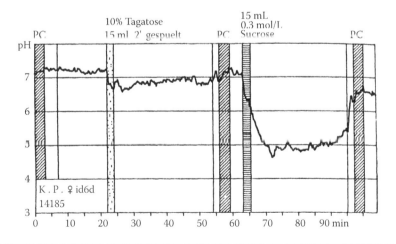

Figure 13.7   Measurement of oral pH in adapted dental trial.

SCFAs. In the mammalian hind gut, acetate, propionate and butyrate account for at least 83% of SCFAs produced and are present in a nearly constant molar ratio 60:25:15 (Velázquez et al. 1996).

In pig studies, D-tagatose altered the composition and in vitro fermentation for 0–4 hours of colonic samples from pigs adapted for 17 days and with 1% D-tagatose added to the samples. The adapted pigs showed 46 mol% of butyrate, compared to a normal mole% of 17, resulting from in vitro fermentation for 0–4 hours of colonic samples from pigs fed the sucrose control diet (Johansen and Jensen 1997).

Concentrations of butyrate in the cecum and colon of pigs were increased in a dose-response manner from ingestion of D-tagatose. Similarly, 12-hour in vitro incubation of intestinal samples from slaughtered pigs also showed a dose-response production of butyrate from ingested D-tagatose (Jensen and Laue 1998) (Figure 13.8).

Increased in vivo concentrations of butyrate were seen in portal vein blood from both adapted and unadapted pigs. The appearance of butyrate in the portal vein of unadapted pigs showed an adaption of butyrate production within 12 hours of the experimental period (Jensen and Laue 1998).

None of the above mentioned studies in pigs, which sample colon contents and blood samples from the portal vein, can be performed in humans for ethical reasons. Pigs and humans, however, have similar gastrointestinal tracts, and qualitatively the same types of indigenous intestinal bacteria. The pig, therefore, serves as a good model for humans in digestion and fermentation. To perform human studies, one has to rely on in vitro studies with D-tagatose added to human fecal slurries.

In human subjects fed 10 g D-tagatose three times daily throughout a 13-day test, the rate of in vitro fermentation (added level of D-tagatose at 1%) was higher with fecal samples of adapted volunteers than unadapted volunteers, and similarly the mole% of butyrate was high (35% versus 22%) in a 4-hour incubation. After 48 hours, the unadapted fecal incubation also showed increased mol% of butyrate. D-tagatose ingestion was characterised by changes in microbial population density and species. Pathogenic bacteria (such as Coliform bacteria) were reduced, and specific beneficial bacteria (such as Lactobacilli and lactic acid bacteria) were increased. The

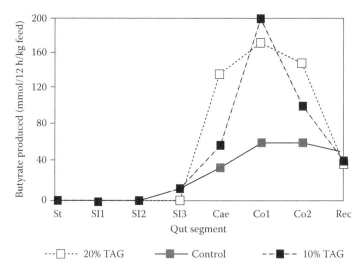

Figure 13.8 Twelve hour *in vitro* production of butyrate in pig gut segment slaughtered six hours after feeding.

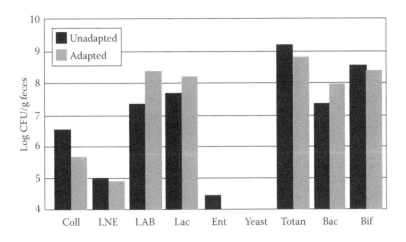

Figure 13.9   Influence of D-tagatose on bacterial composition in human faeces.

enrichment of Lactobacilli is consistent with the screening of pure culture intestinal bacteria, which indicates a high frequency of D-tagatose fermentation in the tested *Lactobacillus* strains (Jensen et al. 1998) (Figure 13.9).

The intestines of both pigs and humans harbor bacteria which produce butyrate as the fermentation end product. D-tagatose is a rather rare sugar for intestinal bacteria, and is only fermented by a limited number of genera of bacteria, Enterococci, Lactobacilli and, obviously, some butyrate-producing intestinal bacteria. Other fibers or undigested carbohydrates are either not a substrate for butyrate-producing bacteria, or competition for the delivered substrate favors growth of other bacteria. A reasonable explanation for the butyrate-inducing effect of D-tagatose is that butyrate-producing bacteria is favored over most other bacteria when D-tagatose is supplied. The adaptation seen with ingestion of D-tagatose in both pigs and human volunteers, is probably due to selection of butyrate-producing bacteria in the colon. The in vivo absorption study in pigs showed that the adaptation or selection of bacteria takes place within 12 hours. Similarly, the in vitro study with human feces indicates selection of bacteria within 48 hours of incubation.

Many studies have documented an important role of butyrate in the colon, because it is the major and preferred fuel for the colonic epithelium (Roediger 1980) and plays an important role in the control of proliferation and differentiation of colonic epithelial cells (Johnson 1995). In contrast to the above effects on the normal epithelium, butyrate arrests growth of neoplastic colonocytes (Hague et al. 1996) and also inhibits the preneoplastic hyperproliferation induced by tumor promotors in vitro (Velázquez et al. 1996).

The Lactobacilli and lactic acid bacteria are important inhabitants of the intestinal tract of man and animals with functional benefits including maintenance of the normal microflora, pathogen interference, exclusion and antagonism, immunostimulation and immunomodulation, anticarcinogenic and antimutagenic activities, deconjugation of bile acids and lactase presentation in vivo (Klaenhammer 1998). Many studies on probiotic bacteria have shown that it is very difficult for these bacteria to colonise the human colon, that is, after stopping the supply of probiotic bacteria in the diet, they disappear. This makes it more obvious to selectively feed the Lactobacillus already present in the colon.

Based on the above documented butyrate stimulation and selection of bacteria in the colon, D-tagatose is a promising prebiotic food bulk sweetener.

Human Tolerance

Extensive human clinical testing has been conducted on D-tagatose to assure safety and tolerance (Bucmann et al. 1997; Buemann et al. 1998b,1998c, 1998d; Lee and Storey 1998; Donner et al. 1998, 1999; Jensen and Buemann 1998). Testing included a 14-day trial with 30 g given in a single daily dose, an eight-week trial with 75 g daily given as three 25 g doses, and a 12-month trial with 45 g daily doses given as three 15 g doses. Because D-tagatose is malabsorbed, gastrointestinal effects similar to those of other undigested sugars would be expected. Unabsorbed molecules in the large intestine retain colonic fluid and increase water content in the feces, contributing to laxation or diarrhea, while fermentation by the microflora produces gas, leading to an increase in flatulence. As seen in clinical trials with tagatose consumption, a 30 g bolus dose was well-tolerated by many subjects, and when gastrointestinal symptoms did occur, they were consistent with the type of effects produced by an undigested compound.

Tolerance to a 30 g divided dose of D-tagatose was investigated in two studies (Jensen and Buemann 1998; Donner et al. 1999). Results demonstrated that most subjects tolerate this level of ingestion well. An analysis of the data available for D-tagatose indicates that a bolus dose of 20 g or a divided daily dose of 30 g is well-tolerated, with some sensitive individuals experiencing gastrointestinal symptoms including mild flatulence, laxation, diarrhea and bloating. The gastrointestinal effects that were observed in all the clinical studies are consistent with those expected when consuming a malabsorbed compound, and there is recognition of inter-individual variability in this response. It is expected that individuals will self-regulate their consumption of this type of food product based on their own gastrointestinal response to the consumption.

Physical Properties

Sweetness

In order to determine the level of sweetness of D-tagatose relative to sucrose, a paired comparison test was conducted on D-tagatose and sucrose using six trained judges. This sweetness equivalency taste test was based on difference threshold methodology (Lawless and Heyman 1998). The sweetness level for D-tagatose was determined by means of a linear regression plot of the fraction of times that D-tagatose solutions were chosen as sweeter than 10% sucrose versus the percent D-tagatose concentration. The point at which judges could not distinguish which solution was sweeter (at 50% probability) resulted in equivalent sweetness of the D-tagatose solution to the 10% sucrose solution. This point was 10.8% D-tagatose. Based on this sweetness equivalence taste test, D-tagatose was determined to be 92% ($10/10.8 \times 100$) as sweet as sucrose in a 10% aqueous solution (Table 13.4).

Flavor Enhancer

In order to determine whether D-tagatose has specific flavor enhancing properties, several tests were performed. These included subjective qualitative evaluation in various diet beverages and Quantitative Descriptive Analysis (QDA) profiling in cola beverages sweetened with aspartame-acesulfame potassium (acesulfame K) blends, and with sucralose. The purpose was to quantify the relative sweetness and mouth-feel enhancing effects of D-tagatose.

Table 13.4   Relative Sweetness of Tagatose and Other Sugars and Polyols

Relative Sweetness	
Sucrose	1
Fructose	0.8–1.7
Tagatose	0.92
Xylitol	0.8–1.0
Glucose	0.5–0.8
Erythritol	0.50
Sorbitol	0.4–0.7
Lactitol	0.4

At practical use levels, a combination of D-tagatose and aspartame provides impressive sweetness synergy. D-tagatose is also synergistic with aspartame-acesulfame K combinations. Consistent changes in flavor attributes have been observed across all sweetener systems. D-tagatose speeds sweetness onset times, and reduces bitterness in blends. The mouth-feel characteristics are improved, for example, mouth drying is significantly reduced, the sweet aftertaste is significantly reduced, and the bitter aftertaste is reduced. The sensory contributions of tagatose were found to be universally beneficial.

From the subjective evaluation of aspartame-acesulfame K sweetened lemon/lime soft drinks containing 0.2% D-tagatose, a fresher and cleaner flavor profile with more depth of flavor was found. A cleaner aftertaste with no bitterness and a fuller, more syrupy mouth-feel results. A subjective evaluation of aspartame-acesulfame K sweetened cola soft drinks containing 0.2% D-tagatose found lemon flavor notes to be enhanced. A more balanced flavor and sweetness resulted. Less lingering sweetness, and enhanced mouth-feel were produced.

Crystal Form

D-tagatose is a white crystalline powder with an appearance very similar to sucrose. The main difference is the crystal form. D-tagatose has a tetragonal bipyramid form, illustrated below (Figure 13.10).

Crystallization from aqueous solution results in anhydrous crystals in an α-pyranose form, with a melting point of 134°C–137°C. In solution, D-tagatose mutarotates and establishes an equilibrium of 71.3% α-pyranose, 18.1% β-pyranose, 2.6% α-furanose, 7.7% β-furanose and 0.3% keto-form (Wolf and Breitmaier 1979).

Hygroscopicity

D-tagatose is a nonhygroscopic product similar to sucrose (Figure 13.11). This means that D-tagatose will not absorb water from its surrounding atmosphere under normal conditions and does not require special storage.

Figure 13.10   Tetragonal bipyramid crystal form of D-tagatose.

Water Activity

Water activity influences product microbial stability and freshness. D-tagatose exerts a greater osmotic pressure and, hence, lower water activity than does sucrose at equivalent concentrations. The effect on water activity is very similar to fructose (same molecular weight).

Solubility

D-tagatose is very soluble in water and similar to sucrose (Figure 13.12). This makes it suitable for use in applications where it is substituted for sucrose. The same amounts produce nearly the same sweetness. D-tagatose is suitable for applications like hard-boiled candy, ice cream, chocolate, soft drinks and cereals. In comparison with polyols, D-tagatose is more soluble than erythritol (36.7% w/w at 20°C) and less soluble than sorbitol (70.2% w/w at 20°C).

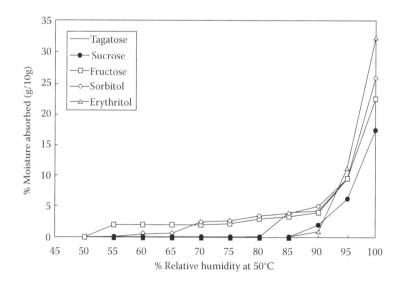

Figure 13.11   Moisture absorbance at different relative humidities.

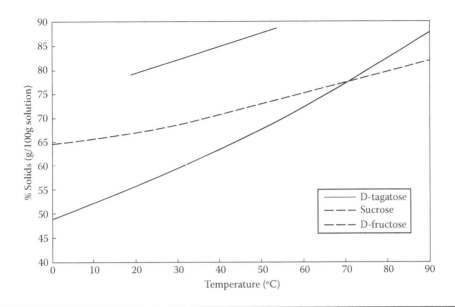

Figure 13.12   Solubility of tagatose, sucrose and fructose in water at different temperatures.

Viscosity

D-tagatose solutions are lower in viscosity than sucrose solutions at the same concentrations, but slightly higher than fructose and sorbitol (180 cP at 70% w/w and 20°C) (Figure 13.13).

Heat of Solution

D-tagatose has a cooling effect stronger than sucrose and slightly stronger than fructose (Table 13.5).

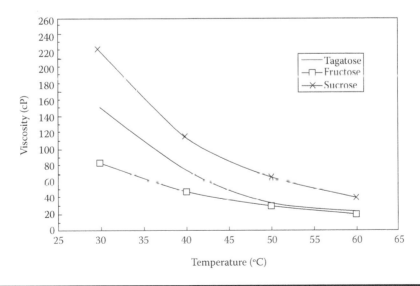

Figure 13.13   Viscosity at 70% w/w concentration and different temperatures.

Table 13.5 Heat of Solution of Tagatose, Sucrose, and Fructose at Different Temperatures

	Heat of Solution [kJ/kg]	
	20°C	40°C
Sucrose	−18.2	−23.9
Fructose	−37.7	−62.3
Tagatose	−42.3	−84.1
Erythritol	−79.4	—
Sorbitol	−111	—

Chemical Properties

As a ketohexose, D-tagatose is a reducing saccharide and very active chemically. D-tagatose takes part in the Maillard reaction which leads to a distinct browning effect. It also decomposes (caramelization) more readily than sucrose at high temperatures. At very low and high pH, D-tagatose is less stable and converts to various compounds. D-tagatose can, however, be used satisfactorily in many different applications at high temperature when process time is kept short. It is relatively straightforward to produce hard-boiled candy under reduced pressure with D-tagatose.

Applications

Confectionery in General

Given its low caloric value, its bulking ability, and the same sweetness as sucrose, D-tagatose is well suited for confectionery products. In terms of flavor profile, confectionery products with D-tagatose will be close to products produced with sucrose.

Chocolate

Chocolate with reduced calories can favorably be produced using D-tagatose as a substitute for sucrose. The sensoric profiles are very similar and calorie reductions of 20%–25% are obtainable. Chocolate with D-tagatose can be produced using standard process equipment, used in the industry today.

Hard-Boiled Candies and Wine Gum

Non-cariogenic hard-boiled candies with good flavor profile and stability can be produced with D-tagatose. When exposed to high temperatures, D-tagatose seems to have a behavior similar to that of fructose. D-tagatose has a low glass transition temperature and promotes the Maillard browning effect. Cooked under vacuum and in combination with other sweeteners (50%–50%) D-tagatose is suitable for hard-boiled candies and wine gum.

Fondant

As D-tagatose has a good ability to crystallize, it is well suited for producing fondants for low calorie pralines.

Chewing Gum

D-tagatose is suited for use in non-cariogenic chewing gum, both for kernels and dragee.

Fudge

The ability of D-tagatose to crystallize makes it suitable for making fudge. As D-tagatose caramelizes at a low temperature, the characteristic caramel flavor is easily obtained.

Caramel

Caramel produced with D-tagatose has a smooth and soft consistensy. Browning and caramel flavor occurs due to the low temperature of caramelizing. This makes D-tagatose suitable for caramel in chocolates.

Ice Cream

D-tagatose can replace sucrose in a one-to-one ratio in ice-cream with good results.

Soft Drinks

In soft drinks it has been found that D-tagatose shows significant synergistic effects at even very low doses with combinations of intense sweeteners. When blended with intense sweeteners, D-tagatose improves flavor and mouth-feel. It has also been found that D-tagatose is able to stabilize the sweetness and flavor profile in soft drinks sweetened with blends of aspartame and acesulfame K. As aspartame degrades, sweetness synergy with acesulfame K is lost and the balance is shifted towards acesulfame K, which has a more bitter taste. Adding as little as 0.2% D-tagatose to the sweetener blend provides an element of stable sweetness and the D-tagatose is able to mask the bitter taste of acesulfame K, thereby prolonging the shelf life of the soft drink.

Breakfast Products

The prebiotic effect of D-tagatose can be well exploited in applications like cereals with a dosage level of 15%, granola bars, and fruit preparations in yogurts at a dose level of 3%.

In conclusion, tagatose has three major areas of application:

1. *Low Calorie Bulk Sweetener.* Based on a significantly lower caloric value than sucrose, yet having a similar level of sweetness and similar physical bulking structure, D-tagatose can be used as a low-calorie bulk sweetener, replacing sugar and other sweeteners in different applications at different dose levels.
2. *Prebiotic Sweetener.* Replacing sucrose in various applications and adding a prebiotic benefit.
3. *Flavor Enhancer.* D-tagatose added to sweetening systems based on potent sweeteners or high intensity sweeteners such as aspartame alone or combinations of aspartame and acesulfame K improves the flavor profile and the mouth-feel.

Conclusion

With its broad range of properties, D-tagatose has a unique application profile and is set to be one of the major low calorie bulk sweeteners of the future.

References

Bar, A. 1990. Factorial calculation model for the estimation of the physiological caloric value of polyols. In *Proc Int Symposium on caloric evaluation of carbohydrates,* ed. N. Hosoya, 209–252. Tokyo: The Japan Assoc. Dietetic & Enriched Foods.

Bar, A., Lina, B.A.R., de Groot, D.M.G., de Bie, B., and Appel, M.J. 1999. Effect of D-tagatose on liver weight and glycogen content of rats. *Reg Toxicol and Pharmacol* 29:S11–S28.

Beadle, J.R., Saunders, J.P., and Wajda, T.J. 1989. Process for manufacturing tagatose. US 5,002,612.

Buemann, B., Toubro, S., Raben, A., and Astrup, A. 1997. Replacement of dietary sucrose by D-tagatose, a stereoisomer of D-fructose, increases energy intake. Internal report dated December 16, 1997.

Buemann, B.,Toubro, S., and Astrup, A. 1998a. D-tagatose, a stereoisomer of D-fructose increases hydrogen production in humans without affecting 24-hour energy expenditure, or respiratory exchange ratio. *J Nutr* 128:1481–1486.

Buemann, B., Nielsen, P., Toubro, S., and Astrup, A. 1998b. Tagaphag. Buffet study on voluntary food intake after D-tagatose. Internal preliminary report.

Buemann, B., Toubro, S., and Astrup, A. 1998c. Human gastrointestinal tolerance to D-tagatose. *Reg Toxicol Pharmacol* 29:571–577.

Buemann, B., Toubro, S., Raben, A., and Astrup, A. 1998d. Human tolerance to a single, high dose of D-tagatose. *Reg Toxicol Pharmacol* 29:566–570.

Buemann, B., Toubro, S., Holst, J.J., Rehfeld, J.F., Bibby, B.M., and Astrup, A. 2000. D-tagatose, a stereoisomer of D-fructose, increases blood uric acid concentration. *Metabol Clin Exp* 49(8):969–976.

Crouzoulon, G. 1978. Les proprietes cinetiques du flux d'entree du fructose a travers la bordure en brosse du jejunum du rat. *Arch Int Physiol Biochim* 86:725–740.

D'Amico, S., Claverie, P., Collins, T., Georlette, D., Gratia, E., Hoyoux, A., Meuwis, M.A., Feller, G., and Gerday, C. 2002. Molecular basis of cold adaption. *Phil Trans R Soc Lond* 357:917–925.

Donner, T., Wilber, J.F., and Ostrowski, D. 1996. D-tagatose: A novel therapeutic adjunct for non-insulin dependent diabetes. *Diabetes* 45(suppl. 2):125a.

Donner, T., Wilber, J.F., and Ostrowski, D. 1999. D-tagatose, a novel hexose: Acute effects on carbohydrate tolerance in subjects with and without Type 2 diabetics. *Diabetes, Obesity and Metabolism* 1(5):285–291.

Donner, T., Wilber, J.F., and Ostrowski, D. 1998. D-tagatose effects upon CHO and lipid metabolism in normal and diabetic subjects. Internal protocol and preliminary results.

European Parliament. 2010. http://www.europarl.europa.eu/sides/getDoc.do?pubRef=-//EP//TEXT+TA+P7-TA-2010-0222+0+DOC+XML+V0//NL&language=NL. See Amendment 197.

Gekas, V., and Lopez-Leiva, M. 1985. Hydrolysis of lactose: A literature review. *Process Biochem* 20:2–11.

Hague, A., Butt, A.J., and Paraskeva, C. 1996. The role of butyrate in human colonic epithelial cells: An energy source or inducer of differentiation and apoptosis. *Proc Nutr Soc* 55:937–943.

Hirst, E.L., Hough, L., and Jones, J.K.N. 1949. Composition of the gum of Sterculia setigera: Occurrence of D-tagatose in nature. *Nat* 163(4135):177.

Jensen, B.B., and Buemann, B. 1998. Preliminary report from a 14-days microbiological adaptation study with D-tagatose. Internal report.

Jensen, B.B., Buemann, B., and Bertelsen, H. 1998. D-tagatose change the composition of the microbiota and enhances butyrate production by human fecal samples. International symposia on pro-biotics and prebiotics. Kiel, Germany, 11–12 June.

Jensen, B.B., and Laue, A. 1998. Absorption of D-tagatose and fermentation products of tagatose from the gastrointestinal tract of pigs. Internal report.

Johansen, H.N., and Jensen, B.B. 1997. Recovery of energy as SCFA after microbial fermentation of D-tagatose. *Int J Obes* 21(suppl 2):S50 (abstr.).

Johnson, I.T. 1995. Butyrate and markers of neoplastic change in the colon. *Eur J Cancer Preven* 4:365–371.

Jørgensen, J., Hansen, O.C., and Stougaard, P. 2004. Enzymatic conversion of D-galactose to D-tagatose: Heterologous expression and characterisation of a thermostable L-arabinose isomerase from Thermoanaerobacter mathranii. *Appl Micro Biotechnol* 64:816–822.

Klaenhammer, T.R. 1998. Functional activities of Lactobacillus probiotics: Genetic mandate. *Int Dairy Journal* 8:497–505.

Lardy, H.A. 1951. The influence of inorganic ions on phosphorylation reactions. In *Phosphorous metabolism*, eds. W.D. McElroy and B. Glass, 477–499. Baltimore, Maryland: John Hopkins.

Lærke, H. N., and Jensen, B.B. 1999. D-tagatose has a low small intestine digestibility, but a high fermentability in the large intestine of pigs. *J Nutr* 129:1002–1009.

Lawless, H.T., and Heymann, H. 1998. Physiological and psychological foundations of sensory function. In *Sensory evaluation of food*, 28–31. New York: Chapman & Hall.

Lee, A., and Storey, D.M. 1998. The gastrointestinal response of young adults following consumption of sucrose, lactitol, and tagatose incorporated into plain chocolate. The University of Salford. Internal report.

Levin, G.V., Zehner, L.R., Saunders, J.P., and Beadle, J.R. 1995. Sugar substitutes: Their energy values, bulk characteritics, and potential health benefits. *Am J Clin Nutr* 62(suppl):1161S–1168S.

Petersen, S.U. 2006. Tagatose. Sweeteners and Sugar Alternatives in Food Technology, 262–294.

Roediger, W.E.W. 1980. Role of anaerobic bacteria in the metabolic welfare of the colonic mucosa in man. *Gut* 21:793–798.

Rognstad, R. 1982. Pathway of gluconeogenesis from tagatose in rat hepatocytes. *Arch Biochem Biophys* 218(2):488–491.

Saunders, J.P., Zehner, L.R., and Levin, G.V. 1999. Disposition of D-[U-^{14}C]Tagatose in the rat. *Reg Toxicol Pharmacol* 29:S46–S56.

Sigrist-Nelson, K., and Hopfer, U. 1974. A distinct D-fructose transport system in isolated brush border membrane. *Biochim Biophys Acta* 367:247–254.

Troyano, E., Martinez-Castro, I., and Olano, A. 1992. Kinetics of galactose and tagatose formation during heat-treatment of milk. *Food Chem* 45:41–43.

University of Zurich. 1996. Telemetric evaluation of D-tagatose provided by MD Foods Ingredients Amba, with regard to the product's qualification as being "safe for teeth." Zurich: Dental Institute, Clinic of Preventive Dentistry, Periodontology and Cariology Bioelectronic Unit.

US Food and Drug Administration (US FDA). 2010. Health Claims: Dietary Noncariogenic Carbohydrate Sweeteners and Dental Caries. Code of Federal Regulations, Sec. 101.80. US Government Printing Office. http://www.accessdata.fda.gov/scripts/cdrh/cfdocs/cfcfr/CFRSearch.cfm. Accessed October 29, 2010.

Velázquez, O.C., Lederer, H.W., and Rombeau, J.L. 1996. Butyrate and the colonocyte: Implications for noplasia. *Dig Dis Sci* 41(4):727–739.

Wolff, G.J., and Breitmeier, E. 1979. C-NMR-Spektroskopishe bistimmung de keto-form in währiger lösungen de fructose, D-sorbose und D-tagatose. *Chem Z* 103:232.

Chapter 14

Less Common High-Potency Sweeteners

A. Douglas Kinghorn and Cesar M. Compadre

Contents

Introduction

In this chapter, recent progress on the study and use of a number of less commonly encountered naturally occurring and synthetic sweet substances will be described. Several of the natural sweeteners have served as template molecules for extensive synthetic modification. Greater emphasis will be provided for those compounds that have commercial use as a sweetener in one or more countries, and more extensive coverage will be given to those sweet compounds for which there is recent published information. Following the format of the earlier versions of this chapter (Nabors and Inglett 1986; Kinghorn and Compadre 1991, 2001), substances that modify the sweet taste response will also be mentioned.

Naturally Occurring Compounds

Glycyrrhizin

Glycyrrhizin, which is also known as glycyrrhizic acid, is an oleanane-type triterpene glycoside for which its use as a sweetener has been reviewed (Nabors and Inglett 1986; Fenwick et al. 1990; Kinghorn and Compadre 2001; Kitagawa 2002). This compound (Figure 14.1), a diglucuronide of the aglycone, glycyrrhetinic acid, is extracted from the rhizomes and roots of licorice (*Glycyrrhiza glabra* L., Fabaceae) and other species in the genus *Glycyrrhiza*. In its natural form, the compound occurs in *G. glabra* in yields of 6%–14% as a mixture of metallic salts. Well established procedures are available for the extraction and purification of glycyrrhizin from the plant. Conversion of glycyrrhizin to ammoniated glycyrrhizin (by treatment of a hot-water extract of licorice root with sulfuric acid, followed by neutralization with dilute ammonia) results in a more water-soluble compound that is reasonably stable at elevated temperatures (Fenwick et al. 1990; Kinghorn and Compadre 2001). Glycyrrhizin has been rated as approximately 50–100 times the sweetness of sucrose, although it has a slow onset of sweet taste and a long aftertaste. Ammoniated glycyrrhizin has similar hedonic properties to glycyrrhizin. Its sweetness potency, which is about 50 times that of sucrose, is increased in the presence of sucrose (Kinghorn and Compadre 2001). There have been several attempts to modify the sensory parameters of glycyrrhizin by synthetically modifying its carbohydrate units, and some additional naturally occurring glycoside analogues of this compound are also highly sweet, such as apioglycyrrhizin and araboglycyrrhizin (Mizutani et al. 1994; Kitagawa 2002). It has been found that the monoglucuronide of glycyrrhizic acid (MGGR; Figure 14.1) is 941 times sweeter than sucrose and is sweeter than several other glycyrrhizin monoglycosides (Mizutani et al. 1994; Kinghorn and Compadre 2001). MGGR may be produced from glycyrrhizin by enzymatic hydrolysis using a microbial enzyme from

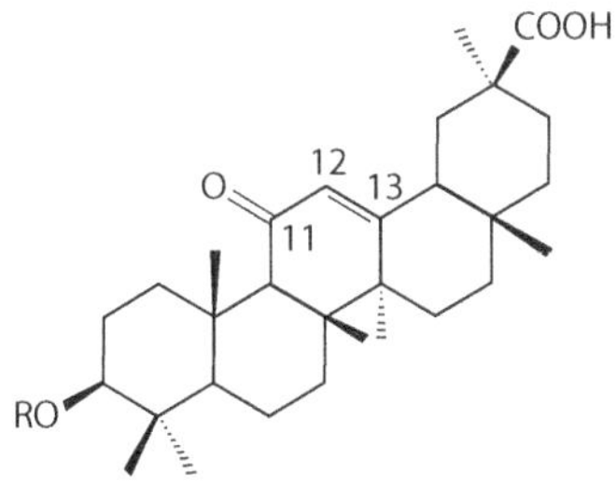

	R
Glycyrrhizin	β-glcA^2-β-glcA
MGGR	β-glcA

Figure 14.1 Structures of glycyrrhizin and MGGR (GlcA = D-glucuronopyranosyl).

Cryptococcus magnus and is now used commercially in Japan to sweeten dairy products such as yogurt, chocolate milk, and soft drinks flavored with fruits (Kinghorn and Compadre 2001). Other derivatives of glycyrrhizin that are sweeter than the parent compound and MGGR are esters of 3-*O*-β-glycyrrhetinic acid with L-aspartyl dipeptide derivatives and with α-amino acids (Kinghorn and Compadre 2001).

Partially purified *Glycyrrhiza* extracts, which contain at least 90% w/w glycyrrhizin, are used widely in Japan for sweetening and flavoring foods, beverages, oriental medicines, cosmetics, and tobacco (Kinghorn and Compadre 2001). In the United States, ammoniated glycyrrhizin, monoammonium glycyrrhizinate, and licorice extract are included in the generally recognized as safe (GRAS) list of approved natural flavoring agents, but not as approved sweeteners (Anonymous 1985). There are many applications for ammoniated glycyrrhizin as a flavorant, flavor modifier, and foaming agent. Although it is useful for incorporation into confectionery and dessert items, ammoniated glycyrrhizin is only used in carbonated beverages that are not too acidic, because this substance tends to precipitate at pH levels less than 4.5 (Kinghorn and Compadre 2001).

There is a very extensive literature on biological activities of glycyrrhizin other than its sweetness, as exemplified by its antiallergic, anti-inflammatory, antitussive, and expectorant actions. Unfortunately, the widespread use of glycyrrhizin and ammoniated glycyrrhizin by humans has been found to lead to pseudoaldosteronism, which is manifested by hypertension, edema, sodium retention, and mild potassium diuresis (Chamberlain and Abolnik 1997; de Klerck et al. 1997). The 11-oxo-$\Delta^{12,13}$-functionality in ring C of glycyrrhetinic acid has been attributed as the part of the molecule responsible for this untoward activity (Kinghorn and Compadre 2001). Glycyrrhetinic acid is known to inhibit the renal enzyme 11β-hydroxysteroid dehydrogenase, which catalyzes the inactivation of cortisol to cortisone (de Klerck et al. 1997). The Ministry of Health in Japan has issued a caution stipulating that glycyrrhizin consumption should be limited to 200 mg/day when used in medicines. Similarly, the Dutch Nutrition Information Board has advised against daily glycyrrhizin intake in excess of 200 mg, corresponding to about 150 g of licorice confectionery. Glycyrrhizin, at a level of 0.5%–1%, has been shown to inhibit plaque formation mediated by *Streptococcus mutans*, a cariogenic bacterial species. As a consequence, it has been suggested that glycyrrhizin is suitable for wider use as a vehicle and sweetener for medications used in the oral cavity (Kinghorn and Compadre 2001).

Periandrins I–IV are additional naturally occurring oleanane-type triterpene glycosides that were isolated by Hashimoto and colleagues from the roots of *Periandra dulcis* Mart. (Fabaceae) (Brazilian licorice) in the early 1980s. These compounds are 90–100 times sweeter than sucrose, but occur in the plant in low yields and are somewhat difficult to purify from bitter substances with which they co-occur (Kinghorn et al. 2010). A fifth compound in this series, periandrin V, which varies structurally from periandrin I as a result of the substitution of a terminal β-D-glucuronic acid unit by a β-D-xylopyranosyl moiety, exhibits about twice the sweetness potency of the other periandrins (Kinghorn et al. 2010). While it was at one time thought that glycyrrhizin is the sweet principle of the leaves of *Abrus precatorius* L. (Fabaceae), five cycloartane-type triterpene glycosides, abrusosides A-E, proved to be responsible for this sweetness. Although these compounds are pleasantly sweet and their water-soluble ammonium salts were rated as up to 100 times the sweetness potency of sucrose, they produce a long-lasting sweet sensation and occur in the leaves of *A. precatorius* in less than 1% w/w combined yield (Choi et al. 1989; Kinghorn et al. 2010).

Mogrosides

Mogrosides IV and V are the principal sweet cucurbitane-type triterpene glycoside constituents of the dried fruits of the Chinese plant *lo han guo*. This species, a member of the family Cucurbitaceae, was accorded the binomial *Momordica grosvenorii* Swingle in 1941, which was then changed to *Thladiantha grosvenorii* (Swingle) C. Jeffrey in 1979. However, more recently, the plant name has been changed again to *Siraitia grosvenorii* (Swingle) C. Jeffrey. Chemical studies on this plant did not begin until the 1970s (Kinghorn and Compadre 2001).

The structures of mogrosides IV and V (Figure 14.2) were established by Takemoto and colleagues after extensive chemical and spectroscopic studies (Kinghorn and Compadre 2001). The most abundant sweet principle of *lo han guo* dried fruits is mogroside V, which may occur at concentration levels of more than 1% w/w. Mogroside V is a polar compound, because it contains five glucose residues, which readily permit its extraction into either water or 50% aqueous ethanol. Aqueous solutions containing mogroside V are stable when boiled (Kinghorn and Compadre 2001). Mogrosides

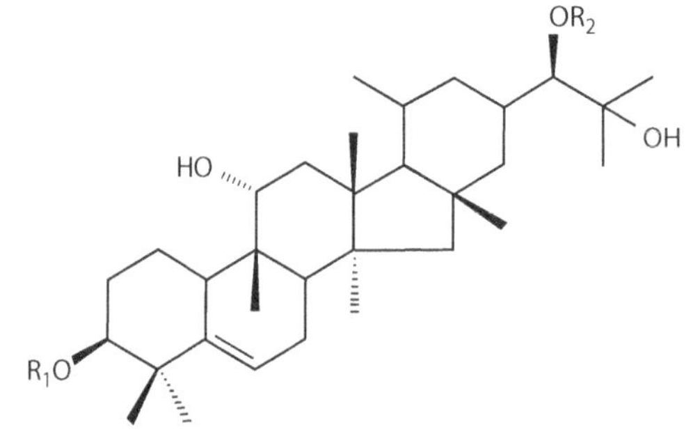

	R_1	R_2
Mogroside IV	β-glc^6-β-glc	β-glc^2-β-glc
Mogroside V	β-glc^6-β-glc	β-glc^2-β-glc $\mid^6$ β-glc

Figure 14.2 Structures of mogroside IV and mogroside V (Glc = D-glucopyranosyl).

IV and V have been rated as being in the ranges 233–392 and 250–425 times sweeter than sucrose by human taste panels, depending on the concentrations at which they were tested (Kinghorn et al. 2010). Several other minor sweet and nonsweet cucurbitane-type triterpenoid glycoside constituents have been reported from both the dried and fresh fruits of *S. grosvenorii* (Matsumoto et al. 1990; Si et al. 1996; Jia and Yang 2009). Of these, siamensoside I was rated as much as 563 times sweeter than sucrose at a concentration of 0.010% w/v (Matsumoto et al. 1990). In a recent study, mogroside V was demonstrated as being a constituent of the sweet-tasting ripe *S. grosvenorii* fruits, with other bitter-tasting cucurbitane glycosides occurring in the unripe fruits (Li et al. 2007).

Extracts made from *lo han guo* fruits have long been used by local populations in Kwangsi province in the southern region of the People's Republic of China, for the treatment of colds, sore throats, and minor stomach and intestinal complaints (Kinghorn and Compadre 1991). Products made from *lo han guo* are available in Japan as a sweetener in food and beverages. The fruits of *S. grosvenorii* are also available in Chinese medicinal herb stores in several other countries including the United States. A major corporation in the United States filed a patent application for the use of sweet-tasting juice from *S. grosvenorii* and other *Siraitia* species some years ago (Fischer et al. 1994). There is now substantial patent literature on the potential food and beverage applications of *lo han guo* extracts, particularly from mainland China.

Some safety studies on mogroside V and extracts of *S. grosvenorii* have been conducted. In earlier work, mogroside V was not nonmutagenic when tested in a forward mutation azaguanine assay with *Salmonella typhimurium* strain TM677 (Kinghorn and Compadre 2001). Also, mogroside V produced no mortalities when administered by oral intubation at doses up to 2 g/kg in acute toxicity studies in mice, and an aqueous extract of *lo han kuo* fruits exhibited an LD_{50} in mice of more than 10 g/kg (Kinghorn and Compadre 2001). More recently, extracts of *S. grosvenorii* were not correlated with any symptoms of toxicity in separate safety studies in rats and dogs (Qin et al. 2006; Marone et al. 2008). There appear to have been no adverse reactions among human populations who have ingested aqueous extracts of *lo han guo*, which would be expected to contain substantial quantities of mogroside V. Therefore, *S. grosvenorii* fruit extracts containing mogroside V are worthy of wider application for sweetening purposes in the future because of their apparent safety, in addition to favorable sensory, stability, solubility, and economic aspects. In early 2010, information appeared on the Internet indicating that GRAS status has been accorded by the US Food and Drug Administration (US FDA) to certain extracts of *S. grosvenorii* containing mogroside V.

Phyllodulcin

Phyllodulcin is produced from its naturally occurring glycosidic form by enzymatic hydrolysis when the leaves of *Hydrangea macrophylla* Seringe var. *thunbergii* (Seibold) Makino (Saxifragaceae) and other species in this genus are crushed or fermented. Phyllodulcin (Figure 14.3) is a dihydroisocoumarin and was isolated initially in 1916 and structurally characterized in the 1920s. In 1959, this sweet compound was found to have 3*R* stereochemistry (Kinghorn and Compadre 2001). In recent work, phyllodulcin, from the unprocessed leaves of its plant of origin, has been found to be a 5:1 mixture of the *R* and *S* forms (Yoshikawa M. et al. 1999). Phyllodulcin has been detected in the leaves of *H. macrophylla* subsp. *serrata* var. *thunbergii* at levels as high as 2.36% w/w (Kinghorn and Compadre 2001). Several patented methods are available for the purification of phyllodulcin. In one such procedure, after initial extraction from the plant with methanol or ethanol, hydrangenol (a nonsweet analog of phyllodulcin) and pigment impurities were removed after pH manipulations and extraction with chloroform. Phyllodulcin was then extracted selectively in high purity at pH 10 with a nonpolar solvent.

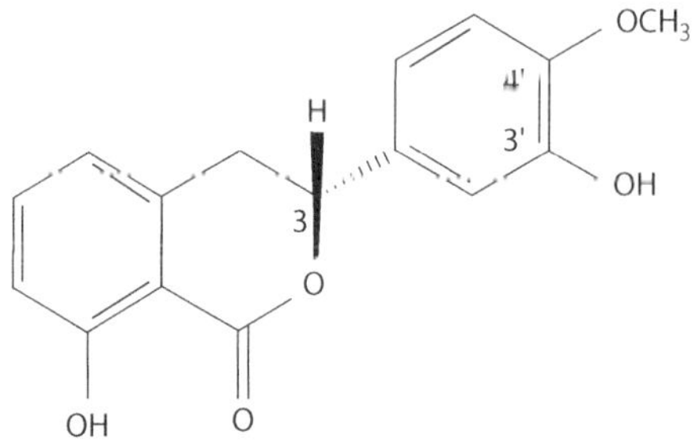

Figure 14.3 Structure of phyllodulcin.

The relative sweetness of phyllodulcin has been variously reported as 400 and 600–800 times sweeter than sucrose, although the compound exhibits a delay in sweetness onset and a licorice-like aftertaste (Kinghorn and Compadre 2001). There have been extensive attempts to modify the phyllodulcin structure to produce compounds with improved sensory characteristics. As a result of such investigations, it has been established that the 3-hydroxy-4-methoxyphenyl (isovanillyl) unit of phyllodulcin must be present for the exhibition of a sweet taste, but the phenolic hydroxy group and the lactone function can be removed without losing sweetness (Kinghorn and Compadre 2001; Bassoli et al. 2002). To date, the phyllodulcin derivatives that have been produced synthetically seem to have limitations in terms of their water solubility, stability, and/or sensory characteristics (Kinghorn and Compadre 2001).

The fermented leaves of *H. macrophylla* var. *thunbergii* (Japanese name *amacha*) are used in Japan to produce a sweet tea that is consumed at *Hamatsuri*, a Buddhist religious festival (Kinghorn and Compadre 2001). This preparation is listed in volume XIII of the *Japanese Pharmacopeia* and is used also in confectionery and other foods for its cooling and sweetening attributes (Yoshikawa M. et al. 1999). Pure phyllodulcin has been found to be nonmutagenic in a forward mutation assay and also not acutely toxic to mice when administered by oral intubation at up to 2 g/kg (Kinghorn and Compadre 2001). The metabolic fate of phyllodulcin on oral absorption has been studied in male Sprague-Dawley rats, and the structures of a number of urinary metabolites were determined (Yasuda et al. 2004). The low solubility in water and the sensory shortcomings of phyllodulcin that have been mentioned, would seem to limit the prospects of this natural product from becoming more widely used as a sweetener in the future.

Sweet Proteins

Thaumatin

Thaumatins I and II are the major sweet proteins obtained from the arils of the fruits of the West African plant *Thaumatococcus daniellii* (Bennett) Benth. (Marantaceae), with there being altogether six different thaumatin molecules now known (thaumatins I, II, III, and a, b, and c) (Kurihara 1992; Kinghorn and Compadre 2001; Crammer 2008). Thaumatin I, composed of 207 amino acid residues, of molecular weight 22209 daltons, has a relative sweetness of between 1600 and 3000 when compared with sucrose on a weight basis (Kurihara 1992). Thaumatin protein (which is known by the trade-name of Talin® protein) was reviewed comprehensively in the First Edition of *Alternative Sweeteners*, in terms of botany, production, biochemistry, physical characteristics, sensory parameters, sweetness synergy with other substances, applications (including flavor potentiation and aroma enhancement effects), safety assessment, cariogenic evaluation, and regulatory status (Higginbotham 1986). In addition, thaumatin has been subjected to review more recently (Lord 2007). The literature on thaumatin continues to be extensive and includes studies on crystallization (Tomcova and Smatanova 2007; Asherie et al. 2008b), solubility (Asherie et al. 2008a), investigations on sites in the molecule responsible for the sweet taste of this protein (Ohta et al. 2008; Ide et al. 2009), and methods for the production of this sweet substance using recombinant strains of microorganisms (Masuda and Kitibatake 2006). Thaumatin protein extract from *T. daniellii*, when sterilized with electron beam irradiation, has been evaluated for its subacute toxicity in Sprague-Dawley rats, and found to be without any adverse effects, with NOAEL (no observed adverse effect level) values estimated as 2502 and 2889 mg/kg/day for male and for female rats, respectively (Hagiwara et al. 2005). Thaumatin protein showed a lack of teratogenicity when given orally to rats, as well as a lack of mutagenic potential when tested against six strains of *Salmonella typhimurium* and *Escherichia coli* WP2. In addition, thaumatin protein does not seem to cause oral sensitization in normal human subjects (Crammer 2008).

Talin® protein was permitted initially as a food additive in Japan in 1979 (Higginbotham 1986). Despite the fact that Talin® protein has also been approved as a sweetener in Australia, Israel, and the European Union (Crammer 2008), it now appears that the major use of this product in the future will be as a flavor enhancer. In the United States, thaumatin has been accorded GRAS status as a flavor adjunct for a number of categories, including milk products, fruit juices and ices, poultry, egg products, fish products, processed vegetables, sweet sauces, gravies, instant coffee and tea, and chewing gum (Kinghorn and Compadre 2001).

Monellin

The sweet protein isolated from the fruits of another African plant, *Dioscoreophyllum cumminsii* (Stapf) Diels (Menispermaceae), has been called "monellin," a substance containing two polypeptide chains of molecular weight 11086 daltons (Kinghorn and Compadre 1991; Kurihara 1992). The A- and B-chains consist of 44 and 50 amino acid residues, respectively. When tasted individually, the A- and B-chains of monellin are not sweet, and the protein must exist in its native conformation in order to exhibit sweetness (Kohmura et al. 2002). Isolation procedures on buffered aqueous extracts of *D. cumminsii* fruit pulp enable 3–5 g of protein to be purified per kilogram of fruit, with a sweetness intensity relative to 7% w/v sucrose of 1500–2000 times. Monellin is somewhat costly to produce, and its plant of origin is difficult to propagate. In addition, the compound has a slow onset of taste, along with a persistent aftertaste, and its sweet effect is both thermolabile and pH sensitive. No toxicological data appear to have been published for this compound (Kinghorn and Compadre 2001).

Despite the fact that monellin is not used commercially as a sweetener, there has been continued interest in this compound in the scientific literature, particularly in regard to its crystal structure (Hobbs et al. 2007), folding (Kimura et al. 2005; Patra and Udgaonkar 2007), recombinant production (Chen et al. 2005, 2007), and molecular mechanism of sweet taste (Kohmura et al. 2002, Esposito et al. 2006).

Other Sweet Proteins

Mabinlins I and II are sweet proteins produced by the seeds of *Capparis masakai* Levl. (Capparidaceae), a plant that grows in Yunnan province in the People's Republic of China, and used in traditional Chinese medicine to treat sore throats and as an antifertility agent. Children are reported to chew the seed meal of *C. masakai* both as a result of its sweet taste and because it imparts a sweet taste to water drunk later (Kinghorn and Compadre 2001). In early work, mabinlin II was found to be more heat stable than mabinlin I, and to have a molecular weight of 14000 on sodium dodecyl sulfate-polyacrylamido gel electrophoresis (Kurihara 1992). Other proteins from *C. masakai* are mabinlins I-1, III, and IV, with mabinlin II having been determined as about 100 times sweeter than sucrose, when determined on a weight basis. The sweet taste of mabinlin II persists on treatment at 80°C for one hour (Crammer 2008). Interest in mabinlin II continues to focus on its crystal structure (Li et al. 2008), and this substance has been produced by cloning and DNA sequencing (Nirasawa et al. 1996). To date, no pharmacological studies appear to have been performed on any of the mabinlin proteins.

Pentadin is a sweet protein with an estimated molecular weight of about 12000 daltons and a sweetness potency of about 500 times sweeter than sucrose on a weight basis, which was reported from the fruits of the African plant *Pentadiplandra brazzeana* Baillon (Pentadiplandraceae) about 20 years ago (Compadre and Kinghorn 2001). More recently, a second sweet protein, brazzein, has been described from the same plant source as pentadin (Ming and Hellekant 1994). Brazzein consists of 54 amino acid residues with a molecular weight of 6473 daltons. This substance was rated as 2000 times and 500 times sweeter, respectively, than 2% w/v and 10% w/v sucrose, with a more sucrose-like temporal profile than other protein sweeteners. The sweetness is not destroyed by storage at 80°C for four hours (Ming and Hellekant 1994). In subsequent work, four disulfide bonds in brazzein were determined by mass spectrometry and amino acid sequencing of the cysteine-containing residues present (Kohmura et al. 1996). The compound has been synthesized by a solid-phase method, and the product was identical to natural brazzein (Izawa et al. 1996). Brazzein has been investigated also in terms of electrophysiological effects on the chorda tympani nerve in the rhesus monkey (Jin et al. 2003). Its binding locus was found to be the cysteine-rich region of subunit T1R3 at the sweetness receptor (Jiang et al. 2004), and more recently the key amino acid regions involved with multi-point binding at the T1R2-T1R3 human receptor have been probed (Assadi-Porter et al. 2010). This sweet protein may be produced through heterologous expression using the bacterium, *Escherichia coli* (Assadi-Porter et al. 2008). Accordingly, brazzein has considerable promise for future commercialization as a naturally occurring sweetening agent, because of its favorable taste profile and thermostability (Hellekant 2007).

The fruits of *Curculigo latifolia* Dryand. (Hypoxidaceae) have afforded two sweet proteins, namely, curculin (Yamashita et al. 1990; Kurisawa 1992) and neoculin (Nakajima et al. 2008). Curculin will be discussed later in this chapter in the section on sweetness-enhancing agents. Neoculin is characterized by both its inherent sweetness and an ability to convert sourness to sweetness. It is a heterodimer of a basic subunit of curculin plus an acidic and glycosylated amino acid subunit of 113 residues (Shirasuka et al. 2004). Neoculin is sweet at weakly acidic pH and interacts with the hT1R2-hT2R3 human sweet receptor (Nakajima et al. 2008). This sweet protein is reported to be about 500 times sweeter than sucrose (Nakajima et al. 2006), and has been produced by a recombinant method using *Aspergillus orizae* (Nakajima et al. 2006).

Miscellaneous Highly Sweet Plant Constituents

Hernandulcin

Hernandulcin (Figure 14.4) is a bisabolane sesquiterpene, which was isolated initially as a minor constituent of a petroleum ether-soluble extract from the aerial parts of the herb *Lippia dulcis* Trev. (Verbenaceae), collected in Mexico.

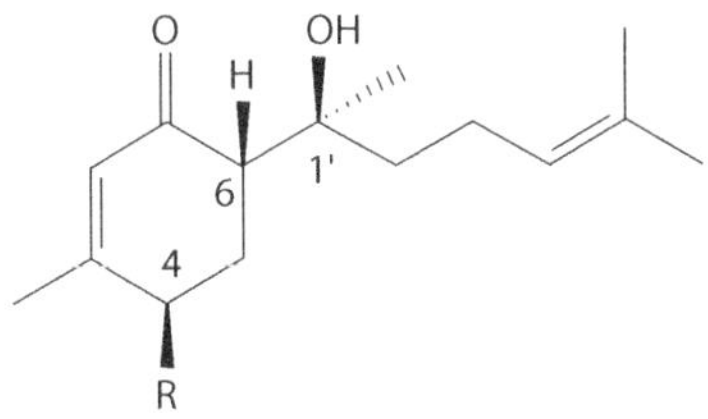

Figure 14.4 **Structures of hernandulcin and 4β-hydroxyhernandulcin.**

Lippia dulcis was known to be sweet by the Aztec people, according to the Spanish physician Francisco Hernández, who wrote a monograph entitled *Natural History of New Spain* between 1570 and 1576. The *L. dulcis* sweet constituent was named in honor of Hernández and was rated as 1000 times sweeter than sucrose on a molar basis when assessed by a taste panel (Kinghorn and Compadre 2001). Racemic hernandulcin was synthesized by a directed aldol condensation from two commercially available ketones and the naturally occurring (6S,1′S)-diastereomer was produced in the laboratory from (R)-limonene (Kinghorn and Compadre 2001). It has been concluded by analog development that the C-1′ hydroxy and the C-1 carbonyl groups of hernandulcin represent the AH and B groups in the Shallenberger model of sweetness, and the C-4′, C-5′ double bond is a third functionality necessary for the exhibition of a sweet taste (Kinghorn and Compadre 2001). In the last few years, additional synthetic routes have been proposed for (+)-hernandulcin (Kim et al. 2003; Gatti 2008). A recollection of *Lippia dulcis* from Panama afforded (+)-hernandulcin in a high yield at the flowering stage (0.15% w/w). Also obtained from this Panamanian sample was a second sweet substance, (+)-4β-hydroxyhernandulcin (Figure 14.4), in which the 4β-OH group provides a possible linkage position for sugars and other polar moieties to synthesize more water-soluble hernandulcin analogs (Kinghorn and Compadre 2001). Recently, six new bisabolane-type sesquiterpenoids were reported from the aerial parts of *Lippia dulcis*, with four of these bearing hydroperoxy groups, but these were not evaluated for sweetness because of the small quantities purified (Ono et al. 2006). Racemic hernandulcin was not a bacterial mutagen and not acutely toxic for mice in preliminary safety studies. The high sweetness potency of hernandulcin is marred by an unpleasant aftertaste and some bitterness, and the molecule is thermolabile (Kinghorn and Compadre 2001).

Rubusoside

Tanaka and coworkers have determined the *ent*-kaurene diterpene rubusoside (Figure 14.5) to be responsible for the sweet taste of the leaves of *Rubus suavissimus* S. Lee (Rosaceae), a plant indigenous to southern regions of the People's Republic of China.

Rubusoside is extractable from the plant with hot methanol and occurs in high yield in the leaves of *R. suavissimus* (more than 5% w/w), and is based on the aglycone, steviol (*ent*-13-hydroxykaur-16-en-19-oic acid) (Kinghorn and Compadre 2001). When evaluated at a concentration of 0.025%, rubusoside was rated as possessing 114 times the sweetness of sucrose, although

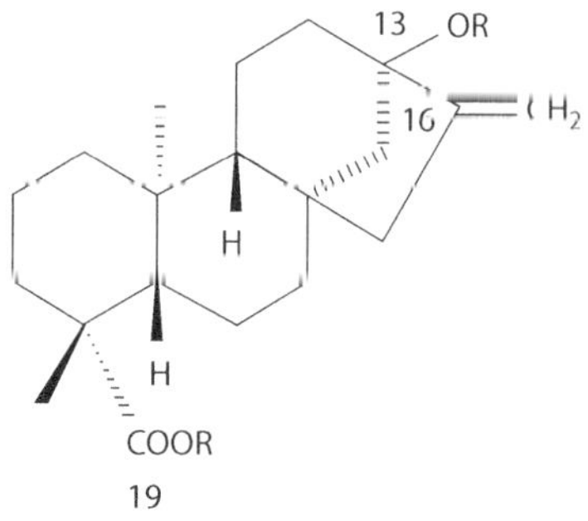

Figure 14.5   Structure of rubusoside (Glc = D-glucopyranosyl).

its quality of taste sensation was marred by some bitterness. Several minor analogs of rubusoside have been isolated and characterized from *R. suavissimus* leaves, with some found to taste sweet and others were bitter or neutral-tasting (Kinghorn and Compadre 2001). A considerable amount of work has been performed on the 1,4-α-transglucosylation of rubusoside, using a cyclodextrin glucanotransferase-starch (CGTase) system produced from *Bacillus circulans* to produce improved sweeteners based on this parent compound (Ohtani et al. 1991).

A sweet tea called *tian-cha*, prepared from the leaves of *R. suavissimus* is consumed as a summer beverage in Guangxi province in the People's Republic of China. Also, during festivals, local populations mix aqueous extracts with rice to make cakes. Furthermore, the tea made from the leaves of *R. suavissimus* has been used in folk medicine to treat diabetes, hypertension, and obesity. *Rubus suavissimus* leaves have begun to be used as a "health-giving" food ingredient in Japan because in addition to rubusoside and the other minor sweet principles, antiallergic ellagitannins are also present (Kinghorn and Compadre 2001). In an acute toxicity experiment on rubusoside, the LD_{50} was established as about 2.4 g/kg when administered orally to mice. In a subacute toxicity study, rubusoside was incorporated into the diet of mice for 60 days at a dose of one-tenth the LD_{50}, and no distinct toxicity or side effects were observed (Kinghorn and Compadre 2001).

In recent work, rubusoside has been detected by ultra-high performance LC-MS as a constituent of the leaves of *Stevia rebaudiana* (Bertoni) Bertoni (Asteraceae) (Gardana et al. 2010).

Baiyunoside

Baiyunoside (Figure 14.6) is a labdane-type diterpene glycoside based on the aglycone, (+)-baiyunol, which was first isolated in 1983 by Tanaka and coworkers from a plant used in Chinese traditional medicine, namely, *Phlomis betonicoides* Diels (Labiatae) (Kinghorn and Compadre 2001).

This butanol-soluble compound was found to be about 500-fold sweeter than sucrose and to possess a lingering aftertaste lasting more than one hour. Synthetic routes are available for both (±) and (+)-baiyunol, and a general glycosylation procedure has been developed for baiyunol (Kinghorn and Compadre 2001). The Nishizawa group synthesized a large number of baiyunoside analogs, some of which were found to be as sweet or sweeter than the parent compound (Yamada and Nishizawa 1992). No safety studies appear to have been performed thus far on baiyunoside.

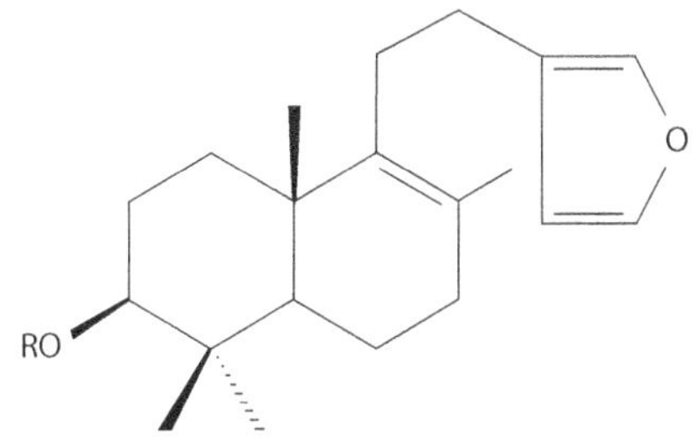

R = β-glc²-β-xyl

Figure 14.6 Structure of baiyunoside (Glc = D-glucopyranosyl; xyl = D-xylopyranosyl).

Steroidal Saponins

Osladin is a steroidal saponin constituent of the fern *Polypodium vulgare* L. (Polypodiaceae), which was isolated and structurally characterized by Herout and coworkers in 1971 (Compadre and Kinghorn 2001). The stereochemistry of the aglycone of osladin was defined by Havel and Cerny in 1975, although the configuration of the rhamnopyranosyl unit at C-26 was not determined. However, when the compound assigned as osladin by Herout et al. was synthesized by Yamada and Nishizawa, it was not found to be sweet. It turned out that the C-22S, C-25R, C-26S stereochemistry originally proposed for osladin required reassignment as C-22R, C-25S, C-26R (Figure 14.7). Moreover, the original sweetness intensity value for osladin relative to sucrose was revised downward from 3000 to 500 (Yamada and Nishizawa 1995).

A related steroidal saponin, polyposide A, was isolated as the major sweet principle of the rhizomes of *Polypodium glycyrrhiza* D.C. Eaton (Polypodiaceae). This plant is known by the

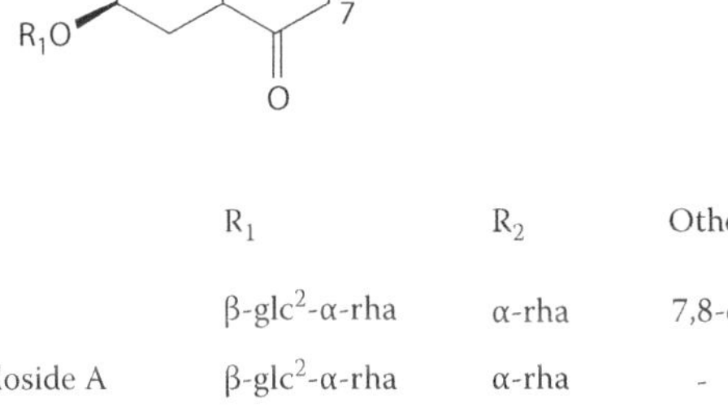

	R_1	R_2	Other
Osladin	β-glc²-α-rha	α-rha	7,8-dihydro
Polypodoside A	β-glc²-α-rha	α-rha	-

Figure 14.7 Structures of osladin and polypodoside A (Glc = D-glucopyranosyl; rha = L-rhamnopyranosyl).

common name of "licorice fern" and is a North American species native to the Pacific Northwest. Polypodoside A was originally assigned as the $\Delta^{7,8}$ analog of osladin, with the configurations of the sugar substituents being determined using spectroscopic methods. The aglycone of polypodoside A was identified as the known compound, polypodogenin, a compound for which the relative configuration was proposed by Czech workers (Kinghorn and Compadre 2001).

A need was felt, however, to re-examine polypodoside A (Figure 14.7), given the uncertainty in the stereochemistry of osladin referred to previously, and synthetic interconversion work also led to a structural revision from C-22S, C-25R, C-26S to C-22R, C-25S, C-26R for this sweet *P. glycyrrhiza* constituent (Nishizawa et al. 1994). Polypodoside A was found to be non-mutagenic and not acutely toxic for mice when dosed by oral intubation at up to 2 g/ kg. In subsequent sensory tests using a small human taste panel, polypodoside A was assessed as exhibiting 600 times the sweetness intensity of a 6% sucrose solution but also revealed a licorice-like off-taste and a lingering aftertaste. Thus, the potential of polypodoside A for commercialization is marred by its relative insolubility in water, its sensory characteristics, and difficulties in collecting *P. glycyrrhiza* rhizomes. Despite the fact that osladin has now been subjected to total synthesis, it is probable that this compound will have similar limitations to polypodoside A in terms of its potential commercial prospects, as previously noted (Kinghorn and Compadre 2001).

Pterocaryosides A and B

Pterocarya Paliurus Batal. (Juglandaceae) is a plant of which the leaves are used by local populations in Hubei province, People's Republic of China, to sweeten foods. Two sweet-tasting secodammarane saponins, designated pterocaryosides A and B (Figure 14.8), were isolated from an extract of the leaves and stems of *P. paliurus*. These compounds differ structurally in only the nature of the attached sugar (a D-quinovose unit in pterocaryoside A compared with a L-arabinose unit in pterocaryoside B). Pterocaryosides A and B were shown not to be toxic in bacterial mutagenesis and mouse acute toxicity tests and were rated as about 50 and 100 times sweeter than 2% w/v sucrose, respectively (Kennelly et al. 1995).

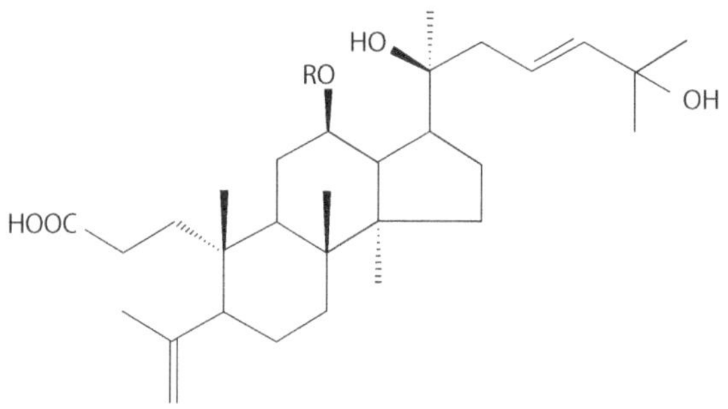

	R
Pterocaryoside A	β-qui
Pterocaryoside B	α-ara

Figure 14.8 Structures of pterocaryosides A and B (Qui = D-quinovopyranosyl; ara = L-arabinopyranosyl).

Pterocaryosides A and B, in being the first sweet-tasting secodammarane sweeteners to have been discovered, represent a new sweet-tasting chemotype and may serve as useful lead compounds in the future for synthetic optimization (Kinghorn and Compadre 2001). Additional work on *Pterocarpa (Cyclocarya) paliurus* carried out in the People's Republic of China has led to the isolation of several sweet dammarane glycosides (Yang et al. 1992).

Synthetic Compounds

In addition to the well-known synthetic sweeteners described in other chapters of this book, activity in this area has yielded some of the most intensely sweet compounds known to mankind and has improved our understanding of the structural requirements for the sweet-tasting response.

Oximes

Perillartine (Figure 14.9), the α-*syn*-oxime of perillaldehyde, has been known to be highly sweet since 1920 and is reported to be up to 2000 times sweeter than sucrose. In contrast, perillaldehyde itself (Figure 14.9), the major constituent of the volatile oil of *Perilla frutescens* (L.) Britton (Labiatae), is only slightly sweet. Perillartine is used commercially in Japan as a replacement for maple syrup or licorice for the sweetening of tobacco, but more widespread use of this compound for sweetening has been restricted by a limited solubility in water, an appreciably bitter taste, as well as a menthol-licorice off-taste that accompanies sweetness (Kinghorn and Compadre 2001).

The intense sweetness and structural simplicity of perillartine have promoted the synthesis of numerous analogs (Kinghorn and Compadre 2001). This work has not only led to a better understanding of the functional groups in compounds of the oxime class that confer sweetness and bitterness, but has also led to the development of several improved sweet compounds. One of the most promising is SRI oxime V (Figure 14.9). This compound is 450 times sweeter than sucrose on a weight basis and exhibits much improved water solubility when compared with perillartine. SRI oxime V has no undesirable aftertaste and is stable above pH 3. This substance was shown not to be a bacterial mutagen in the Ames assay and exhibited a LD_{50} of 1 g/kg in the rat after a single oral dose. The compound is readily absorbed and metabolized, with excretion nearly

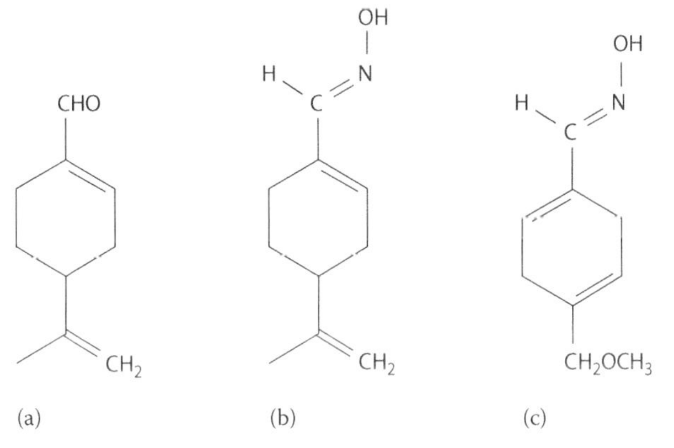

Figure 14.9   Structures of (a) perillaldehyde, (b) perillartine, (c) SRI oxime V.

quantitative within 48 hours after administration to the rat, dog, and rhesus monkey (Kinghorn and Compadre 2001).

The major metabolites of SRI oxime V are products resulting from the oxidation of the methoxymethyl or the aldoxime moieties, and those occurring after thioalkylation and glucuronidation. Subchronic toxicity tests on this compound conducted in rats with a diet containing 0.6% SRI oxime V for eight weeks revealed no apparent toxic effects. It has been suggested that SRI oxime V shows such promise as an artificial sweetener, that a chronic toxicity evaluation is warranted (Kinghorn and Compadre 2001).

Urea Derivatives

Dulcin (*p*-ethoxyphenylurea) has been known to be sweet for more than a century. The compound is about 200 times sweeter than sucrose and was briefly marketed as a sucrose substitute in the United States. Commercial use of this compound was discontinued after it was found to be toxic to rats at a low dose. Dulcin has also been found to be mutagenic. Another group of sweet ureas of more recent interest are the carboxylate-solubilized *p*-nitrophenyl derivatives, which were discovered by Peterson and Muller (Kinghorn and Compadre 2001). Suosan, the sodium salt of *N*-(*p*-nitrophenyl)-*N′*-(β-carboxyethyl)-urea (Figure 14.10), is representative of this series and has been reported to be about 350 times sweeter than sucrose, although it has significant bitterness. Other compounds in this class are even sweeter than suosan (Kinghorn and Compadre 2001).

Structure-sweetness relationships have been investigated for the sweet-tasting arylureas (Kinghorn and Compadre 2001). Combination of the structures of cyanosuasan and aspartame has led to the development of superaspartame (Figure 14.11) with a sweetness potency of 14000 times that of sucrose (Tinti and Nofre 1991). The observation that a replacement of the ureido moiety (NHCONH) with a thioureido moiety (NHCSNH) increases sweetness potency, has led to the development of compounds such as the thio derivative of superaspartame, which is reported to be 50000 times sweeter than sucrose.

The sweetness potentiation induced by sulfur has been reviewed by Roy (1992) and revisited by Spillane and Thompson (2009). Structure-activity relationship studies in the suosan sweetener series have been extended to include additional replacements for the carboxyl group. Tetrazole analogs have been prepared and were found to be sweet. Both the urea and thiourea tetrazolyl analogs, however, exhibited reduced potency compared with the carboxyl-containing compounds (Kinghorn and Compadre 2001).

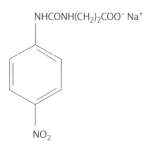

Figure 14.10   Structure of suosan.

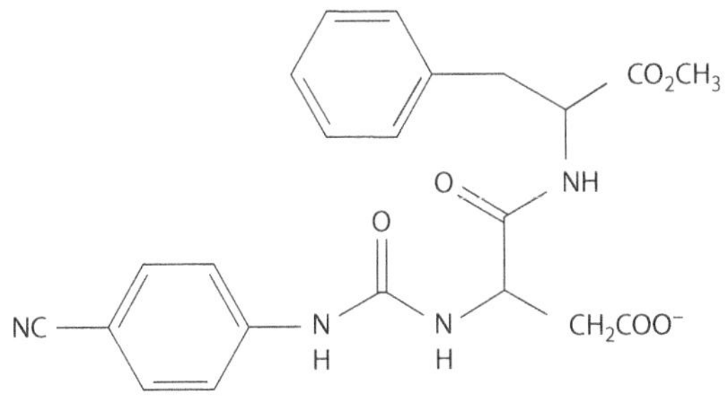

Figure 14.11 Structure of superaspartame.

N-Alkylguanidines

Nofre and coworkers reported the synthesis of a series of (phenylguanidino)- and {[1-phenylamino) ethyl]amino}-acetic acid derivatives with varying sweetness (Kinghorn and Compadre 2001). This series includes carrelame, lugduname, and sucrononic acid (Figure 14.12), the three most potent sweet compounds known, with reported sweetnesses on a molar basis of 230000, 200000, and 200000 times 2% sucrose solution, respectively (Nofre et al. 2002).

Although these compounds are not used commercially, because of their intense sweetness, they have been used to model the binding to the heterodimeric T1R2/T1R3 sweet taste receptor (Walters 2002; Temussi 2006). It also has been suggested that these compounds could have veterinary use, for example, to sweeten animal medications or stimulate animal appetite, for which approved sweeteners such as aspartame, saccharin, and acesulfame potassium are much less effective (Nofre et al. 2002).

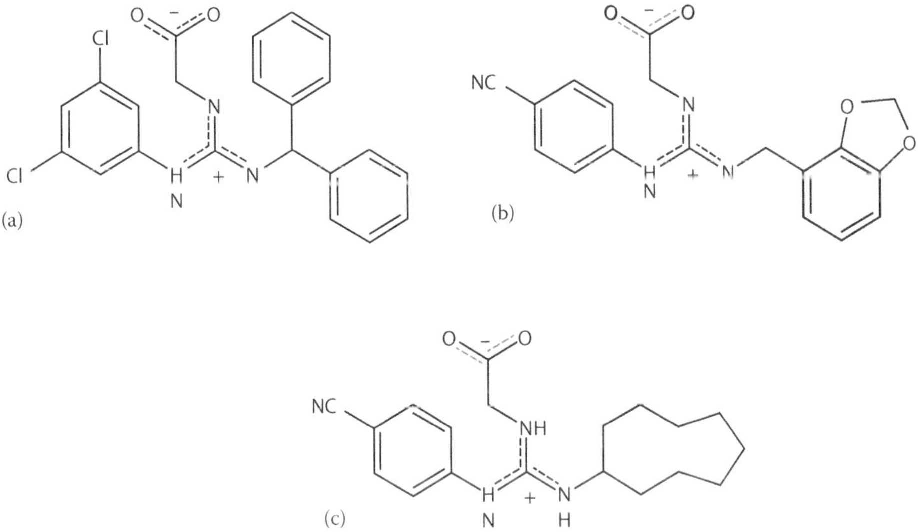

Figure 14.12 Structures of (a) carrelame, (b) lugduname, and (c) sucrononic acid in their zwitterionic forms.

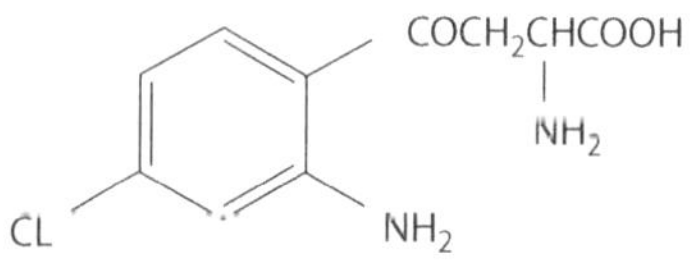

Figure 14.13 Structure of 3-(4-chloroanthraniloyl)-ᴅʟ-alanine.

Miscellaneous Compounds

Tryptophan Derivatives

The sweetness of derivatives of the amino acid tryptophan was discovered by Kornfield and coworkers in 1969, when it was observed that racemic 6-trifluoromethyl-tryptophan has an intensely sweet taste. Additional studies demonstrated that these compounds are sweet when in the ᴅ-form, with 6-chloro-ᴅ-tryptophan being some 1000 times sweeter than sucrose. The ʟ-form of this compound is tasteless but has been found to produce antidepressant activity. It was reported by Finley and Friedman that racemic N'-formyl and N'- acetyl derivatives of kynurenine, an intermediate in the metabolism of tryptophan, are approximately 35 times sweeter than sucrose and elicit an immediate sweet taste on contact with the tongue. The 4-chloro derivative of kynurenine [3-(4-chloroanthraninoyl)-ᴅʟ-alanine] (Figure 14.13) has been reported as being 80 times sweeter than sucrose and possessing no significant aftertaste or off-flavor (Kinghorn and Compadre 2001).

Trihalogenated Benzamides

2,4,6-Tribromobenzamidines, which are substituted at the C-3 position by a carboxyalkyl or a carboxyalkoxy group, are intensely sweet. For example, 3-(3-carbamoyl-2,4,6-tribromophenyl) propionic acid (Figure 14.14) was rated as 4000 times sweeter than sucrose. The compound has a slow onset and a slightly lingering aftertaste, as well as some bitterness. Within this compound class, the intensity of sweet taste depends on the chain length of the carboxyalkyl or carboxyalkoxy group. The acute toxicities of several tribromobenzamides have been determined in mice, and the results were comparable with analogous data obtained for saccharin and aspartame (Kinghorn and Compadre 2001).

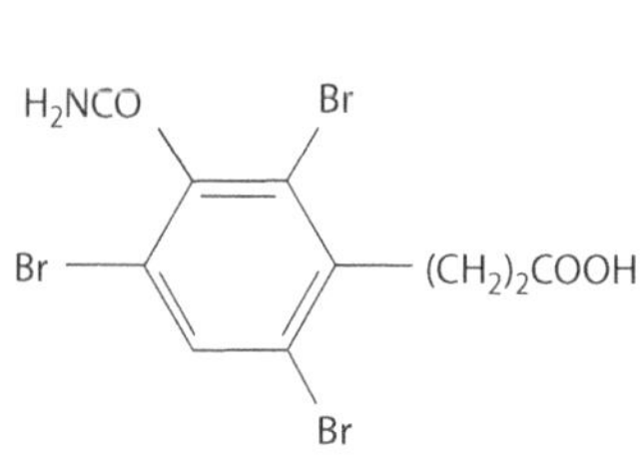

Figure 14.14 Structure of 3-(3-carbamoyl-2,4,6-tribromophenyl)propionic acid.

Sweetness Modifying Substances

Sweetness Inducers and Enhancers

Sweet Proteins

Miraculin is a tasteless basic glycoprotein constituent of the fruits of *Richardella dulcifica* (Schumach. & Thonning) Baehni [formerly *Synsepalum dulcificum* (Schumach. & Thonning) DC. (Sapotaceae) (miracle fruit)], which has the propensity of making sour or acidic materials taste sweet (Kinghorn and Compadre 2001). Native miraculin occurs as a tetramer of a large polypeptide unit, constituted by 191 amino acid residues, with a carbohydrate content of 13.9% and an overall molecular weight of 24600 daltons. The purification, biochemistry, and biological properties of this glycoprotein have been subjected to review (Kurihara 1992; Kinghorn et al. 2010). Miracle fruit concentrate was at one point commercially available in the United States as an aid in dieting but was removed from the market because US FDA approval as a food additive was never obtained (Kinghorn and Compadre 2001). This taste-modifying protein has been expressed in both transgenic lettuce (Sun et al. 2006) and transgenic tomatoes (Hirai et al. 2010).

Curculin, a sweet-tasting protein from *Circuligo latifolia* (Hypoxidaceae), mentioned earlier in this chapter, also possesses sweetness-enhancing effects. The sweet taste of curculin disappears a few minutes after being held in the mouth, but a sweet taste occurs on the subsequent application of water to the mouth. The primary structure of curculin monomer contains 114 amino acids, of molecular weight 12491, with native curculin being a dimer of two monomeric peptides connected through two disulfide bridges (Kurihara 1992). Since curculin is stable at 50°C for a year, studies directed toward the commercialization of this sweet-tasting/sweetness-enhancing protein have been suggested (Kinghorn and Compadre 2001).

Other Sweetness Enhancers

A number of plant constituents have sweetness-enhancing properties, including cynarin, chlorogenic acid, and caffeic acid (Kinghorn and Compadre 2001). Arabinogalactin (larch gum) is capable of enhancing the sweetness potency and taste qualities of saccharin, cyclamate, and protein sweeteners such as thaumatin and monellin (Kinghorn and Compadre 2001). Five oleanane-type triterpene glycoside esters, strogins 1–5, were described from the leaves of *Staurogyne merguensis* Wall. (Acanthaceae). Three of these compounds, strogins 1, 2, and 4, elicit a sweet taste in water when held in the mouth, in similar manner to miraculin and curculin, whereas two of the compounds in this series, strogins 3 and 5, were inactive in this regard (Kinghorn and Compadre 2001). Recently, two flavanones have been found to have activity as sweetness enhancers, namely, hesperetin (Ley et al. 2008) and homoeriodictyol (Ley et al. 2007), which are both constituents of *Eriodictyon californicum* (H. & A.) Torr. (Hydrophyllaceae) (Liu et al. 1992). The latter compound, for example, affords a 6% enhancement in the sweetness of 5% sucrose solution when evaluated at 100 ppm in water (Ley et al. 2007).

Sweetness Inhibitors

Phenylalkanoic Acids

Substituted phenoxyalkanoic acids are potent inhibitors of the sweet taste response without disrupting the taste cell membranes, and their effects are immediately reversible. Lactisole, the

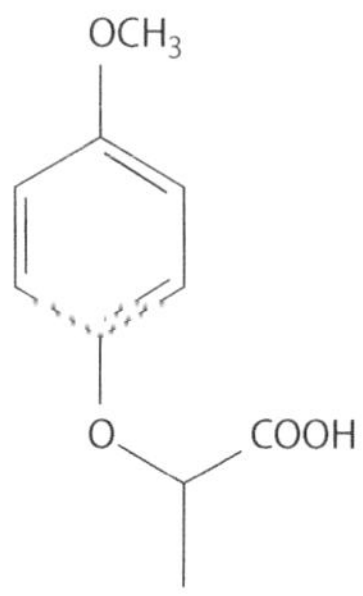

Figure 14.15   Structure of 2-(4-methoxyphenoxyl)propionic acid.

sodium salt of 2-(4-methoxyphenoxy)propionic acid (Figure 14.15) is commercially available and is used to modulate excessive sweetness in formulated products (Kinghorn and Compadre 2001). This compound has been found to be a natural constituent of roasted Colombian Arabica coffee beans and has been granted GRAS status for use in confectionery and frostings, soft candies, and snack products, for use up to levels of 150 ppm. A 100 ppm solution of this compound is able to reduce the sweetness of a 12% sucrose solution to the perceived level of 4% sucrose. It also inhibits the sweetness of other nonsucrose bulk and intense sweeteners, but it does not have an impact on salty, bitter, or sour tastes (Lindley 1991; Johnson et al. 1994). Lactisole inhibits sweetness by interacting with the transmembrane domain of the T1R3 sweet receptor (Jiang et al. 2005).

Interestingly, the structural modification of suosan (Figure 14.10) led to the discovery that *N*-(4-cyanophenyl)-*N′*-[(sodiosulfo)methyl]urea (Figure 14.16) inhibits the sweet taste of a variety of sweeteners. High-potency sweeteners with a slow onset and lingering sweet taste were the least inhibited. This compound can also antagonize the bitter taste responses to caffeine and quinine, although it has no effect on the sour (citric acid) or salty (NaCl or KCl) taste responses (Muller et al. 1992).

Arylalkylketones

Several arylalkylketones and arylcycloakylketones have also been discovered to inhibit the sweet taste of sucrose and other bulk and intense sweeteners. One such compound, the commercially available 3-(4-methoxylbenzoyl)propionic acid (Figure 14.17), is capable of reducing the sweetness of 40% w/v aqueous sucrose by over a six-fold margin when present at a 2% w/w concentration at pH 7, compared with when it is absent in the formulation. This compound and its analogs are recommended for use in soft puddings, infused vegetables, and other food products (Kinghorn and Compadre 2001).

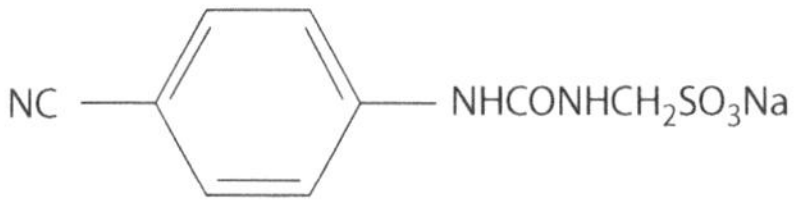

Figure 14.16   Structure of *N*-(4-cyanophenyl)-*N′*-[(sodiosulfo)methyl]urea.

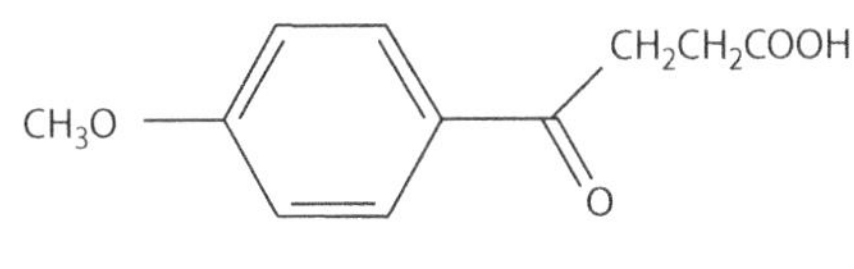

Figure 14.17   Structure of 3-(4-methoxybenzoyl)propionic acid.

Triterpene Glycoside Sweetness Inhibitors

Considerable progress has been made in the characterization of plant-derived triterpene glycosides, particularly from three species, *Gymnema sylvestre* R. Br. (Asclepiadaceae), *Ziziphus jujuba* P. Miller (Rhamnaceae), and *Hovenia dulcis* Thunb. (Rhamnaceae) (Kinghorn et al. 2010). The parent compounds from these three plants are, respectively, gymnemic acid I (Figure 14.18), zizyphin (Figure 14.19), and hoduloside I (Figure 14.20). Other species that have been found also to contain triterpene glycoside sweetness inhibitors are *Gymnema alternifolium* (Lour.) Merr. (Asclepiadaceae), *Stephanotis lutchuensis* Koidz. var. *japonica* (Asclepiadaceae), and *Styrax japonica* Sieb. et Zucc. (Styracaceae) (Kinghorn et al. 2010).

Some 20 analogs of gymnemic acid I (a group of oleanane-type triterpene glycosides) have been reported from *G. sylvestre* leaves, with most of these compounds being less potent as sweetness-inhibitory ("antisweet") substances than gymnemic acid I itself (Kinghorn and Compadre 2001). In addition, a sweetness-inhibitory peptide of 35 amino acids, gurmarin, has been isolated from the leaves of *G. sylvestre* (Kamei et al. 1992). An additional nine oleanane-type triterpene glycoside sweetness inhibitors in the alternoside series have been reported, making 19 compounds in this series in all (Yoshikawa K. et al. 1999; Kinghorn and Compadre 2001). The sitakisosides are another series of antisweet oleanane-type triterpene glycosides reported from the stems of *Stephanotis lutchinensis* var. *japonica* (Kinghorn and Compadre 2001). Jegosaponins A-D are a final group of oleanane-type triterpene glycosides to have

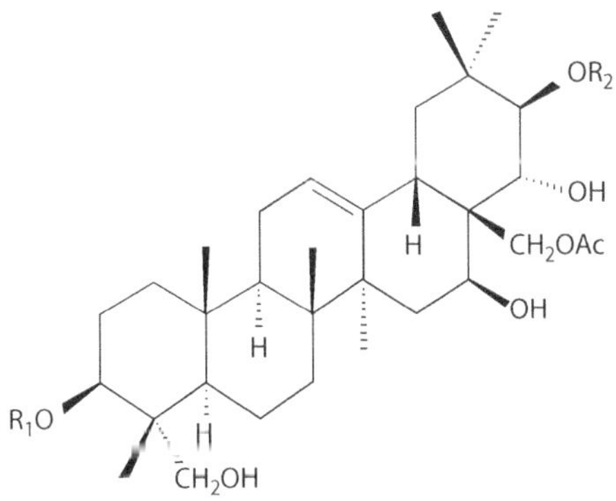

R₁	R₂
β-glcA	tga

Figure 14.18   Structure of gymnemic acid I (GlcA = D-glucuronopyranosyl; tga = tiglic acid).

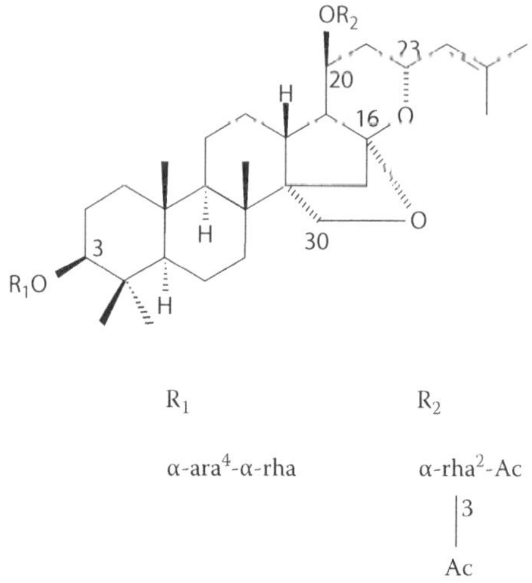
OR2
23
20
H
16
O
O
3
H
30
R1O
H
R1
R2
α-ara4-α-rha
α-rha2-Ac
3
Ac

Figure 14.19 Structure of ziziphin (Ara = L-arabinopyranosyl; rha = L-rhamnopyranosyl).

been reported, but these are all less potent as sweet-inhibitory agents than gymnemic acid 1 (Yoshikawa et al. 2000).

About 10 sweetness-inhibitory dammarane-type triterpenoid sweetness inhibitors have now been isolated and purified from the leaves of *Z. jujuba*, with all of these being equal to or less potent than ziziphin itself (Kinghorn and Compadre 2001). Several dammarane-type sweetness inhibitors have also been reported in the leaves of *H. dulcis*, with the parent compound, hoduloside I, being one of the most active. However, ziziphin and hoduloside I are somewhat less potent than gymnemic acid I as sweetness-inhibitory agents (Kinghorn and Compadre 2001).

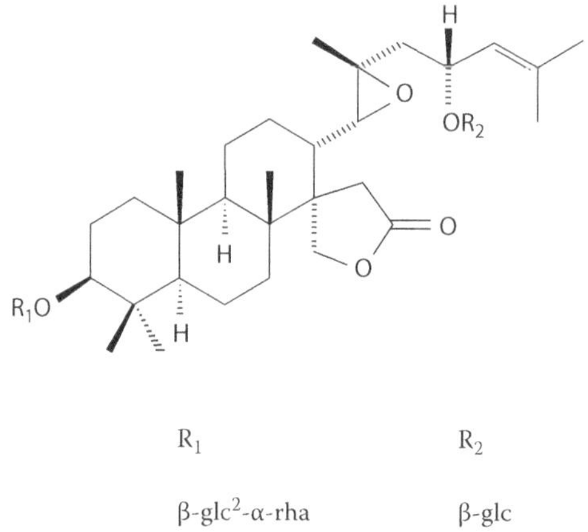
H
O
OR2
O
H
O
R1O
H
R1
R2
β-glc2-α-rha
β-glc

Figure 14.20 Structure of hoduloside I (Glc = D-glucopyranosyl; rha = L-rhamnopyranosyl).

References

Anonymous. 1985. GRAS status of licorice (*Glycyrrhiza*), ammoniated glycyrrhizin, and monoammonium glycyrrhizinate. *Fed Reg* 50(99):21043–21045.

Asherie, N., Ginsberg, C., Blass, S., Greenbaum, A., and Knafo, S. 2008a. Solubility of thaumatin. *Crys Growth Des* 8:1815–1817.

Asherie, N., Ginsberg, C., Greenbaum, A., Blass, S., and Knafo, S. 2008b. Effects on protein purity and precipitant stereochemistry on the crystallization of thaumatin. *Crys Growth Des* 8:4200–1207.

Assadi-Porter, F.M., Maillet, E.L., Radek, J.T., Quijada, J., Markley, J.L., and Max, M. 2010. Key amino acid residues involved in multi-point binding interactions between brazzein, a sweet protein, and the T1R2-T1R3 human sweet receptor. *J Mol Biol* 398:584–589.

Assadi-Porter, F.M., Patry, S., and Markley, J.L. 2008. Efficient and rapid protein expression and purification of high disulfide containing protein brazzein in *E. coli*. *Protein Exp Purif* 58:263–268.

Bassoli, A., Merlini, L., and Moroni, G. 2002. Isovanillyl sweeteners. From molecules to receptors. *Pure Appl Chem* 74:1181–1187.

Chamberlain, J.J., and Abolnik, I.Z. 1997. Pulmonary edema following a licorice binge. *West J Med* 167:184–185.

Chen, Z., Cai, H., Li, Z., Liang, X., and Shangguan, X. 2007. Expression and secretion of a single-chain sweet protein monellin in *Bacillus subtilis* by sacB promoter and signal peptide. *Appl Microbiol Biotechnol* 73:1377–1381.

Chen, Z., Cai, H., Lu, F., and Du, L. 2005. High-level expression of a synthetic gene encoding a sweet protein, monellin, in *Escherichia coli*. *Biotechnol Lett* 27:1745–1749.

Choi, Y.-H., Hussain, R.A., Pezzuto, J.M., Kinghorn, A.D., and Morton, J.F. 1989. Abrusosides A-D, four novel sweet-tasting triterpene glycosides from the leaves of *Abrus precatorius*. *J Nat Prod* 52:1118–1127.

Crammer, B. 2008. Recent trends of some natural sweet substances from plants. In *Selected topics in the chemistry of natural products*, ed. R. Ikan, 189–208. World Scientific: Singapore.

de Klerck, G.J., Nieuwenhuis, M.G., and Beutler, J.J. 1997. Hypokalaemia and hypertension associated with the use of liquorice flavoured chewing gum. *Br Med J* 314:731–732.

Esposito, V., Gallucci, R., Picone, D., Saviano, G., Tancredi, T., and Temussi, P.A. 2006. The importance of electrostatic potential in the interactions of sweet proteins with the sweet taste receptor. *J Mol Biol* 360:448–456.

Fenwick, G.R., Lutomski, J., and Nieman, C. 1990. Liquorice, *Glycyrrhiza glabra* L.—composition, uses, and analysis. *Food Chem* 38:119–143.

Fischer, C.M., Harper, H.J., Henry, Jr., W.J., Mohlenkamp, Jr., M.K., Romer, K., and Swaine, Jr., R.L. 1994. Sweet beverages and sweetening compositions. *PCT Int Appl* WO9418855.

Gardana, C., Scaglianti, M., and Simonetti, P. 2010. Evaluation of steviol and its glycosides in *Stevia rebaudiana* leaves and commercial sweetener by ultra-high-performance liquid chromatography-mass spectrometry. *J Chromatogr A* 1217:1463–1470.

Gatti, F.G. 2008. Enantiospecific synthesis of (+)-hernandulcin. *Tetrahedron Lett* 49:4997–4998.

Hagiwara, A., Yoshino, H., Sano, M., Kawabe, M., Tamaro, S., Sakaue, K., Nakamura, M., Tada, M., Imaida, K., and Shirai, T. 2005. Thirteen-week feeding study of thaumatin (a natural proteinaceous sweetener), sterilized by electron beam irradiation, in Sprague-Dawley rats. *Food Chem Toxicol* 43:1297–1302.

Hellekant, G. 2007. Brazzein. In *Sweeteners*, 3rd edition, ed. R. Wilson, 47–50. Oxford, U.K.: Blackwell.

Higginbotham, J.D. 1986. Talin protein (thaumatin). Nabors, L.O., and G.E. Inglett. 1986. In *Alternative sweeteners*, eds. L.O. Nabors and R.C. Gelardi, 103–34. New York: Marcel Dekker.

Hirai, T., Fukukawa, G., Kakuta, H., Fukuda, N., and Ezura, H. 2010. Production of recombinant miraculin using transgenic tomatoes in a closed cultivation system. *J Agric Food Chem* 58:6096–6101.

Hobbs, J.R., Munger, S.D., and Conn, G.L. 2007. Monellin (MNEI) at 1.15 Å resolution. *Acta Crystallogr* F63:162–167.

Ide, N., Sato, E., Ohta, K., Masuda, T., and Kitabatake, N. 2009. Interactions of the sweet-tasting proteins thaumatin and lysozyme with the human sweet-taste receptor. *J Agric Food Chem* 57:5884–5890.

Izawa, H., Ota, M., Kohmura, M., and Ariyoshi, Y. 1996. Synthesis and characterization of the sweet protein brazzein. *Biopolymers* 39:95–101.

Jia, Z., and Yang, X. 2009. A minor, sweet cucurbitane glycoside from *Siraitia grosvenorii*. *Nat Prod Commun* 4:769–772.

Jiang, P., Cui, M., Zhao, B., Z. Liu, Z., Snyder, L.A., Benard, L.M.J., Osman, R., Margolskee, R.F., and Max, M. 2005. Lactisole interacts with the transmembrane domains of human T1R3 to inhibit sweet taste. *J Biol Chem* 280:15238–15246.

Jiang, P., Ji, Q., Liu, Z., Snyder, L.A., Benard, L.M.J., Margolskee, R.F., and Max, M. 2004. The cysteine rich region of T1R3 determine responses to intensely sweet proteins. *J Biol Chem* 279:45068–45075.

Jin, Z., Danilova, V., Assadi-Porter, F.M., Markley, J.L., and Hellekant, G. 2003. Monkey electrophysiological and human psychophysical responses to mutants of the sweet protein brazzein: Delineating brazzein sweetness. *Chem Senses* 28:491–498.

Jizba, J., Dolejs, L., Herout, V., and Sorm, F. 1971. Structure of osladin—the sweet principle of the rhizomes of *Polypodium vulgare* L. *Tetrahedron Lett* 1329–1332.

Johnson, C., Birch, G.G., and MacDougall, D.B. 1994. The effect of the sweetness inhibitor 2-(4-methoxyphenoxy)propanoic acid (sodium salt) (Na-PMP) on the taste of bitter-sweet stimuli. *Chem Senses* 19:349–358.

Kamei, K., Takano, R., Miyasaka, A., Imoto, T., and Hara, S. 1992. Amino acid sequence of sweet taste suppressing peptide (gurmarin) from the leaves of *Gymnema sylvestre*. *J Biochem* 111:109–112.

Kennelly, E.J., Cai, L., Long, L., Shamon, L., Zaw, K., Zhou, B.-N., Pezzuto, J.M., and Kinghorn, A.D. 1995. Novel highly sweet secodammarane glycosides from *Pterocarya paliurus*. *J Agric Food Chem* 43:2602–2607.

Kim, J.H., Lim, H.J., and Cheon, S.H. 2003. A facile synthesis of (6S,1'S)-(+)-hernandulcin and (6S,1'R)-(+)-epihernandulcin. *Tetrahedron* 59:7501–7507.

Kimura, T., Uzawa, T., Ishimori, K., Morishma, I., Takahashi, S., Konno, T., Akiyama, S., and Fujisawa, T. 2005. Specific collapse followed by slow hydrogen-bond formation of β-sheet in the folding of single-chain monellin. *Proc Natl Acad Sci USA* 102:2748–2753.

Kinghorn, A.D., Chin, Y.-W., Pan, L., and Jia, Z. 2010. Natural products as sweeteners and sweetness modifiers. In *Comprehensive natural products chemistry —II*, ed. R. Verpoorte, 269–315. Oxford, U.K.: Elsevier.

Kinghorn, A.D., and Compadre, C.M. 1991. Less common high-potency sweeteners. In *Alternative sweeteners*, 2nd edition (revised and expanded), eds. L.O. Nabors and R.C. Gelardi, 151–171. New York: Marcel Dekker.

Kinghorn, A.D., and Compadre, C.M. 2001. Less common high-potency sweeteners. In *Alternative sweeteners*, 3rd edition (revised and expanded), eds. L.O. Nabors and R.C. Gelardi, 209–234. New York: Marcel Dekker.

Kitagawa, I. 2002. Licorice root. A natural sweetener and important ingredient in Chinese medicine. *Pure Appl Chem* 74:1189–1198.

Kohmura, M., Ota, M., Izawa, H., Ming, D., Hellekant, G., and Ariyoshi, Y. 1996. Assignment of the disulfide bonds in the sweet protein brazzein. *Biopolymers* 38:553–556.

Kohmura, M., Mizukoshi, T., Nio, N., Suzuki, E.-I., and Ariyoshi, Y. 2002. Structure-taste relationships of the sweet protein monellin. *Pure Appl Chem* 74:1235–1242.

Kornfeld, E.C., Sheneman, J.M., and Suarez, T. 1969. Nonnutritive sweeteners. German Pat. DE-1969-1917844, Apr. 8, 22 pp.

Kurihara, Y. 1992. Characteristics of antisweet substances, sweet proteins, and sweetness-inducing proteins. *Crit Rev Food Sci Nutr* 32:231–252.

Ley, J.P., Blings, M., Paetz, S., Kindel, G., Freiherr, K., Krammer, G.E., and Bertram, H.-J. 2008. Enhancers for sweet taste from the world of non-volatiles. Polyphenols as taste modifiers. In *Sweetness and sweeteners*, eds. D.K. Weerasinghe and G.E. DuBois, ACS Symposium Series 979, 400–409. New York: Oxford University Press.

Ley, J., Kindel, G., Paetz, S., Reiss, T., Haug, M., Schmidtmann, R., and Krammer, G. 2007. Use of hesperetin for enhancing the sweet taste. *PCT Int Pat Appl.* WO2007014879 A1.

Li, D., Ikeda, T., Huang, Y., Liu, J., Nohara, T., Sakamoto, T., and Nonaka, G.-I. 2007. Seasonal variation of mogrosides in Lo Han Guo (*Siraitia grosvenorii*) fruits. *J Nat Medicines* 61:307–312.

Li, D.-F., Jiang, P., Zhu, D.-Y., Hu, Y., Max, M., and Wang, D.-C. 2008. Crystal structure of mabinlin II: A novel structural type of sweet proteins and the main structural basis for its sweetness. *J Struct Biol* 162:50–62.

Lindley, M.G. 1991. Phenoxyalkanoic acid sweeteners. In *Sweeteners: Discovery, molecular design, and chemoreception*, eds. D.E. Walters, F.T. Othoefer, G.E. Dubois, Symposium Series No. 450, 251–260. Washington, DC: American Chemical Society.

Liu, Y.-L., Ho, D.K., Cassady, J.M., Cook, V.M., and Baird, W.M. 1992. Isolation of potential cancer chemopreventive agents from *Eriodictyon californicum*. *J Nat Prod* 55:357–363.

Lord, J. 2007. Thaumatin. In *Sweeteners*, 3rd edition, ed. R. Wilson, 127–134. Oxford, U.K.: Blackwell.

Marone, P.A., Borzelleca, J.F., Merkel, D., Heimbach, J.T., and Kennepohl, E. 2008. Twenty eight-day dietary toxicity study of Lo Han fruit concentration in Hsd:SD rats. *Food Chem Toxicol* 46:910–919.

Masuda, T., and Kitabatake, N. 2006. Developments in microbiological production of sweet-tasting proteins, thaumatin and lysozyme. *Foods Food Ingred J Japan* 211:782–791.

Matsumoto, K., Kasai, R., Ohtani, K., and Tanaka, O. 1990. Minor cucurbitane-glycosides from fruits of *Siraitia grosvenori* (Cucurbitaceae). *Chem Pharm Bull* 38:2030–2032.

Ming, D., and Hellekant, G. 1994. Brazzein, a new high-potency thermostable sweet protein from *Pentadiplandra brazzeana* B. *FEBS Lett* 355:106–108.

Mizutani, K., Kuramoto, T., Tamura, Y., Ohtake, N., Doi, S., Nakaura, N., and Tanaka, O. 1994. Sweetness of glycyrrhetic acid 3-*O*-D-monoglucuronide and the related glycosides. *Biosci Biotech Biochem* 58:554–555.

Muller, G.W., Culbertson, J.C., Roy, G., Zeigler, J., Walters, D.E., Kellogg, M.S., Schiffman, S.S., and Warwick, Z.S. 1992. Carboxylic acid replacement structure-activity relationships in suosan-type sweeteners. A sweet taste antagonist 1. *J Med Chem* 15:1747–1751.

Nabors, L.O., and Inglett, G.E. 1986. A review of various other alternative sweeteners. In *Alternative sweeteners*, eds. L.O. Nabors, and R.C. Gelardi, 309–323. New York: Marcel Dekker.

Nakajima, K.-I., Asakura, T., Maruyama, J.-I., Morita, Y. Oike, H., Shimizu-Ibuka, A., Misaka, T. et al. 2006. Extracellular production of neoculin, a sweet-tasting heterodimeric protein with taste-modifying activity, by *Aspergillus oryzae*. *Appl Environ Microbiol* 72:3716–3723.

Nakajima, K.-I., Asakura, T., Misaka, T., and Abe, K. 2008. Neoculin as a novel sweet protein with taste-modifying activity. *Curr Top Biotechnol* 4:75–82.

Nirasawa, S., Masuda, Y., Nakaya, K., and Kurihara, T. 1996. Cloning and sequencing of a cDNA encoding a heat-stable sweet protein, mabinlin II. *Gene* 181:225–257.

Nishizawa, M., Yamada, H., Yamaguchi, Y., Hatakayama, S., Lee, I.-S., Kennelly, E.J., Kim, J., and Kinghorn, A.D. 1994. Structure revision of polypodoside A: Major sweet principle of *Polypodium glycyrrhiza*. *Chem Lett* 1555–1558; erratum *ibid.* 1579.

Nofre, C., Glaser, D., Tinti, J.-M., and Wanner, M. 2002. Gustatory responses of pigs to sixty compounds tasting sweet to humans. *J Anim Physiol Anim Nutr* 86:90–96.

Ohta, K., Masuda, T., Ide, N., and Kitabatake, N. 2008. Critical molecular regions for elicitation of the sweetness of the sweet-tasting protein, thaumatin I. *FEBS J* 275:3644–3652.

Ohtani, K., Aikawa, Y., Ishikawa, H., Kasai, R., Kitahata, S., Mizutani, K., Doi, S., Nakaura, M., and Tanaka, O. 1991. Further study on the 1,4-α-transglucosylation of rubusoside, a sweet steviol-bisglucoside from *Rubus suavissimus*. *Agric Biol Chem* 55:449–453.

Ono, M., Tsuru, T., Abe, H., Eto, M., Okawa, M., Abe, F., Kinjo, J., Ikeda, T., and Nohara, T. 2006. Bisabolane-type sesquiterpenes from the aerial parts of *Lippia dulcis*. *J Nat Prod* 69:1417–1420.

Patra, A.K., and Udgaonkar, J.B. 2007. Characterization of the folding and unfolding reactions of single-chain monellin: Evidence for multiple intermediates and competing pathways. *Biochem* 46:11727–11743.

Qin, X., Xiaojian, S., Ronggan, L., Yuxian, W., Zhunian, T., Shouji, G., and Heimbach, J. 2006. Subchronic 90-day oral (gavage) toxicity study of a Luo Han Guo mogroside extract in dogs. *Food Chem Toxicol* 44:2106–2109.

Roy, G. 1992. A review of sweet taste potentiation brought about by divalent oxygen and sulfur incorporation. *Crit Rev Food Sci Nutr* 31:59–77.

Shirasuka, Y., Nakajima, K.-I., Asakura, T., Yamashita, H., Yamamoto, A., Hata, S., Nagata, S., Abo, M., Sorimachi, H., and Abe, K. 2004. Neoculin as a new taste-modifying protein occurring in the fruit of *Curculigo latifolia*. *Biosci Biotechnol Biochem* 68:1403–1407.

Si, J., Chen, D., Chang, Q., and Shen, L. 1996. Isolation and determination of cucurbitane-glycosides from fresh fruits of *Siraitia grosvenorii*, *Zhiwu Xuebao* 38:489–494.

Spillane, W.J., and Thompson, E.F. 2009. The effect on taste upon the introduction of heteroatoms in sulphamates. *Food Chem* 114:217–225.

Sun, H.-J., Cui, M.-L., Ma, B., and Ezura, H. 2006. Functional expression of the taste-modifying protein, miraculin, in transgenic lettuce. *FEBS Lett* 580:620–626.

Temussi, P. 2006. The history of sweet taste: Not exactly a piece of cake. *J Mol Recogn* 19:188–199.

Tinti, J.-M., and Nofre, C. 1991. Design of sweeteners. A rational approach. In *Sweeteners: Discovery, molecular design, and chemoreception*, eds. D.E. Walters, F.T. Othoefer, G.E. Dubois, Symposium Series No. 450, 88–99. Washington, DC: American Chemical Society.

Tomcova, I., and Smatanova, I.K. 2007. Copper co-crystallization and divalent metal salts cross-influence effect: A new optimization tool improving crystal morphology and diffraction quality. *J Crys Growth* 306:383–389.

Walters, D.E. 2002. Homology-based model of the extracellular domain of the taste receptor T1R3. *Pure Appl Chem* 74:117–123.

Yamada, H., and Nishizawa, M. 1992. Syntheses of sweet-tasting diterpene glycosides, baiyunoside and analogs. *Tetrahedron* 48:3021–344.

Yamada, H., and Nishizawa M. 1995. Synthesis and structure revision of intensely sweet saponin osladin. *J Org Chem* 60:386–397.

Yamashita, H., Theerasilp, S., Aiuchi, T., Nakaya, K., Nahamura, Y., and Kurihara, Y. 1990. Purification and complete amino acid sequence of a new type of sweet protein with taste-modifying activity, curculin. *J Biol Chem* 265:15770–15775.

Yang, D.J., Zhong, Z.C., and Xie, Z.M. 1992. Studies on the sweet principles from the leaves of *Cyclocarya paliurus*. *Acta Pharm Sin* 27:841–844.

Yasuda, T., Kabaya, S., Takahashi, K., Nakazawa, T., and Ohsawa, K. 2004. Metabolic fate of orally administered phyllodulcin in rats. *J Nat Prod* 67:1604–1607.

Yoshikawa, K., Hirai, H., Tanaka, M., and Arihara, S. 2000. Antisweet natural products. XV. Structures of jegosaponins A-D from *Styrax japonica* Sieb. et Zucc. *Chem Pharm Bull* 48:1093–1096.

Yoshikawa, M., Murakami, T., Ueda, T., Shimoda, H., Yamahara, J., and Matsuda, H. 1999. Development of bioactive functions in *Hydrangea Dulcis Folium*. VII. Absolute stereostructures of 3S-phyllodulcin, 3R- and 3S-phyllodulcin glycosides, and 3R- and 3S-thunberginol H glycosides from the leaves of *Hydrangea macrophylla* Seringe var. *thunbergii* Makino. *Heterocycles* 50:411–4118.

Yoshikawa, K., Takahashi, K., Matsuchika, K., Arihara, S., Chang, H.-S., and Wang, J.-D. 1999. Antisweet natural products. XIV. Structures of alternosides XI-XIX from *Gymnema alternifolium*. *Chem Pharm Bull* 47:1598–1603.

REDUCED-CALORIE SWEETENERS

Chapter 15

Erythritol

Peter de Cock*

Contents

Introduction

Erythritol is a unique member of the polyol family. Polyols or polyhydric alcohols are polyhydroxyl compounds typically produced through the hydrogenation of their parent reducing sugars. These compounds do not have an aldehyde group, therefore, they do not undergo a Maillard reaction and are relatively stable to heat and changes in pH. Erythritol is a small linear sugar alcohol with four carbon atoms (Figure 15.1). It belongs to the acyclic alcohols, a group of sugar alcohols. Its small molecular size gives erythritol its extraordinary physicochemical, nutritional, and physiological

---

* Milda E. Embuscado and Sakharam K. Patil did not participate in the revision but their chapter in the book's 3rd edition was the basis of this revised chapter.

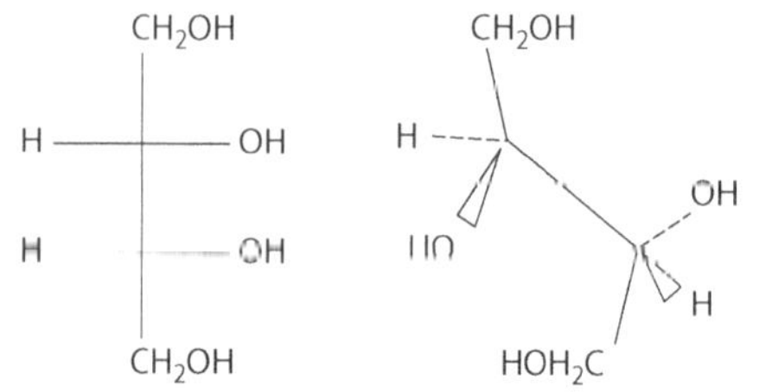

Figure 15.1 **Molecular structure and chemical formula of erythritol.**

properties, such as a zero calorie value and good digestive tolerance that is higher compared with all other polyols. These properties will be discussed in more detail later in this chapter.

Erythritol is a naturally occurring substance and the only polyol commercially produced through a natural fermentation process. It is found in a variety of foods such as grapes, pears, melons, mushrooms and in fermented products such as soy sauce, sake, and wines (Table 15.1). The concentration of erythritol in foods and fermented products can range from just a few milligram up to 1500 mg per liter or kilogram (Perko and de Cock 2006). Erythritol is a minor component of blood and amniotic fluids of cows and other mammals (Roberts et al. 1976). It can also be found in the human body, for instance, in the lens tissue (Tomana et al. 1984), in cerebrospinal fluid (Servo et al. 1977), in seminal plasma (Storset et al. 1978), and in blood serum (Budavari 1996). Erythritol is the main polyol in the human urine with a concentration ranging from 10–30 mg/L (Goossens and Gonze 1996).

Production

Erythritol was first isolated from the algae *Protococcus vulgaris* (now named *Apatococcus lobatus*) in 1852 by Lamy who named the substance phycit (Lamy 1852). Later in 1900, erythritol was also isolated from the algae *Trentepohlia jolithus* (Bamberger and Landsiedl 1900). In 1943, a German patent was granted to Reppe and Schnabel for the synthesis of erythritol from 2-butene-1,4-diol (Budavari 1996). Erythritol can also be synthesized from periodate-oxidized starch (Jeanes and Hudson 1955) or dialdehyde starch (Otey et al. 1961). Basically, there are two main methods of preparing erythritol: through chemical synthesis (e.g., reduction

Table 15.1 **The Natural Occurrence of Polyols in Various Foods**

Foods	Erythritol Content
Wine	130–300 mg/L
Sherry wine	70 mg/L
Sake	1550 mg/L
Soy sauce	910 mg/L
Miso bean paste	1310 mg/kg
Melons	22–47 mg/kg
Pears	0–40 mg/kg
Grapes	0–12 mg/kg

of meso-tartrate or oxidation of 4,6-0-ethylidene-D-glucose) and through fermentation. The chemical process is complex and costly because it involves several intricate steps to obtain the final product and the starting material is expensive. Fermentation is a simpler process requiring only a few steps and, although still costly, it is less expensive and the initial substrate is readily available.

The commercial production of erythritol uses a completely biotechnological process using enzymes and osmophilic yeasts or fungi (Bornet 1994; Kasumi 1995; Goossens and Gonze 1996; Röper and Goossens 1993; Goossens and Röper 1994, 1996). All other polyols are prepared by the catalytic hydrogenation of a precursor; for example, xylose from xylan is used to manufacture xylitol. *Moniliella, Trigonopsis,* or *Torulopsis* are some of the microorganisms that can convert glucose to erythritol in relatively high yields. The basic process is outlined in Figure 15.2. Wheat starch or cornstarch is the usual starting material. These are hydrolyzed primarily to glucose and other carbohydrates in lower concentration. An inoculum of the osmotolerant microorganism is added to the substrate, which ferments glucose to a mixture of erythritol and minor amounts of glycerol and ribitol. One of the yeasts used, *Moniliella,* can thrive at high sugar concentration and at the same time produce erythritol. This characteristic of *Moniliella* is advantageous because a concentrated substrate can be used as a starting material, and glucose can be added continuously to the fermentation tank without adversely affecting the growth of the microorganism and its efficiency to produce erythritol. The resulting concentrated fermentation broth facilitates the isolation of erythritol. After inactivation, the production microorganism is separated from the fermentation broth and impurities are removed through several purification steps. Erythritol is then crystallized and dried to yield crystals with over 99% purity.

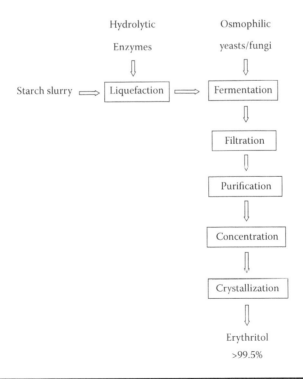

Figure 15.2   Commercial production of erythritol.

Physicochemical and Functional Properties

Sweetness

Tables 15.2 and 15.3 summarize the properties of erythritol, other polyols, and sucrose. The sweetness of erythritol is around 60% that of sucrose. Its sweetness profile is similar to sucrose with slight acidity and bitterness but with no detectable aftertaste. Most polyols have sweetness levels less than sucrose except xylitol that has about the same sweetness intensity and maltitol only slightly less. Perhaps even more important than sweetness intensity is the sweetness quality; the persistence of sweetness, presence or absence of aftertaste, mouthfeel characteristics, and how closely the overall sweetness profile matches the gold standard, sucrose. These quality factors will determine the acceptability and success of any alternative sweetener.

The level of sweetness of erythritol can easily be increased through blending with conventional high potency sweeteners such as aspartame and sucralose or with natural high potency sweeteners like rebaudioside A (rebiana). Rebiana is the best tasting steviol glycoside extracted from the leaves of *Stevia rebaudiana Bertoni*. Table 15.4 shows how relatively low amounts of a high potency sweetener can boost the sweetness intensity of erythritol through quantitative synergy. This is especially useful when replacing sugar pound-for-pound in solid food applications or for uses of erythritol as a tabletop sweetener.

Table 15.2   Properties of Erythritol Compared with Other Polyols and Sucrose

Sugar	Sweetness (Sucrose = 1)	Heat of Solution (cal/g)	Cooling Effect Crystals	Viscosity at 25°C	Hygroscopicity
Erythritol C4 MW = 122	0.6	−43	Very cool	Very low	Very low
Xylitol C5 MW = 152	1.0	−36	Very cool	Very low	Medium
Mannitol C6 MW = 182	0.6	−29	Cool	Low	Low
Sorbitol C6 MW = 182	0.6	−26	Cool	Medium	High
Maltitol C12 MW = 344	0.9	−19	None	Medium	Medium
Isomalt C12 MW = 344	0.4	−9	None	High	Low
Lactitol C12 MW = 344	0.3	−14	Slightly cool	Very low	Medium
Sucrose C12 MW = 342	1.0	−4	None	Low	Medium

Table 15.3 Properties of Erythritol Compared with Other Polyols and Sucrose

Sugar	Melting Point (°C)	Tg (°C)	Solubility g/100g H_2O (25°C)	Heat Stability (°C)	Acid Stability
Erythritol	121	−42	37	>160	2–12
Xylitol	94	−22	64	>160	2–10
Mannitol	167	−39	20	>160	2–10
Sorbitol	99	−5	70	>160	2–10
Maltitol	152	47	60	>160	2–10
Isomalt	96/145	49	25	>160	2–10
Lactitol	92/124	22	57	>160	>3
Sucrose	190	−32	67	Decomposes at 160	Hydrolyzes under acidic and alkaline conditions

Cooling Effect

All polyols exhibit negative heats of solution. Energy is needed to dissolve the polyol crystals, thus they absorb the surrounding energy, resulting in a lowering of the temperature or a cooling of the solution. This property is likewise observed when the dry powder of polyol is dissolved in the mouth. This creates a cooling sensation in the mouth. The degree of cooling depends on the magnitude of the heat of solution and the speed of dissolution. As shown in Table 15.2, the cooling effect ranges from none to very cool. It appears that this cooling effect is only slight or not perceived at all with heat of solution values higher than −20 cal/g. The high cooling effect of erythritol can be advantageous in food and pharmaceutical formulations containing mint or menthol. The cooling sensation is also beneficial in pharmaceutical preparations requiring soothing effects (e.g., lozenges, cough drops, throat medication, and breath mints).

Table 15.4 Percent Synergy for Mixtures of Bulk and Intense Sweeteners

Blend	Bulk Sweetener–Intense Sweetener Sweetness Contribution Ratio[a]					
	1–99	5–95	15–85	85–15	95–5	99–1
Erythritol–Aspartame (% synergy)[b]	−3	−7	10	30[c]	25[c]	24[c]
Erythritol–Acesulfame K (% synergy)[b]	12	8	19[c]	32[c]	31[c]	27[c]

[a] Expected sweetness (sucrose equivalent value SEV) = 10.
[b] % synergy = 100 × [[SE mixture/SE (100% erythritol + 100% intense sweetener) ÷ 2] −1], where SE is the panel's sweet intensity rating. The equation is based on Carr et al. 1993.
[c] Significant at $P < 0.05$.

Solubility and Hygroscopicity

Erythritol is moderately soluble in water unlike sucrose and other polyols (xylitol, sorbitol, maltitol, and lactitol), which are highly soluble in water. Mannitol and isomalt are less soluble in water than erythritol. The low affinity of erythritol for water is reflected in its sorption isotherm (Figure 15.3). Also the isotherm of sucrose is shown for comparative purposes. The sorption isotherms were determined using an SGA-100 water sorption analyzer (VTI Corporation, FL). The change in weight of a completely dried erythritol powder was monitored at 25°C when the relative humidity was increased from 5%–95% and then back to 5% in 5% steps. Erythritol went into solution above 90% relative humidity, whereas sucrose went into solution at a lower relative humidity (around 84%). To determine the exact deliquescence point, erythritol was exposed to relative humidities between 88%–94% in a stepwise fashion. Results show that erythritol starts to deliquesce between 92%–94% relative humidities. When erythritol was exposed to decreasing relative humidities, the sample lost most of its moisture, especially when the relative humidity reached 85%. The adsorption and desorption isotherm profile of erythritol is quite different from that of sucrose (Figure 15.3). As the relative humidity was reduced to 75%, sucrose lost some of

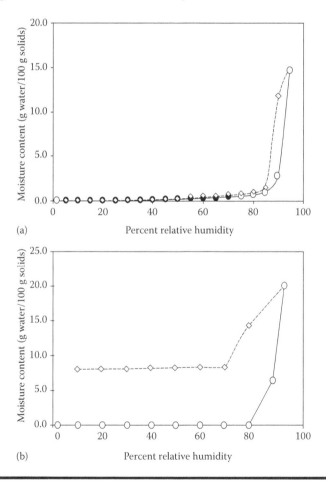

(a)

(b)

Figure 15.3 Sorption isotherms of (a) erythritol and (b) sucrose at 25°C. (Courtesy of VTI Corporation, Hialeah, FL). o—o, Adsorption; ◊- -◊, desorption.

the moisture it absorbed at higher relative humidity. At about 75% relative humidity, the moisture content of sucrose stabilized and remained the same even when the relative humidity was reduced down to 5%. The adsorption and desorption properties of erythritol have important implications in product formulation and storage.

The relatively low solubility of erythritol limits its applications to certain types of food preparations. This property is desirable, however, in applications where moisture pickup of the product on storage should be held at a minimum (e.g., fruit pieces, fruit bars, pastries, powder applications). An increased rate of moisture absorption for these types of products usually results in loss in quality (softening, stickiness) or even microbiological deterioration. When erythritol is used as an ingredient or as a coating material, the moisture pickup of the product is retarded or is held to a minimum. Replacement of hygroscopic sugars with erythritol may also have a desirable impact on the choice of packaging material. In place of expensive multilayered laminates, a simple and less expensive packaging material may suffice to protect the product. In addition, a product coated with erythritol is more stable when directly exposed to relative humidities between 85%–90% compared with products coated with sugars. It is also worth mentioning that the presence of erythritol to control the water activity (Aw) inhibited the growth of *Staphylococcus aureus* even at high Aw levels (0.92–0.94) (Vaamonde et al. 1986). Other solutes, such as sodium chloride, potassium chloride, sucrose, glucose, sodium lactate, and sodium acetate, required a lower Aw to inhibit the growth of *S. aureus*.

Other Characteristics

Erythritol is stable at high temperatures and at a wide pH range (Table 15.3). Like other polyols, it does not undergo a Maillard or browning reaction because it does not have a reactive aldehyde group. This is beneficial in food applications, where maintaining the delicate flavor and the intrinsic qualities of the products is important. Food products with sugars and proteins or amino acids may undergo browning discoloration and develop a bitter taste when processed at high temperature.

Physiological and Health Properties

Metabolic Fate

The metabolic pathway of erythritol is different from sugars and other polyols. Erythritol is rapidly absorbed in the small intestine through passive diffusion because of its small molecular volume. Glucose and sucrose are also readily absorbed in the small intestine after which they are fully metabolized to produce energy and carbon dioxide. Erythritol, on the other hand, is not metabolized at all and excreted unchanged in the urine (Bernt et al. 1996). This also means that erythritol does not have any impact on blood glucose or insulin levels as was confirmed in healthy people (Bornet et al. 1996a) and in patients with diabetes (Ishikawa et al. 1996).

Figure 15.4 shows the rate of $^{13}CO_2$ excretion and the cumulative $^{13}CO_2$ excretion after ingestion of 25 g of ^{13}C-labeled glucose, lactitol, or erythritol. There was no detectable $^{13}CO_2$ excretion for erythritol, whereas for glucose the maximum level was reached after 2–3.5 hours of oral administration (Hiele et al. 1993). The rate of $^{13}CO_2$ for lactitol was slower than for glucose, and it reached its peak six hours after ingestion. The H_2 excretions of ^{13}C-labeled glucose, lactitol, or erythritol are shown in Figure 15.5. The H_2 excretions, indicative for fermentation in the colon,

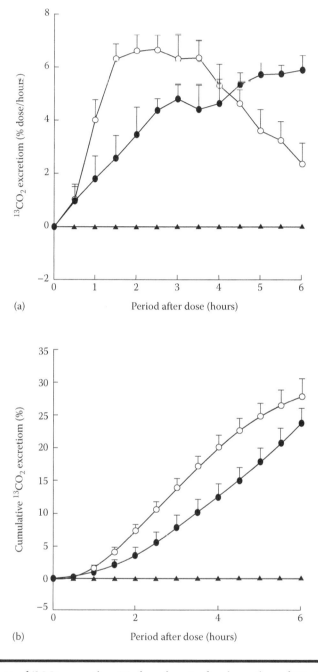

(a)

(b)

Figure 15.4 **(a) Rate of $^{13}CO_2$ excretion (% dose/hour) after ingestion of 25 g ^{13}CO-labeled glucose (o), lactitol (•), or erythritol (▲). (b) Cumulative amount of excretion (% administered dose) after $^{13}CO_2$ ingestion of 25 g ^{13}CO-labeled glucose (o), lactitol (•), or erythritol (▲) in human volunteers. (From Hiele et al.,** *Br J Nutr* **69:169–176, 1993.)**

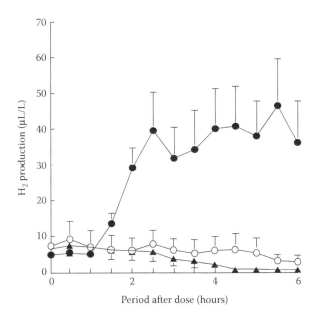

Figure 15.5 H₂ excretion after ingestion of 25 g ¹³CO-labeled glucose (○), lactitol (●), or erythritol (▲) in healthy volunteers. (From Hiele et al., *Br J Nutr* 69:169–176, 1993.)

for erythritol and glucose were low compared with lactitol. Lactitol produced a significant amount of H₂ two hours after ingestion. Hiele et al. (1993) concluded that erythritol is not metabolized by fecal flora. The duration of his fermentation experiments, however, was only six hours.

In order to investigate over a longer period of time whether fresh human intestinal microbiota is able to adapt its enzyme activities to erythritol, a 24 hours lasting fermentation was carried out under well-standardized *in vitro* conditions. For comparison, maltitol, lactulose and a blank (fecal inoculum only) were incubated as well. Fermentation patterns were established by following total gas production, hydrogen accumulation, changes in pH value, Short chain fatty acids (SCFA) production and substrate degradation. Taking all fermentation parameters into account, erythritol turned out to be completely resistant to bacterial degradation within 24 hours, thus excluding an adaptation within that period. Since under *in vivo* conditions more easily fermentable substrates enter the colon continuously, it seems very unlikely that erythritol will be fermented *in vivo* (Arrigoni et al. 2005).

Caloric Value

Erythritol is not metabolized systemically and about 90% of the ingested erythritol is excreted unchanged with the urine (Munro et al. 1998). The small amount of unabsorbed erythritol reaches the large intestine where it is not fermented by the colonic microorganisms (Hiele et al. 1993; Arrigoni et al. 2005). Hence the energy value of erythritol is zero.

Many years ago, some countries like Canada and Australia conservatively estimated the energy value for erythritol at 0.2 kcal/g based on the assumption that the 10% that reaches the colon indeed is fermented. Currently, most countries use a zero energy value for erythritol. Table 15.5 lists the regulatory permitted energy values for all polyols to be used for purposes of nutritional labeling in the United States, the EU and Japan.

Table 15.5   Energy Value of Polyols for Nutritional Labeling (kcal/g)

Polyol	USA	EU	Japan
Erythritol	0	0	0
Xylitol	2.4	2.4	3
Sorbitol	2.6	2.4	3
Sorbitol syrups	2.6	2.4	3
Mannitol	1.6	2.4	2
Maltitol	2.1	2.4	2
Maltitol syrups	3	2.4	2.3–3.4
Isomalt	2	2.4	2
Lactitol	2	2.4	2

Digestive Tolerance

Normal intake of foods formulated with polyols is typically well-tolerated. Excessive intake of polyols, however, can cause digestive side effects such as flatulence and laxation. The latter may be caused by local high concentrations of the non-absorbed polyol fraction resulting in high local osmolality in the large intestine and watery feces. Erythritol however, unlike all other polyols, is readily absorbed and therefore prevents the accumulation of unabsorbed products in the large intestine. Table 15.6 lists the key published digestive tolerance studies conducted with erythritol in adults using an acute bolus dose in a liquid or repeated exposures spread over the day from various foods. The maximum dose that is still well-tolerated is about 40 g when consumed as a bolus liquid dose or 80 g when consumed spread over the day.

The available data clearly demonstrate that erythritol has a much lower potential to cause laxative effects as compared to other polyols, or even compared to sugars like lactose or tagatose. The laxation threshold for erythritol is generally from 2-fold (compared to xylitol and isomalt) to 4-fold (compared to sorbitol and mannitol) greater than other polyols. The high laxation threshold for erythritol enables its use in products such as diet beverages, where high exposures may occur as a bolus over a short period of time.

Tooth-Friendliness

Noncariogenicity and nonacidogenicity are important properties of polyols. The criteria used to assess the noncariogenic property of a substance are its nonfermentability by oral microorganisms, nonacidogenic property, and the absence of glucan formation (Schiweck and Ziesenitz 1995). Fermentation of sugars by oral bacteria produces acid which causes demineralization of tooth enamel below a critical pH. For a substance to be classified as nonacidogenic, the critical pH value in humans as measured by plaque pH telemetry should be equal to or greater than 5.7.

Erythritol, like other polyols, is tooth-friendly. It is not used as a substrate for lactic acid production or for plaque polysaccharide synthesis by oral streptococci (Kawanabe et al. 1992). The acid production of *Streptococcus mutans*, a plaque-forming bacterium, in 3% solutions of erythritol or other polyols under controlled *in vitro* conditions is shown in Figure 15.6. Erythritol

Table 15.6 Summary of Erythritol Digestive Tolerance Studies in Adults Following Acute Bolus or Repeated Exposures

Delivery Vehicle	Dose Type	Number of Subjects	NOEL[a] for Laxation (g)	NOEL for Laxation (g/kg/bw)	Reference
Beverage	Single, Bolus	n = 6	<64	<1	Bornet et al. 1996a
Beverage	Single, Bolus	n = 5	20[b]	N/A	Ishikawa et al. 1996
Beverage–Jelly	Single, Bolus	n = 38	41 F; 43 M	0.80 F; 0.66 M	Oku and Okazaki 1996
Beverage	Single, Bolus	n = 65	50[b]	0.78	Storey et al. 2007
Beverage	Continuous, 14d	n = 11	20[b]	N/A	Ishikawa et al. 1996
Food (Chocolate)	Single, Bolus	n = 24	52–60[b]	0.8	Bornet et al. 1996b
Food/Beverage	Continuous, 7d	n = 12	~79[b]	1.0	Tetzloff et al. 1996

[a] No Observed Effect Level.
[b] Highest dose tested.

and xylitol did not support acid production by this microorganism for more than 36 hours of incubation.

The effects of six months use of erythritol, xylitol and sorbitol were investigated in a cohort of 136 teenage subjects assigned to the respective polyol groups or to an untreated control group (Mäkinen et al. 2005). The daily use of the polyols was 7.0 g in the form of chewable tablets, supplemented by twice-a-day use of a dentifrice containing those polyols. The use of erythritol and xylitol was associated with a statistically significant reduction ($p < 0.001$ in most cases) in the plaque and saliva levels of mutans streptococci. The amount of dental plaque was also significantly reduced in subjects receiving erythritol and xylitol. Such effects were not observed in other experimental groups. These *in vivo* studies were supported by cultivation experiments in which xylitol, and especially erythritol, inhibited the growth of several strains of mutans streptococci. The investigators concluded that erythritol and xylitol may exert similar effects on reducing some risk factors of dental caries, although the biochemical mechanism of the effects may differ (Mäkinen et al. 2005).

Antioxidant Properties

Polyols are excellent scavengers of hydroxyl radicals because of the many hydroxyl groups that they possess. This has, however, only limited potential in animals or humans as polyols are poorly absorbed, with the exception of erythritol. Erythritol is highly bioavailable and, in addition, it is not metabolized and circulates throughout the body where it potentially can exercise its radical scavenging activity.

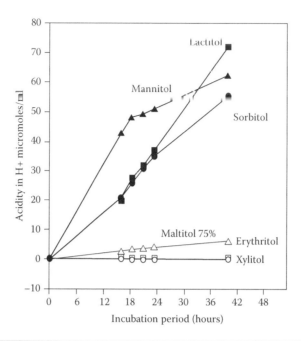

Figure 15.6   Development of the acidity by *Streptococcus mutans* strains growing on a 3% solution of erythritol or another polyol under controlled *in vitro* conditions. Cultures with an initial bacteria count of 3 × 10⁹ were incubated in test tubes at 37°C in a biological buffering system. Acidity is expressed as total H+ concentration in micromoles/ml accumulating over time period. (From Goossens, J., and Gonze, M., *Advances in sweeteners*, ed. T.H. Grenby, 150–186. Gaithersburg, MD: Aspen Publishers, 1996.)

Den Hartog et al. (2009) evaluated the antioxidative and endothelium-protective properties of erythritol. Vascular health is strongly dependent on endothelial state, which regulates vascular tone and inflammatory responses. High glucose-induced endothelial cell apoptosis, likely via reactive oxygen species, has a pivotal role in diabetes-associated vascular diseases, including atherosclerosis. In competition assays, erythritol was shown to be an excellent hydroxyl radical scavenger (inert towards superoxide radicals) and inhibited chemical-induced hemolysis. Erythritol reacted with hydroxyl radicals forming erythrose and erythrulose by carbon-bound hydrogen abstraction. In streptozotocin-induced diabetic rats, erythritol had endothelium-protective effects and erythrose was found in urine. These endothelial protective effects of erythritol cannot be explained by hydroxyl radical scavenging only. More research is needed to elucidate the mechanism of action and to study these effects in humans.

Safety and Regulatory Status

Erythritol has been part of our diet for as long as mankind exists as it occurs naturally at low levels in foods such as grapes, pears, mushrooms and fermented foods (Table 15.1). The consumption of erythritol from its natural occurrence in foods is currently estimated to range from 25 mg/person/day in the United States to 106 mg/person/day in Japan.

The safety of erythritol is well-documented. Numerous safety studies following US FDA's Redbook guidelines have been conducted and published to establish the safety of erythritol. On the basis of the entire safety data base, food safety experts concluded that erythritol is safe for its intended use in foods (Bernt et al. 1996; Munro et al. 1998). Many health authorities worldwide have also reviewed the safety of erythritol and concluded that it is safe (WHO 1999; US FDA 2001; SCF 2003). Currently, erythritol is authorized for use in foods in more than 50 countries around the world including Japan, the United States, Canada, Mexico, Brazil, Argentina, those in the EU, Turkey, Russia, China and various other countries in Asia, Australia and New Zealand. Erythritol is also used as an excipient in pharmaceutical products and included in the European and Japanese Pharmacopeia.

Food and Pharmaceutical Applications

Diet or calorie-reduced beverages are the most popular applications for erythritol. Use of high potency sweeteners in such beverages typically results in lack of mouthfeel and is associated with certain unwanted off-tastes like astringency, bitterness and irritant. Erythritol is capable at relatively low concentrations (1%–3%) to modify the sweetness profile to become more sugar-like by adding mouthfeel and masking some of the unwanted off-tastes.

Erythritol is also often used in tabletop sweeteners because it resembles the taste, appearance and crystallinity of sucrose, without adding the calories. Like in diet beverages, it is mostly used in combination with high potency sweeteners and can likewise modify their flavor profile to become more sugar-like.

Also in dairy applications, erythritol is used, for example, in yogurt or custard to reduce sugar and/or calories without compromising too much on taste and mouthfeel. In ice cream, erythritol is not only used to reduce calories but also to improve the texture and spoonability due to its freezing point depression.

In chewing gum, erythritol adds freshness because of its high cooling effect. Its medium solubility results in a cooling effect and sweetness release that may last longer in comparison to high solubility sweeteners. In chewing gum centers and sticks erythritol can provide a softer texture and greater flexibility. The latter can be useful during processing where it reduces the risk that the chewing gum rope will break after extrusion and cooling.

Chocolate with 30% less calories (required in the EU for use of a calorie-reduced claim) can easily be made with erythritol using traditional manufacturing processes. Other polyols do not provide enough calorie reduction to achieve such a claim through simple sugar replacement. The cooling effect of erythritol combines well with mint-flavored chocolate. If cooling is unwanted it can be masked by using erythritol in conjunction with, for example, inulin which has a positive heat of solution. New patent-pending technology has been developed to reduce the cooling effect of erythritol through co-melting it with specific hydrocolloids, carrageenan in particular (Vercauteren 2008).

Various types of candies can be made with erythritol. Some can contain erythritol as the sole sweetener such as lozenges, resulting in an energy reduction of 90% or more. It can be used to provide a high cooling effect in sherbet-filled stamped hard candies, or "sandwiched" in between two layers in deposited hard candies. Excellent sugarfree fudge and fondant can be formulated with erythritol with a quality similar to conventional sugar-containing versions. The medium solubility and crystallization behavior (extremely high speed of crystallization) of erythritol limit its use in gum confectionery. It can, however, function very well as graining sugar in this application, which is also due to its low hygroscopicity.

In bakery applications, compared to sucrose, erythritol exhibits a different melting behavior, more compact dough, less color formation, and better moisture control. Erythritol is often used in combination with maltitol to leverage its higher sweetness and humectancy properties. This may also result in an improved baking stability (regular and finer crumb structure) and a softer end product like is the case with cake.

Pharmaceutical applications include a wide range of solid and liquid formulations where the taste masking properties of erythritol are of great advantage when using bad tasting actives. Because erythritol is not metabolized by the human body, it is used as an inert excipient and carrier of drugs in capsules. Erythritol also provides good flowability and stability to powder formulations.

Erythritol is the only bulk sweetener without calories. It has unique physicochemical and sensorial properties as well as various potential health benefits. Erythritol creates important new opportunities in reducing sugar and energy intake, highly welcome in our overweight society.

References

Arrigoni, E., Brouns, F., and Amado, R. 2005. Human gut microbiota does not ferment erythritol. *Br J Nutr* 94(5):643–646.

Bamberger, M., and Landsiedl, A. 1900. Erythritol in Trentepohlia jolithus. *Monatshefte fuer Chemie* 21:571–573.

Bernt, W.O., Borzelleca, J.F., Flamm, G., and Munro, I.C. 1996. Erythritol: A review of biological and toxicological studies. *Regul Toxicol Pharmacol* 24:S191–S197.

Bornet, F.J.R. 1994. Undigestible sugars in food products. *Am J Clin Nutr* 59S:763S–769S.

Bornet, F.R.J., Blayo. A., Dauchy. F., and Slama, G. 1996a. Plasma and urine kinetics of erythritol after oral ingestion by healthy humans. *Regul Toxicol Pharmacol* 24:S280–S286.

Bornet, F.R.J., Blayo, A., Dauchy, F., and Slama, G. 1996b. Gastrointestinal response and plasma and urine determinations in human subjects given erythritol. *Regul Toxicol Pharmacol* 24:S296–S302.

Budavari, S., ed. 1996. The Merck Index. Whitehouse Station, NJ: Merck Research Laboratories, pp. 624, 763–764, 912, 979, 1490–1491, 1517–1518, 1723–1724.

Carr, T.B., Pecore, S.D., and Gibes, K.M. 1993. Sensory methods for sweetener evaluation. In *Flavor measurement*, eds. C.T. Ho and C.H. Manley, 219–237. New York: Marcel Dekker, Inc.

Den Hartog, G.J.M., Boots, A.W., Adam-Perrot, A., Brouns, F., Verkooijen, I.W.C.M., Weseler, A.R., Haenen, G.R.M.M., and Bast, A. 2009. Erythritol is a sweet antioxidant. *J Nut* 26(4):449–458.

Goossens, J., and Röper, H. 1994. Erythritol: A new sweetener. *Food Sci Tech Today* 8:144–148.

Goossens, J., and Röper, H. 1996. Erythritol: A new sweetener. *Confec Prod* 62:4, 6–7.

Goossens, J., and Gonze, M. 1996. Nutritional properties and applications of erythritol: A unique combination? In *Advances in sweeteners*, ed. T.H. Grenby, 150–186. Gaithersburg, MD: Aspen Publishers.

Hiele, M., Ghoos, Y., Rutgeerts, P., and Vantrappen, G. 1993. Metabolism of erythritol in humans: Comparison with glucose and lactitol. *Br J Nutr* 69:169–176.

Ishikawa, M., Miyashita, M., Kawashima, Y., Nakamura, T., Saitou, N., and Modderman, J. 1996. Effects of oral administration of erythritol on patients with diabetes. *Regul Toxicol Pharmacol* 24:S303–S308.

Jeanes, A., and Hudson, C.S. 1955. Preparation of meso-erythritol and d-erythronic lactone from periodate-oxidized starch. *J Org Chem* 20:1565–1568.

Kasumi, T. 1995. Fermentative production of polyols and utilization for food and other products in Japan. *Jpn Agric Res* 29:49–55.

Kawanabe, J., Hirasawa, M., Takeuchi, T., Oda, T., and Ikeda, T. 1992. Noncariogenicity of erythritol as a substrate. *Caries Res* 26:258–362.

Lamy, A. 1852. Ueber einige Bestandtheile des Protococcus vulgaris. *J Prakt Chem* 57(1):21–28.

Mäkinen K.K., Saag M, Isotupa K.P., Olak J, Nõmmela R, Söderling E, and Mäkinen P.L. 2005. Similarity of the effects of erythritol and xylitol on some risk factors of dental caries. *Caries Res* 39(3):207–215.

Munro, I.C., Bernt, W.O., Borzella. J.F., Flamm, G., Lynch, B.S., Kennepohl, E., Bär, E.A., and Modderman, J. 1998. Erythritol: An interpretive summary of biochemical, metabolic, toxicological and chemical data. *Food Chem Toxicol* 36:1139–1174.

Oku, T., and Okazaki, M. 1996. Laxative threshold of sugar alcohol erythritol in human subjects. *Nutr Res* 16(4):577–589.

Otey, F.H., Sloan, J.W., Wilham, C.A., and Mehltretter, C.L. 1961. Erythritol and ethylene glycol from dialdehyde starch. *Industrial Eng Chem* 53:267–268.

Perko, R., and de Cock, P. 2006. Erythritol. In *Sweeteners and sugar alternatives in food technology*, ed. H. Mitchell, 149–176. Oxford, UK: Blackwell Publishing Ltd.

Roberts, G.P., MacDiarmid, A., and Gleed, P. 1976. The presence of erythritol in the fetal fluid of fallow deer. *Res Vet Science* 20:154–256.

Röper, H., and Goossens, J. 1993. Erythritol, a new raw material for food and non-food applications. *Starch/Stärke* 45:400–405.

SCF. 2003. Opinion of the Scientific Committee on Food: Erythritol. Scientific Committee on Food. European Commission, Health and Consumer Protection Directorate-General. SCF/CS/ADD/EDUL/215 Final. Opinion expressed 23 March.

Schiweck, H., and Ziesenitz, S.C. 1995. Physiological properties of polyols in comparison with easily metabolisable saccharides. In *Advances in sweeteners*, ed. T.H. Grenby, 56–84. Gaithersburg, MD: Aspen Publishers.

Servo, C., Pabo, J., and Pitkaenen, E. 1977. Gas chromatographic separation and mass spectrometric identification of polyols in human cerebrospinal fluid and plasma. *Acta Neurol Scand* 56:104–110.

Storey, D.M., Lee, A., Bornet, F., and Brouns, F. 2007. Gastrointestinal tolerance of erythritol and xylitol ingested in a liquid. *Eur J Clin Nutr* 61:349–354.

Storset, P., Stooke, O., and Jellum, E. 1978. Monosaccharides and monosaccharide derivatives in human seminal plasma. *J Chrom* 145:351–357.

Tetzloff, W., Dauchy, F., Medimagh, S., Carr, D., and Bär, A. 1996. Tolerance to subchronic, high-dose ingestion of erythritol in human volunteers. *Regul Toxicol Pharmacol* 24:S286–S295.

Tomana, M., Prchal, J.T., Cooper Garner, L., Skala, H.W., and Barker S.A. 1984. Gas chromatographic analysis of lens monosaccharides. *J Lab Clin Med* 103:137–142.

US FDA. 2001. Agency Response Letter GRAS Notice No. GRN 000076.

Vaamonde, G., Scarmato, G., Chirife, J., and Parada, J.L. 1986. Inhibition of Staphylococcus aureus C-243 growth in laboratory media with water activity adjusted using less usual solutes. *Lebensm-Wiss u-Technol* 19:403–404.

Vercauteren, R.L.M. 2008. Reducing the sensory cooling effect of polyols. PCT patent application WO 2008/125344 Al.

WHO. 1999. Summary and conclusions of the fifty-third meeting of the Joint WHO/FAO Expert Committee on Food Additives in Rome, June 1–10, 1999. WHO Technical Report Series 896:18–22.

Chapter 16

Maltitol Syrups and Polyglycitols

Ronald C. Deis

Contents

Introduction

The terms "polyglycitols," "polyglucitols," and hydrogenated starch hydrolysates (HSH) have been used interchangeably, depending on the reference source. In reality, these terms can encompass a wide array of products including maltitol syrups, sorbitol syrups and hydrogenated glucose syrups, but polyglycitols and maltitol syrups are differentiated in the food industry by their maltitol content. Products containing more than 50% maltitol on a dry basis are referred to as "maltitol syrups," while those containing less than 50% maltitol are "polyglycitols." According to *FCC* 7 (*Food Chemicals Codex* 2010), the acceptance criteria for maltitol syrups are no less than 50.0% by weight dry basis (db) maltitol and no more than 8.0% by weight (db) sorbitol.

Polyglycitol is listed under "hydrogenated starch hydrolysate" in FCC 7, and is described as a concentrated aqueous solution or a dried powder, and as a mixture of sorbitol, maltitol, maltitriol, and hydrogenated polysaccharides, with neither the sorbitol nor maltitol fractions comprising 50% of the sample. The range of these products is equivalent, from a molecular weight distribution standpoint, to the range of corn syrups, corn syrup solids and maltodextrins currently on the market. Because of their level of sweetness, the bulk of the products found in the market today are maltitol syrups, but polyglycitols are well-used for their higher molecular weight stability and non-reactivity with other ingredients (colors, flavors, enzymes, etc.) as carriers and flow aides. In total, these food ingredients serve a number of functional roles, including use as bulk sweeteners, viscosity or bodying agents, humectants, crystallization modifiers, cryoprotectants and rehydration aids. They also can serve as sugar-free carriers for flavors, colors and enzymes.

Production

Maltitol syrups and polyglycitols are produced by the partial hydrolysis of corn, wheat, tapioca or potato starch and catalytic hydrogenation of the hydrolysate at high temperature under pressure. By varying the conditions and extent of hydrolysis, the relative occurrence of various mono-, di-, oligo-, and polymeric hydrogenated saccharides in the resulting product can be obtained. The proportion of these hydrogenated species distinctly affects the functional properties and chemistry of the particular maltitol syrup or polyglycitol. For maltitol syrups, the starting point is typically a high maltose syrup and for polyglycitols a lower dextrose equivalent (DE) syrup or maltodextrin.

In the United States, some changes within the industry have occurred within the past ten years—Corn Products International acquired SPI Polyols, Inc., which had previously acquired Lonza Group Ltd.'s polyols business, and Cargill, Inc. acquired Cerestar, Inc. Maltitol syrups are produced and marketed now in the United States by three major manufacturers: Corn Products International (MALTISWEET® and HYSTAR®) Roquette America, Inc. (LYCASIN®), and Cargill, Inc. (Maltidex™). Polyglycitols are produced and marketed in the United States by Corn Products International (STABILITE® and HYSTAR). The products produced by these manufacturers differ in viscosity, sweetness, and hygroscopicity, depending on their composition.

Physical Characteristics

Maltitol syrups and polyglycitols are colorless and odorless and generally available as 70%–85% solids syrups, composed of variable relative amounts of hydrogenated saccharides, characterized by their degree of polymerization (DP). Higher molecular weight polyglycitols, derived from maltodextrins, are also available in a powdered form. As the % maltitol (DP-2) increases in these syrups, the sweetness increases and the average molecular weight (MW) decreases. The effect of molecular weight changes is similar to that seen in corn syrup (Figure 16.1). Typical structures, along with their precursors, are shown in Figure 16.2. Maltitol syrups and polyglycitol syrups can be easily blended to change the change the sweetness, viscosity, humectancy or other properties of the product. In order to further change their properties, high potency sweeteners, soluble fibers, resistant maltodextrins, or glycerin have also been blended into these products, either to change properties or as a delivery system.

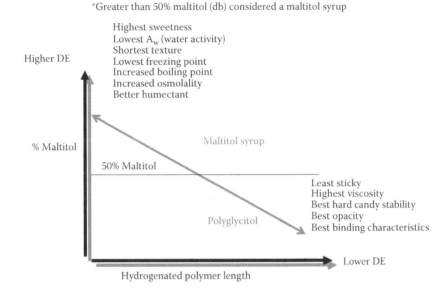

*Greater than 50% maltitol (db) considered a maltitol syrup

Higher DE

Highest sweetness
Lowest A_w (water activity)
Shortest texture
Lowest freezing point
Increased boiling point
Increased osmolality
Better humectant

% Maltitol

50% Maltitol

Maltitol syrup

Least sticky
Highest viscosity
Best hard candy stability
Best opacity
Best binding characteristics

Polyglycitol

Lower DE

Hydrogenated polymer length

Figure 16.1 Polyglycitols containing ≥50% dry basis maltitol are "maltitol syrups." Those containing <50% dry basis maltitol are "polyglycitols." As noted, relative molecular weight affects colligative properties similarly to corn syrups.

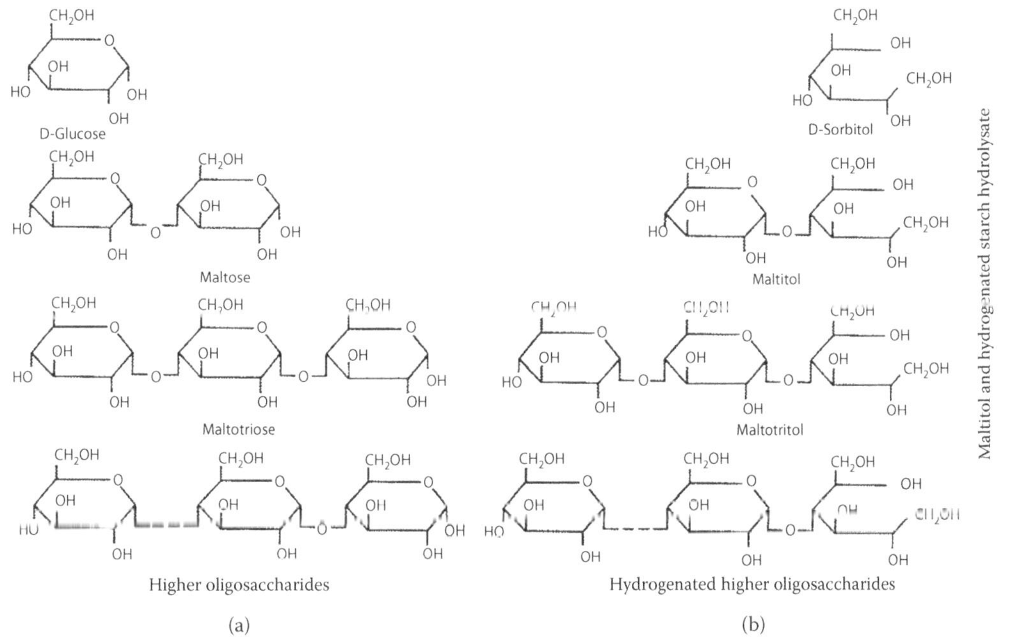

D-Glucose

Maltose

Maltotriose

Higher oligosaccharides

(a)

D-Sorbitol

Maltitol

Maltotritol

Hydrogenated higher oligosaccharides

(b)

Maltitol and hydrogenated starch hydrolysate

Figure 16.2 (a) Chemical structures for monosaccharide, disaccharide, and polysaccharide polyol precursors. (b) Chemical structures for the polyol components of maltitol syrups and polyglycitols.

Applications

The physical properties of maltitol syrups and polyglycitols make them a valuable aid in a variety of applications. Because of the similarity to corn syrup, these products can be used as a substitute in most applications where corn syrups and maltodextrins are used. Advantages over corn syrup include increased humectancy, often a more controlled MW distribution, and non-reducing characteristics, which means less browning at higher temperatures and over time. Also, in the case of the higher molecular weight polyglycitols, this means that they can serve as non-reactive carriers for colors, flavors, enzymes, and other sensitive ingredients.

Confectionery Products

Multiple uses are possible in confectionery products by replacing sugar and corn syrup with maltitol syrups or polyglycitols. Detailed formulations for preparing caramels, gummy bears, jelly beans, hard candy, taffy, nougats, butterscotch, marshmallows, and chewing gum are available from suppliers. As one would expect, maltitol syrups provide more sweetness but a lower average molecular weight leading to less resistance to "cold flow," which is the loss of candy shape (and increased tackiness) as the candy picks up moisture.

Hard candies made with maltitol syrups and polyglycitols should be packaged while warm in a moisture-resistant container. During preparation, the boiling temperature must reach 160°C–170°C under vacuum in order to reach a moisture level below 1% (Billaux 1991). Also, flavors added to these candies must be non-aqueous. In the case of sucrose or corn syrup candy, moisture pickup on the surface leads to crystallization of sucrose (graining), which minimizes the candy against further moisture pick-up. Because maltitol syrups and polyglycitols do not crystallize, however, moisture collection at the surface causes a sticky layer of solubilized syrup that will eventually cause the candy to partially dissolve (Sicard and Leroy 1983). Maltitol syrups and polyglycitols cannot be used as a replacement for sugar in chocolates or in pressed tablets where moisture would be especially detrimental to the product (Moskowitz 1991).

Because of the high molecular weight portion of these products, they resist crystallization, even at low temperatures and high concentrations, which makes them particularly advantageous for use in chewy candies. The crystallization of other components present in the formulation, such as isomalt, sorbitol, mannitol and xylitol, is also prevented—maltitol syrups or polyglycitols are frequently used in isomalt hard candies to control the rate of isomalt crystallization.

Another advantage of maltitol syrups and polyglycitols is that they do not have reducing groups, thus minimizing Maillard browning reactions. Because of this and their resistance to heat and acid, it is possible to manufacture, at high temperatures with acidic ingredients, sweets that remain bright and colorless. Because these reducing groups are less reactive, higher molecular weight polyglycitols are excellent carriers for colors, flavors, enzymes and other sensitive ingredients.

Also, maltitol syrups and polyglycitols are nutritive sweeteners that provide 40% to 90% of the sweetness of sugar, depending on the maltitol content. Unlike sugars, however, they are not readily fermented by oral bacteria, and so are used to formulate sugarless products that do not promote dental caries. To make a candy noncariogenic, however, it is necessary to also remove other sugar-containing ingredients in the product and replace them with a suitable alternative. This may include, for example, replacing milk powder with caseinates and milk fat. To manufacture a product matching the sweetness of sugar-sweetened candies may require the addition of a high potency non-cariogenic sweetener such as aspartame, sucralose, acesulfame potassium or a stevia-based sweetener. Maltitol syrups and polyglycitols blend well with other sweeteners and can mask unpleasant off-flavors such as bitter notes. Using the multiple sweetener approach generally allows a better sweetness profile than using sweeteners alone.

Other Applications

Maltitol syrups and polyglycitols can be used as a replacement for sugar and corn syrup in a variety of sugar free and no sugar added frozen desserts because they will not form crystals. They also act as mild freezing point depressants to increase freeze/thaw stability. These cryoprotective properties protect protein fibers from damage due to ice crystal formation and thermal shock (e.g., freezer burn). For this reason, polyglycitols can be used as a cryoprotectant glaze in seafood and other products to increase shelf life in frozen storage.

Because of their excellent humectancy, maltitol syrups and polyglycitols are also used extensively to partially replace sugar and corn syrup in baked goods, a broad range of other foods, medicinal syrups (e.g., pediatric medicines, cough syrups), dentifrices and mouthwashes. Formulations for the preparation of these products may be obtained from the manufacturer.

Metabolic Aspects

Upon ingestion, maltitol syrups and polyglycitols are enzymatically hydrolyzed to sorbitol, glucose, and maltitol. Only 10% of maltitol may be converted into monosaccharides that are absorbed through the intestinal mucosa. Maltitol is excreted mainly as gas, also in the feces and urine (Modderman 1993). The digestion of the oligo- and polysaccharides is close to 90% (Beaugerie et al. 1989), although the rate of enzymatic hydrolysis might be a function of the chain length of the component polyol. The unabsorbed products of maltitol syrup and polyglycitol hydrolysis reach the lower digestive tract where they are metabolized by naturally occurring colonic bacteria. Evidence of colonic fermentation has been shown using the hydrogen breath test (Wheeler et al. 1990). The absorbed glucose and sorbitol are taken up into the bloodstream. Sorbitol is converted to fructose, which is then utilized through the glycolytic pathway. A review of the metabolism of sorbitol can be found in Chapter 20 on Sorbitol and Mannitol.

Cariogenicity

It is widely accepted that prolonged exposure of the teeth to acid produces dental caries. Sugars and starches are fermented by oral bacteria, producing organic acids that can solubilize tooth enamel and result in decay. Because of obvious ethical restrictions in conducting most tests of cariogenicity in human subjects, screening for proper conditions for cariogenic potential is conducted.

"Cariogenic potential" was defined at the Scientific Consensus Conference on Methods for the Assessment of the Cariogenic Potential of Foods as, "the ability of a food to foster caries in humans under conditions conducive to caries formation" (Working Group Consensus Report 1986). This group also reached an agreement on a line of testing that could be used to establish that a food had either no cariogenic potential or low cariogenic potential. Consensus is that "foods assessed by two recommended plaque acidity test methodologies that result in pH profiles statistically equivalent to those generated by sorbitol would be deemed as possessing no cariogenic potential" (Working Group Consensus Report 1986). Demineralization of tooth enamel definitely occurs below a pH of 5.5; between 5.5 and 5.7 is a transition range where some demineralization may begin.

The United States Food and Drug Administration (US FDA) has authorized the use of the "does not promote tooth decay" health claim for sugar-free food products sweetened with polyols (US Food and Drug Administration 1996). The regulation provides that "when fermentable carbohydrates are present in the sugar alcohol-containing food, the food shall not lower plaque pH below 5.7 by bacterial fermentation either during consumption or up to 30 minutes after consumption, as measured by the indwelling plaque test found in 'Identification of Low Caries Risk Dietary Components,' T.N. Imfeld, Volume 11, *Monographs in Oral Science* (1983)" (Imfeld 1983).

The relative acidogenicity of test products may be predicted based on the component makeup of the compounds. Theoretically, maltitol syrups and polyglycitols with higher hydrogenated saccharides (molecules DP4 and larger) would be more likely to result in the release of free glucose following hydrolysis, which is readily fermentable by oral bacteria (Abelson 1989). Because maltitol syrups and polyglycitols differ in their saccharide profile, studies have been conducted to examine the cariogenicity of each product. Although maltitol syrups and polyglycitols with higher concentrations of higher DP saccharides show a greater drop in pH than those with lower DP fractions, each has been shown to remain above the 5.7 "safe for teeth" value (Abelson 1989; Moscowitz 1991).

The American Dental Association (ADA) has recognized the usefulness of polyols as alternatives to sugars and as a part of a comprehensive program including proper dental hygiene. In October 1998, the ADA's House of Delegates approved a position statement acknowledging the "Role of Sugar-Free Foods and Medications in Maintaining Good Oral Health" (ADA 1999).

Laxation

The ingestion of maltitol syrups and polyglycitols several times a day on an empty stomach of unadapted subjects can result in a laxative effect. This is true of most polyols due to their incomplete absorption and resultant increased osmotic pressure. This effect increases with an increase in the relative amount of sorbitol in the product. Additionally, the digestive system appears to adapt with a decrease in symptoms such as flatulence and diarrhea after repeated daily consumption (Moskowitz 1991). Persons in the 90th percentile for consumption of Lycasin-containing products only consume 1.1–2.6 g/day (US Food and Drug Administration 1984). It is recommended that if the ingestion of 50 g or more is foreseeable, the statement "excess consumption may have a laxative effect" should be used.

Caloric Content

The components of maltitol syrups and polyglycitols are slowly and incompletely absorbed, allowing a portion of the product to reach the large intestine, thereby reducing the carbohydrate available for metabolism. Therefore, unlike sugar that contributes four calories per gram, the caloric contribution of maltitol syrups and polyglycitols is not more than three calories per gram (Federation of American Societies for Experimental Biology 1994). For a product to qualify as "reduced calorie" in the United States, it must have at least a 25% reduction in calories. Maltitol syrups and polyglycitols may, therefore, be of use in formulating reduced calorie food products.

The lower caloric value of polyols is recognized in other countries. The European Union, for example, has provided a Nutritional Labeling Directive stating that all polyols, including maltitol syrups and polyglycitols, have a caloric value of 2.4 calories per gram (EEC 1990).

Suitability in Diabetic Diets

Control of blood glucose, lipids and weight are the three major goals in diabetes management today. Because of their slow and incomplete absorption, maltitol syrups and polyglycitols have a reduced glycemic potential relative to glucose for individuals with and without diabetes (Wheeler et al. 1990). This property permits their use as a reduced calorie alternative to sugar. Doses of 45 to 90 g/day are well tolerated without glycemic effect in either diabetic or nondiabetic subjects (Tacquet 1984). The reduced caloric value (75% or less that of sugar) of maltitol syrups and polyglycitols is also consistent with the objective of weight control.

The American Diabetes Association acknowledges the lower caloric value of polyols but cautions that their use may not contribute to a significant reduction in total calories or carbohydrate content of the daily diet. The calories and carbohydrate from maltitol syrup and polyglycitol sweetened products should be accounted for in the meal plan (Geil 2008). Although studies have shown a reduced glycemic response in comparison to glucose, maltitol syrups, and polyglycitols, other ingredients in the food product may have the potential to affect blood glucose levels. Recognizing that diabetes is complex and requirements for its management may vary between individuals, the usefulness of maltitol syrups and polyglycitols should be discussed between individuals and their physicians.

Toxicity

A broad range of safety studies in man and animals, including long term feeding (Dupas et al. 1984), multigeneration reproduction/development (Leroy and Dupas 1984), and teratology studies (Dupas and Siou 1984) have shown no evidence of any adverse effects from maltitol syrups and polyglycitols. The results of these studies have added to the substantial body of information establishing the safety of maltitol syrups and polyglycitols (Modderman 1993).

The Joint Food and Agriculture Organization/World Health Organization Expert Committee on Food Additives (JECFA) has reviewed the safety information and concluded that maltitol syrups and polyglycitol syrups are safe (JECFA 1993, 1998). The JECFA established an acceptable daily intake (ADI) for maltitol syrup of "not specified," meaning no limits are placed on its use. JECFA defines "not specified" as: "on the basis of available scientific data, the total daily intake of a substance arising from its use at levels necessary to achieve the desired effect, does not, in the opinion of the Committee, represent a hazard to health." Many small countries that do not have their own agencies to review food additive safety, often adopt the JECFA's decisions. In 1984, the Scientific Committee for Food of the European Union evaluated maltitol syrups and also concluded it was not necessary to set a numerical ADI for maltitol syrups (Commission of the European Communities 1985). In 2009, the European Food Safety Authority (EFSA) evaluated polyglycitols and concluded that "there are no indications of a safety concern for the proposed uses and use levels" (EFSA 2009).

Regulatory Status

In the United States, Generally Recognized as Safe (GRAS) petitions for HSH products (maltitol syrups and polyglycitols) have been accepted for filing by the US FDA. Once a GRAS affirmation petition has been accepted for filing, manufacturers are allowed to produce and sell foods containing these sweeteners in the United States. Products from the HSH family are approved in many other countries, including Canada, Japan and Australia. In the EU, maltitol and maltitol syrups (E965) are authorized for food use at *quantum satis* levels in a range of food products identified in Directive 89/10//EC. Based on the 2009 EFSA evaluation of polyglycitols, these products could be listed in the Directive sometime in 2011.

References

Abelson, D. 1989. The effect of hydrogenated starch hydrolysates on plaque pH *in Vivo. Clinical Preventive Dentistry* 11(2):20–23.

American Dental Association. 1999. Position statement on the role of sugar-free foods and medications in maintaining good oral health. http://www.ada.org/1874.aspx. Accessed June 07, 2011.

Billaux, M.S., Flourie, B., Jacquemin, C., and Messing, C. 1991. Sugar alcohols. In *Handbook of sweeteners*, eds. S. Marie, and J.R. Piggott, Blackie and Son.

Beaugerie, L., Flourié, B., Franchisseur, C., Pellier, P., Dupas, H., and Rambaud, J.C. 1989. Absorption intestinale et tolérance clinique au sorbitol, maltitol, lactitol et isomalt. *Gastroenterol Clin Biol* 13, 102 (abstr).

Commission of the European Communities. 1985. Reports of the Scientific Committee for Food Concerning Sweeteners. Sixteenth Series. Report EUR 10210 EN. Office for Official Publications of the European Communities.

Dupas, H., Leroy, P., and D'Alayer, C. 1984. 24-Month Safety Study of Lycasin® 80/55 on Rats. (Roquette Freres, Lestrem, France). Report available from US FDA Dockets Management Branch; specify Docket No. 84G-0003, op cit, Report E-52.

Dupas, H., and Siou, G. 1984. Lycasin® 80/55: Teratogenic Potential Study in Rats. (Roquette Freres, Lestrem, France). Report available from US FDA Dockets Management Branch; specify Docket No. 84G-0003, op cit, Report E-53.

European Economic Community Council (EEC). 1990. Directive on food labeling. Official Journal of the European Communities. No. L 276/40.

European Food Safety Authority. 2009. Scientific Opinion on the use of Polyglycitol Syrup as a food additive. *EFSA Journal* 7(12):1413 (21pp).

Federation of American Societies for Experimental Biology. 1994. The evaluation of the energy of certain polyols used as food ingredients (unpublished).

Food Chemicals Codex. Seventh Edition. 2010. The United States Pharmacopeial Convention. United Book Press, Inc.

Geil, P.B. 2008. Choose your foods: exchange lists for diabetes: the 2008 revision of exchange lists for meal planning. *Diabetes Spectrum* 21(4):281–283.

Imfeld, T. 1983. Identification of low caries risk dietary components, *Monographs in oral science*, Vol. 11, ed. H.M. Myers, Karger AG.

Joint FAO/WHO Expert Committee on Food Additives. 1993. Evaluation of certain food additives and contaminants: Maltitol and maltitol syrup. Forty-first report. WHO Technical Report Series 837, pp. 16–17.

Joint FAO/WHO Expert Committee on Food Additives. 1998. Polyglycitol Syrup. Prepared at the 51st JECFA (1998) and published in FNP 52 Add 6.

Leroy, P., and Dupas, H. 1984. Lycasin 80/55: Three Generation Reproduction Toxicity Studies. (Roquette Freres, Lestrem, France). Report available from US FDA Dockets Management Branch, specify Docket No. 84G-0003, op cit, Report E-33.

Modderman, J.P. 1993. Safety assessment of hydrogenated starch hydrolysates. *Regul Toxicol Pharmacol* 18:80–114.

Moskowitz, A. 1991. Maltitol and Hydrogenated Starch Hydrolysate. In *Alternative sweeteners,* 2nd edition, eds. L. Nabors and R. Gelardi, Marcel Dekker.

Sicard, P.J., and Leroy, P. 1983. Mannitol, sorbitol and lycasin: Properties and food applications. In *Developments in sweeteners,* 2nd edition, ed. T.H. Grenby, Elsevier-Applied Science.

Tacquet, A. 1984. Study on the Clinical and Biological Tolerance of Polysorb Lycasin 80/33 in the Human Being. (Calmette Hospital, d'Lille, France). Report available from US FDA, Dockets Management Branch; specify Docket No. 84G-0003, Report E-35.

US Food and Drug Administration. 1984. Roquette Corp.; Filling of petition for affirmation of GRAS status (hydrogenated glucose syrup). Federal Register, Vol. 49, No. 39:7153.

US Food and Drug Administration. 1996. Food Labeling: Health Claims; Sugar Alcohols and Dental Caries, Final Rule. Federal Register, Bul. 61 No. 165:43433.

Wheeler, M.L., Fineberg, S.E., Gibson, R., and Fineberg, N. 1990. Metabolic response to oral challenge of hydrogenated starch hydrolysate versus glucose in diabetes. *Diabetes Care* 13:733–740.

Working Group Consensus Report.1986 Integration of methods. *J Dent Res* 65(special issue):1537–1539.

Chapter 17

Isomalt

Anke Sentko and Jörg Bernard

Contents

Introduction

Since the early 1990s, the sugar free confectionery market has developed from a niche market to mainstream. This was possible because there was no longer any need to make sacrifices in taste and quality thanks to the sugar replacement ingredients available. Polyols (synonym for sugar alcohols) are nutritive sweeteners, which replace sugars like sucrose, high fructose corn syrup or glucose syrup cup by cup. On top of the replacement of sugars, polyols provide additional benefits to final products. Toothfriendly, low or reduced glycemic and calorie reduced confectionery, baked goods, and pharmaceutical products are possible nowadays. Sugar-free bulk sweeteners are sweeteners that give body and texture to a product, as well as a sweet flavor. Ideally, they do not cause any aftertaste and provide the same functions as sucrose and glucose. Their metabolism in the gastrointestinal tract and their biochemical changes in the mouth, however, differ from sucrose and glucose.

The ideal sweetener should be chemically and biologically stable for an indefinite period of time and provide the same properties to a product as sucrose or glucose do. Its processing parameters should be similar to those of sucrose or glucose, so that existing equipment can be used without requiring major changes. In addition, the finished products should have practically the same taste and appearance as those of traditional products, have an excellent shelf life, and be readily accepted by the consumer.

This chapter presents the polyol isomalt as a sugar-free bulk sweetener. Isomalt is a sugar replacer with properties and characteristics similar to sucrose from a food application point of view. Isomalt is odorless, crystalline, and non-hygroscopic (Sträter 1986, 1988). Isomalt is also available as an aqueous solution. It is a non-reducing sugar alcohol of a disaccharide type. Unlike sucrose, however, it is extremely stable with respect to chemical and enzymatic hydrolysis. It cannot be fermented by a large number of yeasts and other microorganisms found in nature. Isomalt was developed and is manufactured and marketed by BENEO-Palatinit GmbH, a wholly owned subsidiary of Südzucker AG (Germany). Meanwhile, some other producers have also become active in the market. For certain applications, speciality types of isomalt have been developed by BENEO such as isomalt ST, isomalt HC, isomalt GS, isomalt DC and isomalt LM. These different types will be discussed in this chapter in the section "Applications and Product Development." Because a lot of properties are similar for all isomalt types, isomalt ST (i.e., isomalt standard, the best known type) has been selected to describe the properties of isomalt in general.

Production

Isomalt is the only bulk sweetener derived exclusively from sucrose. It is manufactured in a two-stage process in which sugar is first transformed by enzymatic transglucosidation into isomaltulose, a reducing disaccharide (6-O-α-D-glucopyranosyl-D-fructose). Isomaltulose is then hydrogenated into isomalt. Isomalt is composed of 6-O-α-D-glucopyranosyl-D-sorbitol (1,6-GPS) and 1-O-α-D-glucopyranosyl-D-mannitol dihydrate (1,1-GPM dihydrate) (Gau et al. 1979; Schiweck 1980).

The GPS/GPM ratio depends on the isomalt variant: In isomalt ST this is approximately 1:1, in isomalt GS approximately 3:1. In aqueous systems, GPS forms anhydrous crystals, whereas GPM has two molecules of water of crystallization. Additional drying of isomalt ST leads to isomalt LM, a low moisture variant with a water content of less than 0.5%. In a

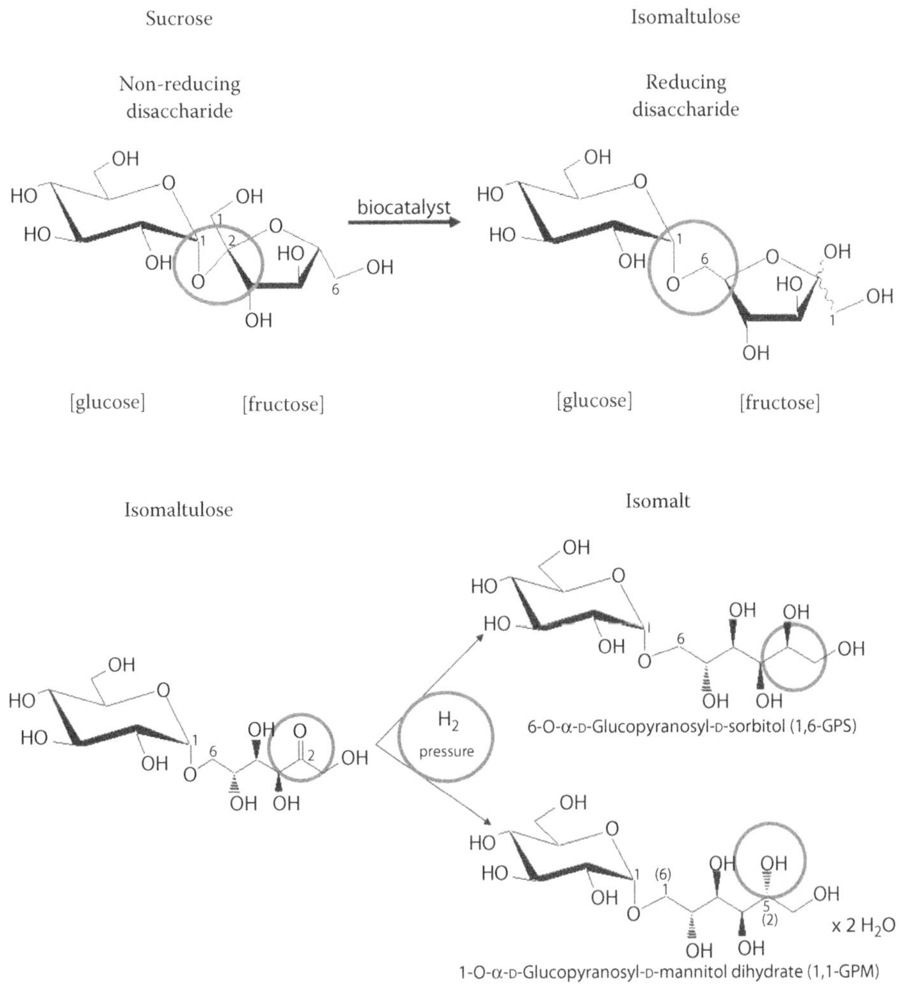
Sucrose
Non-reducing disaccharide
Isomaltulose
Reducing disaccharide
biocatalyst
[glucose]
[fructose]
[glucose]
[fructose]
Isomaltulose
Isomalt
H2
pressure
6-O-α-D-Glucopyranosyl-D-sorbitol (1,6-GPS)
1-O-α-D-Glucopyranosyl-D-mannitol dihydrate (1,1-GPM)
x 2 H2O

Figure 17.1 Production of isomalt from sucrose.

further production step, isomalt ST or isomalt GS can be milled or micronized in order to create a powder-fine product which can be used as such in food applications or as the basic raw material for isomalt DC, an agglomerated variant. The production process is illustrated in Figure 17.1.

Sensory Properties

Sweetness and Taste

The sweetening power of isomalt lies between 0.45 and 0.6 compared with that of sucrose (=1.0). Figure 17.2 shows that the sweetening power is a function of concentration (i.e., it increases with increasing isomalt concentration) (Paulus and Fricker 1980). There is no difference between the sensorially tested and theoretically determined curves.

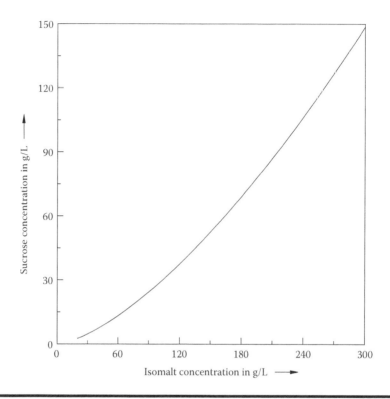

Figure 17.2 Isosweet aqueous solutions of isomalt ST and sucrose, determined sensorially and theoretically.

Isomalt has a pure sweet taste similar to sucrose without any aftertaste. Furthermore, it reinforces flavor transfer in foods. Synergistic effects occur when isomalt is combined with other sugar alcohols (e.g., xylitol, sorbitol, mannitol, maltitol syrup, hydrogenated starch syrup) and with high-intensity sweeteners (e.g., acesulfame potassium, aspartame, sucralose, cyclamate, or saccharin).

In addition, isomalt tends to mask the bitter metallic aftertaste of some intense sweeteners (Schiweck 1980).

Cooling Effect

Several crystalline sugar alcohols, used as sugar substitutes, have a high heat of solution (Cammenga et al. 1996; Cammenga and Stepphuhn 1993; Jasra and Ahluwalia 1982). This results in a cooling sensation in the mouth when these sugar alcohols are consumed in a crystalline state. Although this mouth-cooling effect is a desirable feature for peppermint or menthol products, it is considered atypical in many other products, such as baked products and chocolate. The heat of solution of isomalt ST lies between the values of GPS and GPM and depends on the actual water content of isomalt. Figure 17.3 shows that, compared with other sugar alcohols, isomalt has a low heat of solution, which is comparable to that of sucrose. Crystalline isomalt, therefore, does not produce a cooling effect.

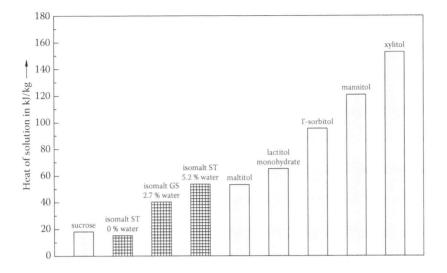

Figure 17.3   Heat of solution (kJ/kg) of sugar and sugar alcohols.

Physico-chemical Properties

Physical Properties

The sorption isotherm (Weisser et al. 1982; see Figure 17.4) of crystalline isomalt shows the typical behavior of a non-hygroscopic material: a nearly constant water content over a broad range of water activity. At 25°C and up to a relative humidity (R.H.) of 85% any moisture is incorporated in isomalt as crystal water. Isomalt remains, under these conditions, palpably dry. The crystal water is bound strongly within the crystal lattice and can only be released under drastic conditions. The low hygroscopicity with respect to free water means

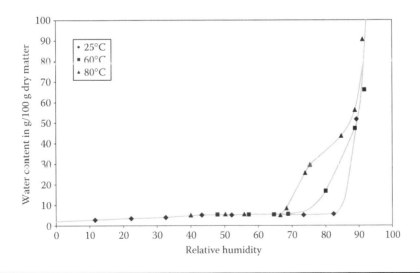

Figure 17.4   Sorption isotherm for isomalt ST at three temperatures (25°C, 60°C, and 80°C).

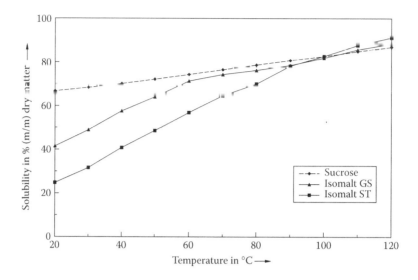

Figure 17.5 Solubility of isomalt ST and isomalt GS in water compared to sucrose as a function of temperature.

that isomalt can be stored easily and distributed without much special care. Furthermore, this explains why products exclusively or mainly based on isomalt tend not to be sticky and to have a long shelf life.

As depicted in Figure 17.5, the solubility of isomalt is much lower than that of sugar at 20°C, namely 24.5 g/100 g solution for isomalt ST and 41.5 g/100 g solution for isomalt GS compared with a value of 66.7 g/100 g solution for sugar (Bollinger 1987a; Schiweck 1980). According to the requirements of the process and the final product the most suitable isomalt type can be selected. On the other hand, the solubility of isomalt increases substantially with rising temperature in contrast to sucrose. Thus, the solubility of all isomalt types is comparable to sucrose at most processing temperatures (greater than 85°C).

The crystals of GPS and GPM or GPM dihydrate, respectively, form a eutectic system with complete miscibility in the liquid phase and immiscibility in the solid phase. The minimum melting temperature corresponds to the eutectic composition of 50% GPM and 50% GPS. Therefore, isomalt shows no distinct melting temperature, but a melting range depending on the exact ratio of GPS/GPM. If GPM dihydrate crystals are heated to temperatures above 100°C then the crystal water is released. The Differential Scanning Calorimetric (DSC) curve in Figure 17.6 illustrates the temperature ranges of both water release and melting of isomalt with a low water content. Thus, the crystal water content of isomalt ST can be adjusted to any value from 5% (m/m) downwards according to the demands of the final product.

The viscosity of aqueous isomalt solutions does not differ significantly from that of corresponding sucrose solutions in a temperature range between 5°C and 90°C (Schiweck 1980). No special engineering requirements need to be taken into consideration. The viscosity of a cooked isomalt mass depends on the shear velocity and can be higher or lower than a sucrose/corn syrup melt. For hard candy production, for example, the cooked isomalt mass shows a lower viscosity than a sucrose/corn syrup melt at the same temperature. The colligative properties (freezing point depression, water activity, boiling point elevation) of aqueous solutions of isomalt are similar to those of sucrose. Nevertheless, to obtain the required water content

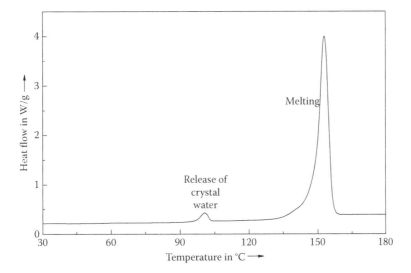

Figure 17.6   DSC curve of isomalt.

in hard candies, isomalt is usually boiled to a higher temperature than sucrose/corn syrup for two reasons: the final water content in isomalt hard candies is lower than that in sucrose/corn syrup hard candies and due to the content of oligosaccharides, the boiling point elevation of sucrose/corn syrup mixtures is lower than that of sucrose or isomalt. During the cooking process, the candy mass is significantly superheated with respect to its thermodynamic boiling point.

Normally vacuum is applied in the cooking process of hard candies to reduce the temperature needed. The cooking curves in Figure 17.7 display the final water content of isomalt

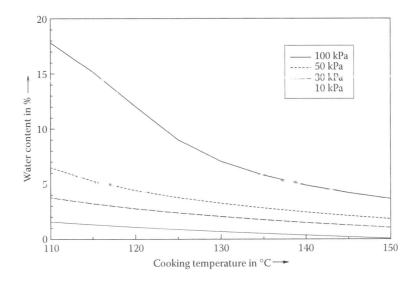

Figure 17.7   Cooking curves of isomalt.

Table 17.1 Physical-Chemical Properties of Isomalt

	Isomalt[a]	*GPM*	*GPS*	*Sucrose*
Molecular mass in g/mol	—	344,324 Anhydrous 380,356 Dihydrate	344,324	342,303
Crystal group	—	Orthorhombic $P2_1P2_1P2_1$ [1]	Monoclinic $P2_1$ [2]	Monoclinic
Specific rotation	91.5° [3]	88.7° [3]	92.9° [3]	66.5°
Melting temperature / range in °C	140 … 155 [6]	168 [4]	166 [4]	186 ± 4 [3]
Heat of fusion in kJ/mol	46.5 [4]	55.0 [4]	56.36 [4]	46.41 [3]
Heat of solution in kJ/mol	5.29 Isomalt ST, 0% water [4] 19.52 Isomalt ST, 5.2% water [6] 14.25 Isomalt GS, 2.7% water [6] −12.48 Amorphous glass [4]	30.02 Crystal dihydrate [4] 6.30 Crystal anhydrous [4] −12.90 Amorphous glass [4]	10.39 Crystal [4] −10.99 Amorphous glass [4]	6.22 Crystal [7] −13,79 Amorphous [7]
Glass transition temperature in °C	63 [8]	65 [4]	50 [4]	65 [3]
Heat capacity in J/(mol K)	480 Crystal [5] 490 Amorphous [5]	370 Crystal anhydrous [5] 410 Amorphous [5]	670 Crystal [5] 610 Amorphous [5]	425.8 Crystal [3] 490.2 Amorphous [3]

[1]: Lindner and Lichtenthaler 1981; [2]: Lichtenthaler and Lindner 1981; [3]: Bubník et al. 1995; [4]: Cammenga et al. 1996; [5]: Zielasko 1997; [6]: Südzucker CRDS, internal data; [7]: Gehrich 2002; [8]: Raudonus et al. 2000.

[a] Physical data of isomalt depend on actual ratio of GPM/GPS and on water content.

melts as a function of cooking temperature at various pressures (Mende 1990). In Table 17.1, some physical-chemical characteristics of isomalt, GPS and GPM have been collected and compared to sucrose (Bubnik et al. 1995; Cammenga et al. 1996; Gehrich 2002; Lichtenthaler and Lindner 1981; Lindner and Lichtenthaler 1981; Raudonus et al. 2000; Südzucker AG Mannheim/Ochsenfurt CRDS 2010; Zielasko 1997).

Chemical Properties

Isomalt is resistant to chemical degradation because of its very stable 1-6 bond between the mannitol or sorbitol moiety and the glucose moiety. When the crystalline substance is heated above the melting point or the aqueous solution above the boiling point, no changes in the molecular structure are observed. The caramelization and other discoloration during the melting, extrusion, or baking processes are significantly inhibited compared to sucrose. In general, reactions of isomalt with other ingredients in the formulation (e.g., with amino acids to produce Maillard reactions) are impaired compared to sucrose. The stability of isomalt during acid hydrolysis was measured at 100°C in 1% hydrochloric acid (Irwin 1990a). Figure 17.8 illustrates that under these conditions sucrose is hydrolyzed in less than five minutes. In contrast, isomalt is completely split after five hours. Also in an alkaline milieu, isomalt shows a much higher resistance than sucrose. Furthermore, isomalt is resistant to enzymatic hydrolysis. Most microorganisms found in foods are unable to use isomalt as a substrate (Bollinger 1987b; Emeis and Windisch 1960; Schiweck 1980).

Physiological Properties

As mentioned in the preceding section, the remarkable feature of isomalt is the very stable glycoside bond in GPS and GPM. Compared with sucrose, this difficult-to-split glycoside bond gives isomalt the following properties after oral intake (Nilsson and Jägerstad 1987):

■ Only about 50% of isomalt is converted into available energy
■ Isomalt is non-cariogenic (does not promote dental caries)
■ Isomalt provides a very low glycemic and insulinemic response

Caloric Value

Humans are only able to absorb and, subsequently, to metabolize monosaccharides from the digestive tract. Because of the stable glycoside bond, isomalt is hardly hydrolyzed and absorbed in the

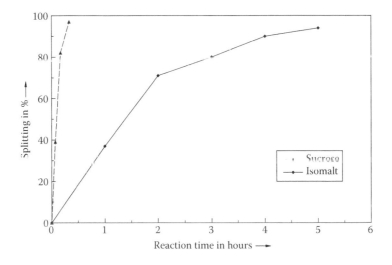

Figure 17.8 Hydrolysis of isomalt and sucrose versus time in a 1% HCl solution at 100°C.

small intestine. Like many dietary fibers, it is mainly fermented in the colon. This results in a lower energy conversion (Berschauer and Spengler 1987; Grupp and Siebert 1978; Life Sciences Research Office 1994; Nilsson and Jägerstad 1987; Sträter 1988; Thiébaud et al. 1984). Because of individual nutrition labeling regulations, the energy value used for food labeling purposes can vary from country to country. Some countries follow the science based route and accept ingredient specific energy conversion factors (e.g., the United States and Canada). Other countries decided on average values for the group of polyols (e.g., European Union). The energy value for isomalt used for food labeling in Japan, for example, is 2.0 kcal/g (food) or 1.2–1.5 kcal/g ("special food for patient"); in the United States and Canada, it is 2.0 kcal/g; and in the European Union it is 2.4 kcal/g (Livesey et al. 2000).

Dental Health

Tooth decay (caries) is the result of acid affecting the tooth enamel. The acids dissolve calcium from the teeth (demineralization) and cavities occur. The acid is produced by microorganisms adhering to the surface of the tooth, forming plaque, when fermentable carbohydrates are available for metabolism by the plaque micoflora. Although these microorganisms easily ferment carbohydrates like sugar, glucose or starch into decay-causing acids, they cannot convert isomalt. The critical pH value in the plaque at which the dental enamel may dissolve is 5.7 or less. Studies, including human plaque pH-telemetry studies, have shown that consumption of isomalt and products containing only isomalt as the bulk sweetener does not result in a plaque pH less than 5.7 (Ciardi et al. 1983; Gehring and Hufnagel 1983; Imfeld 1983; Karle and Gehring 1979). Isomalt is included in the list of non-cariogenic carbohydrate sweeteners in the context of the US Food and Drug Administration (US FDA) health claim approval on carbohydrate sweeteners and dental caries (US FDA 1996).

Low Glycemic and Insulinemic Response

A number of scientific studies have shown that insulin and blood glucose levels in humans increase only very slightly after oral intake of isomalt (Drost et al. 1980; Grupp and Siebert 1978; Livesey 2003; Thiébaud et al. 1984). This means sharp increases in blood glucose, as occur after sucrose or glucose/available starch intake (especially when between-meal snacks and sweets are eaten), can be avoided with isomalt intake. Sydney University measured a relative glycemic response (equivalent to the glycemic index) of 2, based on a gram per gram approach (the Official Website of the Glycemic Index and GI Database 2011). The effect of isomalt in patients with type 2 diabetes was examined in a human intervention study over a period of 12 weeks with 30 g of isomalt instead of higher glycemic carbohydrates. While the test diet was accepted and tolerated very well, significant reductions were observed for glycosylated hemoglobin, fructosamine, fasting blood glucose, insulin, proinsulin, C-peptide, insulin resistance (HOMA-IR) and oxidized LDL. Routine blood measurements and blood lipids were unchanged. The study demonstrated that 30 g isomalt significantly improved the metabolic control of diabetes (Holub et al. 2009).

These results are in line with a recent meta-analysis by Livesey et al. (2008) in which a significant improvement of fasting blood glucose and glycated proteins for interventions with lower GI and GL diets was demonstrated.

Gastrointestinal Tolerance

Sugar alcohols and a number of other carbohydrates (e.g., polydextrose, dietary fiber, lactose) have low or no digestibility. Their digestion process is characterized by a low or no hydrolysis and absorption in the small intestine and fermentation by the gut microflora in the large intestine. Consumption of large amounts may result in a laxative effect. This is due to a water-binding effect (osmotic effect) of non-absorbed nutrients. It is not possible to give a specified numeric threshold limit (g/day) for sugar alcohols such as isomalt as a general rule because this tolerance depends on many parameters (Grabitske and Slavin 2008; Lee et al. 2001; Paige et al. 1986; Spengler et al. 1979a, 1979b, 1987):

- The form in which it is ingested. In a liquid food, the intolerance is higher than in a solid food.
- The individual sensitivity. The intensity of perception (if disturbing or not, etc.) varies from person to person.
- The moment of consumption. Differences in tolerance even exist for the same person from day to day. A person's tolerance is affected by diet (low or high in low digestible carbohydrates) and psychological well-being (emotions, prejudices).
- The adaptation to sugar alcohols or other low or non-digestible carbohydrates. Frequent consumption results in a higher tolerance.

In the case of an individual feeling disturbed due to a physiological overload situation of low or non-digestible carbohydrates, it is recommended to decrease the consumption and/or eat smaller portions spread over the day. This, in many cases, leads to good acceptance.

Like fermentable dietary fibers, isomalt is fermented to short chain fatty acids by the gut microflora. *In vitro*, isomalt was metabolized in several bifidobacteria strains and yielded high butyrate concentrations. A four week human intervention study (double blind, placebo controlled, cross over) with a controlled basal diet enriched with 30 g isomalt or sucrose per day demonstrated a significant increase in the bifidobacteria. The authors concluded that isomalt is to be considered a prebiotic carbohydrate that might contribute to a healthy luminal environment of the colonic mucosa (Gostner et al. 2006).

Toxicological Evaluations

Extensive toxicological and metabolic studies have been conducted that prove conclusively the safety of isomalt (Smits-Van Prooije et al. 1990; Waalkens-Berendsen et al. 1989, 1990a, 1990b). The results of these studies have been summarized in a World Health Organization report prepared by the Joint FAO/WHO Expert Committee on Food Additives (WHO Food Additive Series 1987). This report concludes by assigning isomalt an acceptable daily intake (ADI) "not specified," the safest rating assigned to any evaluated food substance.

Applications

On the basis of its physical and chemical properties, isomalt is a suitable sugar replacer in many areas of the food and pharmaceutical industries. Existing processing equipment can be used for all applications without requiring major changes. Only formula and process parameter modifications are recommended to optimize processes and products. The spectrum of food applications is broad and is listed in Table 17.2.

The main applications and the use of different isomalt variants and types are described in the following sections.

High-boiled Candies

High-boiled candies made with isomalt ST or isomalt HC can be stamped, filled, pulled, combed, or molded. Hard candies with a very good shelf life will be obtained if the water content in the finished product is less than 2%. These candies are very stable against water absorption (Figure 17.9). Compared with candies based on sucrose/corn syrup, only minor changes in the production

Table 17.2   Overview of Isomalt Variants Including Applications

	Isomalt ST	Isomalt GS	Isomalt LM	Isomalt DC	Isomalt HC	Isomalt ST/GS Liquid
	Crystalline/ Powder	Crystalline/ Powder	Crystalline/ Powder	Agglomerated	Liquid	Liquid
	Universal	Good solubility	Low moisture	Direct compressible		
High boiled candies	X				X	X/–
Low boiled candies	X	X				–/X
Gummies/ellies	X	X				X/X
Coated products	X	X				X/X
Chewing gum – coating – center	X X	X X				X/X
Compressed tablets				X		
Baked goods	X	X				
Icings	X	X				–/X
Fillings (fat based)	X	X				–/X
Chocolate			X			
Jam	X	X				X/X
Ice Cream	X	X				

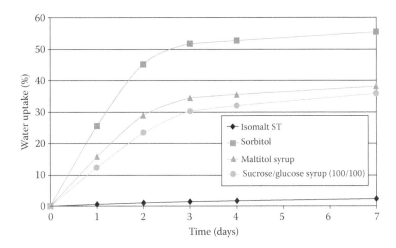

Figure 17.9   Changes in weight of candies by the water absorption storage test at 25°C, 80% R.H. after 7 days without packaging, based on initial weight of candies. (From Raudonus, J.W.J., Untersuchung des Einflusses polymerer Zusätze bei der Herstellung von Isomalt-Hartkaramellen. Diploma Thesis. Hohenheim University, 1999.)

process (batch or continuous) are required. The minor changes are necessary because the following characteristics of isomalt differ from sucrose/corn syrup as already explained in the section "Physical Properties."

- Lower solubility
- Higher cooking temperature
- Lower viscosity of the melt
- Higher specific heat capacity

Low-boiled Candies

Low-boiled candies consist of crystals and a non-crystallizing continuous phase. Isomalt GS can replace sugar in the non-crystallizing phase, isomalt ST the crystals. The water content of the low boiling affects the crystallization and should be in the range between 6% and 10%. Stability of the shape and stickiness are often critical in low-boiled candy production. The low hygroscopicity of isomalt minimizes stickiness during cooling, forming, cutting, and wrapping. In addition, it prevents the soft-boiled mass from sticking to the wrapper. Seeding the low boiling mass with powdered isomalt ST after the boiling process results in increased form stability and reduces stickiness too.

Coated Products

Both isomalt GS and isomalt ST are applied for coated products, often with chewing gum pellets. Isomalt GS can also be used at low process temperature. The coating procedure depends on the type of coating equipment and on the type of center. Isomalt can replace sucrose in all sorts of pan-coating (Fritzsching 1993; Willibald-Ettle 1992).

■ Hard coating with isomalt suspension or solution, possibly in combination with isomalt powder
■ Soft coating with isomalt in combination with another non-recrystallizing sugar substitute
■ Chocolate coating with isomalt chocolate

The low hygroscopicity of isomalt leads to products with an excellent shelf life. Packaging in carton boxes is sufficient in order to maintain stability of the coating and elaborate packaging can be avoided. Isomalt ST and isomalt GS can be used for coating purposes. The greater amount of GPS in isomalt GS improves the crunchiness of a coating made with isomalt GS caused by differences in crystal structure between GPM and GPS. Because the solubility of GPS is higher than the solubility of GPM, isomalt GS has a higher aqueous solubility than isomalt ST. Therefore, a coating with isomalt GS dissolves faster and improves the sweetness impact and the flavor release. Isomalt GS coating stands out due to its stability against abrasion experienced during coating; this reduces damage and results in well-defined corners and edges of coated products. The hard coating process with isomalt GS requires considerably shorter coating time compared to other polyols.

Summary of isomalt benefits in coating:

■ Excellent storage stability due to its low hygroscopicity
■ Colored coatings based on isomalt have an exceptional appearance, smooth surface and brilliant colors
■ Natural taste
■ Isomalt coatings are very stable against cracking and chipping
■ Isomalt requires only short coating times
■ Low process temperature with isomalt GS
■ Single syrup process

Chewing Gum

Chewing gum sticks, pellets, or gum balls can be manufactured with powder-fine isomalt, either ST or GS. Its low solubility causes isomalt to remain crystalline in the chewing gum mass, which leads to a softer/smoother product texture. Chewing gums containing 15%–45% isomalt retain their texture (flexibility, softness, chew) almost completely for a long time period (Südzucker AG Mannheim/Ochsenfurt CRDS 2010) of at least 12 months (Figure 17.10). Due to its low hygroscopicity and its anti-sticking quality isomalt can be used as a dusting or rolling agent in chewing gum production. In this application it can reduce the risk of rope breaking after extrusion and cooling.

Isomalt containing chewing gums are characterized by a longer lasting flavor release and sweetness. The addition of flavored isomalt crystals to the gum base is an interesting new approach in sugar-free products to extend the long-lasting flavor release even more.

Generally, isomalt provides a balance of rigidness and flexibility to the chewing gum which can be used to produce specific shaped products like sticks, pellets, or balls.

Compressed Tablets/Lozenges

Isomalt DC is an agglomerated product (Figure 17.11) that can be applied as raw material for chewable tablets and lozenges by direct compression due to its improved compressibility (Dörr and

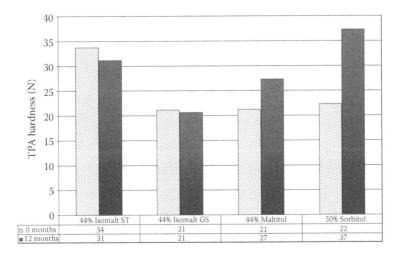

	44% Isomalt ST	44% Isomalt GS	44% Maltitol	50% Sorbitol
0 months	34	21	21	22
12 months	31	21	27	37

Figure 17.10   Texture-profile of chewing gum centers with isomalt ST, isomalt GS, sorbitol and maltitol after storage for 12 months at 30°C, 70% R.H. with packaging.

Willibald-Ettle 1996), as depicted in Figure 17.12, and its free flowability. A very small variance of the tablet weight and a uniform dosing of low concentrated ingredients can be obtained with isomalt DC without the use of further excipients. Depending on the GPS to GPM ratio in isomalt DC, low soluble tablets (using isomalt DC 100) or fast soluble tablets (using isomalt DC 200) can be produced. Isomalt tablets have a long shelf life and the low hygroscopicity of isomalt helps to protect moisture-sensitive ingredients as demonstrated in Figure 17.13.

Isomalt tablets exhibit a sugar-comparable sensorial sensation with a lower sweetness. Isomalt DC can be used both in fruity or mint formulations, as its mild sweetness and low cooling can be combined with any flavor.

Figure 17.11   Isomalt DC 100. (From Beneo Palatinit GmbH, *Food Market Technol* 16(4):6–10, 2002.)

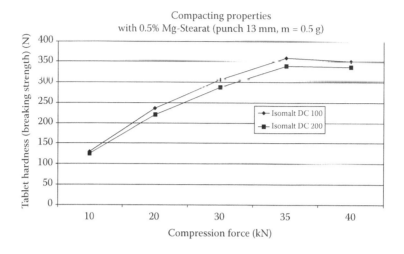

Figure 17.12 Tablet hardness—compression force—profile of isomalt DC 100, DC 200. (From R&D Department Südzucker AG, Germany.)

Chocolate

Isomalt is used in reduced-calorie and diabetic chocolate products because of its sugar-like neutral taste, snap, melting behavior and negligible cooling effect (Bollinger and Keme 1988). The normal formulation for a sucrose chocolate can be used for an isomalt chocolate (mild and bittersweet). Isomalt LM has a lower water content (less than 0.5%) than isomalt ST. This allows higher temperatures during conching. Isomalt chocolate is both sugar free and calorie reduced. Because the major source of energy in chocolate is fat and not sugar, a combination of isomalt and fat substitutes and/or bulking agents is recommended to further lower the calorie content of chocolate (Bollinger and Keme 1988; Irwin 1990b)

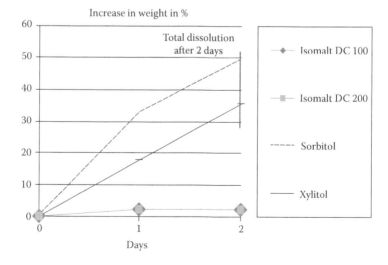

Figure 17.13 Isomalt outperforms sorbitol and xylitol in shelf life stability based on initial weight of tablets; storage at 25°C, 80% R.H., 48 h; as raw material.

Technological benefits of isomalt LM in chocolate
- Low water content allowing standard process temperatures
- Stable flow properties
- Low yield stress values, enrobing possible
- Standard viscosity
- Fineness adjustable; resulting in a smooth melting chocolate
- Sugar-like sensorial profile; sweet, not bitter, no aftertaste, no cooling effect
- Good snap
- Mild sweetness for full sensation of cocoa flavor

Baked Products

Isomalt is a useful ingredient in the formulation of "light" or sugar-free baked goods (Brack et al. 1986a, 1986b; Seibel et al. 1986; Willibald-Ettle and Keme 1992). The low solubility of isomalt, its low hygroscopicity, and its reduced browning reaction have to be considered during formulation development. Only minimal modifications in formulations and processing methods are required for baked goods made with isomalt. The final baked products made with isomalt have a sugar-like taste and a long shelf life (Willibald-Ettle 1992). Isomalt wafers and cookies absorb a lower amount of water than sugar formulations. They stay crispier during storage, therefore, even if the sugar is only partly replaced. The texture of wafers and extruded food products can be improved by adding isomalt to the mixture (Bollinger and Steinhage 1989). For texture improvement, isomalt can be added to sweet, and to salted products.

Technological advantages of isomalt in baked goods
- No reducing capacity → almost no Maillard reaction
- Reduced caramelization (heat stable up to 180°C, as raw material)
- Low hygroscopicity
- Low water activity (like sucrose)
- Low water binding capacity
- Excellent shelf life of cookies (texture)

Jam and Fruit Spreads

Isomalt GS is used to prepare different types of jam and fruit spreads. Due to its acid stability, isomalt GS has a very good gelling behavior for fruit spread, that is even better than sucrose based preparations (Stüber 2003). Isomalt GS spreads retain low syneresis during storage. Figure 17.14 illustrates the well performing gelling and low syneresis performance compared to sucrose or fructose formulations. The browning reaction of these spreads is comparable to that of jam made with sugar. The taste of isomalt GS containing fruit spread is very fruity with no aftertaste. If a low dry substance between 25% and 30% of the final product is desired, a pleasant taste profile in combination with an intense sweetener can be obtained.

Cereals and Bars

Isomalt can partially or completely replace sugar in originally high sugar containing formulations, with up to 40% sugar, of breakfast cereals and corresponding products (Beneo Palatinit GmbH

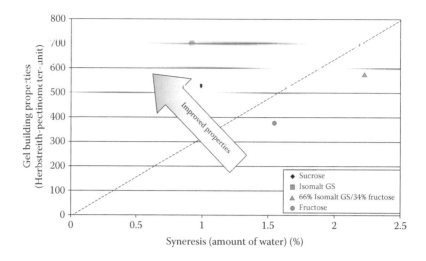

Figure 17.14   Strawberry fruit spread 2:1 with isomalt GS, fructose, isomalt + fructose and sorbitol – gel builder properties versus syneresis properties. (From Stüber, S., Untersuchungen zur Verwendung von Isomalt umd anderen Zuckeraustauschstoffen bei der Herstellung von Diät-Konfitüre. Diploma Thesis, Universität Gießen, 2003.)

2002). Isomalt resembles sugar in body building, glass transition temperature and hygroscopicity. Therefore the substitution of sugar by isomalt ST in breakfast cereals and other cereal products results in similar product properties with respect to taste, bite, bowl-life and color at a one to one replacement in terms of weight. Besides its physiological properties, especially the low solubility, low hygroscopicity, and missing Maillard reaction, isomalt helps to improve the shelf life and texture of cereal products.

Isomalt ST can be used as a binder for crunchy or soft granola/cereals bars as well as for cereals. For soft texture, combinations with a type of maltitol syrup are recommended. Isomalt rounds off the formulation by improving the taste profile and reducing stickiness. Also the shelf life is positively affected due to the low hygroscopicity of isomalt.

Pharmaceutical Applications

Nowadays sugar substitutes such as mono- and disaccharide alcohols are frequently used in pharmaceutical applications. The main reasons to choose sugar alcohols are:

- The suitability for diabetics as no significant increase in body glucose, insulin or lactic acid concentration arises unlike with conventional saccharides
- The chemical stability as they are not subject to the Maillard reaction and therefore more stable than related saccharides
- Their sweet taste
- That they are non-cariogenic
- The non-animal origin

From a regulatory point of view monographs on isomalt are described in current editions of the major pharmacopoeias, the European Pharmacopeia and the Pharmacopeia of the United States.

Isomalt properties that are of particular interest for pharmaceutical applications are:

- Low hygroscopicity enhancing the climate stability of pharmaceutical formulations containing moisture sensitive active ingredients
- Excellent compactibility and therefore reducing the compression force during roller compaction and tableting processes

Different grades of pharmaceutical grade isomalt are marketed under the brand name galenIQ™ by BENEO-Palatinit. Those are mainly used for the manufacture of:

- Chewable, orodispersible and disintegrating tablets
- Capsule fillings
- High boiled lozenges
- Pan coatings for tablets

Scientific literature about the compaction properties of polyols as filler binders for disintegrating tablets comparing direct compressible types of lactitol, sorbitol, xylitol, mannitol, and isomalt reveals that isomalt is one of the most suitable polyols for tablet manufacture as displayed in Figure 17.15 (Bolhuis et al. 2009a). Agglomerated isomalt types show good flowability, moderate lubricant sensitivity and good compaction properties. The latter effect is caused by an early fragmentation of the agglomerates during the compaction process, producing clean-lubricant free particles and a high surface area for bonding. The different GPM/GPS ratios of the agglomerated isomalt types had no significant effect on the compaction properties (Bolhuis et al. 2009a).

Isomalt can also be used in multiparticulate drug delivery systems as the starter core for pharmaceutical coating. Their numerous technological, physiological and therapeutical advantages over single unit dosage forms make them very useful both in drug carrier design and drug development (Digenis 1994; Nastruzzi et al. 2000; Roy and Shahiwala 2009). Physical characteristics of isomalt pellet cores such as mechanical strength, shape and size proved that the inert cores were adequate for further processing. *In vitro* drug dissolution test results demonstrated that saccharose

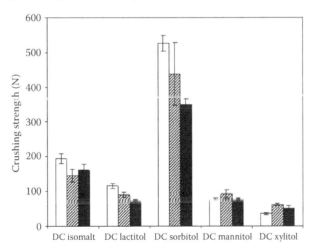

Figure 17.15 **Crushing strength of tablets, compressed at 20 kN from different polyols for direct compaction. The tablets were unlubricated (white bar), or mixed with 0.5% magnesium stearate for 2 min (gray bar) or 30 min (black bar). (From Bolhuis et al., *Drug Dev Ind Pharm* 35(6):671–677, 2009b.)**

based and isomalt type pellet cores demonstrated similar drug release profiles. Moreover drug dissolution from isomalt pellet cores proved to be much less sensitive to the change in osmolality of the dissolution media in comparison to non soluble pellet cores (Kallai et al. 2010).

Regulatory Status

Isomalt is approved in more than 80 countries worldwide, including those in NAFTA, those in the EU, Japan, Australia, New Zealand, East European countries, and in many Asian and South American countries. First approvals for use as a sugar replacer were given in 1983/84 in Switzerland and the United Kingdom followed by countries all over the world. In most European countries, isomalt is approved as a food additive. A Generally Recognized As Safe (GRAS) affirmation petition was accepted for filing by the US FDA in 1990. Since then, isomalt has been marketed in the United States as a GRAS substance. In Japan isomalt is accepted as a food, in China it has new resource food approval. Isomalt is included in the General Standard on Food Additives (GSFA) of the Codex Alimentarius for use in food in general (Table 3 of the GSFA, updated to the 32nd Session of the Codex Alimentarius Commission in 2009). The specification relevant for international trade is the JECFA specification prepared at the 69th meeting (FAO JECFA Monographs 2008).

References

Beneo Palatinit GmbH 2002. Sugar reduced breakfast cereals can now be enjoyed. *Food Market Technol* 16(4):6–10.

Berschauer, F., and Spengler, M. 1987. Energetische Nutzung von Palatinit. *Dtsch Zahnärztl Z* 42:145–150.

Bolhuis, G.K., Engelhart, J.J., and Eissens, A.C. 2009a. Compaction properties of isomalt. *Eur J Pharm Biopharm* 72(3):621–625.

Bolhuis, G.K., Rexwinkel, E.G., and Zuurman, K. 2009b. Polyols as filler-binders for disintegrating tablets prepared by direct compaction. *Drug Dev Ind Pharm* 35(6):671–677.

Bollinger, H. 1987a. Palatinit® (Isomalt) – ein kalorienreduzierter Zuckeraustauschstoff – Technologische und physiologische Eigenschaften. *Gordian* 87/5:92–95.

Bollinger, H. 1987b. Palatinit® (Isomalt) – ein kalorienreduzierter Zuckeraustauschstoff – Technologische und physiologische Eigenschaften. *Gordian* 87/6:111–114.

Bollinger, H., and Keme, T. 1988. Kalorienreduzierten Zuckeraustauschstoff Palatinit® (Isomalt) verwenden – Saccharosefreie Schokolade herstellen. *Zucker-und Süßwaren Wirtschaft* 41(1):23–27.

Bollinger, H., and Steinhage, H. 1989. The production of low-calorie cereal extrudates. *Zucker- und Süßwaren Wirtschaft* 42(2):48–52.

Brack, G., Seibel, W., and Bretschneider, F. 1986a. Backtechnische Wirkung des Zuckeraustauschstoffes Palatinit (Isomalt): Feinteige ohne Hefe. *Getreide, Mehl Und Brot* 40(9):269–274.

Brack, G., Seibel, W., and Bretschneider, F. 1986b. Backtechnische Wirkung des Zuckeraustauschstoffes Palatinit (Isomalt): Massen mit Aufschlag. *Getreide, Mehl Und Brot* 40(10):302–306.

Bubník, Z., Kadlec, P., Urban, D., and Bruhns, M. 1995. *Sugar Technologists Manual*. 8th Edition. Berlin: Bartens Verlag.

Cammenga, H.K., Figura, L.O., and Zielasko, B. 1996. Thermal behaviour of some sugar alcohols. *J Thermal Anal* 47:427–434.

Cammenga, H.K., and Stepphuhn, I.D. 1993. Polymorphic status of sorbitol: Solution calorimetry versus DSC. *Thermochimica Acta* 229:253–256.

Ciardi, J., Bowen, W., Rolla, G., and Nagorski, K. 1983. Effect of sugar substitutes on bacterial growth, acid production and glucan synthesis. *J Dent Res* 62:182.

Digenis, G.A. 1994. The in vivo behavior of multiparticulate versus single unit dosage formulations. In *Multiparticulate Oral Drug Delivery*, I. Ghebre-Sellassie (Ed.), pp. 333–355. New York: Marcel Dekker, Inc.

Dörr, T., and Willibald-Ettle, I. 1996. Evaluation of kinetics of dissolution of tablets and lozenges consisting of saccharides and sugar alcohols. *Pharm Ind* 58(10):947–952.

Drost, H., Spengler, M., Gierlich, C., and Jahnke, K. 1980. *Palatinit ein neuer Zuckeraustauschstoff: Untersuchungen bei Diabetikern vom Erwachsentyp. Aktuelle Endokrinolgie*, Vol. 2, 171. Stuttgart, New York: Georg Thieme Verlag.

Emeis, C.C., and Windisch, S. 1960. Palatinosevergärung durch Hefen. *Z Zuckerindustrie* 10(5):248–250.

FAO JECFA Monographs. Isomalt (5). 2008.

Fritzsching, B. 1993. Isomalt, a sugar substitute ideal for the manufacture of sugar-free and calorie-reduced confectionery. *Food Ingr Eur, Conf Proc 1993*, 371–377.

Gau, W., Kurz, J., Müller, L., Fischer, E., Steinle, G., Grupp, U., and Siebert, G. 1979. Analystische Charakterisierung von Palatinit. *Z Lebensm Unters Forsch* 168:125–130.

Gehrich, K. 2002. *Phasenverhalten einiger Zucker und Zuckeraustauschstoffe*. PhD Thesis, TU Braunschweig.

Gehring, F., and Hufnagel, H. 1983. Intra-und extraorale pH-Messungen an Zahnplaques des Menschen nach Spülungen mit einigen Zucker-und Saccharoseaustauschstoff-Lösungen. *Oralprophylaxe* 5:13–19.

Gostner, A., Blaut, M., Schaffer, V., Kozianowski, G., Theis, S., Klingeberg, M., Dombrowski, Y., et al., 2006. Effect of isomalt consumption on faecal microflora and colonic metabolism in healthy volunteers. *Br J Nutr* 95(1):40–50.

Grabitske, H.A., and Slavin, J.L. 2008. Low-digestible carbohydrates in practice. *J Am Diet Assoc* 108(10):1677–1681.

Grupp, U., and Siebert, G. 1978. Metabolism of hydrogenated palatinose, an equimolar mixture of alpha-D-glucopyranosido-1,6-sorbitol and alpha-D-glucopyranosido-1,6-mannitol. *Res Exp Med (Berl)* 173(3):261–278.

Holub, I., Gostner, A., Hessdorfer, S., Theis, S., Bender, G., Willinger, B., Schauber, J., Melcher, R., Allolio, B., and Scheppach, W. 2009. Improved metabolic control after 12-week dietary intervention with low glycaemic isomalt in patients with type 2 diabetes mellitus. *Horm Metab Res* 41(12):886–892.

Imfeld, T.N. 1983. Identification of Low Caries Risk Dietary Components, 123–127. Basel: Verlag S Karger.

Irwin, W.E. 1990a. Reduced calorie bulk ingredients: Isomalt. *Manuf Confect* 70(11):55–60.

Irwin, W.E. 1990b. Sugar substitute in chocolate. *Manuf Confect* 70(5):150–154.

Jasra, R.V., and Ahluwalia, J.C. 1982. Enthalpies of solution, partial molal heat capacities and apparent molal volumes of sugars and polyols in water. *J Sol Chem* 11:325–338.

Kallai, N., Luhn, O., Dredan, J., Kovacs, K., Lengyel, M., and Antal, I. 2010. Evaluation of drug release from coated pellets based on isomalt, sugar, and microcrystalline cellulose inert cores. *AAPS Pharm Sci Tech* 11(1):383–391.

Karle, E.J., and Gehring, F. 1979. Kariogenitätsuntersuchungen von Zuckeraustauschstoffen an xerostomierten Ratten. *Dtsch Zahnärztl Z* 34(7):551–554.

Lee, A., Livesey, G., Storey, D., and Zumbé, A. 2001. International Symposium on Low-Digestible Carbohydrates – Consensus Statements. *Br J Nutr Supplement* 85(3):S5.

Lichtenthaler, F.W., and Lindner, H.J. 1981. The preferred conformations of glycosalditols. *Liebigs Ann Chem* 2372–2383.

Life Sciences Research Office. 1994. *The Evaluation of the Energy of Certain Sugar Alcohols used as Food Ingredients*. Bethesda, MD: Federation of the American Society for Experimental Biology.

Lindner, H.J., and Lichtenthaler, F.W. 1981. Extended zigzag conformation of 1-O-D-alpha-glucopyranosyl-D-mannitol. *Carbohyd Res* 93:135–140.

Livesey, G. 2003. Health potential of polyols as sugar replacers, with emphasis on low glycaemic properties. *Nutr Res Rev* 16(2):163–191.

Livesey, G., Buss, D., Coussement, P., Edwards, D., Howlett, J., Jonas, D., Kleiner, J., Müller, D., and Sentko, A. 2000. Suitability of traditional energy values for novel foods and food ingredients. *Food Control* 11(4):249–289.

Livesey, G., Taylor, R., Hulshof, T., and Howlett, J. 2008. Glycemic response and health – a systematic review and meta-analysis: Relations between dietary glycemic properties and health outcomes. *Am J Clin Nutr* 87(1):258S–268S.

Mende, K. 1990. *Thermodynamische und rheologische Untersuchungen an Saccharidlösungen und -schmelzen.* PhD Thesis. TU Berlin.

Nastruzzi, C., Cortesi, R., Esposito, E., Genovesi, A., Spadoni, A., Vecchio, C., and Menegatti, E. 2000. Influence of formulation and process parameters on pellet production by powder layering technique. *AAPS Pharm Sci Tech* 1(2):E9.

Nilsson, U., and Jägerstad, M. 1987. Hydrolysis of lactitol, maltitol and Palatinit by human intestinal biopsies. *Br J Nutr* 58(2):199–206.

Official Website of the Glycemic Index and GI Database. 2011. University of Sydney. Available from http://www.glycemicindex.com. Accessed June 06, 2011.

Paige, D.M., Bayless, T., and Davis, L. 1986. *The Evaluation of Palatinit Digestibility* (Revised Report II). Baltimore, MD: J. Hopkins Univ.

Paulus, K., and Fricker, A. 1980. Zucker-Ersatzstoffe Anforderungen und Eigenschaften am Beispiel von Palatinit. *Z Lebensmitteltechnologie Und Verfahrenstechnik* 31(3):128–132.

Raudonus, J., Bernard, J., Janßen, H., Kowalcyk, J., and Carle, R. 2000. Effect of oligomeric or polymeric additives on glass transition, viscosity and crystallization of amorphous isomalt. *Food Res Int* 33:41–51.

Raudonus, J.W.J. 1999. Untersuchung des Einflusses polymerer Zusätze bei der Herstellung von Isomalt-Hartkaramellen. Diploma Thesis. Hohenheim University.

Roy, P., and Shahiwala, A. 2009. Multiparticulate formulation approach to pulsatile drug delivery: Current perspectives. *J Control Release* 134(2):74–80.

Schiweck, H. 1980. Palatinit – Herstellung, technologische Eigenschaften und Analytik palatinithaltiger Lebensmittel. *Alimenta* 19:5–16.

Seibel, W., Brack, G., and Bretschneider, F. 1986. Backtechnische Wirkung des Zuckeraustauschstoffes Palatinit (isomalt): Feinteige mit Hefe. *Getreide, Mehl Und Brot* 40(8):239.

Smits-Van Prooije, A.E., de Groot, A.P., Dreef-Van der Meulen, H.C., and Sinkeldam, E.J. 1990. Chronic toxicity and carcinogenicity study of isomalt in rats and mice. *Food Chem Toxicol* 28(4):243–251.

Spengler, M., Schmitz, H., and Biermann, C. 1979a. Vergleich der Palatinit-Toleranz gegenüber der Sorbit-Toleranz gesunder Erwachsener nach 14-tägiger oraler Gabe. Bayer AG Pharmabericht 8449. Bayer AG, Pharma Bericht 8449 v.23.5.1979.

Spengler, M., Sommer, J., and Schmitz, H. 1979b. Vergleich der Palatinit-Toleranz gegenüber der Sorbit-Toleranz gesunder Erwachsener nach einmaliger oraler Gabe in aufsteigender Dosierung. Bayer AG Pharmabericht 8449. Bayer AG, Pharma Bericht 8457 v.13.6.1979.

Spengler, M., Somogyi, J.C., Pletcher, E., and Boehme, K. 1987. Tolerability, Acceptance and Energetic Conversion of Isomalt (Palatinit(R)) in Comparison with Sucrose. *Akt Ernähr* 12(6):210–214.

Sträter, P.J. 1986. Palatinit – Technological and Processing Characteristics. *Food Sci Technol* 17:217–244.

Sträter, P.J. 1988. Palatinit (isomalt), an energy-reduced bulk sweetener derived from saccharose. In *Low-Calorie Products,* eds. G.G. Birch and M.G. Lindley, p. 36. London: Elsevier Applied Science.

Stüber, S. 2003. Untersuchungen zur Verwendung von Isomalt umd anderen Zuckeraustauschstoffen bei der Herstellung von Diät-Konfitüre. Diploma Thesis, Universität Gießen.

Südzucker AG Mannheim/Ochsenfurt CRDS. Internal Data. 2010.

Thiébaud, D., Jacot, E., Schmitz, H., Spengler, M., and Felber, J.P. 1984. Comparative study of isomalt and sucrose by means of continuous indirect calorimetry. *Metabolism* 33(9):808–813.

US Food and Drug Administration (US FDA). 1996. Food Labeling: Health Claims; Sugar Alcohols and Dental Caries 21 CFR Part 101. *Federal Register,* 61(No. 165).

Waalkens-Berendsen, D.H., Koeter, H.B., Schluter, G., and Renhof, M. 1989. Developmental toxicity of isomalt in rats. *Food Chem Toxicol* 27(10):631–637.

Waalkens-Berendsen, D.H., Koeter, H.B., and Sinkeldam, E.J. 1990a. Multigeneration reproduction study of isomalt in rats. *Food Chem Toxicol* 28(1):11–19.

Waalkens-Berendsen, D.H., Koeter, H.B., and van Marwijk, M.W. 1990b. Embryotoxicity/teratogenicity of isomalt in rats and rabbits. *Food Chem Toxicol* 28(1):1–9.

Weisser, H., Weber, J., and Locin, M. 1982. Water vapour sorption isotherms of sugar substitutes in the temperature range 25 to 80°C. *ZFL* 33(2):89–97.

WHO Food Additive Series. 1987. Isomalt. In *Toxicological Evaluation of Certain Food Additives and Contaminants*, 20, p. 207. Cambridge: Cambridge Universtiy Press.

Willibald-Ettle, I. 1992. Isomalt: A sugar substiute. Its properties and applications in confectioncry and baked goods. Presented at Practical Confectionery Course, William Angliss College, Melbourne, Australia.

Willibald-Ettle, I., and Keme, T. 1992. Recent findings concerning isomalt in different applications. *Int Food Ingred Mag* 1:17–21.

Zielasko, B. 1997. Ermittlung physikalisch-chemischer Daten von Isomalt und seinen Komponenten. PhD Thesis, TU Braunschweig.

Chapter 18

Maltitol

Malcolm W. Kearsley and Navroz Boghani

Contents

Introduction

Maltitol powder is a member of the family of bulk sweeteners and although first produced in 1940 has only been used in food products for about 25 years in Europe and the United States and for a longer period in Japan. As non hygroscopic grades became available in the early 1980s (Hirao et al. 1983), its use has grown and it is now indispensible for the manufacture of specific high quality "sugar free" and "no added sugar" products as a replacement for sucrose specifically and sugars generally. Like other polyols, maltitol is promoted as being safe for the teeth, suitable for diabetics, having a low glycemic index (GI) and reduced calorie; all categories of interest to food manufacturers looking to gain sales and marketing advantage for their products. Maltitol finds particular use in chocolate and baked goods where it functions as probably the best replacement for sugar. The range of sugar free products, however, is still very limited compared with traditional sugar or sugar/glucose based products. There are many and varied reasons given for this including cost, market demand, difficulty of manufacture, and availability of raw materials. Additionally, maltitol is used in coatings for tabletted gum where it gives a highly desirable crunchy texture and a glossy surface. The laxative effect of polyols generally is often quoted as the major reason for their lack of market penetration. It does seem likely, though, that the demand and manufacture of sugar free foods generally will increase in the future and maltitol is set to play a key role in meeting this demand.

Of the permitted sugar alcohols, maltitol offers the closest approximation to the properties of sugar (as shown in Table 18.1) and the closely related maltitol syrups can replace glucose syrups,

Table 18.1 Comparison of Maltitol and Sucrose

	Sucrose	Maltitol
Molecular Weight	342	344
Sweetness	1.00	0.90
Solubility at 22°C	67%	65%
Melting Point (°C)	168–170	144–152
Heat of solution (cal/g)	−4.3	−5.5
ERH for water uptake (20°C)	84%	89%
Calories (kcal/g)	4.0	2.4 (EU)
		2.1 (USA)
		2.0 (Japan)

ERH = Equilibrium Relative Humidity.

thereby offering a complete "sugar replacement" package. This combination gives a good starting point for new product development although in some circumstances specific polyols offer better quality options than simple sugar and glucose replacement.

There is an abundance of general and patent literature available on maltitol production, properties and applications and the reader is directed specifically to two reviews by Kato and Moskowitz (2001) and Kearsley and Deis (2006).

Production

The manufacturing process for maltitol was pioneered by Hayashibara in Japan, which subsequently licensed the technology to Towa (Japan), Cerestar (now Cargill) (EU) and Roquette (EU and the United States). Additionally, a number of Chinese manufacturers also claim to make maltitol powder. In common with the majority of polyols, the disaccharide maltitol is manufactured by the catalytic hydrogenation of the appropriate reducing sugar and for maltitol this is either maltose or more usually very high maltose glucose syrup. The reactive aldehyde groups in each case are reacted with hydrogen to give stable alcohol groups and by changing only the reactive reducing group, the polyol retains much of the parent sugar's structure, bulk, and functionality and simultaneously gains other beneficial properties. This makes it an almost perfect sugar replacer.

Hydrogenation

The hydrogenation conditions to manufacture maltitol are similar to those used to make other polyols. The maltose or maltose syrup is reacted with hydrogen at high temperature (typically 100°C–150°C) and high pressure (typically 100–150 bar) in the presence of a suitable catalyst. Catalysts include Raney Nickel, supported nickel, molybdenum, palladium, and platinum. Reaction times are typically in the order of 1–2 hours depending on the conditions.

The maltose or maltose syrup used in the hydrogenation process is highly purified to prevent poisoning of the catalyst and this is achieved using ion exchange and carbon treatment. Depending on the catalyst used in the process, the hydrogenated product may then be subject to a further ion exchange treatment to remove dissolved catalyst. The raw material for the manufacture of maltose is starch, and while this can be from any source, maize (corn) and tapioca starches are most widely used commercially.

There are several related routes which can be used to make maltitol (α(1–4) glucosylsorbitol) but all share a common first stage in which the starch is liquefied typically by cooking to about 110°C in the presence of a heat stable α-amylase enzyme. A second cook to 135°C may then be carried out to ensure all the starch is gelatinized before proceeding to the next stage, where specifically maltose is produced from the starch using a combination of saccharifying enzymes including β-amylase and pullulanase. The latter is added to specifically hydrolyse (1–6) linkages in the starch and thereby open up the starch granules to allow the β-amylase greater access. While it may be possible under specific conditions to totally convert starch to maltose, owing to time and financial constraints the industrial process does not attempt this, but rather makes a very high maltose syrup containing typically 85%–95% maltose on a dry basis. The maltitol manufacturer now has several options, most of which have been the subject of patents at one time or other:

1. Maltose can be recovered from the maltose syrup by crystallization and after redissolving in water the high purity maltose can then be hydrogenated to give predominantly maltitol. Alternatively, industrial scale liquid chromatography can be used to create a purified maltose stream, which after concentration by evaporation, can then be hydrogenated. Maltitol powder can then be recovered by aqueous or melt crystallization and typically has a purity in excess of 99%.

2. The maltose syrup can be hydrogenated and after application of liquid chromatography, maltitol powder can be recovered from the purified maltitol stream by aqueous or melt crystallization. Again purity is in excess of 99%.

3. If the maltose content of the syrup after hydrolysis of the starch is sufficiently high, then after hydrogenation the maltitol powder can alternatively be recovered by melt crystallization. Purity varies but typically is around 92%–93% with the balance being mainly sorbitol and maltotriitol.

In aqueous crystallization, a saturated solution of a solute is cooled and seeded with crystals of the solute. The solute crystallizes out in pure form and is recovered by filtration or centrifugation from the mother liquor and dried.

In melt crystallization, water is progressively evaporated from the high maltitol syrup until the solids content exceeds about 95%. The polyol is not in solution under these conditions, but is effectively a high solids melt. After seeding, the molten maltitol is then bled from the cooker and solidified. Drying, milling and screening complete the process to give maltitol powder. Melt crystallization does not result in purification of the solute stream.

The EU Sweeteners in Food Regulations specify a minimum 98% maltitol content for maltitol powder (on a dry basis). Dry products containing less than 98% maltitol are defined as dried maltitol syrups. In the United States, *Food Chemicals Codex* (*FCC*) guidelines require the maltitol content of maltitol powder to be not less than 92% and not more than 100.5% maltitol calculated on a dry basis.

Owing to the additional processes which need to be carried out on the starting material (very high maltose syrup) maltitol is significantly more expensive than sucrose or glucose syrup and typically twice the price of sorbitol powder.

Availability

Packaging is usually in 25, 500, and 1000 kg bags, or in bulk (about 20 mt).

Structure

The structure of maltose and its subsequent conversion to maltitol by hydrogenation is shown in Figure 18.1. During hydrogenation only the terminal glucose residue reacts with hydrogen and is converted to sorbitol.

Physical and Chemical Properties

It is important when replacing sucrose and/or glucose with polyols that the food manufacturer selects a polyol whose properties most closely mimic the sugar(s) to be replaced. Optimum results will, however, only be obtained when the disaccharide sucrose is replaced by a disaccharide polyol whose properties are most similar to those of sucrose. This product is maltitol powder and "one for one" replacement of sucrose is feasible in many applications.

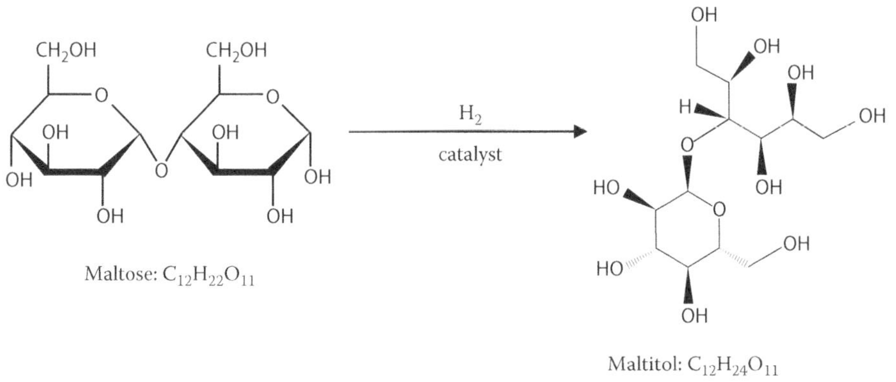

Figure 18.1 Hydrogenation of maltose to maltitol.

Chemical Reactivity

In common with other polyols, maltitol does not take part in Maillard browning reactions (Kearsley 1978). Normally this is a positive attribute as browning is undesirable, but in some applications (e.g., caramel manufacture), browning is an essential part of the manufacturing process. Sugar free caramels require addition of caramel color and flavor when sugar free products are made. It is often more appropriate to make a "no added sugar" claim than a "sugar free" claim for caramels so that milk products containing lactose can remain in the formulation.

Compressibility

Maltitol powder can be used in direct compression applications to make tablets, but these are not as hard as sorbitol based products. The method of manufacture of the maltitol will affect the hardness of the tablet, with melt crystallized material giving a harder product owing to its greater surface area and lower maltitol assay. Particle size is also important. Additionally, tablets made with maltitol do not require addition of high potency sweeteners, which may be an advantage with regard to a "cleaner" label on the product. Maltitol powder generally commands a substantially higher price than sorbitol powder, however, and this has to be balanced against any potential labelling benefits.

Cooling Effect (Heat of Solution)

The solid forms of some sugars and related products have a noticable heating or cooling effect when placed on the tongue. The effect, known as the "heat of solution," can be positive (warming effect) or more usually in the case of polyols, negative (cooling effect). The smaller the particle size of the product, the more quickly it will dissolve and the more pronounced the effect.

In some food products, a cooling effect is desirable, for example, a mint-flavored directly compressed tablet, while in others, for example, chocolate, it is not. Sucrose has a negative heat of solution of about −4 cal/g and maltitol has a value closest to this (−5.5 cal/g) compared with other polyols again confirming maltitol as the product of choice to replace sucrose in foods.

Humectancy

Maltitol powder has little functionality as a humectant in its main applications (gum, bakery products, and chocolate).

Hygroscopicity

The hygroscopic tendencies of food ingredients can have very serious consequences during production and storage (shelf life) of the food itself and also may affect how the ingredient itself is stored. Maltitol is one of the least hygroscopic of the polyols and does not begin to absorb moisture until the atmospheric relative humidity exceeds 80%. It can, therefore, be handled in most countries without the need for air conditioning in the manufacturing plant. In the gum industry, this property is important because the maltitol coated tablets retain their crunchy texture and glossy appearance for a longer period of time compared with tablets coated using other polyols.

Molecular Weight

Pure maltitol powder has a molecular weight of 344. When replacing sugar(s) with polyols, it is important to match the molecular weight of the sugar(s) and the polyol(s) to ensure the colligative properties remain unchanged. Molecular weight can play a significant role in influencing the overall texture and functionality of the finished product. Specifically it influences viscosity, freezing point depression, boiling point elevation and osmotic pressure of the product in which it is used.

Freezing point depression is an important factor when making ice cream for example, where traditionally, sucrose, glucose syrups and maltodextrins are used to provide functionality and sweetness. As in all applications, replacement of like with like will give the best results. Thus sucrose can be replaced with maltitol powder, glucose syrup with maltitol syrup and maltodextrin with polyglycitol. Osmotic pressure becomes important when products containing maltitol are consumed as any undigested maltitol may draw water from the body into the gut leading to osmotic diarrhea.

Solubility

This property is fundamental to the manufacture of almost all confections. Traditional confections (such as hard or chewy candy) in their most basic form are supersaturated sugar/glucose solutions, where either crystallization (grained) is required to some degree or not at all (ungrained). Through many years of use, confectioners have utilized the solubility of sucrose to create the standards of appearance, taste, texture, shelf life and sweetness that we now come to expect from these types of products.

Polyols such as erythritol, isomalt, and mannitol are least soluble and crystallize most readily, while maltitol has a relatively high solubility (in line with sucrose) of about 175 g per 100 g of water at 25°C. Maltitol finds application in gum where it is used to form a crystalline coating around the gum tablet. In crystalline form maltitol is less hygroscopic than sucrose. Maltitol syrups can be used in combination with maltitol to control its crystallization.

Sweetness

Sweetness is one of the most important properties of any sugar or sugar replacer. We generally like sweet foods and, despite manufacturers making foods with reduced sweetness, the demand for the sweet taste is still one of the most sought after. Maltitol has a sweetness of about 90% that of sucrose so usually does not need an addition of high potency sweeteners to boost its sweetness.

Sweetness of a carbohydrate is related to its molecular weight and structure and the chemical conversion of a reducing sugar to its corresponding polyol by the process of hydrogenation normally results in an increase in sweetness. While maltose, with its $\alpha(1-4)$ linked glucose residues, increases quite dramatically in sweetness after hydrogenation, cellobiose, with its $\beta(1-4)$ linked glucose residues, decreases in sweetness after hydrogenation. Lactose also increases in sweetness on hydrogenation but to a much lower extent than maltose (Kearsley et al. 1980).

It is immediately evident that maltitol is unique among the permitted polyols in that hydrogenation of maltose more than doubles the sweetness of the reducing sugar. This anomalously high sweetness has been attributed to a conformational interaction, mediated by intramolecular hydrogen bonding, between the sugar residue and the aglycone (Birch and Kearsley 1977).

Similarly hydrogenation of a high maltose syrup would be expected to yield an increase in sweetness in direct relation to its maltose content.

Physiological Properties

Caloric Value

In the EU, maltitol (in common with all other permitted polyols except erythritol) has a caloric value of 2.4 kcal/g (10 kJ/g) (EECC 1990) while in the United States, it has a value 2.1 kcal/g (LSRO 1994). This value was confirmed by the Life Sciences Research Office (LSRO) in 1999. In Japan, a value of 1.8 kcal/g was initially proposed (JMHW 1991) but this was reviewed in 1996 and changed to 2 kcal/g. In practice, this makes calculation of the caloric value of a food containing maltitol very difficult as a food made in the EU would apparently be less caloric when consumed in the United States or Japan. Different polyols will be digested in different ways and while there may be some similarities between different polyols giving the same end caloric value, it should not be assumed that this will be the case. Some polyols will have a higher value than 2.4 kcal/g and some values will be lower. The lower value compared with sugars is due to the relatively poor absorption of polyols in the small intestine. While some polyols will be hydrolysed and/or absorbed at this site, what is not absorbed will find its way to the large intestine where it will be fermented to volatile fatty acids or if the gut is overloaded excreted in the feces. Fatty acids are absorbed by the body and contribute about 2 kcal/g to the general calorie pool.

Owing to their different composition, it is most unlikely that all polyols would have the same caloric value. In some countries (e.g., the United States and Japan), this difference is recognized and different polyols have different caloric values. In this respect the values used in the United States and Japan are probably a truer reflection of the value for maltitol.

Maltitol can be used in foods to make reduced calorie and low calorie claims if appropriate reductions are made (e.g., 30% reduction in calories for reduced calorie in the EU, 25% in the United States).

Dental Aspects

Dental caries result from prolonged exposure of tooth enamel to acid which is produced by fermentation of dietary sugars by the oral bacteria. Dextran, the main component of plaque, is produced by microorganisms such as *Streptococcus Mutans*, and is responsible for the adherence of the microorganisms to the teeth. In common with other polyols, maltitol is either not fermented or fermented to a much lower extent than sugars by the oral bacteria so no acid is produced and maltitol is described as non-cariogenic (Ziesentiz and Siebert 1987). While maltitol may cause a slight fall in pH at the tooth surface it does not normally cause a fall in pH below 5.7. If the pH falls below 5.7, this can lead to decalcification of the tooth enamel and eventually to tooth decay or dental caries. Maltitol finds best use as a sweetener in those foods which remain in the mouth for a considerable time, for example hard candy, gum or chocolate and therefore provides a means of making such products safe for the teeth. This is the basis for a health claim in the United States— US Code of Federal Regulations, Title 21:101.80 (example: "Frequent eating of foods high in sugars and starches as between-meal snacks can promote tooth decay. The sugar alcohol maltitol used to sweeten this food may reduce the risk of dental caries."). While maltitol is non-cariogenic, other components of the food may not be. For example, chocolate contains milk which in turn contains lactose, so even if the traditional sucrose is replaced with maltitol the chocolate may still be cariogenic.

Finished products can be tested at certain recognised institutes to ensure they do not cause a lowering of the plaque pH and, if the product passes the test, the ToothFriendly logo may be used on the packaging. This testing involves continuous measurement of pH at the tooth surface using an in-dwelling pH electrode and if pH does not fall below 5.7 within 30 minutes of taking a food in the mouth, it passes the test and may be certified as "Safe for Teeth." While there are *in vitro* methods which can be used for rapid screening, only the *in vivo* test takes into account factors such as salivary flow, the buffering capacity of saliva and the swallowing of sugars or polyols dissolved in the saliva. Several studies have shown that maltitol per se is not cariogenic (Firestone et al. 1980; Matsuoka 1975; Rundgen et al. 1980). Detailed information on the cariogenicity of foods generally is available in the literature (JDR 1986).

The US Food and Drug Administration (US FDA) has authorized a "does not promote tooth decay" claim for sugar free foods containing polyols and thus maltitol (US FDA 1996).

Diabetic Suitability

Critical in the management of diabetes is control of blood glucose and weight. Use of polyols generally and maltitol specifically to replace sugar in the diet can be used to advantage in both of these areas owing to its reduced effect on blood glucose and its reduced calorie value (Wolever 2002; Wolever et al. 2002). Maltitol is metabolized differently by the body compared with traditional sugars leading to reduced blood glucose and insulin response levels after ingestion while simultaneously still providing the same bulk as traditional sugars. In addition to confectionery, maltitol can also be used in a wide range of baked goods for diabetics, giving products which are almost indistinguishable from the traditional offering. The metabolism of maltitol has been extensively discussed in the literature (Dwivedi 1986).

Glycemic Index

The concept of Glycemic Index (GI) was developed over twenty years ago by David Jenkins at the University of Toronto, as a tool to allow diabetics to better manage their diets, and is defined as the incremental area under the blood glucose response curve of a 50 g digestible carbohydrate portion of a test food, expressed as a percentage of the response to the same amount of carbohydrate from a standard food (typically glucose) taken by the same subject (FAO/WHO 1997). It is a method of ranking foods according to the extent to which they raise blood glucose levels after consumption.

Foods containing carbohydrates which break down quickly after ingestion, giving a fast and high blood glucose response, have the highest GI values, while foods containing carbohydrates which break down slowly after ingestion, giving a slow and low blood glucose response, have the lowest GI values. These are placed on a scale where glucose is given a GI of 100:

Low GI = 55 and below
Medium GI = 56–69
High GI = 70 and above

In addition to being important in the control of diabetes, low GI diets have also been promoted for weight loss (although more research is required in this case) and in improving the body's sensitivity to insulin.

Maltitol is considered to be low GI (about 35) (Livesey 2003).

Laxative Effects

Maltitol has been an important feature of sugar-free confectionery for many years and has proved extremely useful in reducing calories, lowering blood glucose response and formulating tooth-friendly products. In common with other polyols, there has always been concern over excessive consumption leading to laxative side effects. Humans do not have the necessary enzymes to fully metabolize maltitol in the small intestine and maltitol is therefore described as a low digestible carbohydrate. This may be viewed as a direct consequence of the non-cariogenic nature of maltitol where the bacteria in the mouth also do not possess the necessary enzymes to fully metabolize maltitol. It is important to note that any maltitol not broken down in the upper gastrointestinal tract can, at certain concentrations, lead to an osmotic imbalance and/or fermentation by bacteria in the lower gut, causing some digestive discomfort in the form of flatulence and diarrhea. If the maltitol is broken down but too much sorbitol generated, then this could lead to a similar effect. Maltitol and polyols generally are not unique in this respect. Fiber, for example, is also a low digestible carbohydrate and over consumption of fiber will have similar effects; over consumption of simple sugars such as sucrose and fructose can also have the same effect. Many excellent papers have been published over the years concerning the laxative effects of polyols and related products (Livesey 2001).

There is generally inconsistency in published data for the tolerance of polyols generally and maltitol specifically. It is possible, though, to draw certain conclusions and guidelines for the tolerance and ingestion levels of maltitol and to make recommendations on the maximum level which can be consumed first as a single dose and secondly on a daily basis without causing significant digestive issues in the majority of people.

Polyols have a wide range of molecular weight and other characteristics that can affect their digestion, absorption and fermentation (in the large intestine) and therefore their gastrointestinal tolerance.

Disaccharide sugar alcohols such as maltitol, are slowly digested by enzymes in the small intestine. This process is usually incomplete but there will be active absorption of any glucose generated and simultaneous passive absorption of sorbitol generated by the enzymic (maltase) hydrolysis. Any unhydrolyzed or non-absorbed maltitol passes to the colon where it is fermented by bacteria (Grabitske and Slavin 2009). If large doses of maltitol are ingested, the system becomes overloaded, both in terms of maltitol hydrolysis by maltase and the amount of sorbitol being too much for the passive diffusion through the gut wall and into the blood stream. The maltase enzyme does not have time to hydrolyze all the maltitol as the hydrolysis of maltitol is much slower than maltose (Ziesentiz and Siebert 1987). Excess maltitol passes into the lower gut with any unabsorbed sorbitol, where they are fermented producing gas (and in larger quantities they upset the water balance producing osmotic laxation). Generally larger quantities of disaccharide polyols can be tolerated compared with monosaccharide polyols, as the former are less osmotically active (and half of maltitol is of course glucose which is well tolerated in the body) (Sicard and Le Bot 1988).

Data from the literature indicates a maximum of 30 g of maltitol can be consumed as an individual portion without laxation in the majority of adults and a maximum of 50 g of maltitol can be consumed per day. Generally, an individual can consume larger doses of maltitol per day than other polyols without discomfort and larger still doses of maltitol syrups and polyglycitols. Body weight is also important with regard to the laxative effect of maltitol and it has been reported that the maximum single dose in both men and women is 0.3 g/kg body weight (no laxative effect) with a daily maximum of 0.8 g/kg body weight (Koizumi et al. 1983).

There is very little published information regarding the laxative potential of maltitol (or in fact any of the polyols) in children. This has been remedied to a certain extent in a recently published study, "Short-term digestive tolerance of chocolate formulated with maltitol in children," in which the authors found that maltitol was better tolerated than previously reported. The study indicated that maltitol is relatively well tolerated by children at up to 15 g/day and its use could be extended to include new sugar free confectionary and food products (Thabuis et al. 2010).

These values should be treated as guidelines when developing new products using maltitol as other factors such as gender, age, body weight and health can also affect tolerance. Maltitol, in common with other polyols, is usually better tolerated in solid foods and when taken with other food ingredients such as fat or protein.

A further consideration is that prolonged consumption of maltitol (polyols) leads to adaptation with consequent improved tolerance (Livesey 2001). After four to five days on a relatively high intake of maltitol, the laxative effects tend to be reduced and may in some cases disappear.

The US FDA requires that foods that may result in a daily ingestion of 50 g of sorbitol be labeled with the statement, "Excess consumption may have a laxative effect" (CFR 2010). Such a label for maltitol, which is metabolized into sorbitol, therefore may be required (Newsome 1993).

The Codex Alimentarius recommends that if a food provides a daily intake of sugar alcohols in excess of 20 g, there should be a statement on the label to the effect that the food may have a laxative effect. This is not mandatory, however, just a guideline. If polyols are used within sensible guidelines they can improve a formulation without causing problems.

Applications in Foods

Some of the properties of the maltitol powder, particularly those related to crystal morphology (e.g., compressibility), change depending on the method of manufacture, but generally melt crystallized and aqueous crystallized maltitol can be used interchangeably with minor processing changes in most cases. The main exception is in pan coating where differences in crystallization rates and coating hardness may be found between the grades. There is, additionally, little difference in sweetness between the grades. The milling and screening processes give a range of particle size in the powders and these are used in different applications.

Maltitol can be used in the production of "diet," "light" and "reduced" (sugar or calorie) foodstuffs and it is now possible to buy a wide range of such items to meet the demands of an increasingly overweight and obese population. Maltitol based "sugar free" confectionery items include chewing gum, tabletted mints and related products, and chocolate. In some confectionery categories, for example gum, products are predominantly sugar free (mainly due to the very desirable non-cariogenic properties of the polyols), while in other categories, for example, chocolate, sugar free accounts for only a very small proportion of sales. It does seem likely, though, that the manufacture of sugar free confectionery will increase as the demand for "reduced calorie" and "reduced sugar" foods increases (Zumbe et al. 2001). Similarly maltitol based "no added sugar" baked goods, particularly biscuits, are now well established although small in volume.

Maltitol, in common with other polyol powders, is white in color but there the similarity ends. All polyols are not the same and have different properties. It is important when replacing sugar and/ or glucose with polyols that the manufacturer selects a polyol whose properties most closely mimic the sugar(s) to be replaced. Sucrose can be replaced directly in confectionery by several of the permitted polyols. Optimum results will, however, only be obtained when the disaccharide sucrose is replaced by a disaccharide polyol whose properties are most similar to sucrose. This product is maltitol powder. Similarly the optimum replacement for a glucose syrup in a food is generally a maltitol syrup.

It is important that the formulator understands the differences between different polyols so that the correct polyol can be selected for a particular application. Sweetness, solubility, and cooling effect are amongst the most important properties of polyols when considering their use in foods.

Through many years of use, confectioners have utilized sucrose to create the standards of appearance, taste, texture, shelf life, and sweetness that we now come to expect from these types of products. To create these products in sugar-free form, the formulator must understand the functional requirements of the application. Is the product going to be grained? Is it going to be a low moisture system? Are there any handling, storage, or processing concerns? What is the expected shelf life of the product? Once the formulator knows the answer to these questions, formulator can then decide which polyol or polyols to use in the formulation.

Maltitol is one of the most soluble of the polyols and most similar to sucrose. The texture and flavor release in a confection made using this product will be most similar to a sucrose based product and this allows its use as an effective replacement for sucrose. Maltitol also crystallizes in a similar way to sucrose and in crystalline form is less hygroscopic than sucrose.

Specific Applications of Maltitol

Chocolate

Although not present in large volumes there are "sugar free" and "no added sugar" chocolate products on the market in the EU and the United States. In the EU, it is not permissible to partly replace the sugars in chocolate with maltitol but it is permissible to completely replace all added sugar with maltitol and call the product a "no added sugar (milk) chocolate with sweeteners." In the United States, if any polyols are added to chocolate which has a standard of identity it may no longer be called "chocolate" and would have to be called "chocolate flavored" or some other name.

Very acceptable sugar free chocolate can be made with any of the permitted polyols. These include: sorbitol, mannitol, isomalt, lactitol, xylitol, and maltitol and although each polyol brings benefits to the chocolate, there are more often than not associated negative effects. Sorbitol and mannitol are highly cooling, lack sweetness and bring laxative issues. Xylitol is almost as sweet as sugar and needs no addition of high potency sweeteners but its strong cooling effect is not desirable. Isomalt and lactitol suffer from lack of sweetness and high potency sweeteners need to be included in any formulation with these products. Additionally they have relatively low laxation thresholds, particularly lactitol. Maltitol with its anhydrous crystalline form, low hygroscopicity, high melting point, low cooling effect, high sweetness (no additional high potency sweetener needed), and stability can be used to replace sugar with none of these drawbacks and is the product of choice in this application. Substitution of sucrose with maltitol requires negligible processing changes. A large particle size maltitol powder is preferable to avoid needing excess additional fat in the formulation but otherwise the manufacturing process is identical to that where sucrose is used.

General Candy

While maltitol can be used to make the complete range of sugar free candies it now finds limited application in this confectionery group, as there are more appropriate polyols with better functionality. Hard candy in Europe is now almost exclusively isomalt based while in the United States, many hard candies are made using maltitol syrup. Tabletted products (e.g., compressed mints) are sorbitol based, and caramels and chewy candy are maltitol syrup based.

Sugar Free Panning

Soft Panning

Traditionally sugar and glucose syrup are used to give a soft coating to products such as jelly beans. These can be replaced by a combination of maltitol powder and maltitol syrup. It is important that the maltitol in the maltitol syrup does not crystallize, as this will lead to surface hardening of the coated product.

Hard Panning

Maltitol with properties most similar to sugar, enables a similar process to that when using sugar and also produces a coating most similar to the original material. There are many patents covering coating with maltitol and the user should ensure that the process they use does not infringe upon these. For pellet gum, maltitol is the polyol of choice for coating, owing to its crunchiness and it giving a glossy surface.

Dairy Applications

Maltitol can be used to replace sucrose in conventional full sugar ice cream. It has approximately the same molecular weight as sucrose and, therefore, gives the same freezing point depression as sucrose and hardness of the ice cream. Optimum results, however, appear to be found when an appropriately formulated maltitol syrup is used to replace both sucrose and glucose syrup. Maltitol in the maltitol syrup provides the sweetness and the higher molecular weight polymers in the maltitol syrup replace the glucose syrup. The no added sugar ice cream, therefore, has a solids content and freezing point profile identical to the full sugar ice cream, as well as a similar sweetness profile (Bordi et al. 2004).

Similarly, the sucrose and glucose syrup in drinkable yogurts and flavored milks can be replaced by maltitol, or better still a maltitol syrup, to give the product a more rounded sweetness profile. To avoid possible laxative effects, it is important to ensure the polyol content of the product remains below about 20 g per serving.

Bakery Applications

In bakery applications, sucrose and glucose syrups are the traditional sweeteners, so the consumer has been very much conditioned to these products as the gold standard with respect to product quality. As in confectionery and dairy applications, the molecular weight and solubility of the sweetener system is very important, because these properties not only affect the sweetness, but also other factors including the starch gelatinization temperature, water immobilization, protein denaturation, and the bulk of the product. These characteristics in turn affect the spread, the volume, the baking characteristics, and the stickiness of the biscuit. Many of the lessons learned from working with confectionery can be extended to baked goods (and also dairy products) because confectionery items such as caramels, creams, marshmallow, nougats and chocolate are also components of many baked goods. Since many of the physical and chemical characteristics of maltitol are similar to sucrose, maltitol can replace sucrose directly in most baked goods applications and maltitol or polyglycitols are available to replace glucose syrups or maltodextrins. Maltitol, maltitol syrups and polyglycitols do not participate in Maillard browning and the products will therefore brown less during processing. Sweetness may need to be adjusted, but the textural properties of the finished product should remain unchanged. In very sweet products, close attention should be given to the grams of polyol per serving because consumers are quite likely to consume more than one serving. Again, a limit of about 20 g of polyols per serving should ideally be used. For this reason, reduced sugar baked goods should be considered as a healthy alternative to full sugar baked goods rather than attempting to replace all of the sugar.

Additionally, maltitol can be used to partially or totally replace the fat in baked goods (Bakal et al. 1993).

Legal status

Maltitol powder is permitted for food and pharmaceutical use in most countries. Some countries, for example, those in the EU and the United States, have their own legal requirements and purity criteria for these products, but where these do not exist, the appropriate Codex Alimentarius specifications are usually adopted. All are essentially similar and have the same objective—to present the consumer with high quality food ingredients. In the EU, polyols are considered as food additives and their use in foods is controlled by the Sweeteners in Food Regulations. Maltitol (and maltitol syrups) have been given the E number E965 and are authorized for food use at *quantum satis* for the range of food products listed in the regulations. Depending on the country, this range typically includes confectionery, baked goods, ice cream and desserts, and fruit preparations. Beverages are not usually permitted to contain maltitol and its derivatives owing to possible over consumption leading to laxative side effects. In the EU, maltitol cannot be used in foods in conjunction with sugars unless the polyol is present for a technological function other than sweetness or a 30% reduction in calories results from the combination. In the United States, the restrictions on the use of polyols in combination with sugars do not apply.

In the United States, polyols are considered either a food additive or "Generally Regarded As Safe" (GRAS) by the US FDA. Maltitol (and maltitol syrups) are both self-affirmed GRAS.

Safety

Extensive toxicological testing has shown that maltitol is safe for consumption and the Joint FAO/WHO Expert Committee of Food Additives (JECFA) has given it an Acceptable Daily Intake (ADI) "not specified" (FAO/WHO 1980, 1985; SCF 1985).

Conclusion

In food applications, selection of the most appropriate polyol(s) can be based on an understanding of the functionality of sugar and/or glucose syrup in a particular product and matching the properties of the polyol to the sugar(s) they are to replace. In some foods, this polyol will be maltitol while in other cases it will be a different polyol. There will be an optimum polyol for each application.

Of all the permitted polyols, maltitol offers unique opportunities to the new product formulator owing to its physical and chemical properties. The main advantages and properties of maltitol when used in food products are given below:

- Bulk sweetener with a clean, sweet taste (about 90% of the sweetness of sucrose).
- White crystalline powder.
- Reduced calorie compared with traditional sugars.
- Suitable for diabetics.
- Safe for teeth.
- Low glycemic index and low insulin response.
- Not hygroscopic.
- Heat stable.
- The polyol of choice for no-added sugar/sugar free chocolate and baked goods.

References

Bakal, A.I., Nanbu, S., and Muraoka, T. 1993. Foodstuffs containing maltitol as sweetener or fat replacement. *European Patent* 039:299 B1.

Birch, G.G., and Kearsley, M.W. 1977. Some human physiological responses to the consumption of glucose syrups and related carbohydrates, *Die Starke* 29(10):348–352.

Bordi, P., Cranage, D., Stokols, J., Palchak, J., and Powell, L. 2004. Effect of polyols versus sugar on the acceptability of ice cream among a student and adult population. *Foodservice Research International* 15:41–50.

Code of Federal Regulations (CFR). 2010. Title 21, Section 184.1835(e). http://www.accessdata.fda.gov/scripts/cdrh/cfdocs/cfcfr/CFRSearch.cfm?fr=184.1835. Accessed October 29, 2010.

Dwivedi, B.K. 1986. Polyalcohols: Sorbitol, mannitol, maltitol and hydrogenated starch hydrolysates. In *Alternative Sweeteners*, L. O'Brien, and R.C. Gelardi (Eds.), NY: Marcel Dekker.

European Economic Community Council (EECC). 1990. Council directive: Nutrition labelling for foodstuffs. *Official Journal of the European Communities* No. L276/41.

FAO/WHO. 1980. Twenty Fourth Report of the Joint FAO/WHO Expert Committee on Food additives, Rome 1980. WHO Technical Report Series No. 653.

FAO/WHO. 1985. Report Prepared by the 29th Meeting of the Joint FAO/WHO Expert Committee on Food Additives. *Toxicological Evaluation of Certain Food additives and Contaminants*. June 3–12, 1987, pp. 179–206. Geneva, Switzerland: Published by Cambridge University Press on behalf of WHO.

FAO/WHO. 1997. Carbohydrates in Human Nutrition, Report of the joint FAO/WHO Expert Consultation, Rome, 14–18 April.

Firestone, R., Schmid, R., and Muhlemann, H.R. 1980. The effects of topical applications of sugar substitutes on the incidence of caries and bacterial agglomerate formation in rats. *Caries Research* 14:324.

Grabitske, H.A., and Slavin, J.L. 2009. Gastrointestinal effects of low-digestible carbohydrates. *Critical Reviews in Food Science and Nutrition* 49:327–260.

Hirao, M., Hijija, H., and Miyaka, T. 1983. Anhydrous crystals of maltitol and whole crystalline hydrogenated starch hydrolysate mixture solid containing the crystals and process for the production and use thereof. US Patent 4,408,041.

Japanese Ministry of Health and Welfare (JMHW). 1991. *Evaluation of the Energy Value of Indigestible Carbohydrates in Special Nutritive Foods.* Official Notice of Japanese Ministry of Health and Welfare, no. Ei-shin 71, Tokyo.

Journal of Dental Research (JDR). 1986. Scientific Consensus Conference on the Methods for assessment of the Cariogenic Potential of Foods. Special Issue, 1473–1543.

Kato, K, and Moskowitz, A.H. 2001. Maltitol. In *Alternative Sweeteners*, 3rd edition, L. O. Nabors (Ed.). NY: Marcel Dekker.

Kearsley, M.W. 1978. The control of hygroscopicity, browning and fermentation in glucose syrups. *Journal of Food Technology* 13(4):339–348.

Kearsley, M.W., and Deis, R.C. 2006. Maltitol and maltitol syrups. In *Sweeteners and Sugar Alternatives in Food Technology*, H. Mitchell (Ed.). Blackwell Publishing.

Kearsley, M.W., Dziedzic, S.Z., Birch, G.G., and Smith, P.D. 1980. The production and properties of glucose syrups. *Starke* 32(7):244–247.

Koizumi, N., Fujii, M., Ninomiya, R., Inoue, Y., Kagawa, T., and Tsukamoto, T. 1983. Studies on transient laxation effects of sorbitol and maltitol1: estimation of 50% effective does and maximum non-effective dose. *Chemosphere* 12(1):45.

Life Sciences Research Office (LSRO). 1994. In *The Evaluation of the Energy of Certain Sugar Alcohols Used as Food Ingredients.* J.M. Talbot, S.A. Anderson, and K.D. Fisher (Eds.). Bethesda, MD: Prepared for the Calorie Control Council, Atlanta, Georgia, Federation of American Societies for Experimental Biology (FASEB).

Livesey, G. 2001. Tolerance of low digestible carbohydrates – a general view. *British Journal of Nutrition* 85(Suppl. 1):S7–S16.

Livesey, G. 2003. Health potential of polyols as sugar replacers, with emphasis on low glycemic properties. *Nutrition Research Reviews* 16:163–191.

Matsuoka, K. 1975. On the possibility of maltitol and lactitol and se-58 as non-cariogenic polysaccharides. *Nihon University Dental Journal* 49:334.

Newsome, R. 1993. *Sugar Substitutes in Low-Calorie Foods Handbook*, A.B. Altschul (Ed.), p. 159. Marcel Dekker, Inc.

Rundgen, J., Koulourides, T., and Encson, T. 1980. Contribution of maltitol and lycasin to experimental enamel demineralisation in the human mouth. *Caries Research* 14:67.

Scientific Committee for Food (SCF). 1985. Commission of the European Communities. Reports of the Scientific Committee for Food Concerning Sweeteners. Sixteenth Series. Report EUR 10210 EN. Office of official Publications of the European Communities, Luxembourg.

Sicard, P.J., and Le Bot, Y. 1988. *From Lycasin to Crystalline Maltitol: A new series of versatile sweeteners.* Presented at the International Conference on Sweeteners – Carbohydrate and Low Calorie, September 22–25. Los Angeles, California: Sponsored by the American Chemical Society.

Thabuis, C., Cazaubiel, M., Pichelin, M., Wils, D., and Guerin-Deremaux, L. 2010. Short-term digestive tolerance of chocolate formulated with maltitol in children. *Int J Food Sci Nutr* 61(7):728–738.

US Food and Drug Administration (US FDA). 1996. Food Labelling: Health Claims: Sugar Alcohols and Dental Caries, Final Rule. Federal Register, 61, No. 165:43433, August 23.

Wolever, T.M.S. 2002. Dietary carbohydrates in the management of diabetes: Importance of source and amount. *Endocrinology Rounds* 2(5).

Wolever, T.M.S., Piekarz, A., Hollands, M., and Younker, K. 2002. Sugar alcohols and diabetes: A review. *Canadian Journal of Diabetes* 26(4):356–362.

Ziesentiz, S.C., and Siebert, G. 1987. Polyols and other bulk sweeteners. In *Developments in Sweeteners 3*, T. Grenby (Ed.). Elsevier-Applied Science Publishers.

Zumbe, A., Lee, A., and Storey, D. 2001. Polyols in confectionery: The route to sugar free, reduced sugar and reduced calorie confectionery. *British Journal of Nutrition* 85(Suppl. 1):S31–S45.

Chapter 19

Lactitol

Christos Zacharis and Julian Stowell[*]

Contents

---

[*] Paul H. J. Mesters, John A. Van Velthuijsen, and Saskia Brokx did not participate in the revision but their chapter in the book's 3rd edition was the basis of this revised chapter.

History

Lactitol is a bulk sweetener. It is a sweet-tasting sugar alcohol derived from lactose (milk sugar) by reduction of the glucose part of this disaccharide. It has not been found in nature and was described in the literature for the first time by Senderens in 1920 (Senderens 1920). The first useful preparation was made by Karrer and Büchi in 1937 (Karrer and Büchi 1937). The Chemicals Abstracts Service (CAS) registry number of lactitol (also called lactit, lactositol, and lactobiosit) is 585-86-4. The molecular weight of lactitol monohydrate ($C_{12}H_{24}O_{11}.H_2O$) is 362.34 and the chemical structure of lactitol, 4-0-(β-D-galactopyranosyl)-D-glucitol, is shown in Figure 19.1.

Preparation

The principles of the preparation of lactitol are general knowledge. The industrial process is a hydrogenation of a 30%–40% lactose solution at about 100°C with a Raney nickel catalyst (van Velthuijsen 1979). The reaction is carried out in an autoclave under a hydrogen pressure of 40 bar or more. On sedimentation of the catalyst, the hydrogenated solution is filtered and purified by means of ion-exchange resins and activated carbon. The purified lactitol solution is then concentrated and crystallized. The monohydrate, as well as the anhydrate and the dihydrate, can be prepared depending on the conditions of crystallization (Wijnman et al. 1983).

Hydrogenation under more severe conditions (130°C, 90 bar) results in partial epimerization to lactulose and partial hydrolization to galactose and glucose and hydrogenation to the corresponding sugar alcohols lactitol, lactulitol, sorbitol, and galactitol (dulcitol) (Saijonmaa et al. 1980).

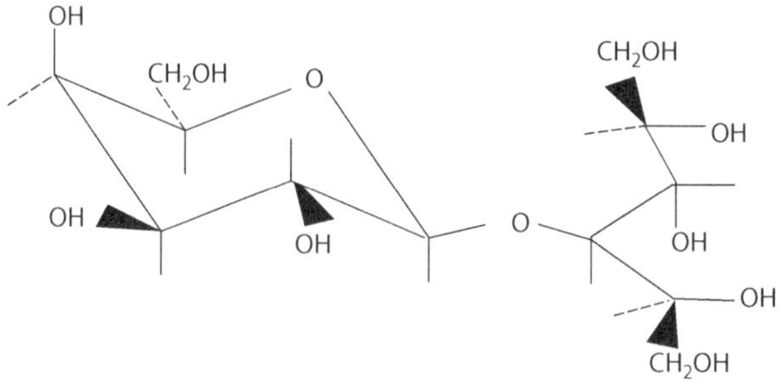

Figure 19.1   The chemical structure of lactitol.

Toxicology

All the required studies for a food additive have been carried out (CIVO-TNO, The Netherlands). There are no deleterious effects of feeding lactitol at levels up to 10% of the diet. The Joint FAO/WHO Expert Committee of Food Additives (JECFA) approved lactitol in April 1983. The Committee allocated an acceptable daily intake (ADI) "not specified" to lactitol (JECFA 1983).

The Scientific Committee on Food of the European Community (SCF-EC) evaluated lactitol and considered it to be a safe product; they stated that "consumption of the order of 20 g per person per day of polyols is unlikely to cause undesirable laxative symptoms" (SCF 1984, 3). The same was stated for isomalt, maltitol, mannitol, sorbitol, and xylitol.

Safety studies in experimental animals included long-term feeding studies at high dietary levels for 2.5 years in rats and for 2 years in mice. The safety data are not summarized here. Full data are described in CIVO-TNO (Zeist, The Netherlands) reports, which are published in the *Journal of the American College of Toxicology* (Sinkeldam et al. 1992). These reports have been evaluated by JECFA, SCF-EC, and the US Food and Drug Administration (US FDA).

Use and Purpose of Lactitol Based on Biological Properties

Lactitol is a sweet-tasting sugar alcohol with interesting nutritional, physiological, and pharmaceutical properties. Lactitol is neither hydrolyzed nor absorbed in the small intestine. It is metabolized by bacteria in the large intestine, where it is converted into biomass, short chain fatty acids, lactic acid, CO_2, and a small amount of H_2. Beneficial bacteria in the large intestine, such as *Bifidobacteria* and *Lactobacillus* spp., use lactitol as a substrate.

Lactitol consumption does not induce an increase of blood glucose or insulin levels, making it a desirable sugar substitute for diabetic patients. Furthermore, it has been demonstrated that its nutritional caloric use is half that of carbohydrates, with a metabolic energy value maximum of 2 kcal/g or maximum 8.4 kJ/g.

Lactitol is very slowly converted to organic acids (lactic acid) by tooth plaque bacteria, so that lactitol does not cause dental caries and satisfies the Swiss regulations as safe for teeth and United States health claim regulations for "does not promote tooth decay." These properties suggest important dietary applications, for example, products for people with diabetes, noncariogenic confections, low-calorie foods, and products with a prebiotic effect.

Caloric Value

Biochemical Aspects

Karrer and Büchi (1937) have studied the action of galactosidase-containing enzyme preparations on the splitting of lactitol into galactose and sorbitol. They found that lactitol is hydrolyzed very slowly by these enzyme preparations. Later studies reported in a German patent (Maizena 1974) have confirmed that lactitol is only slowly split by enzymes at about a tenth the speed at which lactose is split. On the basis of both these *in vitro* studies, a reduced calorie value can be expected.

In patents of Hayashibara (Hayashibara 1971; Hayashibara and Sugimoto 1975), it is claimed that lactitol has no caloric value because it is not digested or absorbed. The intestines of test rabbits not fed for 24 hours beforehand were closed at both ends and injected with a 20% aqueous solution of lactitol or with an equimolar amount of a sucrose solution. After a lapse of several hours, the lactitol or sucrose left in the intestines was determined. Although 85% of the sucrose intake had been lost because of absorption and digestion, all the lactitol was still present.

Metabolic Energy

At the Agricultural University of Wageningen in The Netherlands, a study was carried out (Van Es et al. 1986) determining the energy balance of eight volunteers on diets supplemented with either lactitol or sucrose. In this study, volunteers were kept for four days in a respiratory room. This was done twice: once with 49 g of sugar a day in a diet, the other time, with the sugar replaced by lactitol. The dosage of 50 g lactitol monohydrate was ingested in four to six portions during the day. Intakes of metabolizable energy (ME) were corrected, within subjects, to energy equilibrium and equal metabolic body weight. Further correction of ME intake was made toward equal actometer activity. With regard to the value of ME to supply energy for maintaining the body in energy equilibrium, the energy contribution to the body of lactitol monohydrate was 60% less than for sucrose.

Because the lactitol monohydrate contains 5% water and the results had an inaccuracy of standard error (SE) 10%, the final conclusion is justified that the metabolic energy of lactitol, on the basis of the dry substance, is at most 50% sucrose. Rounding off upward, a caloric value of lactitol in man at 2 kcal/g seems fully justified.

Prebiotic Effects of Lactitol

Lactitol reaches the colon untouched, where it is used as an energy source by the intestinal microflora. The fermentation of lactitol favors the growth of saccharolytic bacteria and decreases the amount of proteolytic bacteria that are responsible for the production of ammonia, carcinogenic compounds, and endotoxins. This can positively influence the health of humans and animals.

In vitro studies show that lactitol stimulates the growth of *Lactobacillus* spp. and *Bifidobacteria*. The growth of proteolytic bacteria such as *Enterobacteria* and *Enterococci* is inhibited (Table 19.1).

Table 19.1 Fermentability of Lactitol, and Glucose by Intestinal Microorganisms[a]

Species	Lactitol			Glucose			Sugar-free Control
	No. Positive[b]	pH[c]	Gas	No. Positive[b]	pH[c]	Gas	pH[c]
Obligate Anaerobic Microorganisms							
Bacteriodes thetaiomicron	7/8	5.6 ± 0.5	0	8	5.3 ± 0.7	0	6.6
Bacteriodes fragilis	5/8	5.6 ± 0.4	0	8	5.1 ± 0.4	0	6.4 ± 0.4
Bacteriodes distasonics	8/8	6.1 ± 0.6	0	8	5.4 ± 0.6	0	6.6 ± 0.3
Bifidobacterium bifidum	1/1	5.5	0	1	4.0	0	6.5
Clostridium perfringens	6/8	5.7 ± 0.6	+++	8	5.1 ± 0.4	+++	6.8 ± 0.4
Lactobacillus acidophilus	7/9	4.7 ± 0.6	0	9	5.0 ± 0.9	0	6.4 ± 0.5

Table 19.1 (Continued) Fermentability of Lactitol, and Glucose by Intestinal Microorganisms[a]

Species	Lactitol			Glucose			Sugar-free Control
	No. Positive[b]	pH[c]	Gas	No. Positive[b]	pH[c]	Gas	pH[c]
Lactobacillus species	2/2	5.3 ± 1.7	0	2	5.0 ± 1.4	0	6.8 ± 0.4
Streptococcus mutans	5/7	5.3 ± 0.4	0	7	4.0	0	6.6 ± 0.5
Clostridium glycolcium	0/1	—	0	1	5.0	+++	6.5
Propionibacterium species	1/2	6.5	0	1	5.5 ± 0.7	0	7.0
Lactobacillus fermentum	0/1	—	0	1	6.0	0	6.5
Facultative Anaerobic and Obligate Anaerobic Microorganisms							
Enterobacter	2/7	7.0	+	7/7	6.8 ± 0.4	+++	7.8 ± 0.4
Klebsiella	14/15	5.6 ± 0.4	+	15/15	5.4 ± 0.4	++	6.3 ± 0.3
E. coli	0/2	—	0	2/2	4.7 ± 0.4	++	7.5
Proteus mirabilis	0/1	—	0	1/1	4.5	+	7.5
Proteus morgani	0/2	—	0	2/2	4.5	++	7.7 ± 0.4
Proteus vulgaris	0/2	—	0	2/2	5.0	+	7.5
Serratia	0/2	—	0	2/2	6.2 ± 0.4	++	7.7 ± 0.4
Citrobacter	0/2	—	0	2/2	5.0 ± 0.7	+	7.7 ± 0.4
Hafnia	0/2	—	0	2/2	4.5	0	7.0
Pseudomonas aereeginosa	0/2	—	0	2/2	—	0	8.5
Candida albicans	0/2	—	0	2/2	6.2 ± 0.4	0	7.7 ± 0.4
Enterococci	8/9	5.7 ± 0.4	0	9/9	4.1 ± 0.2	0	7.0 ± 0.2
Staphylococcus epidermidis	5/10	5.5 ± 0.4	0	10/10	4.8 ± 0.3	0	7.8 ± 0.4
Staphylococcus aureus	4/8	6.3 ± 0.3	0	8/8	4.9 ± 0.5	0	6.5 ± 0.4

Source: From Lebek, G., and Luginbühl, S.P., *Hepatic Encephalopathy. Management with Lactulose and Related Compounds*, H.O. Conn and J. Bircher (Eds.), 271–282. East Lansing, Michigan, 1989.

[a] All experiments done at least in triplicates.
[b] Number of positive strains per number of examined strains.
[c] pH is the mean on day 2.

Inhibition of these microorganisms is caused by the production of short-chain fatty acids, which reduce the pH, and inhibition of the adhesion of these bacteria to the epithelial cell walls (Lebek and Luginbühl 1989; Finney et al. 2007; Drakoularakou et al. 2007; Scevola et al. 1993a, 1993b).

In vivo lactitol is tested both in humans and animals. Stool samples and cecal material was retrieved from the tested subjects to determine the presence of microorganisms, short-chain fatty acid compounds, fecal pH, moisture, and activity of certain enzymes (Felix et al. 1990; Kitler et al. 1992; Ravelli et al. 1995). These studies confirmed the results of the *in vitro* tests. This prebiotic effect can be used to develop new products.

Lactitol in Combination with Probiotics

Recently, a human intervention study was published whereby the effects of lactitol were looked at in combination with *L. acidophilus* NCFM (Ouwenhand et al. 2009) in 51 healthy elderly subjects. This study took place in order to evaluate any potential synbiotic effects in the light of lactitol being a good carbon source for lactobacilli and especially *L. acidophilus* (Kontula et al. 1999). After the intervention, there was a significant increase in fecal *L. acidophilus* NCFM levels in the symbiotic group. In addition, stool frequency was observed to be higher in the synbiotic group, while there were also modest changes in specific biomarkers (i.e., PGE_2, IgA and spermidine). This study suggests that lactitol with *L. acidophilus* NCFM may improve the microbiota composition and the mucosal function in the healthy elderly population via a synbiotic mechanism.

Lactitol and Diabetes

Because lactitol does not have any significant effect on blood glucose or insulin levels, it is suitable for insulin-dependent diabetic patients (type I). It can also be consumed by noninsulin-dependent diabetic patients (type II). This group can limit its diabetic status by taking dietetic measures. Lactitol fits in their diets because of its reduced calorie value compared with sucrose and sorbitol and its similar metabolism to dietary fiber.

Tooth-Protective Properties

Sugars are a major factor in the pathogenesis of dental caries. Oral bacteria convert sugars into polysaccharides that are deposited on the teeth; these plaque sugars are then fermented into acids. The acid demineralizes the enamel and causes cavities.

Dental Studies

Three main types of experiments are used to evaluate the cariogenicity of foods and food ingredients:

Studies *in vitro*
Experiments in laboratory animals
Clinical trials and investigations in human subjects

Studies In vitro

Among the earliest reports are those of Havenaar (Havenaar 1976; Havenaar et al. 1978) on the use of lactitol by oral bacteria, with formation of acids. A number of plaque bacteria were found that could metabolize lactitol as a substrate, including *Streptococcus mutans*, which is known to possess cariogenic activity, and certain strains of *S. sanguis, Bifidobacterium*, and *Lactobacillus*. These first experiments did not establish the speed of fermentation, which because of the limited time that sugars and sweeteners remain in the mouth, is a determinant of their cariogenicity. It was later shown that pH drop was slow, leading to the conclusion that lactitol could be fermented slowly by these microorganisms.

In addition to acid production, another important property of cariogenic bacteria is the capacity to synthesize extracellular polysaccharide from carbohydrate substrates, making dental plaque. No evidence for extracellular polysaccharide synthesis was found.

A more recent study on lactitol *in vitro* comparing five other bulk sweeteners, on incubation with mixed cultures of human dental plaque microorganisms, measuring acid development, insoluble polysaccharide synthesis, and the attack of the acid on enamel mineral, was done by Grenby and Phillips (1989a). The six different sweeteners fell into three groups, with the highest acid generation, polysaccharide production, and enamel demineralization from glucose and sucrose, less from sorbitol and mannitol, and least of all from lactitol and xylitol.

Experiments in Laboratory Animals

The cariogenicity of lactitol was also tested in rats (Van der Hoeven 1986). For this purpose, lactitol was incorporated in a powdered diet, consisting of 50% of a basal diet (SPP, Trouw & Co, Putten, The Netherlands), 25% wheat flour, and 25% of test substance. Lactitol was compared with sorbitol, xylitol, sucrose, and a control consisting of 50% wheat flour and 50% basal diet. The rats were program fed. None of the animals had diarrhea.

Especially in the lactitol group, no significant adverse effects on general health were observed. The results are shown in Table 19.2. Obviously, the substitution of sucrose by lactitol reduces caries significantly. The caries results for sorbitol and xylitol were in accordance with other studies in which these polyols were used.

Table 19.2 Average of Carious Fissure Lesions and Weight Gains in Five Dietary Groups of 12 Rats

Diet	Number of Fissure Lesions[a]		Weight Gain (g)
	B	*C*	
Sucrose	5.75	2.58	82.0
Sorbitol	1.33[b]	0.42[b]	46.0
Lactitol	1.17[b]	0.33[b]	72.2[b]
Wheat flour	0.67[b]	0.25[b]	80.8[b]
Xylitol	0.42[b]	0.00[b]	48.6[c]
Pooled standard error	0.56	0.43	5.4

Source: From Felix et al. *Microbial Ecology Health Dis* 3:259–267, 1990.

[a] Twelve fissures at risk.
[b] $P < .01$, significantly different from the sucrose group.
[c] $P < .0001$ (Tukey's test).

In the second experiment, human food was fed to rats. Shortbread biscuits containing 16.6% sucrose or lactitol were incorporated at 66% in pulverized, blended diets fed to two groups of 21 rats, so that the final level of lactitol in the test diet was 11%. Again, after a period of eight weeks, caries attack showed highly significant reduction when replacing the sucrose in the biscuits with lactitol.

The most recent studies on the dental properties of lactitol in rats tested it at lower levels in the diet and in the form of a finished human food product, rather than as a raw ingredient in a blended animal diet (Grenby and Phillips 1989b). At a level of 16% lactitol or 16% xylitol in the blended diet, the carious scores, lesion counts, and severity of the lesions were so close on the two polyol regimens that they were indistinguishable but significantly beneficial compared with the 16% sucrose group.

Investigations in Man

At the University of Zurich, a special method has been developed for the *in vivo* determination of the plaque pH in humans. After consumption of chocolate or confections in which sucrose is replaced with lactitol, changes in plaque acidification are detected by the electrodes and transmitted electronically to a graph recorder.

The term "zahnschonend" (safe for teeth/friendly to teeth) is used officially in Switzerland when the pH of dental plaque does not drop below 5.7 during a 30 minute period. Professor Muhlemann demonstrated in chocolates that lactitol is "safe for teeth" (Muhlemann 1977; Grenby and Desai 1988).

In Figure 19.2, it is shown that during and after eating 13 g LACTY chocolate (plain lactitol-containing chocolate), the plaque pH is not affected, whereas with a 15 ml 10% sucrose solution, the plaque-pH is reduced to about 4.5, which indicates that sucrose is being fermented into acids by the oral bacteria.

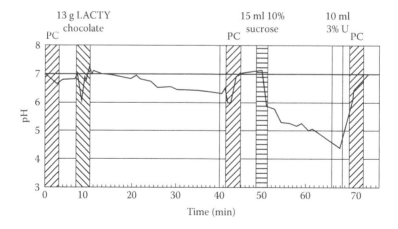

Figure 19.2 Telemetrically recorded pH of 5 days interdental plaque in subject H. H. during and after consumption of 13 g LACTY plain chocolate. A 10% sucrose solution was used as a positive control. PC, 3-min paraffin chewing gum; U, rinsing with 3% carbamide.

Physical Properties

Lactitol is a white crystalline powder available as a monohydrate or anhydrate. In the following sections the most important characteristics of lactitol will be described.

Sweetness

Lactitol is less sweet than sucrose. Lactitol increases in relative sweetness when its concentration is increased (Table 19.3). In most applications, a sweetness of 0.4 times that of sucrose will be found. Lactitol has a clean sweet taste, without an aftertaste. In some products, the lower sweetness will allow the flavor of other ingredients to develop better. In other products, the sweetness will need to be increased. This is possible by using an intense sweetener. To obtain a sweetness equivalent to sucrose, approximately 0.3% aspartame or acesulfame potassium, or approximately 0.15% saccharin needs to be added to lactitol. Combinations with new intense sweeteners on the market will also need to be investigated.

Crystal Forms

Crystallization from aqueous solutions results in nonhygroscopic white monohydrate or dihydrate crystals or hygroscopic anhydrate crystals. The dihydrate has a melting point of 76°C–78°C, and the monohydrate has a more complicated melting behavior. Under the melting microscope, most of the crystals melt at 121°C–123°C, but part of the crystal-bound water evaporates below 100°C, converting this monohydrate into the anhydrous form. The melting point determination of the monohydrate by differential scanning calorimetry, shows the main endothermal peak in the range of 96°C–107°C (water evaporation). The melting point of lactitol anhydrate is 145°C–150°C, determined by differential scanning calorimetry.

Hygroscopicity

Hygroscopicity is related to the amount of water absorbed by a product. Sucrose is not hygroscopic. Only under severe circumstances will sucrose absorb water. The same applies to lactitol monohydrate. This can be seen in Figure 19.3. In contrast, sorbitol has a high level of hygroscopicity. This means that, for example, crisp bakery products will easily become soggy because they absorb water from the air. Lactitol monohydrate is one of the least hygroscopic of the polyols

Table 19.3 Relative Sweetness of Lactitol Solutions Compared with Sucrose Solutions at 25°C

Sucrose Solution Concentration (w/w)	Relative Sweetness of Lactitol (Sucrose = 1.0)
2%	0.30
4%	0.35
6%	0.37
8%	0.39
10%	0.42

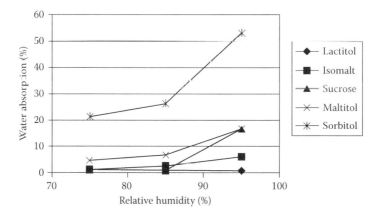

Figure 19.3 **Water activity of lactitol, sucrose, sorbitol and maltitol at 25°C.**

together with mannitol. This makes lactitol suitable for all applications in which water absorption is a critical parameter, like bakery products, tablets, and panned confection.

Water Activity

Water activity influences enzymatic activities, Maillard reactions, fat oxidation, microbial stability, and texture. These elements combined influence the shelf life. The influence of lactitol on water activity is similar to that of sucrose (on a dry solids basis) because the molecular weight of lactitol is almost identical to that of sucrose. Sorbitol, being a much smaller molecule, has a much larger effect on the water activity (Figure 19.4).

Solubility

Solubility of the ingredients is important for many products (e.g., ice cream, hard-boiled sweets). A low solubility can make production processes more difficult (e.g., in hard-boiled sweets, most of

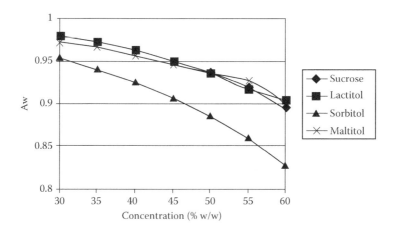

Figure 19.4 **Water activity of lactitol, sucrose, sorbitol, and maltitol at 25°C.**

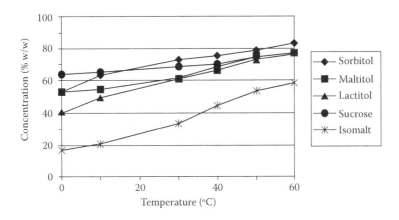

Figure 19.5   Solubility of lactitol and other polyols at different temperatures.

the water will have to be evaporated because more water will have to be added when the solubility is low). The solubility of lactitol at lower temperatures is less than that of sucrose but still good enough not to cause inconveniences during processing (Figure 19.5). The solubility of maltitol is comparable with that of lactitol. Mannitol and isomalt have a lower solubility, which influences processing, mouth-feel, and flavor release.

Viscosity

The viscosity of lactitol in aqueous solution is slightly higher than that of sucrose (Figure 19.6). In the case of hard-boiled sweets, the viscosity of the lactitol-melt is more important. In this case, a combination of lactitol and hydrogenated starch hydrolysate is used (comparable to sucrose and glucose syrup). For this combination, a slightly lower viscosity is found than for its sucrose counterpart, but on cooling down, the same viscosity (or plasticity) will be reached, so lactitol will also be suitable for molded sweets.

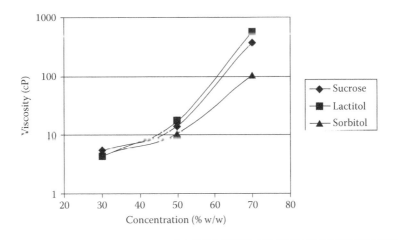

Figure 19.6   Viscosity of polyol solutions at different concentrations at 25°C.

Table 19.4   Heat of Solution of Various Sweeteners

Sweetener	Cal/g
Sorbitol	−26.6
Mannitol	−28.9
Xylitol	−36.6
Lactitol monohydrate	−12.4
Lactitol anhydrate	−6.0
Maltitol	−18.9
Isomalt	−9.4
Sugar	−4.3

Heat of Solution

Lactitol has a cooling effect that is slightly stronger than that of sucrose but far less than that of sorbitol or xylitol (Table 19.4).

Freezing-Point-Depressing Effect

If sugars are replaced in ice cream, a replacer causing a similar freezing-point decrease is required. Lactitol is such a product, as shown in Figure 19.7. This effect is closely related to the molecular weight because this is virtually the same for lactitol and sucrose; the effect is very similar.

Chemical Properties

Lactitol is a polyol with nine OH-groups that can be esterified with fatty acids resulting in emulsifiers (van Velthuijsen 1979) or that can react with propylene oxide (to produce polyurethanes).

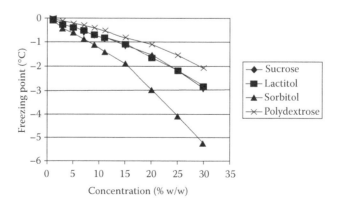

Figure 19.7   Freezing-point-depressing effect.

Table 19.5 Decomposition of a 10% Lactitol Solution under Extreme Conditions of pH and Temperature

pH	% Weight Recovery at 24 h at 105°C			
	Lactitol	*Galactose*	*Sorbitol*	*Not Identified*
2.0	42.9	28.1	21.4	7.6
10.0	98.0	—	—	2.0 (acids)
12.0	98.3	—	—	1.7 (acids)

Because of the absence of a carbonyl group, lactitol is chemically more stable than related disaccharides like lactose. Its stability in the presence of alkali is much higher than that of lactose. The stability of lactitol in the presence of acid is similar to that of lactose. The absence of a carbonyl group also means that lactitol does not take part in nonenzymatic browning (Maillard) reactions.

Lactitol solutions have an excellent storage stability. A 10% lactitol solution in the pH range 3.0–7.5 at 60°C shows no decomposition after one month. After two months at pH 3.0, some decomposition (15%) is detected. At a higher pH this does not occur. With increasing temperature and especially with increasing acidity, hydrolytic decomposition of lactitol is observed. Sorbitol and galactose are the main decomposition products. At high pH lactitol is stable even at 105°C (Table 19.5).

When heated to temperatures of 179–240°C, lactitol is partly converted into anhydrous derivatives (lactitan), sorbitol, and lower polyols.

Applications

Lactitol could potentially play a role in any application in which carbohydrates are used. However, some general guidelines for specific application areas are provided in the following sections.

Lactitol is used in products for its bulking properties. Therefore, on most occasions, a one-to-one replacement will be required. Because of the lower sweetness of lactitol, most products contain a combination of lactitol and intense sweeteners.

Bakery Products

In this area there are many products in which sugar can be replaced. For a large number of bakery products, the crispiness of the product is important. Most bulk sweeteners have the tendency to be hygroscopic, but if lactitol products are handled like sucrose products, crispiness can be assured.

In the soft bakery products (e.g., cakes), we find that lactitol (being nonhygroscopic) often benefits by a small addition of a humectant (e.g., sorbitol). This gives a smooth, moist feel without affecting the good structure and texture obtained by lactitol.

Next to sugar-free bakery products, products low in calories or with functional claims, such as high fiber, are developed. In these products lactitol is a suitable bulk sweetener. In low-calorie products, lactitol is often combined with polydextrose, which is an optimal combination to reduce the calorie content and maintain a high-quality product. High-fiber products claim to have a positive effect on the colon. Because of the beneficial effects that lactitol has on the colon, it will be a suitable sweetener in such products.

Chewing Gum

Sugar-free chewing gum is becoming very popular. The advantage of the sugar-free chewing gum is the tooth-friendliness. Lactitol, like all other polyols, is tooth-friendly and suitable for use in chewing gum. It can be used as a sweetener in the gum base, as rolling compound/dusting powder, and in the panning layer. In the gum base, lactitol gives a more flexible structure, and there is no need to increase the gum base level as is required in sorbitol-containing chewing gum.

As noted above, lactitol and mannitol are the polyols with the lowest hygroscopicity. This makes lactitol an excellent rolling compound. The low hygroscopicity will increase shelf life, especially when stored at high temperatures and humidities. Furthermore, lactitol will improve mouth-feel and, compared to mannitol, lactitol has a better solubility that prevents sandiness of the chewing gum.

A lactitol panned chewing gum is very stable because of the low hygroscopicity.

Chocolate

Lactitol can successfully be applied when producing a sugar-free chocolate. Both lactitol monohydrate and anhydrate can be used for the production of chocolate. The difference between these two crystal forms during processing is the conching temperature. As discussed before, lactitol monohydrate is not hygroscopic, however, it does contain crystal water. When lactitol monohydrate is used, the crystal water will be bound sufficiently to use temperatures of up to about 60°C during the conching stage of chocolate. Above this temperature, the water will come free and the viscosity of the chocolate mass increases. With anhydrous lactitol the conching temperature can be 80°C. Another difference between the two forms is the cooling effect. The anhydrous lactitol has a lower cooling effect than the monohydrate. This will improve the taste of the chocolate, which should have a warm taste.

Evaluation of the sensoric profile of three types of "no added sugar" chocolate, namely milk, white and dark, shows that there is a good synergistic effect between lactitol/maltitol blend in milk chocolate. In particular, 'no added sugar' milk chocolate with lactitol/maltitol blend showed an almost identical sensory profile to the standard (i.e., sugar) milk chocolate when it came to key parameters, such as sweet and milk/caramel flavor as well as significantly enhanced cocoa taste.

Lactitol was also shown to exert excellent creaminess and sweet taste due to the interaction with the milk solids, in particular in the white type chocolate, where the addition of intense sweetener is not necessary.

In dark type chocolate, lactitol was shown to promote the bitterness and the cocoa flavor which are key parameters for this type of chocolate.

As mentioned earlier, the physical properties of the anhydrous form of lactitol are such that the use of lactitol in chocolate applications guarantees a top quality end product, without bringing any complications to processing (Internal unpublished data).

Confectionery

In confectionery, a large range of products is imaginable. A combination of different sugars is often found when looking at sugar confectionery. This is necessary to obtain the desired effect and results. In general, lactitol replaces the sucrose part of recipes, whereas glucose syrup needs to be replaced by hydrogenated starch hydrolysate or maltitol syrup. These latter products are used as crystallization inhibitors. The right ratio differs from product to product, (e.g., hard-boiled sweets need a 70/30 ratio [lactitol/hydrogenated starch hydrolysate], chewies, fruit gums, and pastilles a 40/60 ratio).

Ice Cream

Lactitol shows a combination of physical properties (e.g., freezing-point-depressing effect, hygroscopicity, solubility) that are suitable for ice cream. Not only will the use of lactitol reduce the caloric content of the ice cream, but it will also permit a low fat percentage, allowing an even further reduction in calories.

Tablets

Granulated lactitol is suitable for use in tabletted confections, such as breath fresheners, and as an excipient in pharmaceutical preparations like vitamin tablets. In this application, the low hygroscopicity of lactitol is important. This low hygroscopicity prolongs shelf life and protects active agents against moisture. Figure 19.8 shows that lactitol tablets absorb hardly any water.

In pharmaceutical tablets, direct compressible sorbitol and mannitol are most often used. The advantage of lactitol in this application is the low hygroscopicity compared with sorbitol and the better solubility compared with mannitol. With respect to tablet hardness, lactitol performs in between the two (Figure 19.9). This makes direct compressible lactitol a good alternative.

Preserves

The solution to low-calorie preserves is often found in a low dry solids preserve. To combine this with lactitol will result in an even larger reduction in calories, while maintaining the good flavor and texture.

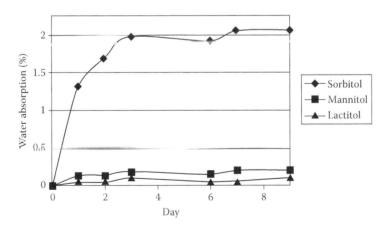

Figure 19.8   The water absorption of tablets made with direct compressible polyols. (Tablet hardness, 150N; temperature, 20°C; RH, 70%).

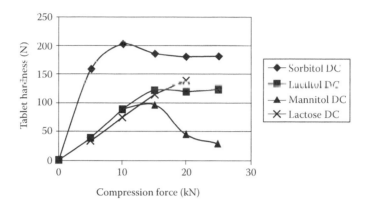

Figure 19.9 **The compression profile of direct compressible polyols on a rotary press (30,000 rpm, 0.5% Mg-stearate, Ø 9 mm, 650 mg).**

Regulatory Status for Lactitol

Argentina	Allowed
Australia	Allowed
Brazil	Allowed
Canada	Allowed
European Union	Lactitol is allowed as a sweetener in all EU countries regulated by the sweeteners in food directive No. 94/35/EC, until such time as the above Directive is fully replaced by Regulation (EC) 1333/2008 of December 16, 2008 regarding food additives. This is expected to take place after the end of 2010.
	Lactitol can also be used as an additive as regulated by the Directive on food additives other than colors and sweeteners No. 95/2/EC. Again, in due course this Directive will be fully replaced by Regulation (EC) 1333/2008 of December 16, 2008 regarding food additives.
	The E-number for lactitol is E 966
Israel	Allowed
Japan	Allowed
Norway	Allowed
Sweden	Allowed
Switzerland	Allowed
United States	The petition to affirm lactitol as GRAS was accepted for filing by the US FDA on May 17, 1993.

Conclusion

Lactitol is a unique, commercially available bulk sweetener. Its physical properties guarantee optimal product performance during processing and storage of the food containing lactitol. Being so similar to sucrose, it is possible to develop new sugar-free and light food products in a short period of time.

References

Drakoularakou, A., Hasselwander, O., Edinburgh, M., and Ouwehand, A.C. 2007. Food science and technology bulletin. *Functional Foods* 3(7):71–80.

Finney, M., Smullen, J., Foster, H.A., Brokx S., and Storey, D.M. 2007. Effects of low doses of lactitol on faecal microflora, pH, short chain fatty acids and gastrointestinal symptomology. *Eur J Nutr* 46:307–314.

Felix, Y.F., Hudson, M.J., Owen, R.W., Ratcliffe, B., Van ES, A.J.H., van Velthuijsen, J.A., and Hill, M.J. 1990. Effect of dietary lactitol on the composition and metabolic activity of the intestinal microflora in the pig and in humans. *Microbial Ecology Health Dis* 3:259–267.

Grenby, T.H., and Desai, T. 1988. A trial of lactitol in sweets and its effects on human dental plaque. *Br Dent J* 164:383–387.

Grenby, T.H., and Phillips, A. 1989a. Studies of the dental properties of lactitol compared with five other bulk sweeteners in vitro. *Caries Res* 23:315–319.

Grenby, T.H., and Phillips, A. 1989b. Dental and metabolic observations on lactitol in laboratory rats. *Br J Nutr* 61:17–24.

Havenaar, R. 1976. Microbial investigations on the cariogenicity of the sugar substitute lactitol. Unpublished report from the University of Utrecht (Netherlands), Department of Preventive Dentistry and Oral Microbiology; June.

Havenaar, R., Huis in't Veld, J.H.J., Backer Dirks, O., and Stoppelarr, D. 1978. Some bacteriological aspects of sugar substitutes. In *Health and Sugar Substitutes. Proceedings of the ERGOB Conference*, B. Guggenheim (Ed.), pp. 192–198. Geneva, Basel: Karger.

Hayashibara, K., and Sugimoto, K. 1975. Containing lactitol as a sweetener. US Patent 3,973,050, 1975.

Hayashibara, K. 1971. Improvements in and relating to the preparation of foodstuffs. UK Patent 1,253,300.

Joint FAO/WHO Expert Committee on Food Additives. 1983. IPCS Toxicological evaluation of certain food additives and contaminants: Lactitol. 27th report Geneva, WHO Food Additives Series, pp. 82–94.

Karrer, P., and Büchi, J. 1937. Reduktionsprodukte von disacchariden: Maltit, lactit, cellobit. *Helvetica Chim Acta* 20:86–90.

Kitler, M.E., Luginbuhl, M., Lang, O., Wuhl, F., Wyss, A., and Lebek, L. 1992. Lactitol and lactulose an in vivo and in vitro comparison of their effects on the human intestinal flora. *Drug Invest* 4:73–82.

Kontula, P., Suihko, M.L., Von Wright, A., and Mattila-Sandholm, T. 1999. The effects of lactose derivatives on intestinal lactic acid bacteria. *J Dairy Sci* 82:249–256.

Lebek, G., and Luginbühl, S.P. 1989. Effects of lactulose and lactitol on human intestinal flora. In *Hepatic Encephalopathy. Management with Lactulose and Related Compounds*, H.O. Conn and J. Bircher (Eds.), 271–282. East Lansing, Michigan.

Maizena GmbH. 1974. Verwendung von Lactit als Zuckeraustauschstoff. *German Patent* 2,133,428.

Muhlemann, H.R. 1977. Gutachten uber die zahnschonenden Eigenschaften von Lacty Schokolade und Lacty Milch schokolade der Firma CCA. Holland, unpublished report from Zahnarztliches Institut der Universitat Zunch, July.

Ouwenhand, A.C., Tiihonen, K., Saarinen, M., Putaala, H., and Rautonen, N. 2009. Influence of a combination of Lactobacillus acidophilus NCFM and lactitol on healthy elderly: Intestinal and immune parameters. *British Journal of Nutrition* 101:367–375.

Ravelli, G.P., Whyte, A., Spencer, R., Hotten, P., Harbron, C., and Keenan, R. 1995. The effect of lactitol intake upon stool parameters and the faecal bacterial flora in chronically constipated women. *Acta Therapeutica* 21:243–255.

Saijonmaa, T., Heikonen, M., Kreula, M., and Linko, P. 1980. Preparation and Characterization of Milk Sugar Alcohol Dihydrate (abstract). Sixth European Chrystallograph Meeting, Barcelona, p. 26.

Scevola, D., Bottari, G., Franchini, A., Guanziroli, A., Faggi, A., Monzillo, V., Pervesi, L., and Oberto, L. 1993a. The role of lactitol in the regulation of the intestinal microflora in liver disease. *Giornale di malattie infettive e parassitaire* 45:906–918.

Scevola, D., Bottari, G., Oberto, L., Monzillo, V., Pervesi, L. and Marone, P. 1993b. Intestinal bacterial toxins and alcohol liver damage: Effects of lactitol, a synthetic disaccharide. *La Clinica Dietologica* 20:297–314.

Senderens, J.B. 1920. Catalytic hydrogenation of lactose. *Comptes Rendus* 170:47–50.

Sinkeldam, E.J., Woutersen, R.A., Hollanders, V.M.H., Til, H.P., Van Garderen-Hoetmer, A., and Baer, A. 1992. Subchronic and chronic toxicity/carcinogenicity feeding studies with lactitol in rats. *J Am Col Toxicol* 11(2):165–188.

The Scientific Committee for Food. 1986. Report on sweeteners, EC-Document III/1316/84/CS/EDUL/27 rev.

van der Hoeven, J.S. 1986. Cariogenicity of lactitol in program-fed rats. *Caries Res* 20:441–443.

Van Es, A.J.H., de Groot, L., and Vogt, J.F. 1986. Energy balances of eight volunteers fed on diets supplemented with either lactitol or saccharose. *Br J Nutr* 56:545–554.

van Velthuijsen, J.A. 1979. Food additives derived from lactose: Lactitol and lactitolpalmitate. *J Agric Food Chem* 27:680–686.

Wijnman, C.F., van Velthuijsen, J.A, and van den Berg, H. 1983. Lactitol monohydrate and a method for the production of crystalline lactitol. *European Patent* 39981.

Chapter 20

Sorbitol and Mannitol

Peter R. Jamieson*

Contents

---

* Anh S. Le and Kathleen Bowe Mulderrig did not participate in the revision but their chapter in the book's 3rd edition was the basis of this revised chapter.

Introduction

Sorbitol and mannitol are low digestible carbohydrates (LDC) which have existed commercially for more than 70 years. They are classified as either "sugar alcohols" or "polyols" and are often used as sugar-free bulking agents, humectants, cross-linking agents, cryoprotectants, and crystallization modifiers in various food, oral care, personal care, pharmaceutical, and industrial applications.

Sorbitol was first discovered by a French Chemist named Joseph Boussingault in 1872. He isolated it from the fresh juice of the mountain ash berries. Mannitol is found in the extrudates of trees, manna ash, marine algae, and fresh mushroom. Although present in natural sources, sorbitol and mannitol were not commercially available until 1937, when they were first manufactured in full production scale by the Atlas Powder Company, Wilmington, Delaware (Mellan 1962).

Production of Sorbitol and Mannitol

Glucose syrups, invert sugar, and other hydrolyzed starches are important raw materials for the manufacture of sorbitol and mannitol. Sorbitol is produced from the catalytic hydrogenation of glucose. The hydrogenation reaction is driven by a catalyst, such as nickel. After the reaction is complete, the catalyst is filtered out and the solution is purified. It is then evaporated to 70% solids and sold as sorbitol solution (Mellan 1962).

Crystalline sorbitol is made by further evaporating the sorbitol solution into molten syrup containing at least 99% solids. The molten syrup is then crystallized by either seeding the melt with sorbitol crystals or utilizing only high shear to force crystallization (i.e., melt crystallization). The resulting material is then tempered or cured to ensure complete and proper crystallization of the sorbitol has occurred. Once cured, the crystallized mass is milled and sifted through various screens to achieve a desired particle size distribution.

Mannitol can be obtained by various ways, such as through fermentation or extraction from specific types of seaweed. However, it is more commonly produced through the hydrogenation of fructose from starch based syrups or sucrose. When these raw material ingredients are hydrogenated, both sorbitol and mannitol are produced. The mannitol is separated from sorbitol on the basis of its lower solubility (22 g/100 g water versus 235 g/100 g water) and is crystallized from the solution. The mannitol crystals are then filtered, dried, and sold as white powder or can be processed further to make free-flowing granules.

Physical and Chemical Properties of Sorbitol and Mannitol

Sorbitol and mannitol are six-carbon, straight-chain polyhydric alcohols, meaning they have more than one hydroxyl group. Both sorbitol and mannitol have six hydroxyl groups and the same molecular formula, $C_6H_{14}O_6$. They are isomers of one another and have different molecular configurations. The difference between sorbitol and mannitol occurs in the planar orientation of the hydroxyl group on the second carbon atom (Figures 20.1 and 20.2). This dissimilarity has a powerful influence and results in an individual set of properties for each isomer.

The major difference between the two isomers is that sorbitol is hygroscopic and mannitol is non-hygroscopic. Sorbitol, therefore, is used as a humectant because of its affinity for moisture, and mannitol is used as a pharmaceutical and nutritional tablet excipient because of its inertness and stability against moisture.

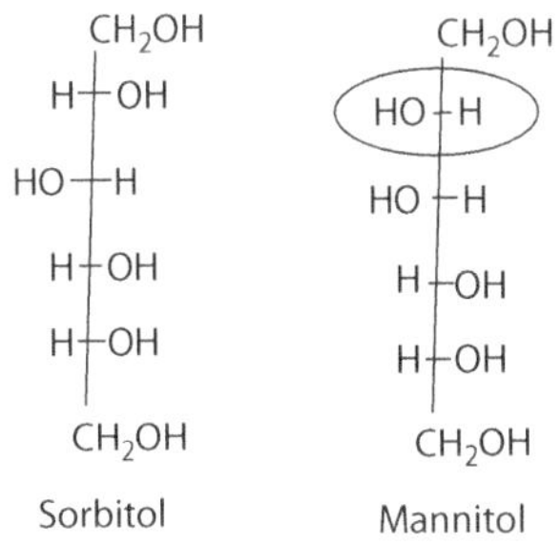

Figure 20.1 The chemical structures of sorbitol and mannitol.

Sorbitol solution is hygroscopic, attracting and releasing moisture under varying humidity conditions, but it does so very slowly. Polyols of lower molecular weight, such as glycerin, tend to gain and lose water more rapidly. Sorbitol provides improved moisture control and is more likely to maintain equilibrium in the surrounding environment. This slower rate of change in moisture content protects the food products in which sorbitol is used, thus maintaining the as-made quality of the product and extending the shelf life (Figure 20.3).

In crystalline form, mannitol and sorbitol can have different crystal types or polymorphs. Typically, most stable polymorphs are preferred commercially, since they often have the highest melting point and lowest hygroscopicity. In addition, working with the most stable polymorph is important and helps to ensure the crystal form being used will not change during processing (i.e., an unstable polymorph converting to the stable polymorph) and subsequently will not affect the finished product (Wadke et al. 1989).

The most stable polymorph of sorbitol is known as gamma (γ) sorbitol and was determined from the work of Atlas Powder Company and University of Pittsburgh researchers in the early 1960s. Today, this particular polymorph of crystalline sorbitol is the most commercially available (DuRoss 1982, 1984). It can be identified by its single melting point (99°C–101°C) and heat

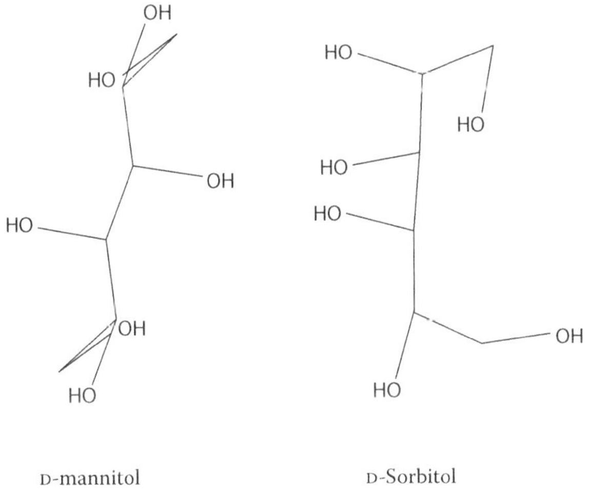

Figure 20.2 The planar configurations of sorbitol and mannitol.

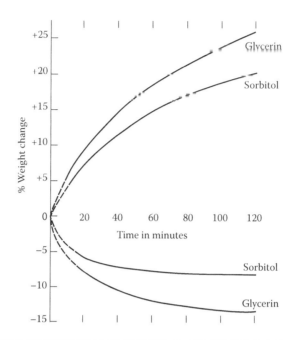

Figure 20.3 **Rates of moisture exchange between air and sorbitol solutions and glycerin. These curves show the rate of moisture gain occurring when solutions of sorbitol or glycerin, which are in hygroscopic equilibrium at 58% relative humidity, are transferred to an atmosphere of 79% relative humidity; also the rate of moisture loss when such solutions are transferred to an atmosphere of 32% relative humidity. Note that the sorbitol solutions neither gain nor lose water as rapidly as do the glycerin solutions.**

of fusion (42 cal/g, assuming 44 cal/g represents a fully crystallized crystalline sorbitol) when evaluated by a differential scanning calorimeter (DSC). The properties of γ-sorbitol and its other polymorphs are contained in Table 20.1 (DuRoss 1984). Mannitol on the other hand, commonly exists in its most stable form which is a (β)-polymorph (Debord et al. 1987).

Sorbitol and mannitol are sweet, pleasant-tasting polyols. Sorbitol is approximately 60% as sweet as sucrose, while mannitol is approximately 50% as sweet. In crystalline form, they both have a negative heat of solution when dissolved in water that results in a cooling sensation when tasted. The heat of solution of sorbitol is −26.5 cal/g (at 25°C), and the heat of solution of mannitol is −28.9 cal/g (at 25°C) (Table 20.2).

Absorption and Metabolism

Sorbitol and mannitol are widely accepted by the food and pharmaceutical industries as nutritive ingredients because of their ability to improve the taste and shelf life of regular foods and special dietary products.

Sorbitol and mannitol are slowly absorbed into the body from the gastrointestinal tract and are metabolized by the liver, largely as fructose, a carbohydrate that is highly tolerated by people with diabetes. Sorbitol is absorbed and metabolized in the liver by way of a pathway located entirely in the cytoplasmic compartment (Allison 1979); this pathway is illustrated in Figure 20.4.

Table 20.1 Physical Properties of the Polymorphic Forms of Crystalline Sorbitol

Property	Form									
	E		α		Δ		β		Γ	
Peak[a] (°C)	75/83		88		89		97.5		101	
Onset of peak[a] (°C)	68		85		86		94.5		99	
Melting range (°C)	77–78.5		90.0–91.5		86.8–88.5		94.2–95.0		99.0–101.6	
Density (g/cm^{-3}) flotation	n.d.[d]		1.550		1.460		1.481		1.527	
H_2O solubility[b]	252		n.d.§		252		216		211	
X-ray interplanar spacingsc (Å) (6 strongest)	d	I/I_0	d	I/I_0	d	I/I_0	d	I/I_0	D	I/I_0
	4.73	100	4.22	100	4.23	100	3.95	100	4.73	100
	3.50	67	3.93	75	3.71	66	5.95	80	3.49	67
	4.04	60	4.33	70	4.77	64	4.66	74	4.04	61
	6.00	45	2.65	51	2.12	53	4.98	50	3.91	53
	7.57	40	3.78	41	2.64	44	4.24	42	4.33	50
	5.07	40	2.77	41	8.85	27	3.48	38	3.50	50

Source: Reproduced with permission from Pharmaceutical Technology, Vol. 8, Number 9, September 1984, pages 50–56. Copyright by Advanstar Communications, Inc. Advanstar Communications, Inc. retains all rights to their article.

[a] As determined by differential thermal analysis (DTA) and differential scanning calorimetry (DSC).
[b] H_2O at 20°C (parts per 100 parts H_2O).
[c] Space group: P2,2,2; Axes (Å): a = 8.677, b = 9.311, c = 9.727.
[d] n.d. = not determined.

The initial steps in the metabolism of sorbitol and mannitol in the liver, their uptake by the liver cells, and their conversion to glucose, are independent of insulin (Allison 1979). Consumption of sorbitol and mannitol, therefore, does not significantly increase blood glucose levels. Subsequently, they have lower glycemic indices than other carbohydrates and can usually be used safely by people with diabetes.

The caloric value of sorbitol is 2.6 kcal/g and the caloric value of mannitol is 1.6 kcal/g in the US. In the European Union (EU), all polyols except for erythritol are 2.4 kcal/g, while Japan considers mannitol and sorbitol as 2.0 kcal/g and 3.0 kcal/g, respectively (Table 20.2). Regardless of the regional caloric variation, they are still lower in calories than traditional sugars (i.e., glucose, sucrose and maltose), which are 4 kcal/g (EEC 1990). As a result, both can be effectively used to achieve reduced calorie and low-calorie claims where permitted.

Like many low digestible carbohydrates, sorbitol and mannitol will reach the lower gastrointestinal (GI) tract intact. Consequently, at elevated concentrations, they can exhibit a laxative effect by creating an osmotic imbalance, drawing water into the lower GI tract, softening

Table 20.2 Characteristics of Sorbitol and Mannitol

	Sorbitol	Mannitol
Chemical formula	$C_6H_{14}O_6$	$C_6H_{14}O_6$
Form	White powder or 70% solution	White powder
Sweetness	60% of sucrose	50% of sucrose
Taste	Sweet/cool	Sweet/cool
Odor	None	None
Noncariogenic	Yes	Yes
Moisture sensitivity	Hygroscopic	Non-hygroscopic
Solubility in H_2O (at 25°C)	235 g/100 g H_2O	23 g/100 g H_2O
Caloric value	2.6 cal/g	1.6 cal/g
Melting point	100°C	164°C
Molecular weight	182	182
Heat of solution (at 25°C)	−26.5 cal/g	−28.9 cal/g
Chemical stability	Stable in air in the absence of catalysts and in cold, dilute acids and alkalis. Sorbitol does not darken or decompose at elevated temperatures or in the presence of amines. It is nonflammable, non-corrosive, and nonvolatile.	Stable in dry state or in sterile aqueous solutions. In solutions, it is not attacked by cold dilute acids or alkalis, or by the atmospheric oxygen in the absence of catalysts. Mannitol does not undergo Maillard reactions.

stool and/or be fermented by the body's microflora leading to flatulence. According to the Code of Federal Regulations (21 CFR 180.25) if the foreseeable daily consumption of mannitol surpasses 20 grams per day or 50 grams per day for sorbitol, then an "excess consumption may cause laxation" statement is required to be listed on a product's label. In the EU, a trigger amount is expressed differently. Mandatory labeling is triggered when polyol comprises 10% of the food product. Furthermore, the Codex Alimentarius: Standard 181 (1991) states that if the food provides a daily intake of sugar alcohols in excess of 20 g per day, there shall be a statement on the label to the effect that the food may have a laxative effect. This is not mandatory – it is a guideline.

Sorbitol and mannitol do not increase the incidence of dental caries, a condition started and promoted by acid conditions that develop in the mouth after eating carbohydrates and proteins. Results from pH telemetry tests show that sorbitol and mannitol do not increase the acidity or lower the pH of the mouth after ingestion. This means that they will not promote tooth decay. For this reason, sorbitol and mannitol are used in oral care and pediatric applications. The usefulness of polyols (e.g., including sorbitol and mannitol) as alternatives to sugars and as part of a comprehensive program including proper dental hygiene has been recognized by numerous authorities, including the American Dental Association.

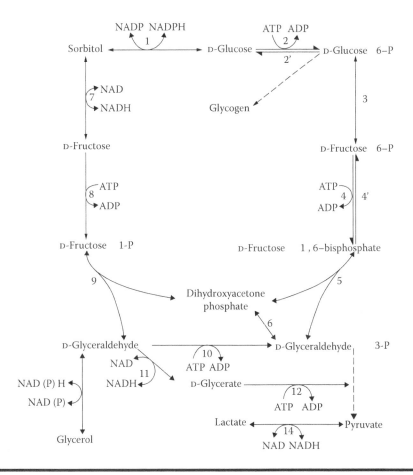

Figure 20.4 Sorbitol and mannitol metabolism. (From Allison, R.G. 1979. Dietary Sugars in Health and Disease III. Sorbitol, pp. 1–51: Report from Life Sciences Research Office Federation of American Societies for Experimental Biology, Bethesda, MD.)

Applications

Sorbitol and mannitol are used in the food, confectionery, oral care, and pharmaceutical industries because of their unique physical and chemical properties. Because of these properties, sorbitol and mannitol perform certain functions that are beneficial within the final product. As mentioned previously, sorbitol and mannitol have different properties from one another because of their different planar orientations. For this reason, they serve different functions in food and pharmaceutical products.

Sorbitol Applications

Chewing Gum

Crystalline sorbitol is widely used as a bulking agent in sugar-free chewing gum. Crystalline sorbitol does not promote dental caries. It provides sweetness and a pleasant cooling effect, which are synergistic with other flavoring agents such as spearmint, peppermint, cinnamon, wintergreen, and fruit flavors. The level of crystalline sorbitol used in a sugar-free chewing gum is typically between 50%–55% by weight. High-intensity sweeteners such as aspartame, acesulfame potassium, and sucralose can be used with crystalline sorbitol to improve the sweetness and flavor release. The size and porosity of the crystalline sorbitol granules are very important in regard to flexibility, chew, cohesion, and smoothness in chewing gum. The particle size distribution of crystalline sorbitol used in sugar-free chewing gum is typically 94.5% through the US Standard Testing Sieve Number 40.

The process of making sugar-free chewing gum is basically a blending operation that uses a horizontal sigma blade mixer. The mixer is equipped with a circulating water-heated jacket. The temperature of the water is heated to between 50°C–55°C. Crystalline sorbitol is mixed with the preheated gum base (the temperature of the gum base is 50°C–55°C) and the humectant solution until the mixture is homogenous. The mixing time is typically about eight to nine minutes. Gum bases are derived from natural sources and are synthetically produced to provide desired chewing properties. The humectant solution is usually a sorbitol solution, glycerin, and/or a maltitol solution. The humectant is used to prevent the sugar-free chewing gum from becoming too dry or stale during storage. The flavoring agent is then added to the homogenous mixture and mixed for an additional three to four minutes. The gum is removed from the sigma blade mixer and shaped into thin sheets by running it through a set of rollers several times. A dusting agent, mannitol powder, is usually used to dust the surfaces of the gum sheets to reduce the tendency of the gum to stick to the rolling, cutting, and wrapping equipment. The thin sheets of gum are then stored in a constant temperature room (25°C and less than 40% relative humidity) for 24 hours. They are then cut and wrapped into individual sticks of chewing gum.

Sugar-Free Hard Candy

Sorbitol hard candies were introduced commercially in the late 1970s and were very popular until the mid 1990s. Sorbitol is used for its noncariogenic properties, sweet cool taste and its ability to readily crystallize. Today, however, most sugar free hard candy is produced using polyols, such as isomalt, maltitol syrups, and polyglycitol syrups, that yield higher quality products without the need for specialized production equipment.

Sugar-free sorbitol hard candy is produced by a batch depositing method. In this method, sorbitol solution is cooked at high temperatures until most of the water is driven off. The molten sorbitol is slowly and evenly cooled to a certain temperature, at which time flavor and a small amount of crystalline sorbitol are added. The crystalline sorbitol is used as a seed that nucleates the melt and starts the crystallization of sorbitol. The melt is deposited into molds and allowed to crystallize further. The sorbitol continues to crystallize and sets up into a hard candy.

The type of sorbitol solution used can control the rate of sorbitol crystallization. In the manufacture of sorbitol solution, mannitol is also produced because it is an isomer. Sorbitol solution 70%, USP, contains a small amount of mannitol (approximately 3%), which can inhibit or control the rate of sorbitol crystallization. Consequently, adjusting the percentage of mannitol up or down can increase or decrease the candy's set time. If the candy manufacturer has an automated batch depositing process and wants the candy to set up quickly, the mannitol level in sorbitol solution can be adjusted down. Alternatively, if the candy manufacturer has a manual process and does not want the candy to set up too quickly for fear of it setting up in the depositor, the mannitol level in sorbitol solution can be adjusted up.

Pressed Mints and Pharmaceutical and Nutritional Tablets

Crystalline sorbitol can be used in various types of tablets. It is most often used in a confectionery tablet such as a pressed mint. Because of its negative heat of solution, sorbitol has a cool taste that almost gives the perception of mint. Used with a mint flavor, this attribute of sorbitol enhances the tablet's flavor. Sorbitol is 60% as sweet as sucrose, so a mint tablet will not be too sweet, which would detract from its breath-freshening properties.

Typically, sorbitol is very compressible and binds other tablet ingredients effectively. However, its compressibility can be improved further by modifying its density or porosity through special processes such as spray drying or agglomeration (DuRoss 1984). Crystalline sorbitol made for tablet applications, commonly have special particle size distributions that make the product free-flowing and suitable for direct compression. The particle size is controlled to optimize the flow characteristics of the granulations in modern high-speed tablet presses (Shangraw et al. 1981).

Sorbitol is also used in pharmaceutical and nutritional tablets. The hygroscopicity of sorbitol can pose a problem to some tablet formulations, because many drug substances are moisture sensitive and degrade if moisture is present. A small amount of sorbitol can be used in the tablet without greatly affecting the moisture pickup. By adding a small amount of sorbitol, a pharmaceutical formulator can make a pleasant tasting, strong tablet without jeopardizing the potency of the active drug substance. In some pharmaceutical tablets, moisture uptake is not a great problem. Sugar-free antacid tablets use sorbitol as the major inactive ingredient to bind the active ingredients effectively.

Confections

Sorbitol is related to sugars, but has a different carbohydrate structure. Despite the chemical relationship, sorbitol modifies or "doctors" the crystallization of sugar by interfering with the crystallizing process. As a result, it influences the rate of crystallization, crystal size, and crystal-syrup balance in sugar-based confections.

In grained confections such as fondants, fudges, aerated mallows, and crèmes, sorbitol functions as one of the key doctors used to modify crystal structures to extend shelf life. In the production of any confection, there is a point at which sucrose crystallization should take place to achieve maximum shelf life. If crystallization takes place on either side of this "optimal point," the crystals will be either too large or too small to achieve the stability required for maximum shelf life. Including sorbitol in the doctor system of a confection will complex the total sucrose/doctor system. This provides the confectioner a broader area in which to crystallize the confection and still be at the optimum point as measured by quality and shelf life. Sorbitol also improves finished texture and moisture retention because of its humectant properties. It is unique because, unlike other doctors, enough sorbitol can be used to extend shelf life without adversely affecting finished texture or taste.

Baked Goods

Sorbitol can be used in both sugar-free and sugar-based baked good products such as cakes, cookies, crèmes, icings and muffins. It is commonly used as a bulking agent, sweetener and shelf life extender. When only sorbitol is used as the bulking agent, a high-intensity sweetener may be needed, because sorbitol is only approximately 60% as sweet as sucrose.

Sorbitol can also be used in sugar-based baked goods to extend the product's shelf life. Sorbitol works as a humectant, attracting moisture from the environment. It holds onto this moisture, maintaining the proper balance and not allowing the baked good to dry out.

Surimi

Sorbitol is used to make surimi products such as imitation crabmeat, shrimp, and lobster. Surimi is fish (pollack) that is processed and frozen. It is then shipped to the imitation seafood manufacturers who color it and mold it into the shape of the seafood they wish to imitate. This product has enjoyed tremendous growth in the US market and has always been extremely popular in the Asian markets.

Because surimi is frozen, shipped, and subjected to freeze-thaw cycling in transit, the primary muscle protein of the fish becomes denatured, resulting in a reduction of gel-forming ability and gel strength. As deterioration of surimi quality progresses, the surimi becomes unsatisfactory for the production of high-quality imitation seafood products. A cryoprotectant is needed to protect the surimi from the freeze-thaw cycle (Lee 1984).

Initially, the Japanese used sucrose as the primary cryoprotectant in surimi, at levels of 6%–8%. The resulting product had good freeze-thaw stability, but the sucrose had imparted an undesirable sweetness to the surimi analog and in some cases caused a significant color change in the surimi paste. These two problems led to the evaluation of sorbitol in blends with sucrose.

Sorbitol proved effective. It lowered the sweetness (60% that of sucrose) and improved the flavor of the surimi analog. It served as an effective cryoprotectant because of its ability to lower the freezing point of water, and it prevented the deterioration of the texture of the surimi because of its ability to maintain proper moisture balance. Crystalline sorbitol can be used alone or in combination with sucrose. Sorbitol solution is also effective as a flavor enhancer and cryoprotectant and can be used in place of crystalline sorbitol/sucrose blends (Yoon and Lee 1984).

Cooked Sausages

Sorbitol can be used in cooked sausages to improve the flavor and method of cooking. The term "cooked sausages" is defined in the USDA Regulation 9 CFR 318.7 and includes products labeled as "franks," "frankfurters," and "knockwurst." It does not include raw sausage products such as "brown and serve" products.

When frankfurters are cooked commercially, such as at a sports stadium or convenience market, the sausage is cooked continuously and can spend a great deal of time on the rotary grill. This can lead to burning or charring of the sausage casing, which makes its appearance unappealing to the customer. It is the sugar or corn syrup used in the cooked sausage that is caramelizing and causing the charring. Sorbitol does not caramelize as do the sugars or corn syrups, therefore, the cooked sausage will not char on the grill.

Sorbitol not only enhances the flavor of the cooked sausage due to its sweetness, but also improves the stability of the sausage's reddish color due to its non-reactivity. Furthermore, it has been shown to reduce the potential for bacterial growth on casings, subsequently increasing shelf life of the frankfurters (Geisler and Papalexis 1971).

The previously mentioned benefits make the cooked sausage more appealing to the consumer. The addition of sorbitol to the cooked sausage also benefits the manufacturer in that it increases the ease of peeling off the sausage casing after the sausage has been made, such as is done in the skinless frankfurter manufacturing process.

Shredded Coconut

The pleasant, sweet taste and hygroscopicity of sorbitol protects shredded coconut from loss of moisture. Because it is nonvolatile, sorbitol has a more permanent conditioning effect than other humectants.

Liquid Products

Sorbitol has many advantages for a wide range of pharmaceutical or functional based liquids such as antacids, antibiotic suspensions, and cough syrups.

Sorbitol solution acts as a bodying agent because of its viscosity of approximately 110 centipoise (cps) at 22°C. It improves the mouthfeel of the finished product by eliminating a watery or thin organoleptic sensation. Sorbitol enhances flavor and imparts a characteristic sweet pleasant taste that is not cloying.

Sorbitol exhibits stability and chemical inertness when in contact with many chemical combinations of ingredients. Sorbitol solution can be used alone or with other liquid vehicle components, if necessary, for enhanced solvent power or special taste and mouthfeel effects. With sorbitol solution as the vehicle, almost any medicament can be used. Sensitive, insoluble active ingredients such as vitamins or antibiotics retain their potency in oral liquids containing sorbitol solution, USP. Sorbitol can also act as a cryoprotectant and therefore lowers the freezing point of water. This ability gives the formulators the means to protect liquid pharmaceutical preparations from low-temperature damage in storage or transit (Courtney 1980).

As discussed previously, sorbitol can serve as a crystallization inhibitor for liquid sugar systems. In any concentrated solution of pure solids or suspension systems, crystallization can occur. When crystallization is localized on the threads of a bottle, the cap may become difficult or impossible to remove. The condition is known as "caplocking." Sorbitol helps eliminate crystallization and its undesirable effects by developing a more complex solids system in syrups that makes them less readily crystallizable.

Oral Care and Toothpaste

Sorbitol is used in toothpaste and other oral care products as the primary sweetener. Sorbitol solution and non-crystallizing sorbitol solution are the primary types of sorbitol used. The role of oral care products is to prevent tooth decay. As mentioned previously, sorbitol does not lower the pH of the mouth, so will not promote the growth of bacteria that can lead to tooth decay. The fact that sorbitol is only 60% as sweet as sucrose is a benefit, because the oral care product will not be too sweet.

Sorbitol is also used because of its humectant properties. Sorbitol is a static humectant, meaning it picks up moisture and retains that moisture over time. It does not have the rapid moisture gain and loss cycle of other polyols, such as glycerin. For this reason, sorbitol is used by itself or in combination with other polyols to maintain the appropriate moisture balance in the toothpaste or mouthwash (Figure 20.3).

Sorbitol is used as a bodying agent to give oral care products the appropriate mouthfeel. The viscosity of sorbitol solution or non-crystallizing solution gives the mouthwash or toothpaste a smooth feel without making them too viscous or syrup like (Figure 20.5)

Special Dietary Foods

Flavor, palatability, and texture are often lost to a noticeable extent when foods are modified for dietary reasons. Sorbitol has been used as an adjunct in these products to make them more

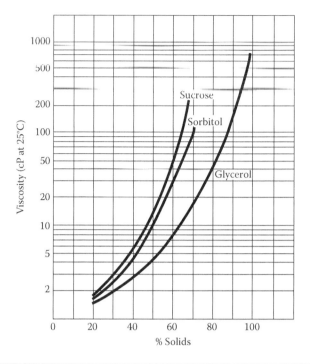

Figure 20.5   Viscosity of solutions of sucrose, sorbitol, and glycerol at various concentrations.

palatable. Its sweet, cool taste and humectant properties lend it to application in many product types. The variety of foods is almost endless and includes dried fruits, granola bars, ice cream, jams, roasted nuts, and pancake syrup, in addition to the products discussed in the previous sections.

Mannitol Applications

Chocolate-Flavored Compound Coatings

Sugar-free chocolates are popular for consumers with dietary restrictions, such as individuals with diabetes. Mannitol can be used to make sugar-free or no-sugar-added chocolate-flavored compound coatings that ultimately enrobe ice creams and confections such as marshmallows and butter creams. Although mannitol is still used today in the manufacture of sugar-free compound coatings, its low laxation threshold and sweetness have limited its use. As a result, the polyol maltitol is considered the current sugar-free bulking agent of choice by coating manufacturers.

Mannitol is used in compound coating manufacture because it is non-hygroscopic. When sugar-free chocolate compound coatings were first manufactured, sorbitol was used as the replacement for sucrose. Sorbitol is hygroscopic and picks up moisture, which proved to be a problem. In the early stages of manufacture, the chocolate is refined and the particle size reduced so that it will have a smooth mouthfeel in the end product. During this step, if too much moisture is introduced from the equipment or environment, the chocolate viscosity will be adversely affected. Because the viscosity is too high, the end product will require more fat than usual to achieve the appropriate texture. However, fat can be one of the most expensive ingredients in chocolate manufacture. For that reason, the chocolate manufacturer wants to use the least amount of fat possible to make the best chocolate at the lowest cost; this is possible with mannitol. It is widely known in the industry that mannitol compound coatings perform better than other sugar-free coatings, because less chocolate is required to successfully enrobe a confection.

Chewing Gum

In chewing gum, mannitol is used in a small percentage. It is used in two areas, both in the gum and on top of the stick of gum as a processing aid. In the gum, mannitol is used as part of the plasticizer system to help maintain the soft texture of the gum. On the surface of stick gum, mannitol is the primary ingredient used as a "dusting agent" during manufacture and wrapping. It is employed because of its non-hygroscopic characteristics, which protects the gum from picking up moisture and becoming sticky, thus preventing it from adhering to the equipment or gum wrapper.

Pharmaceutical and Nutritional Tablets

Mannitol is primarily used in tableting applications as a diluent or filler. A diluent or filler is an inactive ingredient that binds with the active drug substance and helps to compress it into a tablet. As a diluent, mannitol often makes up the bulk of the tablet imparting its properties to the rest of the granulation. Mannitol comes in two forms for tablet applications. These include powdered mannitol for wet granulation tablet manufacture and granular mannitol for direct compression tablet manufacture. The only difference between these products is their particle size distribution. The granular mannitol products contain a larger and more uniform particle size distribution allowing them to be more free flowing. This is important for the direct compression method of tablet manufacture, because the tablet ingredients are simply dry blended before they are introduced into the tablet press hopper.

Mannitol is most often used in chewable tablets because of its pleasant sweet taste and good mouthfeel. Mannitol is 50% as sweet as sucrose and has a cooling effect because of its negative heat of solution. These attributes lend themselves to masking the bitter tastes of vitamins and minerals, herbs, or pharmaceutical actives.

One of the major benefits of mannitol for tableting is its chemical inertness. Mannitol is one of the most stable tablet diluents available. It will not react with other tablet ingredients under any conditions. Many other tablet ingredients, such as lactose or dextrose, undergo the Maillard reaction when combined with active drug substances that contain a free amine group. This reaction causes the tablet to discolor, turning a brownish color. Mannitol does not undergo the Maillard reaction so therefore will not result in tablet discoloration. Mannitol has a high melting point and will not discolor at higher temperatures, as do some other tablet ingredients (Wade and Weller 1994).

Another extremely important benefit of mannitol, is that it is non-hygroscopic. Mannitol is very stable and will not pick up moisture, even under conditions of high temperature and high humidity. Many pharmaceutical active substances are moisture sensitive and will degrade when moisture is present. Mannitol protects the active substance from moisture in the environment so that its potency will not be lost (Figure 20.6).

Special Dietary Foods and Candy

Mannitol can be used in a variety of foods and candies, but its actual use is limited because of its price and laxation potential. It is popular in special dietary foods because of its inert nature and

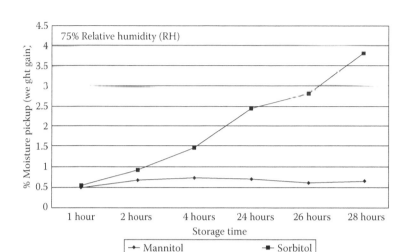

Figure 20.6 Moisture pickup of sorbitol and mannitol.

the fact that it is considered safe for people with diabetes. It is used as a flavor enhancer because of its sweet and pleasantly cool taste. Because of its non-hygroscopic nature, mannitol can be used to maintain the proper moisture balance in foods to increase their shelf life and stability. It is non-cariogenic and can be used in pediatric and geriatric food products because it will not contribute to tooth decay.

Regulatory Status

Sorbitol is Generally Recognized As Safe (GRAS) for use as a direct additive to human food according to the United States Food and Drug Administration's (US FDA) regulation 21 CFR 184.1835 (sorbitol). Mannitol is considered a "food additive" and is permitted in food at specified usage levels stated in 21 CFR 180.25 (mannitol). In regards to dental claims, sugar-free foods sweetened with sorbitol and mannitol may bear the "does not promote tooth decay" health claim in accordance with US regulations 21 CFR 101.80 (U.S. Food and Drug Administration 2009).

The Joint Food and Agriculture Organization/World Health Organization Expert Committee on Food Additives (JECFA) has reviewed the safety data for both sorbitol and mannitol and determined they are safe. The JECFA has established an acceptable daily intake (ADI) for sorbitol and mannitol of "not specified," meaning no limits are placed on its use (JECFA 1982 and 1986).

Sorbitol solution 70%, non-crystallizing sorbitol solution 70%, crystalline sorbitol, and mannitol all have monographs in the United States Pharmacopoeia/National Formulary (USP/NF) and Food Chemical Codex (FCC), as well as the various pharmacopoeias around the world (European, British, Japanese, etc.).

Summary

Sorbitol and mannitol are naturally occurring polyols found in animals and plants. They are present in small quantities in almost all vegetables. They are widely used in specialty foods and pharmaceuticals. Sorbitol and mannitol are important ingredients in sugar-free baked goods, candies, chewing gum, and tablets. The sweet, cool taste of sorbitol and mannitol makes them useful for many taste-masking or sweetening applications. The unique moisture retention properties of sorbitol make it useful in improving the shelf life of food products. The non-hygroscopic nature and chemical inertness of mannitol are attractive benefits for pharmaceutical tablets. Sorbitol and mannitol have a wide variety of uses as evidenced by the tremendous number of applications in which they are found.

References

Allison, R.G. 1979. Dietary Sugars in Health and Disease III. Sorbitol, pp. 1–51: Report from Life Sciences Research Office Federation of American Societies for Experimental Biology, Bethesda, MD.

Courtney, D. 1980. Polyols in creams and lotions. *J Cosmet Toiletr* 95:27–34.

Debord, B., Lefebvre, C., Guyot-Hermann, A.M., Hubert, J., Bouché, R., and Cuyot, J.C. 1987. Study of different crystalline forms of mannitol: Comparative behaviour under compression. *Drug Devl Indust Pharm* 13:1533–1546.

DuRoss, J.W. 1982. Modified crystalline sorbitol. *Manuf Confect* 35–41.

DuRoss, J.W. 1984. Modification of the crystalline structure of sorbitol and its effects on tableting characteristics. *Pharm Technol* 8:50–56.

European Economic Community Council (EEC). 1990. Directive on food labeling. *Off J Eur Commun* No. L 276/40.

Geisler, A.S., and Papalexis, G.C. 1971. Method of preparing a frankfurter product and compositon for use therein. *US Patent* 3,561,978.

Joint FAO/WHO Expert Committee on Food Additives. 1982. Toxicological evaluation of certain food additives: Sorbitol. Twenty-sixth report. WHO Technical Report Series 683, p. 27.

Joint FAO/WHO Expert Committee on Food Additives. 1986. Mannitol. Prepared at the 30th JECFA and published in FNP 37.

Lee, C.M. 1984. Surimi process technology. *Food Technol* 38:69–80.

Mellan, I. 1962. *Polyhydric Alcohols*, pp. 185–202. Washington, DC: McGregor & Werner.

Shangraw, R.F., Wallace, J.W., and Bowers, F.M. 1981. Morphology and functionality in tablet excipients for direct compression. *Pharm Technol* 5:68–78.

US Food and Drug Adminstration. 2009. *Code of Federal Regulations*, Title 21. Washington, DC: US Government Printing Office.

Wade, A., and Weller, P.J. 1994. Mannitol. In *Handbook of Pharmaceutical Excipients,* N.A., Armstrong, G.E., Reier, and D.A., Wadke (Eds.), pp. 294–298. Washington, DC: American Pharmaceutical Association.

Wadke, D.W., Serajuddin, A.T.M., and Jacobson, H. 1989. Preformulation testing. In *Tablets*, H.A. Lieberman, L. Lachman, and J.B. Schwartz (Eds.), Vol. 1, pp. 1–69. New York: Marcel Dekker.

Yoon, K.S., and Lee, C.M. 1984. Assessment of the cryoprotectability of liquid sorbitol for its application in surimi and extruded products. Presented at The 31st AFTC Meeting, Halifax, Nova Scotia.

Chapter 21

Xylitol

Christos Zacharis and Julian Stowell*

Contents

---

* Philip M. Olinger and Tammy Pepper did not participate in the revision but their chapter in the book's 3rd edition was the basis of this revised chapter.

Introduction

Xylitol is a five-carbon polyol with a sweetness similar to that of sucrose (Hyvoenen et al. 1977; Munton and Birch 1985). It is found in small amounts in a variety of fruits and vegetables (Mäkinen and Söderling 1980; Washuett et al. 1973) and is formed as a normal intermediate in the human body during glucose metabolism (Touster 1974). Xylitol has been shown to be of value in the prevention of dental caries because it is not an effective substrate for plaque bacteria (Bär 1988a; Mäkinen 1978, 1988, 1994; Linke 1986; Loesche 1985; Kleinberg 1995). In addition, xylitol is known to inhibit growth, metabolism, as well as polysaccharide production of mutans streptococci (Söderling et al. 2008). Because of its largely insulin-independent metabolic utilization, it may also be used as a sweetener in the diabetic diet and as an energy source in parenteral nutrition (Foerster and Mehnert 1979; Georgieff et al. 1982, 1985). As a sweetening agent, xylitol has been used in human food since the 1960s. Different aspects of its applications, metabolic properties, and

Table 21.1 Some Physical and Chemical Properties of Xylitol

Formula	$C_5H_{12}O_5$ (molecular weight, 152.15)
Appearance	White, crystalline powder
Odor	None
Specific rotation	Optically inactive
Melting range	92°C–96°C
Boiling point	2l6°C (760 mmHg)
Solubility at 20°C	169 g/100 g H_2O; 63 g/100 g solution; sparingly soluble in ethanol and methanol
pH in water (l00 g/liter)	5–7
Density of solution	10%, 1.03 g/ml; 60%, 1.23 g/ml
Viscosity (20°C)	10%, 1.23 cP; 60%, 20.63 cP
Heat of solution	+34.8 cal/g (endothermic)
Heat of combustion	16.96 kJ/g
Refractive index (25°C)	10%, 1.3471; 50%, 1.4132
Moisture absorption (4 days, RT)	60% RH, 0.051% H_2O, 92% RH, 90% H_2O
Relative sweetness	Equal to sucrose; greater than sorbitol and mannitol
Stability	Stable at 120°C, no caramelization; stable also under usual conditions in food processing; caramelization occurs if heated for several minutes near the boiling point

cP = centipoise.

RH = relative humidity.

RT = room temperature.

safety evaluation have been the subject of several previous reviews (Counsell 1978; Hyvoenen and Koivistoinen 1982; Mäkinen 1978; Ylikahri 1979; Mäkinen and Scheinin 1982; Bär 1985, 1986). Its physicochemical properties are summarized in Table 21.1.

Production

Xylitol was first synthesized and described in 1891 by Emil Fischer and his associate (Fischer and Stahel 1891). On a commercial scale it is produced by chemical conversion of xylan (Jaffe 1978; Aminoff et al. 1978). Potential sources of xylan include birch wood and other hardwoods, almond shells, straw, corn cobs, and by-products from the paper and pulp industries. The practical suitability of these raw materials depends, among other factors, on their xylan content, which may vary considerably, and the presence of by-products (e.g., poly- or oligosaccharides), which have to be removed during the production process. In principle, the commercial synthesis of xylitol involves four steps. The first step is the disintegration of natural xylan-rich material and the hydrolysis of the recovered xylan to xylose. The second step is the isolation of xylose from the hydrolysate by means of chromatographic processes to yield a pure xylose solution. Thirdly, xylose is hydrogenated to xylitol in the presence of a nickel catalyst. Alternatively, hydrogenation of the impure xylose solution may be conducted first, followed by purification of the xylitol syrup. Ultimately, xylitol is crystallized in orthorhombic form (Carson et al. 1943).

The synthesis of xylitol by fermentative or enzymatic processes is possible in principle (Onishi and Suzuki 1969; Chen and Gong 1985; Gong et al. 1983; Kitpreechavanich et al. 1985). However, such procedures have so far not been used on a commercial scale. Other approaches to the synthesis of xylitol are of merely scientific interest (Holland and Stoddart 1982; Kiss et al. 1975).

Xylitol is supplied commercially in crystalline, milled, and granulated forms. Crystalline xylitol has a particle size fraction ranging from 150 to 2300 microns. Milled forms of xylitol range in mean particle sizes from approximately 50 to 200 microns. The granulated forms of xylitol are suitable for direct compression.

Properties and Applications

At present, xylitol is used as a sweetener mainly in noncariogenic confectionery (chewing gum, candies, chocolates, gumdrops), and less frequently in dietetic foods (food products for people with diabetes), in pharmaceutical preparations (tablets, throat lozenges, multivitamin tablets, cough syrup), and in cosmetics (toothpaste and mouthwash) (Pepper and Olinger 1988; Krueger 1987, 1988; Pepper 1987; Dodson and Pepper 1985; Imfeld 1984; Bray 1985). Xylitol is used at low levels in selected low-calorie soft drink applications to improve product mouthfeel and sweetness profile. In principle, the manufacture of various baked goods with xylitol is possible. However, if crust formation, caramelization, or nonenzymatic browning is required, the addition of a reducing sugar is necessary. Because xylitol inhibits the growth and fermentative activity of yeast, it is not a suitable sweetener for products containing yeast as a leavening agent (Askar et al. 1987; Varo et al. 1979). Xylitol is also being used as a natural bulk sweetener in a growing range of applications due to its excellent sweetness profile.

Sweetness

Xylitol is the sweetest polyol (Figure 21.1) (Moskowitz 1971; Gutschmidt and Ordynsky 1961, Lindley et al. 1976; Yamaguchi et al. 1970; Lee 1977) At 10% solids (w/w) xylitol is isosweet to sucrose, whereas at 20% solids (w/w) xylitol is about 20% sweeter than sucrose (Munton and Birch 1985). Combinations of xylitol and other polyols, such as sorbitol, create significant sweetness synergisms.

Cooling Effect

The heats of solution of crystalline polyols and sucrose are shown in Figure 21.2. The loss of heat when dissolving polyols in water is much greater than with sucrose. Crystalline xylitol provides a significant cooling effect. This interesting organoleptic property is most notable in sugar-free chewing gum, compressed candies, and chewable vitamins. The cooling effect enhances mint flavor perception, and the presence of xylitol contributes a refreshing coolness.

A cooling effect is obviously not perceived from products in which xylitol is already dissolved (e.g., toothpaste, mouth rinse) or in which it exists in an amorphous form (jellies; boiled, transparent candies).

Due to its high solubility and high negative heat of solution, xylitol has the most pronounced cooling effect among the polyols.

Solubility of Polyols

The solubility of a bulk sweetener has a critical impact on the mouthfeel and texture of the final product. Bulk sweetener solubility also affects the release or onset of flavor and sweetness perception and the release and bioavailability of the active ingredients of pharmaceuticals. Table 21.2 shows the solubility of selected polyols at 20°C. The solubility of xylitol is 2% higher than sucrose at body temperature.

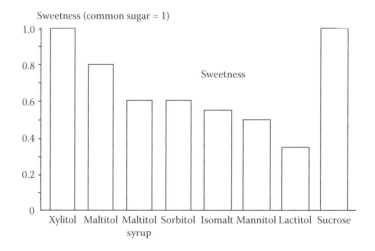

Figure 21.1   Relative sweetness of polyols.

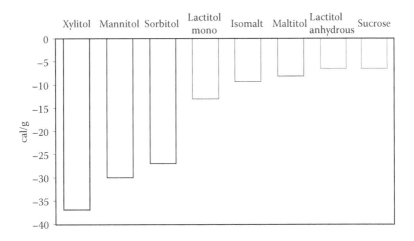

Figure 21.2   Cooling effect of polyols.

Other Characteristics

The viscosity of polyol solutions depends on their molecular weight. Monosaccharide alcohols such as sorbitol and xylitol have a relatively low viscosity, whereas maltitol syrups with a high content of hydrogenated oligosaccharides are fairly viscous.

Xylitol is chemically quite inert because of the lack of an active carbonyl group. It cannot participate in browning reactions. This means that there is no caramelization during heating, as is typical of sugars. Because xylitol does not form Maillard reactants, it is ideally suited to use as an excipient in conjunction with amino active ingredients.

Use of Xylitol in Chewing Gum and Confectionery

Sugar-free chewing gum is the leading worldwide application of xylitol. Xylitol is used to sweeten both stick and pellet (dragee) forms of chewing gum. In addition to the dental benefits attributed to it, xylitol provides a pleasant cooling effect and a rapid onset of both sweetness and flavor, which is especially perceived with mint flavors. Because of its rapid drying and crystallization character, xylitol is often used to coat pellet forms of sugar-free chewing gum.

Table 21.2   Solubility of Sugar Alcohols in Water at 20°C

Sweetener	Solubility (% w/w)
Sucrose	66
Xylitol	63
Sorbitol	75
Maltitol	62
Lactitol	55
Isomalt	28
Mannitol	18

Sugar-free chewing gums usually contain mixtures of polyols and intense sweeteners to ensure a satisfactory sweetness profile over time. Although a few brands are sweetened solely with xylitol, most products contain less than 50% xylitol. The use of xylitol-containing chewing gum in either stick or pellet form as a between-meal supplement has been associated with improvements in oral health. Among the reported benefits are reductions in plaque levels, plaque bacteria levels, and caries increment (Söderling et al. 1988; Soparkar et al. 1989; Birkhed 1994; Gordon et al. 1990; Edgar 1998).

Xylitol is commercially used in many countries, either alone or in combination with other sugar substitutes, in the manufacture of noncariogenic sugar-free or no sugar added confectionery. Although it is, in principle, possible to produce gum arabic pastilles, chocolate, hard candies, ice cream fillings, and other confections with xylitol alone, other sugar substitutes are normally added to optimize sweetness, texture, or shelf life. Sugar-free pectin jellies, which are indistinguishable from conventional sugar/corn syrup jellies, can be produced with a combination of xylitol and hydrogenated starch hydrolysates. Crystalline xylitol, which has an appearance similar to sucrose, is an excellent sugar-free sanding material in conjunction with pectin jellies and other confection forms.

Xylitol can be used either with sorbitol or hydrogenated starch hydrolysate to produce an exceptional sugar-free fondant. In each case, xylitol is applied as the crystalline phase in the production. The resulting fondants exhibit a fine texture and a pleasant cooling effect.

Pharmaceutical and Oral Hygiene Preparations

Xylitol can be used as an excipient or as a sweetener in many pharmaceutical preparations. As in foods, the advantages are suitability for diabetic patients, non-cariogenic properties, palatability and non-fermentability. Cough syrups, tonics, and vitamin preparations made with xylitol can neither ferment nor mold. Because xylitol is chemically inert, it does not undergo Maillard reactions or react with other excipients or active ingredients of pharmaceuticals. Xylitol-sweetened medications can be given to children at night after tooth brushing without any harm to the teeth (Imfeld 1984; Feigal et al. 1981; Souney et al. 1983; Roberts and Roberts 1979; Crawford 1981; Muehlemann et al. 1982).

In tablets, xylitol can be used as a carrier and/or as a sweetener (Kristoffersson and Halme 1978; Laakso et al. 1982). In addition to sweetness, nonreactivity, and microbial stability, xylitol offers the advantages of high solubility at body temperature and a pleasing, cooling effect. As a carrier, xylitol has also been tested in solid dispersions because of its low melting point and stability up to 180°C. In these studies, it was found that solid dispersions of hydrochlorothiazide or *p*-aminobenzoic acid with xylitol, showed a faster release than the micronized drugs (Bloch et al. 1982; Sirenius et al. 1979). Milled xylitol may be granulated and pressed into tablets after flavoring (Gordon et al. 1990). Alternatively, directly compressible grades of xylitol can be used to facilitate tablet production (Morris et al. 1993; Garr and Rubinstein 1990). Coatings with xylitol or mixtures of xylitol and sorbitol (up to 20%) can be made by conventional pan coating and by sintering the surface of compressed tablets in a hot-air stream (Voirol and Etter 1979).

In toothpaste, xylitol may partially or completely replace sorbitol as a humectant. Because of its greater sweetness, xylitol improves the taste of the dentifrice, and in the manufacture of transparent gels, it has been said to exhibit properties slightly superior to those of sorbitol. Furthermore, xylitol can enhance the anticaries effect of a fluoride toothpaste. The inclusion of 10% xylitol in a fluoride toothpaste, for example, resulted in a 12% reduction in decayed/filled surfaces after three years compared with a fluoride-only toothpaste (Sintes et al. 1995). In addition, there is evidence that xylitol exerts a plaque-reducing effect (Mäkinen et al. 1987; Mäkinen et al. 1985a) and that it interferes with bacterial metabolism particularly in the presence of fluoride and zinc ions (Scheie et al. 1989). An inhibitory effect on enamel demineralization has also been postulated (Arends et al. 1984; Smits and Arends 1988). There is also evidence that use of a xylitol-containing dentifrice can result in a significant reduction of *Streptococcus mutans (S. mutans)* in saliva (Svanberg and Birkhed 1995). Because of its overall favorable effects on dental health, xylitol has also been applied in other oral care products, such as in mouth rinses and in artificial saliva (Featherstone et al. 1982; Vissink et al. 1985; Cobanera et al. 1987; Petersson et al. 1989).

Metabolism

In principle, two different metabolic pathways are available for the catabolism of xylitol: (a) direct metabolism of absorbed xylitol in the mammalian organism, mainly in the liver, or (b) indirect metabolism by means of fermentative degradation of unabsorbed xylitol by the intestinal flora.

Indirect Metabolic Utilization

All polyols, including xylitol, are slowly absorbed from the digestive tract, because their transport through the intestinal mucosa is not facilitated by a specific transport system. After ingestion of large amounts, therefore, only a certain proportion of the ingested xylitol is absorbed and enters the hepatic metabolic system through the portal vein blood. A comparatively larger amount of the ingested xylitol reaches the distal parts of the gut, where extensive fermentation by the intestinal flora takes place. Besides minor amounts of gas (H_2, CH_4, CO_2), the end-products of the bacterial metabolism of xylitol are mainly short-chain, volatile fatty acids, (i.e., acetate, propionate, and butyrate) (Grimble 1989; Salyers and Leedle 1983; Cummings 1981). These products are subsequently absorbed from the gut and enter the mammalian metabolic pathways (Ruppin et al. 1980). Acetate and butyrate are efficiently taken up by the liver and used in mitochondria for production of acetyl-CoA. Propionate is also almost quantitatively removed by the liver and yields propionyl-CoA (Cummings et al. 1987; Buckley and Williamson 1977; Skutches et al. 1979).

The production of volatile fatty acids (VFA) is a normal process associated with the consumption of polyols and dietary fibers (cellulose, hemicelluloses, pectins, gums) for which hydrolyzing enzymes are lacking or poorly efficient in the small intestine. Under normal conditions, most of the generated VFA are absorbed from the gut and are further used by established metabolic pathways in animals and man (Cummings 1981; Cummings et al. 1986). For the energetic use of slowly digestible materials, this fermentative, indirect route of metabolism plays an important role, and evidence has been presented that, even under normal dietary conditions (i.e., in the absence of polyols), the contribution of intestinal fermentations to the overall energy balance is significant (McNeil 1984; Flemming and Arce 1986).

Direct Metabolism via the Glucuronic Acid–pentose Phosphate Shunt

In addition to the indirect route of use, a direct metabolic pathway is available for the portion of xylitol that is absorbed unchanged from the gastrointestinal tract (Touster 1974; Demetrakopoulos and Amos 1978; Baessler 1978).

The metabolism of xylitol and its general relationship to the carbohydrate metabolism by means of the pentose phosphate pathway is shown in Figure 21.3. This scheme illustrates how the

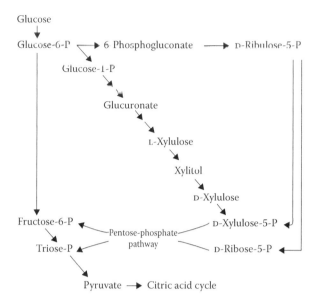

Figure 21.3   Metabolism of xylitol.

transformation of L-xylulose by way of xylitol to D-xylulose links the oxidative branch of the glucuronate pentose phosphate pathway with the nonoxidative branch, which yields glyceraldehyde-3-phosphate and fructose-6-phophate, as well as ribose-5-phosphate for ribonucleotide biosynthesis (Williams et al. 1987; Wood 1986; Baquer et al. 1988). Thus, xylitol can be converted by means of the pentose phosphate pathway to intermediates of the glycolytic pathway, which may either undergo further degradation or transformation to glucose-1-phosphate, a precursor of glycogen (Touster 1974; Demetrakopoulos and Amos 1978; Baessler 1978; Williams et al. 1987). Because gluconeogenesis from exogenous xylitol is associated with the generation of NADH in the cytosol, reoxidation of cytosolic NADH is a necessary step for xylitol use.

The recognition of xylitol as a normal endogenous metabolite in humans originates from observations in patients with a genetic abnormality, essential pentosuria. These persons excrete considerable quantities of L-xylulose in the urine. When incubated in tissue preparations, L-xylulose was found to be metabolized to xylitol by a specific NADPH-linked enzyme, L-xylulose-reductase. This enzyme has subsequently been demonstrated to be deficient in essential pentosuria. Because pentosuric patients excrete 2–15 g of L-xylulose per day, the daily production of xylitol was estimated to be of a similar order of magnitude (Touster 1974; Lane 1985; Soyama and Furukawa 1985).

Estimation of the Caloric Value of Xylitol

If the caloric value of polyols is to be estimated on the basis of their metabolic fate, precise knowledge about the different steps of their digestion and metabolic use is required. In particular it is, for example, essential to know (a) how much of an ingested dose is absorbed directly from the gut, (b) by which pathways the absorbed portion is metabolized in the human organism, (c) which proportion of an ingested dose is fermented by the gut microflora, (d) to what extent the resulting fermentation products are absorbed and metabolically used by the host, and (e) how much of an ingested dose leaves the intestinal tract unchanged with the feces or urine. Although the experimental database on xylitol does not allow one to answer all these questions with sufficient precision, it is possible to obtain reasonable estimates by extrapolation from existing data and studies on other polyols that are also incompletely absorbed and are subject to the same fermentative degradation in the gut.

On the basis of a thorough assessment of numerous *in vitro* and *in vivo* experiments with xylitol and other polyols, it has been estimated that approximately one quarter of an ingested xylitol dose is absorbed from the gastrointestinal tract (FASEB 1994). This portion of xylitol, which is effectively metabolized by means of the glucuronate-pentose phosphate shunt, is energetically fully available and provides about 4 kcal/g. The nonabsorbed three-quarters of the ingested load, however, are almost completely fermented by the intestinal flora. Combining a 78% retention of energy in bacterial fermentation products and a growth yield of 20 g bacteria per 100 g substrate, it has been estimated that about 42% of the energy provided with unabsorbed xylitol is consumed by bacterial metabolism and growth, whereas about 58% of the energy becomes available to the host after absorption of the fermentation end-products (VFA) (Bär 1990). This value is well in line with an estimated 50% energy salvage proposed by a Dutch expert group (Anonymous 1987).

On the basis of these estimates, a metabolizable energy value of about 2.8–2.9 kcal/g may be calculated for xylitol (FASEB 1994). This value is in line with the results of an *in vivo* study in which xylitol was found to be about 60% as effective as glucose in promoting growth (Karimzadegan et al. 1979).

Besides laborious and expensive energy balance trials, only indirect calorimetry allows one to determine the energetic use of substrates in humans. By use of this noninvasive technique, the energetic use of xylitol was examined in 10 healthy volunteers (Mueller-Hess et al. 1975). The results of this study revealed that, during a 2.5-hour period, the cumulated increase in carbohydrate oxidation amounted to only one quarter of that caused by glucose. The total increase in metabolic rate was 52% lower after the xylitol load than after the glucose treatment. This result suggests that xylitol has a caloric value of only about 50% that of glucose. Considering, however, that the absorption (and hence metabolic use) of the bacterial fermentation products may not have been complete within the 2.5-hour experimental period, it is likely that the true caloric value for xylitol is somewhat higher than the proposed 2 kcal/g. The net energy value of 2.4 kcal/g has been assigned to xylitol by the Federation of American Societies for Experimental Biology (FASEB). The net energy is defined as that portion of gross energy intake that is deposited or mobilized in the body to do physical, mental, and metabolic work (FASEB 1994).

Food regulatory authorities are increasingly taking note of the reduced caloric value of polyols, which is by now well supported by a still growing volume of scientific literature. In the EU, a caloric value of 2.4 kcal/g has been allocated for all polyols including xylitol (EC 1990; Gutschmidt and Ordynsky 1961; Lindley et al. 1976; Yamaguchi et al. 1970; Lee 1977; Söderling et al. 1988). On the basis of the 1994 FASEB review, the US Food and Drug Administration (US FDA) acknowledged the xylitol caloric value of 2.4 kcal/g (letter to American Xyrofin Inc from US FDA concerning use of a self-determinated energy value for xylitol 1994).

Dental Benefits

Caries Formation

According to current knowledge, dental caries are caused by bacteria that accumulate in large masses, known as dental plaque, on the teeth in the absence of adequate oral hygiene. Fermentation of common dietary carbohydrates by plaque bacteria leads to the formation of acid end-products. Acid accumulation and a decrease in plaque pH will follow. A decrease in plaque pH caused by bacterial fermentation of carbohydrates may lead to undersaturation of the plaque with respect to calcium and phosphate ions, to demineralization of the tooth enamel, and eventually to formation of a cavity.

Approaches aimed at the prevention or elimination of dental caries include reduction of acid dissolution of tooth mineral and stimulation of enamel remineralization by fluoride, removal of dental plaque by brushing the teeth, and reduction of the availability of fermentable carbohydrates by appropriate dietary habits (Shaw 1987).

For obvious reasons, fermentable sugars cannot be completely eliminated from our daily food supply. From the point of view of caries prevention, this is not even necessary, because a number of studies have indicated that eating sucrose-sweetened foods in moderation and with regular meals, results in little or no caries (Newbrum 1982). On the other hand, experiments with rats, as well as some studies in humans, indicate a strong relationship between the frequency of consumption of sucrose between meals and caries activity (Huxley 1977; Gustaffson et al. 1953). The reason for this observation is that under conditions of frequent consumption of sugary snacks, the plaque pH drops and remains below the critical value of 5.7 for prolonged periods of time (Imfeld and Muehlemann 1978). As a result, periods with demineralization overwhelm the recovery phases, and dental caries may subsequently develop. This suggests that substitution of sucrose and other fermentable sugars by nonfermentable sugar substitutes has considerable impact on caries activity, even if this substitution is limited to snacks and beverages that are consumed between the main meals. Because of its noncariogenic or even cariostatic properties, xylitol is a particularly suitable substance for this purpose (Edgar 1998; Bär 1989; Loesche 1985).

Acidogenicity

Considerable evidence documents the fact that xylitol is not fermented by most oral microorganisms and that exposure of dental plaque to xylitol does not result in a reduction of plaque pH (Bär 1988a; Guelzow 1982; Bibby and Fu 1985; Firestone and Navia 1986; Rekola 1988; Imfeld 1983; Maki et al. 1983). Even after chronic exposure for two years, no adaptation occurred with respect to the ability of the dental plaque to ferment xylitol (Mäkinen et al. 1985a; Gehring, 1977).

The noncariogenicity of xylitol was demonstrated in several rat caries studies in which xylitol was fed in the absence of other readily fermentable dietary sugars. Under these conditions, xylitol did not exhibit any cariogenic potential (Bär 1988a; Van der Hoeven 1986). The cariostatic properties of xylitol were investigated in numerous studies in which rats received xylitol in admixture to a cariogenic (i.e., sucrose-containing) diet. The results of these studies indicate that xylitol tends to reduce the cariogenicity of such diets and that this effect is greater for moderately cariogenic diets with 25% or less fermentable sugars than for diets containing 50% or more sucrose (Bär 1988a; Havenaar et al. 1984; Havenaar 1984b; Karle and Gehring 1987). More pronounced and more consistent reductions of caries formation were observed in rat studies in which xylitol was administered alternately with sucrose-containing meals (i.e., between the cariogenic challenges) (Bär 1988a; Havenaar 1984a; Hefti 1980; Leach and Green 1980). Under these conditions, which mimic the human meal pattern more closely, the caries scores of the xylitol-treated animals were on average 35% less than those of the corresponding controls. In one study it was even found that xylitol may promote the remineralization of early caries lesions of rats (Leach and Green 1980; Leach et al. 1983). In summary, the results obtained in the rat model suggest that xylitol exhibits cariostatic properties.

Efficacy of Xylitol in Human Caries Studies

The clinical benefit of xylitol for caries prevention has been demonstrated in several studies with children and adult human volunteers, in which consumption of xylitol was consistently associated with a significant reduction of the caries increment. In these studies, which were conducted by independent investigators at different locations, the efficacy of xylitol was tested under carefully controlled conditions, with randomized assignment of the subjects to the different treatment groups (Scheinin et al. 1975a, 1975b; Rekola 1986a, 1986b, 1987) and under the less standardized but "real-life" oriented conditions of field studies involving large numbers of caries-prone children (Scheinin et al. 1985; Kandelman et al. 1988; Kandelman and Gagnon 1985). Regardless of the differences in study design, significant reductions of the caries increment were observed in all xylitol-treated groups. Depending on the dose (Kandelman et al. 1988) and frequency of the xylitol administration (Isokangas 1987), this reduction varied between 45% and more than 90%. The dental benefits attributed to xylitol have been shown to extend beyond the study period. When subjects from a Finnish chewing gum study (Isokangas 1987) were evaluated up to five years after the initial study, subjects from the xylitol group continued to demonstrate a reduction in caries rate compared with the control group (Isokangas et al. 1993).

In a further study involving 10-year-old children, xylitol, xylitol/sorbitol, and sorbitol-sweetened gums were evaluated in terms of caries risk and their impact on remineralization. Although each chewing gum group exhibited a reduction in caries risk and an increase in remineralization, the xylitol group had a significantly lower caries risk and higher remineralization than the other gum groups (Mäkinen et al. 1995a, 1995b, 1996). The results support the view that xylitol has an active anticaries effect above that of a mere sugar substitute (Edgar 1998).

It is particularly noteworthy that daily fluoride brushing plus xylitol consumption was found to be more efficient than daily fluoride brushing alone (Scheinin et al. 1985; Kandelman et al. 1988; Kandelman and Gagnon 1987; Kandelman and Gagnon 1988). Considering the different modes of the cariostatic action of fluoride and xylitol, this is not surprising, but it is certainly relevant in terms of a most efficient combination of caries preventive measures (Bär 1989).

Mother and Child Study

A clinical study published in the *Journal of Dental Research* reported on an innovative new caries prevention method by which treatment given to mothers reduced their children's tooth decay by over 70% (Söderling et al. 2000). The aim of the study was to establish whether maternal consumption of sugar free chewing gum sweetened with the naturally occurring sweetener xylitol, could reduce the transmission of harmful *S. mutans* bacteria from mother to child and thereby reduce the risk of dental caries in their children.

Several studies have shown that the earlier that infant's teeth are infected with *S. mutans*, the higher is the infant's future caries risk. Thus if the infection or transmission can be delayed after the so called "window of infectivity," ranging from 19 to 33 months, the child's risk for future caries is decreased considerably. In the mother-child study, mothers from the Ylivieska area in Finland were recruited during pregnancy and sub-divided into three treatment groups. The mothers of the xylitol group received xylitol chewing gum approximately four times per day from 3 to 24 months after the child was born. In the two control groups, the mothers received either fluoride (not expected to affect the transmission, but to protect and strengthen the mothers teeth) or chlorhexidine (expected to decrease the transmission of the oral bacteria) varnish treatments 6, 12 and 18 months post-delivery. Otherwise, normal dietary and oral hygiene practices were maintained during the study.

The children did not chew gum or receive any varnish treatments. Dental examinations of the children were carried out annually. The plaque *S. mutans* levels of the children were assessed at two years of age; the children's risk of *S. mutans* colonization was five-fold higher in the fluoride group and three-fold higher in the chlorhexidine group as compared to the xylitol group. The results of the mother–child study further support the theory that regular xylitol consumption selects for *S. mutans* with reduced adhesion/transmission properties.

The objective of this study was to monitor the caries incidence of the children whose mothers had participated in the original trial. The children were examined annually and any decayed, missing or filled tooth surfaces (DMFS) were scored, providing an "index" of caries incidence.

There was for a significant reduction in caries incidence as a direct result of the mother's consumption of xylitol chewing gum for the first two years of the child's life (71%–74% reduction in caries incidence compared with the two control groups). It is important to remember that only the mothers of these children received any treatment, and this was only during the first two years of the child's life. The children themselves received no intervention during this period.

Interestingly, although the children of mothers in the chlorhexidine group exhibited reduced colonization by *S. mutans* after two years, this did not translate into a reduction in caries incidence after five years. This further supports the theory that xylitol does not exert an antibacterial effect on the oral bacteria, but rather a "modulating" effect on the oral flora as a whole, resulting in a shift towards a less virulent/cariogenic flora.

In a separate study, which was carried out in the Department of the Behavioural Pediatric Dentistry of Okayama Graduate School of Medicine, pregnant women with a high incidence of caries were allocated into two groups. In the first group the pregnant women would be given a Xylitol chewing gum to be consumed at least four times a day on top of basic prevention strategies (i.e., tooth brushing), while the other group were not given a chewing gum. Both groups, however, were given a full professional check up and dental cleaning prior to the initiation of the study (Nakai et al. 2009).

Women in the Xylitol group continued consuming 100% Xylitol chewing gum for 13 months beginning from the 6th month of their pregnancy, so effectively until their children reached nine months of age. Dental professionals of the clinic were examining saliva samples of the children for evidence of *S. mutans* colonization between the ages of 6 months up to the age of 24 months (a period of 15 months since the intervention study took place (6th month of pregnancy).

The results that followed were remarkable, because they did not only confirm the earlier findings from the mother and child studies carried out in the Nordic countries, but also determined that consuming 100% Xylitol chewing gum from the 6th month of pregnancy and for as little as one year, serves as a very early intervention measure minimizing the "window of infectivity" for the new generation with less bad bacteria colonizing the teeth surfaces (Nakai et al. 2009).

Discrimination and Effect against Streptococcus Mutans

Microbiological investigations in caries-active humans, as well as studies in experimental animals, have demonstrated that certain types of bacteria are particularly active in initiating and promoting dental caries. Culprit number one in this respect is *S. mutans*. This microorganism has the ability to adhere strongly to the crowns of the teeth by means of sticky extracellular glucans, to grow and metabolize optimally in a relatively acidic environment, and to live under microaerobic or strictly anaerobic conditions, as met in the depth of pits and fissures (Hamada and Slade 1980; van Houte 1980; Alaluusua et al. 1989; Stecksen-Blicks 1988).

Several studies have suggested that the predominance of *S. mutans* in cariogenic plaque depends on the ability of this bacterium to remain metabolically active even in a relatively acidic environment. In fact, *S. mutans* is more active at pH 5 than at pH 7, whereas many other members of the plaque flora become metabolically inactive under such conditions. This observation suggests that the frequent ingestion of sucrose gives a competitive edge to *S. mutans* over the other plaque microbes. In this way, a *circulus vitiosus* for the formation of a more cariogenic plaque is formed (Scheie et al. 1984; Harper and Loesche 1984; Sgan-Cohen et al. 1987; Maiwald et al. 1982).

In this regard, the use of xylitol as a sucrose substitute becomes an extremely attractive means to control and prevent dental caries for two different reasons. First, no acid is formed from xylitol by the dental plaque. In fact, during and after chewing of a xylitol-sweetened gum, an elevation rather than a decrease of the plaque pH is observed (Maiwald et al. 1982; Graber et al. 1982). Under such conditions, however, the metabolism of *S. mutans* is, as mentioned previously, not optimally active, and other bacteria may successfully compete with *S. mutans*. Second, in addition to this indirect, pH-mediated effect, xylitol appears to inhibit the growth and metabolism of *S. mutans* in a more direct way. Several experiments have shown that the addition of xylitol to a glucose-containing medium reduced the growth of *S. mutans* (Assev et al. 1983; Vadeboncoeur et al. 1983; Assev et al. 1985; Assev and Roella 1986b; Ziesenitz and Siebert 1986; Beckers 1988; Siebert et al. 1988; Forbord and Osmundsen 1992). This inhibition by xylitol appears to be related to the accumulation of xylitol-5-phosphate and xylulose-5-phosphate within the cells (Assev and Scheie 1986; Trahan et al. 1985; Assev and Roella 1986a; Waaler et al. 1985; Pihanto-Leppala et al. 1990; Waher 1992). Probably as a result of the intracellular accumulation of these metabolites, the ability of *S. mutans* to adhere to surfaces is decreased (Söderling et al. 1987; Verran and Drucker 1982), and disintegration of the ultrastructure of the cells may occur (Tuompo et al. 1983).

In line with these results, lower *S. mutans* counts were found in the plaque and/or saliva of xylitol-treated human volunteers (Overmyer et al. 1987; Söderling et al. 1989; Loesche 1984; Mäkinen 1989; Banoczy et al. 1985; Gehring 1977; Söderling et al. 1991). These observations suggest that the formation of xylitol-tolerant strains of *S. mutans* after chronic xylitol exposure does not annihilate its inhibitory effect (Trahan and Mouton 1987; Trahan 1995).

In a recent study by Perez et al. (2010), it was shown that consuming chewing gum that contained 15% xylitol induced the reduction of the levels of *S. mutans* in twelve subjects with high *S. mutans* counts over a period of 30 days (five times a day). Such reduction in *S. mutans* numbers persisted, even after the interruption of the study.

General Effects on Dental Plaque

Because dental plaque plays a crucial role in the formation of tooth decay, many investigators have examined the effects of xylitol on dental plaque. It is well established that frequent consumption of sucrose promotes the growth of a voluminous, sticky dental plaque. Xylitol is not a substrate for oral microorganisms, and it therefore does not favor plaque formation. Consequently, studies comparing the effects of sucrose and xylitol on dental plaque have demonstrated consistently that the plaque weights are lower in the xylitol-treated subjects than in the positive controls consuming sucrose (Bär 1988a). Studies suggest that xylitol inhibits the formation of insoluble glucans and lipoteichoic acid, two products of bacterial metabolism that play an important role in the adhesion and cohesion of dental plaque (Assev et al. 1983; Vadeboncoeur et al. 1983; Assev et al. 1985; Assev and Roella 1986a, 1986b; Ziesenitz and Siebert 1986; Beckers 1988; Siebert et al. 1988; Forbord and Osmundsen 1992; Assev and Scheie 1986; Trahan et al. 1985; Waaler et al. 1985; Pihanto-Leppala et al. 1990; Waher 1992; Söderling et al. 1987).

Chewing gums sweetened with xylitol or sorbitol have been investigated regarding their impact on the formation of plaque. The results of these studies suggest that chewing gums sweetened with xylitol have a greater impact on plaque reduction (Söderling et al. 1988; Soparkar et al. 1989).

A recent study by Splieth et al. (2009) adds to the already existing evidence that the action of xylitol as an ingredient is beneficial for oral health by modifying the dental plaque. When xylitol was compared with sorbitol in lozenges, there was a marked reduction in the plaque acidogenicity, thus awarding xylitol with an additional benefit in caries prevention. In particular, the study showed that the consumption of five lozenges per day for four weeks, reduced the plaque acidogenicity significantly amongst 61 adults.

Dental Endorsements and Claims

The dental benefits provided by xylitol have been recognized by a number of dental associations.

Recently the European Food Safety Authority (EFSA) issued with a positive opinion regarding children and chewing gum that is sweetened with 100% xylitol as part of Article 14 of the European Commission Regulation (EC 2009) 1924/2006 regarding health claims. In particular, the proposed wording for such a claim was published as follows: "Chewing gum sweetened with 100% xylitol has been shown to reduce dental plaque. High content/level of dental plaque is a risk factor in the development of caries in children."

Furthermore, the EFSA has published an opinion whereby a cause and effect relationship was established between the consumption of sugar free chewing gum and reduction of tooth demineralization and a reduction in the incidence of caries. Two to three grams of the "sugar-free" chewing gum, consisting primarily of xylitol, should be consumed for 20 minutes at least three times per day after meals in order to have the claimed effect. From the pooled results of the scientific evidence, it was made known that the xylitol containing sugar free gum was more effective (58% prevention factor; PF), followed by the xylitol/sorbitol containing sugar free chewing gum (52% PF), followed by the sorbitol containing sugar free chewing gum (20% PF) (EFSA 2010).

Evolving Applications

New, beneficial and potentially existing applications of xylitol continue to be either discovered or suggested. These new applications include the following:

1. The use of xylitol as a bulk sweetener
2. The use of xylitol to mediate enhanced satiety
3. The use of xylitol to reduce the potential occurrence of acute otitis media

The effects of xylitol on gastric emptying and food intake when measured in 10 healthy male volunteers suggests that ingestion of a 25 g xylitol preload may be associated with an approximate 25% reduction of caloric intake during a subsequent test meal (Shafer et al. 1987). Other bulk sweeteners did not suppress caloric intake when tested at the same 25 g level. The study suggests that xylitol could be applied at efficacious levels in meal replacement or diet-supporting products to facilitate weight reduction programs.

Xylitol was also applied in yoghurts either alone (Xyl) or in combination with polydextrose (XylPDX). The aim of the study was to evaluate the effects that the ingredients would render, either alone or in combination, to the hunger and energy intake over a period of 10 days to healthy male and female subjects. All the experimental yoghurts (Xyl, PDX and XylPDX) given as pre-loads (90 minutes before the test meal) induced a lower energy intake at the test meal compared with the control yoghurt, despite containing less energy. In addition, when the energy of the yoghurt pre-load was accounted for, the suppression induced by the experimental yoghurts was strong (range of 11%–17%) and highly statistically significant. Yoghurts containing xylitol and polydextrose were shown to induce strong satiating effects compared with a control yoghurt, showing their potential benefit as an ingredient in functional foods for appetite control.

This study has demonstrated that measures of appetite (subjective hunger and *ad libitum* test lunch) were sensitive enough to detect differences between the experimental yoghurts and the control yoghurt following consumption of a single 200 g pre-load (King et al. 2005).

Acute otitis media, an infection of the middle ear, is a common illness that affects a significant number of young children. During 1990, for example, an estimated 24.5 million visits were made to office-based physicians in the US at which the principal diagnosis was otitis media. A study of the effect of xylitol ingestion on acute otitis media with either a xylitol-sweetened chewing gum, syrup, or lozenge, suggests that each xylitol-sweetened material has a positive influence in the reduction of acute otitis media occurrence. Reductions of 40%, 30%, and 20% were reported for the chewing gum, syrup, and lozenge, respectively (Uhari et al. 1996, 1998). The observed benefit is believed to be associated with the ability of xylitol to reduce the growth of *Streptococcus pneumoniae* and thus minimize the attacks of acute otitis media caused by pneumococci (Kontiokari et al. 1995).

In addition, the functionality of xylitol with regards to preventing acute otitis media lies within its ability to inhibit the attachment of both pneumococci and *Haemophilus influenzae* on the nasopharyngeal cells. Attachment of bacteria on nasopharyngeal cells is an important part of the pathomechanism of acute otitis media.

It has been shown that children who consume xylitol either in the form of chewing gum or syrups will experience significantly less episodes of acute otitis media and will require a reduced level of antimicrobial intervention, if required at all. Overall, xylitol is an attractive alternative to prevent acute otitis media in children (Uhari et al. 2001; Vernacchio et al. 2006).

Use of Xylitol as a Sweetener in Diabetic Diets

Historically, the first proposed application of xylitol concerned its use as a sugar substitute for diabetic patients (Mellinghoff 1961). Diabetes mellitus is a chronic metabolic disorder characterized by fasting hyperglycemia and/or plasma glucose levels above defined limits during oral glucose tolerance testing. It is caused either by a total lack of insulin (type I, insulin-dependent diabetes mellitus) or by insulin resistance in the presence of normal or even elevated plasma insulin levels (i.e., by a decreased tissue sensitivity or responsiveness to insulin [type II, non-insulin dependent diabetes mellitus]).

The major goals of dietetic and drug-based management of diabetes mellitus are to achieve normal control of glucose metabolism and glycemia, and thereby to prevent macro- and microvascular complications. The recommended treatment modalities are dietary modification, increased physical activity, and pharmacological intervention with either an oral hypoglycemic agent or insulin.

Modification of the diet is the most important element in the therapeutic plan for diabetic patients, and for some patients with type II diabetes, it is the only intervention needed to control the metabolic abnormalities associated with the disease.

A specific goal of medical nutrition therapy for people with diabetes is the maintenance of as near normal blood glucose levels as possible. This includes balancing food intake with either endogenous or exogenous or oral glucose-lowering medications and physical activity levels. It is the current position of the American Diabetes Association (ADA) that first priority be given to the total amount of carbohydrates consumed rather than the source of the carbohydrate. According to the ADA, the calories and carbohydrate content of all nutritive sweeteners must be taken into account in a meal plan and that all have the potential to affect blood glucose levels. The ADA recognizes, however, that polyols produce a lower glycemic response than sucrose and other carbohydrates and have approximately 2 cal/g compared with 4 cal/g from other carbohydrates (American Diabetes Association 1999).

Even an "ideal" diet plan is worthless if patients do not adhere to it. To increase compliance, it seems therefore appropriate to use low glycemic or non-glycemic sweeteners for the preparation of special diabetic products. Traditionally, one may consider using nonnutritive, intense sweeteners for this purpose. These compounds are noncaloric and have no deleterious effect on diabetic control. Undoubtedly, they are most useful for the sweetening of beverages (soft drinks, coffee, tea). Incorporation of nonnutritive sweeteners into solid foods, however, causes a major technological problem. In normal products, sucrose represents a considerable portion of bulk, and its replacement by nonnutritive sweeteners in special products for diabetic patients requires the addition of bulking agents (i.e., typically fat and/or starch). Because diabetic patients are predisposed to macrovascular disease and are required to restrict their fat intake, products with increased fat content, although carbohydrate-modified, are not recommended. The addition of starch on the other hand, increases the glycemic index of the food and eliminates the glycemic and caloric advantage that one hoped to achieve by using an intense sweetener. To avoid these problems, bulk sugar substitutes like fructose, sorbitol, and xylitol may be used as an appropriate alternative. These sweeteners produce only a slight increase in blood glucose concentration and require only small amounts of insulin for their metabolism in both healthy and diabetic individuals.

The effects of xylitol on blood glucose and insulin levels and its general suitability for inclusion in foods for diabetic patients, have been examined in several acute and subchronic studies with healthy and diabetic volunteers. When xylitol is given orally, no increase in blood glucose levels is observed, even in diabetic patients (Mueller-Hess et al. 1975; Manenti and Della Casa 1965; Huttunen et al. 1975; Yamagata et al. 1965). Similarly, plasma insulin concentrations do not rise at all (Yamagata et al. 1965) or only moderately (Mueller-Hess et al. 1975; Berger et al. 1973) after oral xylitol application in normal and diabetic subjects. These observations indicate that the conversion of xylitol to glucose, which, in principle is possible, is apparently too slow to raise the blood glucose concentration to a significant extent.

In a study involving eight healthy nonobese men, it was observed that ingestion of xylitol caused significantly lower increases in plasma glucose and insulin concentrations compared with the ingestion of glucose. The glycemic index of xylitol was determined to be seven (Natah et al. 1997).

It has been suspected that the obvious advantage of xylitol in terms of blood glucose and insulin requirement may disappear when it is incorporated into a meal. Therefore, a study was conducted in which 30 g xylitol or sucrose was substituted for an equal amount of starch in a meal of a diabetic diet regimen. The results of this investigation demonstrate that the insulin requirement after sucrose is significantly higher than after starch or xylitol (Hassinger et al. 1981).

In an early subchronic study, the application of 45–60 g/day of xylitol had no adverse effects on the metabolic condition of 20 diabetic patients (Mellinghoff 1961). These results were confirmed in a subsequent study in which the effect of 30 g xylitol/day on the carbohydrate and lipid metabolism of 12 well-controlled diabetic patients (type II) was examined for a duration of two to six weeks (Yamagata et al. 1969). The urinary glucose excretion disappeared with the xylitol diet, at least in some of the participating patients. The good tolerance of xylitol was also established in a study with 18 diabetic children (type I) who received 30 g/day of xylitol for four weeks (Foerster et al. 1977).

Even at a dose of 70 g/day administered over a period of six weeks, xylitol was well tolerated by type I and type II diabetic individuals and by healthy nondiabetic controls. Contrary to expectations, however, no significant differences were noted in plasma glycosylated Hb, in fasting or postprandial glucose, and in total urinary glucose (Bonner et al. 1982). In conclusion, these investigations demonstrate that xylitol can safely be incorporated as a sweetening agent in the diabetic diet without any negative effects on their metabolic condition.

Use of Xylitol in Parenteral Nutrition

The aim of parenteral nutrition is to optimize fluid, energy, nitrogen, and electrolyte balance by the infusion of adequate solutions providing energy (e.g., glucose or lipids), amino acids, and mineral salts (Shanbhogue et al. 1987; Schmitz et al. 1985). However, a stable metabolic condition is often difficult to reestablish in severely injured patients because the post trauma metabolism may be seriously disturbed in various respects. In particular, the use of glucose, which is the body's normal energy source, is often impaired, especially in patients with shock, severe trauma, burns, sepsis, and diabetes.

Under the conditions of such a general disturbance of metabolism and of the hormone-regulated control mechanisms, insulin must be administered concomitantly with the glucose to ensure sufficient use of this energy source. However, this procedure requires careful regular monitoring of the blood glucose levels to avoid severe complications. In addition, the infusion of glucose in excess of 0.12–0.24 g/kg/hr may be generally questioned, because higher infusion rates lead neither to a further suppression of the endogenous gluconeogenesis at the expense of physiologically important proteins nor to an increase of the peripheral glucose use (Wolfe et al. 1980). Because with this recommended maximum rate of glucose infusion, the energy requirement of the injured or stressed organism is not completely covered, the infusion of fructose and xylitol, in addition to glucose, has been advocated (Georgieff et al. 1982, 1984, 1985; Doelp et al. 1986). In several studies, the positive influence of such infusion regimens on nitrogen balance and visceral protein synthesis has been demonstrated in stressed or post-trauma patients (Georgieff et al. 1985; Doelp et al. 1986), as well as in an animal model (Georgieff et al. 1984). However, these effects were not always reproducible (Behrendt et al. 1984; Semsroth and Steinbereithner 1985).

More recently, it has been proposed to supplement lipid emulsions with xylitol. In this combination, xylitol is preferred over glucose because it only marginally stimulates insulin secretion and, therefore, does not suppress lipolysis, as demonstrated by a positive adenosine triphosphate (ATP) balance. In this way, xylitol could improve the metabolic situation (Georgieff 1986; Dillier and Leutenegger 1988).

At present, xylitol is used mainly as a glucose substitute in parenteral amino acid solutions, as well as in combination with fructose and glucose in so-called trisugar solutions. These products are frequently applied, particularly in Germany. Solutions of xylitol alone are nowadays rarely used, except in Japan.

Most clinical investigations on the use of xylitol in parenteral solutions have concentrated on the general energetic and amino acid-sparing effects. Other potentially beneficial, metabolic effects have not yet been explored in detail. Because xylitol is metabolized by way of the pentose phosphate pathway, it is conceivable that the levels of phosphorylated nucleotides (ATP) might be enhanced. This could be relevant in reperfused myocardial tissue, as has been suggested by studies with ribose, which is subject to the same catabolic pathway (Dillier and Leutenegger 1988; Zimmer and Gerlach 1978, McAnulty et al. 1987; Zimmer et al. 1987). Other potential advantageous effects include an increase of the 2,3-diphosphoglycerate levels and an NADH-mediated reduction of oxidized glutathione in erythrocytes (Stigge et al. 1973; Vora 1987; Asakura et al. 1970; Pousset et al. 1981; Ukab et al. 1981). This latter effect may be particularly relevant in the tens of millions of people who are affected by different degrees of glucose-6-phosphate dehydrogenase deficiency, which as such, causes little concern, but which predisposes to radical-induced pathology under certain conditions (drug therapy, sepsis, shock). Whether a reported effective treatment of cardiac arrhythmias with xylitol is also related to a shift to a more reduced status is not known (Tanaka and Arimura 1979; Yoshimura et al. 1979).

Although a large number of studies indicate the absence of adverse side effects with parenteral xylitol administration, some authors advise against the use of xylitol because of the possibility of increased serum lactic acid, uric acid, and bilirubin concentrations (Froesch 1978; Korttila and Mattila 1979) and because of the possible deposition of calcium oxalate crystals in the kidney and in the brain (Conyers et al. 1985; Ludwig et al. 1984; Galaske et al. 1986; Rosenstiel 1986). Regarding the occurrence of renocerebral oxalosis in association with xylitol infusions, however, it is noteworthy that in most of these cases the recommended maximum daily dose and/or infusion rate was surpassed. If xylitol is infused at recommended rates (0.25 g/kg/hr or less and 3 g/kg/day or less) (Schroeder 1984) and in combination with glucose and fructose, such complications are not likely to occur more frequently than in patients not treated with xylitol (Pfeiffer et al. 1984; Mori and Beppu 1983; Bär and Loehlein 1987). In general, no adverse side effects are observed if these recommendations are observed. Contrary to fructose, sorbitol, and glucose, a metabolic intolerance to xylitol is not known.

Toxicity and Tolerance

The results of animal tests for acute toxicity have indicated that xylitol is of very low toxicity by all routes of administration. Conventional tests for embryotoxicity and teratogenicity and for adverse effects on reproduction have given entirely negative results. Similarly, both *in vitro* and *in vivo* tests for mutagenicity and clastogenicity have given uniformly negative results (WHO/FAO 1977).

Long-term studies in animals for safety evaluation have included 2-year treatment of rats, mice, and dogs. In these studies, xylitol was tested at a maximum dose level of 20% of the diet. Although the findings of these animal studies generally supported the safety of oral xylitol, some observations required further investigation. These observations were urinary tract calculi in mice and a slight increase in the incidence of adrenal medullary pheochromocytomas in male rats (WHO/FAO 1978).

The results of subsequent studies and additional data from experiments with other polyols and lactose demonstrated, however, that the adverse effects observed in mice and rats are generic in nature and lack significance for safety evaluation in humans because of the species specificity of the underlying mechanisms (Bär 1985, 1986, 1988b; LSRO 1986; JECFA 1997; ILSI 1995).

Human tolerance of high oral doses of xylitol has been investigated in numerous studies with healthy and diabetic volunteers. The results of these studies invariably demonstrate the good tolerance, even of extremely high intakes (up to 200 g/day), of xylitol. Adverse changes of clinical parameters were not observed. The only side effect that was occasionally noted was transient laxation and gastrointestinal discomfort (Mäkinen 1978; WHO/FAO 1977; Foerster et al. 1981; Mäkinen et al. 1981; Culbert et al. 1986; Akerblom et al. 1981). Such effects are generally observed after consumption of high doses of polyols and slowly digestible carbohydrates (e.g., lactose). The slow absorption of these compounds from the gut, and the resulting osmotic imbalance, are considered to be the cause of these effects, which are readily reversible on cessation or reduction of the amounts consumed. With continued exposure, tolerance usually develops (Foerster 1978; Mäkinen and Scheinin 1975).

Regulatory Status

On a supranational level, the Joint FAO/WHO Expert Committee on Food Additives (JECFA) has allocated an acceptable daily intake (ADI) "not specified," the most favorable ADI possible, for xylitol (WHO/FAO 1983) and the Scientific Committee for Food for the European Economic Community (EEC) proposed "acceptance" of this polyol in 1984. On a national level, xylitol is approved for foods, cosmetics, and pharmaceuticals in many countries. Claims such as "noncariogenic" or "safe for teeth" may be applied where permitted.

References

Akerblom, H.K., Koivukangas, T., Punka, R., and Mononen, M. 1981. The tolerance of increasing amounts of dietary xylitol in children. *Int J Vit Nutr Res* (Suppl. 22): 53–66.

Alaluusua, A., Nystroem, M.,Groenroos, L., and Peck L. 1989. Caries-related microbiological findings in a group of teenagers and their parents. *Caries Res* 23:49–54.

American Diabetes Association. 1999. Nutrition recommendations and principles for people with diabetes mellitus. *Diabetes Care* 22(SI):S42–S45.

Aminoff, C., Vanninen, E., and Doty, T.E. 1978. The occurrence, manufacture and properties of xylitol. In *Xylitol*, J.N. Counsell (Ed.), pp 1–9.. London: Applied Science Publishers Ltd.

Anonymous. 1987. De energetische waarde van suikeralcoholen. *Voeding* 48:357–365.

Arends, J., Christoffersen, J., Schuthof, J., and Smits, M.T. 1984. Influence of xylitol on demineralization of enamel. *Caries Res* 18:296–301.

Asakura, T., Adachi, K., and Yoshikawa, H. 1970. Reduction of oxidized glutathione by xylitol. *J Biochem* 67:731–733.

Askar, A., Abd El-Fadeel, M.G., Sadek, M.A., El Rakaybi, A.M.A., and Mostafa, G.A. 1987. Studies on the production of dietetic cake using sweeteners and sugar substitutes. *Dtsch Lebensm Rundschau* 83:389–394.

Assev, S., and Roella, G. 1986a. Further studies on the growth inhibition of *Streptococcus mutans* omz 176 by xylitol. *Acta Path Microbiol Immunol Scand Sect B* 94:97–102.

Assev, S., and Roella, G. 1986b. Sorbitol increases the growth inhibition of xylitol on *Strep. mutans* OMZ 176. *Acta Path Microbiol Immunol Scand Sect B* 94:231–237.

Assev, S., and Scheie, A.A. 1986. Xylitol metabolism in xylitol-sensitive and xylitol-resistant strains of *Streptococci. Acta Path Microbiol Immunol Scand Sect B* 94:239–243.

Assev, S., Waler, S.M., and Roella, G. 1983. Further studies on the growth inhibition of some oral bacteria by xylitol. *Acta Path Microbiol Immunol Scand Sect B* 91:261–265.

Assev, S., Vegarud, G., and Roella, G. 1985. Addition of xylitol to the growth medium of *Streptococcus mutans* omz 176—effect on the synthesis of extractable glycerol-phosphate polymers. *Acta Path Microbiol Immunol Scand Sect B* 93:145–149.

Bär, A. 1985. Safety assessment of polyol sweeteners—some aspects of toxicology. *Food Chem* 16:231–241.

Bär, A. 1986. Toxicological aspects of sugar alcohols—studies with xylitol. In *Low Digestibility Carbohydrates, Proceedings of a Workshop Held at the TNO-CIVO Institutes Zeist*, November 27–28, The Netherlands, pp. 42–50.

Bär, A. 1988a. Sugars and adrenomedullary proliferative lesions: The effects of lactose and various polyalcohols. *J Am Coll Toxicol* 7:71–81.

Bär, A. 1988b. Caries prevention with xylitol, a review of the scientific evidence. *Wld Rev Nutr Diet* 55:183–209.

Bär, A. 1989. Significance and promotion of sugar substitution for the prevention of dental caries. *Lebensm-Wiss u-Technol* 22:46–53.

Bär, A. 1990. Factorial calculation model for the estimation of the physiological caloric value of polyols. In *Proc. Int. Symposium on Caloric Evaluation of Carbohydrates*, N. Hosoya (Ed.), 209–257. The Japan Assoc. Dietetic & Enriched Foods, Tokyo.

Bär, A., and Loehlein, D. 1987. Limited role of endogenous glycolate as oxalate precursor in xylitol-infused patients. *Contr Nephrol* 58:160–163.

Baessler, K.H. 1978. Absorption, metabolism, and tolerance of polyol sugar substitutes. *Pharmacol Ther Dent* 3:85–93.

Banoczy, J., Orsos, M., Pienihaekkinen, K., and Scheinin, A. 1985. Collaborative WHO xylitol field studies in Hungary. IV. Saliva levels of *Streptococcus mutans*. *Acta Odontol Scand* 43:367–370.

Baquer, N.Z., Hothersall, J.S., and McLean, P. 1988. Function and regulation of the pentose phosphate pathway in brain. *Curr Top Cell Regul* 29:265–289.

Beckers, H.J.A. 1988. Influence of xylitol on growth, establishment, and cariogenicity of *Streptococcus mutans* in dental plaque of rats. *Caries Res* 22:166–173.

Behrendt, W., Minale, C., and Giani, G. 1984. Parenterale ernährung nach herzchirurgischen operationen. *Infusionstherapie* 11:316–322.

Berger, W., Goeschke, H., Moppert, J., and Kuenzli, H. 1973. Insulin concentrations in portal venous and peripheral venous blood in man following administration of glucose, galactose, xylitol and tolbutamide. *Horm Metab Res* 5:4–8.

Bibby, B.G., and Fu, J. 1985. Changes in plaque pH in vitro by sweeteners. *J Dent Res* 64:1130–1133.

Birkhed, D. 1994. Cariogenic aspects of xylitol and its use in chewing gums: A review. *Acta Odontal Scand* 25:116–127.

Bloch, D.W., El Egakey, M.A., and Speiser, P.P. 1982. Verbesserung der Auflösungsgeschwin-digkeit von Hydrochlorothiazid durch feste Dispersionen mit Xylitol. *Acta Pharm Technol* 18:177–183.

Bonner, R.A., Laine, D.C., Hoogwerf, B.J., and Goetz, F.C. 1982. Effects of xylitol, fructose and sucrose in types I and II diabetics: A controlled, cross-over diet study on a clinial research center. *Diabetes* 31(Suppl. 2):80A (Abstr.).

Bray, F. 1985. Zuccheri alternativi e loro influenza nel gelato. *Ind Aliment* 24:895–898.

Buckley, B.M., and Williamson, D.H. 1977. Origin of blood acetate in the rat. *Biochem J* 166:539–545.

Carson, J.F., Waisbrot, S.W., and Jones, F.T. 1943. A new form of crystalline xylitol. *J Am Chem Soc* 65:1777–1778.

Chen, L.F., and Gong, C.S. 1985. Fermentation of sugarcane bagasse hemicellulose hydrolysate to xylitol by a hydrolysate-acclimatized yeast. *J Food Sci* 50:226–228.

Cobanera, A., Mopasso, E., White, P., and Cuevas Espinosa, M. 1987. Xylitol-sodium fluoride: Effect on plaque. *J Dent Res* 66:Abstr. 56.

Conyers, R.A.J., Rofe, A.M., Bais, R., James, H.M., Edwards, J.B., Thomas, D.W., and Edwards, R.G. 1985. The metabolic production of oxalate from xylitol. *Int Z Vitam Ernährungsforsch* (Beih.) 28:9–28.

Counsell, J.N. 1978. In *Xylitol*, J.N. Counsell (Ed.), p. 191. London: Applied Science Publishers Ltd.

Crawford, P.J.M. 1981. Sweetened medicine and caries. *Pharm J* 226:668–669.

Culbert, S.M., Wang, Y.M., Fritsche, Jr., H.A., Carr, D., Lantin, E., and van Eys, J. 1986. Oral xylitol in American adults. *Nutr Res* 6:913–922.

Cummings, J.H., Englyst, H.N., and Wiggins, H.S. 1986. The role of carbohydrates in lower gut function. *Nutr Rev* 44:50–54.

Cummings, J.H., Pomare, E.W., Branch, W.J., Naylor, C.P.E., and MacFarlane, G.T. 1987. Short chain fatty acids in human large intestine, portal, hepatic and venous blood. *Gut* 28:1221–1227.

Cummings, J.H. 1981. Short chain fatty acids in the human colon. *Gut* 22:763–779.

Demetrakopoulos, G.E., and Amos, H. 1978. Xylose and xylitol. *World Rev Nutr Diet* 32:96–122.

Dillier, C., and Leutenegger, A. 1988. Vereinfachung der parenteralen Ernährungstherapie durch eine neue, gebrauchsfertige all-in-one-Lösung, presented at the Annual Meeting of the Schweizerische Gesellschaft für Ernährungsforschung, March 5.

Dodson, A.G., and Pepper, T. 1985. Confectionery technology and the pros and cons of using non-sucrose sweeteners. *Fd Chem* 16:271–280.

Doelp, R., Gruenert, A., Schmitz, E., and Ahnefeld, F.W. 1986. Klinische Untersuchungen zur peripher-venösen parenteralen Ernährung. *Beitr Infusionstherapie klin Ernähr* 16:64–76

Edgar, W. 1998. Sugar substitutes, chewing gum and dental caries—a review. *Br Dent J* 184(1):29–31.

EFSA Journal. 2010. 8(10):1775, p. 13.

European Commission (EC). 2009. Regulation (EC) 1024/2009 of 29 October 2009 on the authorisation and refusal of authorisation of certain health claims made on food and referring to the reduction of disease and to children's development and health.

European Council (EC). 1990. Council Directive No. 90/496/EEC of 24 September 1990 on nutrition labelling for foodstuffs. *Off J European Communities* 33 (L276).

Featherstone, J.D.B., Cutress, T.W., Rodgers, B.E., and Dennison, P.J. 1982. Remineralization of artificial caries-like lesions in vivo by a self-administered mouthrinse or paste. *Caries Res* 16:235–242.

Federation of American Societies for Experimental Biology (FASEB). 1994. The Evaluation of the Energy of Certain Sugar Alcohols Used as Food Ingredients. Unpublished.

Feigal, R.J., Jensen, M.E., and Mensing, C.A. 1981. Dental caries potential of liquid medications. *Pediatrics* 68:416–419.

Firestone, A.R., and Navia, J.M. 1986. In vivo measurements of sulcal plaque pH in rats after topical applications of xylitol, sorbitol, glucose, sucrose and sucrose plus 53 mM sodium fluoride. *J Dent Res* 65:44–48.

Fischer, E., and Stahel, R., 1891. Zur Kenntnis der Xylose. *Berichte Dtsch Chem Gesellschaft* 24:528–539.

Flemming, S.E., and Arce, D.S. 1986. Volatile fatty acids: Their production, absorption, utilization, and roles in human health. *Clin Gastroenterol* 15:787–814.

Foerster, H., and Mehnert, H. 1979. Die orale Anwendung von xylit als zucker-austauschstoff in der diät des diabetes mellitus. *Akt Ernährungsmedizin* 4:296–314.

Foerster, H., Boecker, S., and Walther, A. 1977. Verwendung von Xylit als Zuckeraustauschstoff bei diabetischen Kindern. *Fortschr Med* 95:99–102.

Foerster, H., Quadbeck, R., and Gottstein, U. 1981. Metabolic tolerance to high doses of oral xylitol in human volunteers not previously adapted to xylitol. *Int J Vit Nutr Res* (Suppl. 20):67–88.

Foerster, H. 1978. Tolerance in the human adults and children. In *Xylitol*, J.N. Counsell (Ed.), pp. 43–66. London: Applied Science Publishers, Ltd.

Forbord, B., and Osmundsen, H. 1992. On the mechanism of xylitol-dependent inhibition of glycolysis in Streptococcus sobrinus OMZ 126. *Inst J Biochem* 24:509–514.

Froesch, E.R. 1978. Parenterale ernährung: Glucose oder glucoseersatzstoffe? *Schweiz Med Wochenschr* 108:813–815.

Galaske, R.G., Burdelski, M., and Brodehl, J. 1986. Primär polyurisches nierenversagen und akute gelbe leberdystrophie nach infusion von zuckeraustauschstoffen im kindesalter. *Dtsch med Wschr* 111:978–983.

Garr, J.S.M., and Rubinstein, M.H. 1990. Direct compression characteristics of Xylitol. *Inst Pharm* 64:223–226.

Gehring, F. 1977. Mikrobiologische untersuchungen im rahmen der "Turku Sugar Studies." *Dtsch. Zahnärztl Z* 32:84–88.

Georgieff, M. 1986. Zur stoffwechselwirkung von intravenös verabreichter glukose, xylit oder glyzerin während unterschiedlicher dosierung und kombination. *Beitr Infusionstherapie klin Ernähr* 16:103–119.

Georgieff, M., Ackermann, R.H., Baessler, K.H., and Lutz, H. 1982. Die vorteile von xylit gegenüber glucose als energiedonator in rahmen der frühen postoperativen, parenterajen nährstoffzufuhr. *Z Ernährungswiss* 21:27–42.

Georgieff, M., Moldawer, L.L., Bistrian, B.R., and Blackburn, G.L. 1984. Mechanisms for the protein-sparing action of xylitol during partial parenteral feeding after trauma. *Surg Forum* 35:105–108.

Georgieff, M., Moldawer, L.L., Bistrian, B.R., and Blackburn, G.L. 1985. Xylitol, an energy source of intravenous nutrition after trauma. *J Enteral Parenteral Nutr* 9:199–209.

Gong, C.S., Claypool, T.A., McCracken, L.D., Maun, C.M., Ueng, P.P., and Tsao, G.T. 1983. Conversion of pentoses by yeasts. *Biotechnol Bioeng* 25:85–102.

Gordon, J., Cronin, M., and Reardon, R. 1990. Ability of a xylitol chewing gum to reduce plaque accumulation. *J Dent Res* 69:136.

Graber, T.M., Muller, T.P., and Bhatia, V.D. 1982. The effect of xylitol gum and rinses on plaque acidogenesis in patients with fixed orthodontic appliances. *Swed Dent J* 15:41–55.

Grimble G. 1989. Fibre, fermentation, flora, and flatus. *Gut* 30:6–13.

Guelzow, H.J. 1982. Ueber den anaeroben umsatz von palatinit durch mikro-organismen der menschlichen mundhöhle. *Dtsch zalmärztl Z* 37:669–672.

Gustaffson, B., Quensel, C.E., Lanke, L., Lundqvist, C., Grahnen, W., Gonow, B.E., and Krasse, B. 1953. The Vipeholm dental caries study: The effect of different levels of carbohydrate intake on caries activity in 436 individuals observed for 5 years. *Acta Odont Scand* 11:232–363.

Gutschmidt, J., and Ordynsky, G. 1961. Bestimmung des Süssungsgrades von Xylit. *Dtsch Lebensm Rdsch* 57:321–324.

Hamada, S., and Slade, H.D. 1980. Biology, immunology and cariogenicity of Streptococcus mutans. *Microbiol Rev* 44:331–384.

Harper, D.S., and Loesche, W.J. 1984. Growth and acid tolerance of human dental plaque bacteria. *Arch Oral Biol* 29:843–838.

Hassinger, W., Sauer, G., Cordes, U., Beyer, J., and Baessler, K.H. 1981. The effects of equal caloric amounts of xylitol, sucrose and starch on insulin requirements and blood glucose levels in insulin-dependent diabetics. *Diabetologia* 2:37–40.

Havenaar, R., Huis in't Veld, J.H.J., de Stoppelaar, J.D., and Backer Dirks, O. 1984. Anti-cariogenic and remineralizing properties of xylitol in combination with sucrose in rats inoculated with Streptoccus mutans. *Caries Res* 18:269–277.

Havenaar, R. 1984a. The anti-cariogenic potential of xylitol in comparison with sodium fluoride in rat caries experiments. *J Dent Res* 63:120–123.

Havenaar, R. 1984b. Effects of intermittent feeding of sugar and sugar substitutes on experimental caries and the colonization of Streptococcus mutans in rats. In *Sugar Substitutes and Dental Caries*, R. Havenaar (Ed.), pp. 73–79. Utrecht Drukkerij Elinkwijk BV, Basel.

Hefti, A. 1980. Cariogenicity of topically applied sugar substitutes in rats under restricted feeding conditions. *Caries Res* 14:136–140,

Holland, D., and Stoddart, J.F. 1982. A stereoselective synthesis of xylitol. *Tetrahed Lett* 23:5367–5370.

Huttunen, J.K., Mäkinen, K.K., and Scheinin, A. 1975. Turku Sugar Studies XI. Effects of sucrose, fructose, and xylitol diets on glucose, lipid and urate metabolism. *Acta Odontol Scand* 33(Suppl 70):239–245.

Huxley, H.G. 1977. The effect of feeding frequency on rat caries. *J Dent Res* 56:976.

Hyvoenen, L., and Koivistoinen, P. 1982. Food technological evaluation of xylitol. *Adv Food Res* 28:373–403.

Hyvoenen, L., Kurkela, R., Koivistoinen, P., and Merimaa, P. 1977. Effects of temperature and concentration on the relative sweetness of fructose, glucose and xylitol. *Lebensm Wiss Technol* 10:316–320.

International Life Sciences Institute (ILSI). 1995. Low digestible carbohydrates (polyols and lactose): Significance of adrenal medullary proliferative lesions in the rat. Report for the Project Committee on Polyols, ILSI North America.

Imfeld, T., and Muehlemann, H.R. 1978. Cariogenicity and acido-genicity of food, confectionery and beverages. *Pharmacol Ther Dent* 3:53–68.

Imfeld, T. 1984. Zahnschonende halswehlutschtabletten. *Swiss Dent* 5:19–22.

Imfeld, T.N. 1983. Identification of low caries risk dietary components, *Monogr Oral Sci* 11:1983.

Isokangas, P., Mäkinen, K., Tieko, J., and Alanen, P. 1993. Long-term effect of xylitol chewing gum in the prevention of dental caries: A follow-up five years after termination of a prevention program. *Caries Res* 27:495–498.

Isokangas, P. 1987. Xylitol chewing gum in caries prevention. A longitudinal study on Finnish school children. *Proc Fin Dent Soc* 83:1–117.

Jaffe, G.M. 1978. Xylitol—a specialty sweetener. *Sugar y Azucar* 73:36–42.

Joint FAO/WHO Expert Committee on Food Additives (JECFA). 1997. Toxicological significance of pro-liferative lesions of the adrenal medulla in rats fed polyols and other poorly digestible carbohydrates. In *Evaluation of Certain Food Additives and Contaminants*, pp. 8–12. World Health Organization Technical Report Series 868.

Kandelman, D., and Gagnon, G. 1985. Effect on dental caries of xylitol chewing gum; two year results. *J Dent Res* 67:172 (Abstr. 472).

Kandelman, D., and Gagnon, G. 1987. Clinical results after 12 months from a study of the incidence and progression of dental caries in relation to consumption of chewing-gum containing xylitol in school preventive programs. *J Dent Res* 66:1407–1411.

Kandelman, D., Bär, A., and Hefti, A. 1988. Collaborative WHO xylitol field study in French Polynesia. I. Baseline prevalence and 32 months caries increment. *Caries Res* 22:55–62.

Karimzadegan, E., Clifford, A.J., and Hill, F.W. 1979. A rat bioassay for measuring the comparative avail-ability of carbohydrates and its application to legume foods, pure carbohydrates and polyols. *J Nutr* 109:2247–2259.

Karle, E.J., and Gehring, F. 1987. Zur Kariogenität von Mischungen aus Kohlenhydraten und Xylit im kon-ventionellen und gnotobiotischen Tierversuch. *Dtsch Zahnärztl Z* 42:835–840.

King, N.A., Craig, S.A., Pepper, T., and Blundell, J.E. 2005. Evaluation of the independent and combined effects of Xylitol and Polydextrose consumed as a snack on hunger and energy intake over 10d. *Brit J Nutr* 93:911–915.

Kiss, J., D'Souza, R., and Taschner, P. 1975. Präparative Herstellung von 5-Desoxy-L-arabinose, Xylit und d-Ribose aus "Diacetonglucose." *Helv Chim Acta* 58:311–317.

Kitpreechavanich, V., Nishio, N., Hayashi, M., Nagai, S. 1985. Regeneration and retention of NADP (H) for xylitol production in an ionized membrane reactor. *Biotechnol Lett* 7:657–662.

Kleinberg, I. 1995. Oral effects of sugars and sweeteners. *Int Dent J* 35:180–189.

Kontiokari, T., Uhari, M., and Koskela, M. 1995. Effect of xylitol on growth of nasoharyngal bacteria in vitro. *Antimicrob Agents Chemother* 39:1820–1823.

Korttila, K., and Mattila, M.A.K. 1979. Increased serum concentrations of lactic, pyruvic and uric acid and bilirubin after postoperative xylitol infusion. *Acta Anesthesiol Scand* 23:273–277.

Kristoffersson, E., and Halme, S. 1978. Xylitol as an excipient in oral lozenges. *Acta Pharm Fenn* 87:61–73.

Krueger, C. 1987. Zuckerfreie Pralinen—zuckerfreie Füllungen und Schokoladen-massen. *Süsswaren* 11:506–516.

Krueger, C. 1988. Zuckeraustauschstoffe, Arten, technologische, sensorische und ernährungsphysiologische Eigenschaften und Synergie-Effekte. *Zucker-und Süsswaren Wirtschaft* 41(11):360–365.

Laakso, R., Sneck, K., and Kristoffersson, E. 1982. Xylitol and Avicel pH 102 as excipients in tablets made by compression and from granulate. *Acta Pharm Fenn* 91:47–54.

Lane, A.B. 1985. On the nature of l-xylulose reductase deficiency in essential pentosuria. *Biochem Genet* 23:61–72.

Leach, S.A., Agalamanyi, E.A., and Green, R.M. 1983. Remineralisation of the teeth by dietary means. In: *Demineralisation and Remineralisation of the Teeth*, S.A. Leach, and W.M. Edgar (Eds.), pp. 51–73. Oxford: IRL Press Ltd.

Leach, S.A., and Green, R.M. 1980. Effect of xylitol-supplemented diets on the progressional and regression of fissure caries in the albino rat. *Caries Res* 14:16–23.

Lee, C.K. 1977. Structural functions of taste in the sugar series: Taste properties of sugar alcohols and related compounds. *Fd Chem* 2:95–105.

Life Sciences Research Office (LSRO). 1986. Health aspects of sugar alcohols and lactose. Report prepared for the Bureau of Foods. Food and Drug Administration, Washington, DC under Contract No. FDA 223–2020 by the Life Sciences Research Office, Federation of American Societies for Experimental Biology. Bethesda, MD. P. 85.

Lindley, M.G., Birch, G.G., and Khan, R. 1976. Sweetness of sucrose and xylitol, structural considerations. *J Sci Food Agric* 27:140–144.

Linke, H.A.B. 1986. Sugar alcohols and dental health. *Wld Rev Nutr Diet* 47:134–162.

Loesche, W.J. 1984. The effect of sugar alcohols on plaque and saliva level of Streptococcus mutans. *Swed Dent J* 8:125–135.

Loesche, W.J. 1985. The rationale for caries prevention through the use of sugar substitutes. *Int Dent J* 35:1–8.

Ludwig, B., Schindler, E., Bohl, J., Pfeiffer, J., and Kremer, G. 1984. Reno-cerebral oxalosis induced by xylitol. *Neuroradiology* 26:517–521.

Maiwald, H.J., Banoczy, J., Tietze, W., Toth, Z., and Vegh, A. 1982. Die Beeinflussung des Plaque-pH durch zuckerhaltigen und zuckerfreien Kaugummi. *Zahn- Mund-Kieferheilk* 70:598–604.

Mäkinen, K.K. 1970. Biochemical principles of the use of xylitol in medicine and nutrition with special consideration of dental aspects. *Experientia* 30:1–160.

Mäkinen, K.K. 1988. Sweeteners and prevention of dental caries. *Oral Health* 78:57–66.

Mäkinen, K.K. 1989. Latest dental studies on xylitol and mechanisms of action of xylitol in caries limitation. In *Progress in Sweeteners*, T.H. Grenby (Ed.), pp. 331–362. New York: Elsevier.

Mäkinen, K.K. 1994. Sugar alcohols. In Functional Foods, Designer Foods, Pharmafeeds Nutraceuticals, I. Goldberg (Ed.), Chapter 11, pp. 219–241. New York: Chapman & Hall.

Mäkinen, K.K., Bennett, C.A., Hujoel, P.P., Isokangas P.J., Isotupa K., Pape, H.R., and Mäkinen, P.L. 1995b. Xylitol chewing gums and caries rate: As 40-month cohort study. *J Dent Res* 74:1904–1913.

Mäkinen, K.K., Hujoel, P.P., Bennett, C.A., Isotupa, K.P., Mäkinen, P.L., and Allen, P. 1996. Polyol chewing gums and caries rates in primary dentition: A 24-month cohort study. *Caries Res* 30:408–417.

Mäkinen, K.K., and Scheinin, A. 1975. Turku sugar studies VI. The administration of the trial and the control of the dietary regimen. *Acta Odontol Scand* 33:105–127.

Mäkinen, K.K., and Scheinin, A. 1982. Xylitol and dental caries. *Ann Rev Nutr* 2:133–150.

Mäkinen, K.K., and Söderling, E. 1980. A quantitative study of mannitol, sorbitol, xylitol, and xylose in wild berries and commercial fruits. *J Food Sci* 45:367–374.

Mäkinen, K.K., Söderling, E., Haemaelaeinen, M., and Antonen, P. 1985a. Effect of long-term use of xylitol on dental plaque. *Proc Finn Dent Soc* 81:28–35.

Mäkinen, K.K., Söderling, E., Hurttia, H., Lehtonen, O.P., and Luukkala, E. 1985b. Biochemical, microbiologic and clinical comparisons between two dentifrices that contain different mixtures of sugar alcohols. *J Am Dent Assoc* 111:745–751.

Mäkinen, K.K., Söderling, E., Isokangas, P., Tenovuo, J., and Tiekso, J. 1989. Oral biochemical status and depression of Streptococcus mutans in children during 24- to 36-month use of xylitol chewing gum. *Caries Res* 23:261–267.

Mäkinen, K.K., Mäkinen, P.L., Pape, H.R., Allen, P., Bennett, C.A., Isokangas, P.J., and Isotupa, K.P. 1995a. Stabilization of rampant caries: polyol gums and arrest of dentin caries in two long-term cohort studies in young subjects. *Int Dent J* 45:93–107.

Mäkinen, K.K., Söderling, E., and Laeikkoe, I. 1987. Zuckeralkohole (Polyole) als "aktive" Zahnpastenbestandteile. *Oralprophylaxe*. 9:115–120.

Mäkinen, K.K., Ylikahri, R., Mäkinen, P.L., Söderling, E, and Haemaelaeinen, M. 1981. Turku sugar studies XXIII. Comparison of metabolic tolerance in human volunteers to high oral doses of xylitol and sucrose after long-term regular consumption of xylitol. *Int J Vit Nutr Res* (Suppl. 22):29–51.

Maki, Y., Ohta, K., Takazoe, I., Matsukubo, I.Y., Takaesu, Y., Topitsoglou, V., Frostell, G., Manenti, F., and Della Casa, L. 1983. Acid production from isomaltulose, sucrose, sorbitol, and xylitol in suspensions of human dental plaque (short communication). *Carries Res* 17:335–339.

Manenti, F., and Della Casa, L. 1965. Effetti dello xylitolo sull'equilibrio glicidico del diabetico. *Boll Soc Medico-Chirurgica di Modena* 65:1–8.

McAnulty, J.F., Southard, J.H., and Belzer, F.O. 1987. Improved maintenance of adenosine triphosphate in five-day perfused kidneys with adenine and ribose. *Transplant Proc* 19:1376–1379.

McNeil, N.I. 1984. The contribution of the large intestine to energy supplies in man. *Am J Clin Nutr* 39:338–342.

Mellinghoff, C.H. 1961. Ueber die Verwendbarkeit des Xylit als Ersatzzucker bei Diabetikern. *Klin Wochenschr* 39:447.

Mori, S., and Beppu, T. 1983. Secondary renal oxalosis: A statistical analysis of its possible causes. *Acta Pathol Jpn* 33:661–669.

Morris, L., Moore, J., and Schwart, J. 1993. Characterisation and Performance of a New Direct Compression Excipient for Chewable Tablets: Xylitab, American Association of Pharmaceutical Scientist Annual Meeting.

Moskowitz, H.R. 1971. The sweetness and pleasantness of sugars. *Am J Psychol* 84:387–405.

Muehlemann, H.R., and Firestone, A. 1982. Drei neue "zahnscshonende" Präparate. *Swiss Dent* 3:25–30.

Mueller-Hess, R., Geser, C.A., Bonjour, J.P., Jequier, E., and Felbe, J.P. 1975. Effects of oral xylitol administration on carbohydrate and lipid metabolism in normal subjects. *Infusionstherapie* 2:247–252.

Munton, S.L., and Birch, G.G. 1985. Accession of sweet stimuli to receptors. I. Absolute dominance of one
molecular species in binary mixtures. *J Theor Biol* 112:539–551.

Nakai, Y., Shinga-Ishihara, C., Kaji, M., Moriya, K., Murakami-Yamanaka, K., and Takimura, M. 2009.
Xylitol Gum and Maternal Transmission of Mutans Streptococci. *J Dent Res*. Nov 30.

Natah, S., Hussein, K., Tuominen, J., and Koivisto, V. 1997. Metabolic response to lactitol and xylitol in
healthy men. *Am J Clin Nutr* 65:947–950.

Newbrun, E. 1982. Sugar and dental caries: A review of human studies. *Science* 217:418–423.

Onishi, H., and Suzuki, T. 1969. Microbial production of xylitol from glucose. *Appl Microbiol* 15:1031–1035.

Overmyer, C.A., Söderling, E., Isokangas, P., Tenovuo, J., and Mäkinen, K.K. 1987. Oral biochemical status
of children given xylitol gums. *J Dent Res* 66 (Spec. Issue): Abstr. 1551.

Pepper, T., and Olinger, P.M. 1988. Xylitol in sugar-free confections. *Food Techn* 42:98–106.

Pepper, T. 1987. Sugar substitutes, their use in chocolate and chocolate fillings. *Manuf Conf* 67(6):83, 1987.

Perez Trindade Fraga, C., Pinto Alvez Mayer, M., and Delgado Rodrigues, C.R.M. 2010. Use of chewing
gum containing 15% of xylitol and reduction in mutans streptococci levels. *Braz Oral Res* April–Jun
24(2):142–146.

Petersson, L.G., Johansson, M., Joensson, G., Birkhed, D., and Gleerup, A. 1989. Caries preventive effect of
toothpastes containing different concentration and mixture of fluorides and sugar alcohols. *Caries Res*
23:109.

Pfeiffer, J., Danner, E., and Schmidt, P.F. 1984. Oxalate-induced encephalitic reactions to polyolcontaining
infusions during intensive care. *Clin Neuropathol* 3:76–87.

Pihanto-Leppala, A., Soderling, E., and Mäkinen, K. 1990. Expulsion mechanism of xylitol-5-phosphate in
Streptococcus mutans. *Scand J Dent Res* 98:112–119.

Pousset, J.L., Bourn, B., and Cavé, A. 1981. Action antihémolytique du xylitol isolé des écorces de carica
papaya. *J Med Plant Res* 41:40–47.

Rekola, M. 1986a. A planimetric evaluation of approximal caries progression during one year of consuming
sucrose and xylitol chewing gums. *Proc Finu Dent Soc* 82:213–218.

Rekola, M. 1986b. Changes in buccal white spots during 2-year consumption of dietary sucrose or xylitol.
Acta Odontol Scand 44:285–290.

Rekola, M. 1987. Approximal caries development during 2-year total substitution of dietary sucrose with
xylitol. *Caries Res* 21:87–94.

Rekola, M. 1988. Acid production from xylitol products in vivo and in vitro. *Proc Finn Dent Soc* 84:39–44.

Roberts, I.F., and Roberts, G.J. 1979. Relation between medicines sweetened with sucrose and dental disease.
BMJ 2:14–16.

Rosensdel, K. 1986. Nierenversagen nach Infusion von Zuckeraustauschstoffen. *Dtsch med Wschr*
111:1340–1341.

Ruppin, H., Bar-Meir, S., Soergel, K.H., Wood, C.M., and Schmitt, Jr., M.G. 1980. Absorption of short-
chain fatty acids by the colon. *Gastroenterol* 78:1500–1507.

Salyers, A.A., and Leedle, J.A.Z. 1983. Carbohydrate metabolism in the human colon. In *Human Intestinal
Flora in Health and Disease*, D.J. Hentges (Ed.), pp. 129–146. New York: Academic Press.

Scheie, A.A., Arneberg, P., Orstavik, D., and Afserh, J. 1984. Microbial composition, pH depressing capacity
and acidogenicity of 3-week smooth surface plaque developed on sucrose regulated diets in man. *Caries
Res* 18:74–86.

Scheie, A.A., Fejerskov, O., Assev, S., and Roella, G. 1989. Ultrastructural changes in *Streptococcus sobrinus*
induced by xylitol, NaF and $ZnCl_2$. *Caries Res* 23:320–327.

Scheinin, A., Banoczy, J., Szoeke, J., Esztari, I., Pienihaekkinen, K., Scheinin, U., Tiekso, J., Zimmermann,
P., and Hadas, E. 1985. Collaborative WHO xylitol field studies in Hungary. I. Three-year caries activity in institutionalized children. *Acta Odontol Scand* 43:327–347.

Scheinin, A., Mäkinen, K.K., and Ylitalo, K. 1975a. Turku sugar studies V. Final report on the effect of sucrose,
fructose and xylitol diets on the caries incidence in man. *Acta Odontol Scand* 33(Suppl. 70):67–104.

Scheinin, A., Mäkinen, K.K., Tammisalo, E., and Rekola, M. 1975b. Turku sugar studies XVIII. Incidence of dental caries in relation to 1-year consumption of xylitol chewing gum. *Acta Odontol Scand* 33(Suppl. 70):307–316.

Schmitz, J.E., Doelp, R., Altemeyer, K.H., Gruenert, A., and Ahnefeld, F.W. 1985. Parenterale Ernährung: Stoffwechsel und Substrate. *Arzneimit-teltherapie* 3:162–172.

Schroeder, R. 1984. Beachting der Dosierungsgrenzen für Xylit bei der parenteralen Ernährung. *Dtsch med Wschr* 109:1047–1048.

Semsroth, M., and Steinbereithner, K. 1985. Stickstoffmetabolismus und renale Aminosäure-nausscheidung während totaler parenteraler Ernährung hypermetaboler Patienten mit verschiedenen Kohlenhydratregimen. *Infusionstherapie* 12:136–148.

Sgan-Cohen, H.D., Newbrun, E., Huber, R., and Sela, M.N. 1987. The effect of low-carbohydrate diet on plaque pH response to various foods. *J Dent Res* 66:911(Abstr. 22).

Shafer, R., Levine, A., Marlette, J., and Morley, J. 1987. Effects of xylitol on gastric emptying and food intake. *Am J Clin Nutr* 45:744–747.

Shanbhogue, L.K.R., Chwals, W.J., Weintraub, M., Blackburn, G.L., and Bistrian, B.R. 1987. Parenteral nutrition in the surgical patient. *Br J Surg* 74:172–180.

Shaw, J.H. 1987. Medical progress—causes and control of dental caries. *N Engl J Med* 317:996–1004.

Siebert, G., Thim, P., Brenner, H.P., and Ziesenitz, S.C. 1988. Antiacidogenic effects of non-cariogenic bulk sweeteners on fermentation of sucrose by *S. mutans* NCTC 10449. *J Dent Res* 67:696(Abstr. 110).

Sintes, J.L., Escalante, C., Stewart, B., McCool, J.J., Garcia, L., Volpe, A.R., and Triol, C. 1995. Enhanced anticaries efficacy of a 0.243% sodium fluoride/10% xylitol/silica dentifrice: 3-year clinical results. *Am J Dent* 8(5):231–235.

Sirenius, I., Krogerus, V.E., and Leppaenen, T. 1979. Dissolution rate of *p*-aminobenzoates from solid xylitol dispersions. *J Pharm Sci* 68:791–792.

Skutches, C.L., Holroyde, C.P., Myers, R.N., Paul, P., and Reichard, G.A. 1979. Plasma acetate turnover and oxidation. *J Clin Invest* 64:708–713.

Smits, M.T., and Arends, J. 1988. Influence of extraoral xylitol and sucrose dippings on enamel demineralization in vivo. *Caries Res* 22:160–165.

Söderling, E., Alaraeisaenen, L., Scheinin, A., and Mäkinen, K.K. 1987. Effect of xylitol and sorbitol on polysaccharide production by and adhesive properties of Streptococcus mutans. *Caries Res* 21:109–116.

Söderling, E.M., Ekman, T.C., and Taipale, T.J. 2008. Growth inhibition of Streptococcus mutans with low xylitol concentrations. *Curr Microbiol* 56(4):382–385. Epub 5 Jan 2008.

Söderling, E., Isokangas, P., Pienihakkinen, K., and Tenovuoa, J. 2000. Influence of maternal Xylitol consumption on acquisition of mutans streptococci by infants. *J Dnt Res* 79(3):882–887.

Söderling, E., Isokangas, P., Tenovuoa, J., Mustakallio, S., and Mäkinen, K. 1991. Long-term xylitol consumption and mutans streptococci in plaque and saliva. *Caries Res* 25:153–157.

Söderling, E., Isokangas, P., Tenovuoa, J., Mäkinen, K.K., Mustakallio, S., and Maennistoe, H. 1989. Habitual xylitol consumption and Streptococcus mutans in plaque and saliva. *J Dent Res* 68(Spec. Issue): Abstr. 953.

Söderling, E., Mäkinen, K.K., Chen, C.Y., Pape, H.R., and Mäkinen, P.L. 1988. Effect of sorbitol, xylitol or xylitol/sorbitol gum on plaque. *J Dent Res* 67:279(Abstr. 1334).

Soparkar, P., DePaola, P., Vrasanos, S., and Mandel, I. 1989. The effect of xylitol chewing gums on plaque and gingivitis in humans. *J Dent Res* 68:Abstr. 833.

Souney, P.E., Cyr, D.A., Chang, J.R., and Kaul, A.E. 1983. Sugar content of selected pharmaceuticals. *Diabetes Care* 6:231–240.

Soyama, K., and Furukawa, N. 1985. A Japanese case of pentosuria—case report. *J Inher Metab Dis* 8:37.

Splieth, D.H., Alkilzy, M., Schmitt, J., Berndt, C., and Welk, A. 2009. Effect of xylitol and sorbitol in plaque acidogenesis. *Quintessence International* (40)4, April.

Stecksen-Blicks, C. 1988. Lactobacilli and Streptococcus mutans in saliva, diet and caries increment in 8- and 13-year-old children. *Scand J Dent Res* 95:18–26.

Stigge, V., Strauss, D., Roigas, H., and Raderecht, H.J. 1973. Verbesserte Erhaltung der zellulären Konzentration von 2,3-Diphosphoglyzerat im Konservenblut durch Zusatz von Xylit. *Dtsch Ges Wesen* 28:1805–1810.

Svanberg, M., and Birkhed, D. 1995. Effects of dentrifices containing either xylitol and glycerin or sorbitol on Streptococcus mutans in saliva. *Caries Res* 25:74–79.

Tanaka, T., and Arimura, H. 1979. Validity of xylitol for management of cardiac arrhythmia during anesthesia. *J Uoeh* 1:365–368.

Touster, O. 1974. The metabolism of polyols. In *Sugars in Nutrition*, H.L. Sipple, and K.W. McNutt (Eds.), pp. 229–239. New York: Academic Press.

Trahan, L., and Mouton, C. 1987. Selection for Streptococcus mutans with an altered xylitol transport capacity in chronic xylitol consumers. *J Dent Res* 66:982–988.

Trahan, L., Bareil, M., Gauthier, L., and Vadeboncoeu, C. 1985. Transport and phosphorylation of xylitol by a fructose phosphotransferase system in Streptococcus mutans. *Caries Res* 19:53–63.

Trahan, L. 1995. Xylitol: A review of its action on mutans streptococci and dental plaque—the clinical significance. *Int Dent J* 45(Suppl. 1):77–92.

Tuompo, H., Meurman, J.H., Lounatmaa, K., and Linkola, J. 1983. Effect of xylitol and other carbon sources on the cell wall of Streptococcus mutans. *Scand J Dent Res* 91:17–25.

Uhari, M., Kontiokari, T., and Niemela, M. 1998. A novel use of xylitol in preventing acute otitis media. *Pediatrics* 102(4):879–884.

Uhari, M., Kontiokari, T., Koskela, M., and Niemela, M. 1996. Xylitol chewing gum in prevention of acute otitis media: Double blind randomized trial. *BMJ* 313:1180–1184.

Uhari, M., Tapiainen, T., and Kontiokari, T. 2001. Xylitol in preventing Acute Otitis Media. *Vaccine* 19:S144–S147.

Ukab, W.A., Sato, J., Wang, Y.M., and van Eys, J. 1981. Xylitol mediated amelioration of acetyl-phenylhydrazine-induced hemolysis in rabbits. *Metabolism* 30:1053–1059.

Vadeboncoeur, C., Trahan, L., Mouton, C., and Mayrand, D. 1983. Effect of xylitol on the growth and glycolysis of acidogenic oral bacteria. *J Dent Res* 62:882–884.

Van der Hoeven, J.S. 1986. Cariogenicity of lactitol in program-fed rats (short communication). *Caries Res* 20:441–443.

van Houte, J. 1980. Bacterial specificity in the etiology of dental caries. *Int Dent J* 30:305–326.

Varo, P., Westermarck-Rosendahl, C., Hyvoenen, L., and Koivistoinen, P. 1979. The baking behavior of different sugars and sugar alcohols as determined by high pressure liquid chromatography. Lebensm. *Wiss Technol* 12:153–156.

Vernacchio, L., Vezina, R.M., and Mitchell, A.A. 2006. Tolerability of oral xylitol solution in young children: Implications for otitis media prophylaxis. *Int J Pediatric Otorhinolaryngol*, Nov. 6.

Verran, J., and Drucker, D.B. 1982. Effects of two potential sucrose-substitute sweetening agents on deposition of an oral Streptococcus on glass in the presence of sucrose. *Archs Oral Biol* 27:693–695.

Vissink, A., s-Gravenmade, E.J., Gelhard, T.B.F.M., Panders, A.K., and Franken, M.H. 1985. Rehardening properties of mucin- or CMC-containing saliva substitutes on softened human enamel. *Caries Res* 19:212–218.

Voirol, F., and Etter, R. 1979. Verfahren zur Erzeugung einer Deckschicht auf Tablettenkernen. *German Patent Appl* 29 13 555.

Vora, S. 1987. Metabolic manipulation of key glycolytic enzymes: A novel proposal for the maintenance of red cell 2,3-DPG and ATP levels during storage. *Biomed Biochim Acta* 46:285–289.

Waaler, S.M., Assev, S., and Roella, G. 1985. Metabolism of xylitol in dental plaque. *Scand J Dent Res* 93:218–221.

Waher, S. 1992. Evidence for xylitol 5 P production in human dental plaque. *Scand J Dent Res* 100:204–206.

Washuett, J., Riederer, P., and Bancher, E. 1973. A qualitative and quantitative study of sugar-alcohols in several foods. *J Food Sci* 38:1262–1263.

WHO/FAO. 1977. Summary of toxicological data of certain food additives. Twenty-first Report of the Joint FAO/WHO Expert Committee on Food Additives. Geneva, WHO Technical Report Series No. 617, pp. 124–147.

WHO/FAO. 1978. Summary of toxicological data of certain food additives and contaminants, Twenty-second Report of the Joint FAO/WHO Expert Committee on Food Additives. Geneva, WHO Technical Report Series No. 631, pp. 28–34.

WHO/FAO. 1983. Evaluation of certain food additives and contaminants, Twenty-seventh Report of the Joint FAO/WHO Expert Committee on Food Additives, Geneva, WHO Technical Report Series No. 696, pp. 23–24.

Williams, J.F., Arora, K.K., and Longenecker, J.P. 1987. The pentose pathway: A random harvest. *Int J Biochem* 19:749–817.

Wolfe, R.R., O'Donnell Jr., T., Stone, M.D., Richmand, D.A., and Burke, J.F. 1980. Investigation of factors determining the optimal glucose infusion rate in total parenteral nutrition. *Metabolism* 29:892–900.

Wood, T. 1986. Physiological functions of the pentose phosphate pathway. *Cell Biochem Func* 4:241–247.

Yamagata, S., Goto, Y., Ohneda, A., Anzai, M., Kawashima, S., Kikuchi, J., Chiba, M., Maruhama, Y., Yamauchi, Y., and Toyot, T. 1969. Clinical application of xylitol in diabetics. In *Pentoses and Pentitols*, B.L. Horecker, K. Lang, and Y. Takagi (Eds.), pp. 316–325. Berlin: Springer-Verlag.

Yamagata, S., Goto, Y., Ohneda, A., Anzai, M., Kawashima, S., Shiba, M., Maruhama, Y., and Yamauchi, Y. 1965. Clinical effects of xylitol on carbohydrate and lipid metabolism in diabetes. *Lancet ii*:918–921.

Yamaguchi, S., Yoshikawa, T., Ikeda, S., and Ninomiya, T. 1970. Studies on the taste of some sweet substances. *Agric Biol Chem* 34:181–197.

Ylikahri, R. 1979. Metabolic and nutritional aspects of xylitol. *Adv Food Res* 25:159–180.

Yoshimura, N., Yamada, H., and Haraguchi, M. 1979. Anti-arrhythmic effect of xylitol during anesthesia. *Jap J Anesthesiol* 28:841–848.

Ziesenitz, S.C., and Siebert, G. 1986. Nonnutritive sweeteners as inhibitors of acid formation by oral microorganisms. *Caries Res* 20:498–502.

Zimmer, H.G., and Gerlach, E. 1978. Stimulation of myocardial adenine nucleotide biosynthesis by pentoses and pentitols. *Pflügers Arch* 376:223–227.

Zimmer, H.G., Zierhut, W., and Marschner, G. 1987. Combination of ribose with calcium antagonist and b-blocker treatment in closed-chest rats. *J Mol Cell Cardiol* 19:635–639.

CALORIC ALTERNATIVES III

Chapter 22

Crystalline Fructose

John S. White

Contents

Introduction

Crystalline fructose first became widely available for industrial food and pharmaceutical applications nearly twenty-five years ago. It is physically and functionally distinct from other carbohydrates in solubility, freezing point depression, boiling point elevation, water activity, osmotic pressure, Maillard browning and flavor development, flavor enhancement, starch synergy, and metabolism. Most important are its high relative sweetness, unique sweetness intensity profile, and synergy with other sweeteners.

Students of organic chemistry learn very early that sucrose is a disaccharide that can be readily hydrolyzed (inverted) to provide equimolar amounts of two monosaccharides, fructose and dextrose. Although it would seem simple to obtain both fructose and dextrose from the inversion of sucrose, it was only in the mid-1970s that sucrochemical advances enabled commercial quantities of pure fructose to become available in the United States.

It was certainly not the case that food scientists and members of the medical profession were unaware of the increased sweetening power and unique metabolic properties of pure fructose. It made little sense, however, to expend research resources to formulate foods and diets utilizing crystalline fructose without assurance that bulk industrial quantities (i.e., truckloads and railcars) would be available. The late 1980s and early 1990s saw the convergence of a plentiful raw material, proven refining technologies, economies of scale and reduced prices that assured the integration of crystalline fructose into mainstream food and beverage applications.

Manufacture of Crystalline Fructose

Producers have tested many processes and raw materials, seeking an economical manufacturing process for crystalline fructose. In the 1960s, Finnish, French, and German manufacturers, expanding on the pioneering work of the Frenchman Dubrufant (circa 1850), began producing industrial quantities of crystalline fructose from sucrose. Their processes involved the straightforward hydrolysis of sucrose with release of the simple sugars fructose and dextrose. The subsequent separation of the sugars by ion-exclusion chromatography and isolation of the fructose by carefully controlled crystallization, however, extended the total manufacturing time impracticably to more than a week.

The use of inulin, a polyfructose storage polymer isolated from tuberous plants such as dahlias, Jerusalem artichokes and the Hawaiian Ti plant, has also been examined as a raw material for fructose production. Through the controlled hydrolysis of inulin's $\beta(2{\to}1)$ fructosidic linkages, fructofuranose units are freed and converted to the more stable fructopyranose anomer. Despite repeated attempts, however, inulin processes have never been developed to produce crystalline fructose at competitive prices or in quantities required for ingredient use in the food industry.

As the worldwide demand for fructose grew, the need to increase capacity, find cheaper raw materials and develop less energy intensive processes became critical. In 1981, the world's first facility for making crystalline fructose from corn came on stream in Thomson, Illinois. American Xyrofin used liquid dextrose derived from cornstarch as the starting material. Finnish refining technology was used to produce a particularly high quality crystalline fructose with production time reduced to about five days.

The solubility of fructose in methanol or ethanol is very low, approximately 0.07 g per g alcohol, in comparison with water at 4 g per g H_2O. It is for this reason that alcohol became the solvent of choice for many early European crystallization processes. Removal and recovery of the alcohol, coupled with disposal of mixed waste streams, however, has caused alcohol crystallization to largely fall out of favor.

Finnish manufacturers were among the first to experiment with aqueous crystallization. Though aqueous crystallizations are the most successful processes in use today, the Finns and their successors have learned that these processes hold unique challenges beyond the high solubility of fructose in water. Unfavorable crystallization conditions result in the formation of fructose hemihydrate and fructose dihydrate crystalline forms, which are more hygroscopic (moisture absorbing), deliquesce in humid environments and melt at lower temperatures than pure fructose. Difructose dianhydrides also form under certain crystallization conditions, reducing yields and altering physical properties of the crystalline product.

Over time, the following general processing scheme has emerged for the successful manufacture of crystalline fructose (Hanover 1992):

1. Preparation of concentrated fructose feed; generally high fructose corn syrup (HFCS) with more than 90% fructose and more than 90% solids
2. Seeding with anhydrous crystalline fructose or undried crystals
3. Crystallization via batch or continuous process, and aqueous, alcohol or aqueous alcohol solvents
4. Centrifugation and washing of crystals to remove surface impurities
5. Drying via fluidized bed, tray, rotary, box or belt particulate dryer
6. Crystal conditioning
7. Screening
8. Packaging

The Finnish company, Suomen Sakeri Osakeyhtio, received one of the first US patents issued for an aqueous process in 1975 (Forsberg et al. 1975). Subsequent patents describe improved processes for fructose concentration under atmospheric or reduced-pressure conditions; seeding to initiate crystallization; cooling to allow for crystal growth; crystal washing and drying steps; and batch and continuous processes with computer-controlled crystallization cycles (Hanover 1992). Table 22.1 shows the composition of a typical lot of fructose produced to *United States Pharmacopoeia* and *Food Chemicals Codex* specifications.

Existence in Nature

Fructose, levulose and fruit sugar are synonyms for this sweetest of all naturally occurring sugars. Fructose is the chief constituent in honey, historically the most available form of naturally occurring fructose that "until the end of the Middle Ages … was everywhere the sweetener par excellence" (Tannahill 1973). The earliest recorded mentions of sweet-tasting substances reference honey thousands of years before its principal component was isolated and characterized as

Table 22.1 Typical Physical Properties

Appearance	White crystalline powder, forming anhydrous needle-shaped crystals
Empirical formula	$C_6H_{12}O_6$
Molecular weight	180.16
Melting point	102°C–105°C
Density	1.60 g/cm^3
Bulk density (loose)	0.8 g/cm^3
Caloric value	3.7 cal/g
Loss on drying (70°C for 4 hr in vacuum oven)	Less than 0.2%
Residue on ignition	Less than 0.5%
Heavy metals	Less than 5 ppm
Arsenic	Less than 1 ppm
Chloride	Less than 0.018%
Sulfate	Less than 0.025%
Calcium and magnesium	Less than 0.005%
Hydroxymethylfurfural	Less than 50 ppm
Glucose	Less than 0.1%
Assay (dry basis)	98.0%–102.0%
pH in aqueous solution (1 g/10 ml)	5.0–7.0

Note: State-of-the-art instrumentation is now available to permit rapid, precise measurements of fructose purity utilizing liquid chromatography. The reader is referred to the Sweetener Group of Tate & Lyle PLC of Decatur, IL, for this methodology.

fructose. Honey-sweetened foods were popular in Rome, where honey was referred to as "chief among all sweet things." Biblical references to honey are found throughout both the Old and New Testaments. While no single list of constituents can define the wide variety of honeys harvested worldwide, Table 22.2 gives representative values approximating those found in most commercial honeys.

Table 22.2 Constituents of Honey

Fructose	40%
Glucose	35%
Water	18%
Other saccharides	4%
Other substances	3%

Table 22.3   Fructose Contents of Fruits

Fruit	Percent Fructose in Fruit	Percent Fructose in Total Solids
Apple	6.04	37.8
Blackberry	2.15	14.1
Blueberry	3.82	24.0
Currant	3.68	20.8
Gooseberry	3.90	26.3
Grape	7.84	41.0
Pear	6.77	49.9
Raspberry	4.84	17.2
Sweet cherry	7.38	32.9
Strawberry	2.40	25.4

The characterization of fructose as "fruit sugar" stems from its significant presence in fruits and berries. Table 22.3 shows the percentages of fructose in a variety of fruits, based on the entire fruit and in relation to the total fruit solids (Moskowitz 1974).

Physical and Functional Properties

Fructose possesses physical and functional properties that distinguish it from sucrose, dextrose, corn syrups, starches, high-intensity sweeteners and the myriad of other ingredients used in food formulation. The following discussion reviews these unique differences. "Practical Applications of Crystalline Fructose," found later in this chapter, illustrates ways in which these properties may be used to advantage in food products.

Relative Sweetness

The functional property that most distinguishes fructose from other nutritive carbohydrates, is its high relative sweetness. Relative sweetness is a subjective comparison of the peak organoleptic perception of sweetness of a substance, usually in relation to a sucrose reference. Reported relative sweetness values fall in the range of 1.8 times that of sucrose for crystalline fructose and 1.2 times that of sucrose for liquid fructose (White and Parke 1989). It must be emphasized, however, that the relative sweetness of fructose is dependent on the anomeric state of fructose at the time the sweetness comparison is made.

Only the sweetest, β-D-fructopyranose, anomer exists in crystalline fructose. Fructose rapidly mutarotates upon dissolution in water, forming three additional tautomers possessing lower sweetness (Shallenberger 1978). The extent of mutarotation can be determined using optical rotation, gas-liquid chromatography and nuclear magnetic resonance (NMR). These techniques have been employed to determine that at 22°C, the tautomeric equilibrium concentrations of 20% solids fructose in D_2O are as illustrated in Figure 22.1 (Hyvonen et al. 1977).

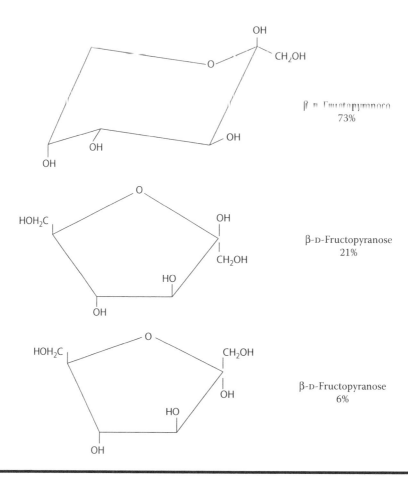

β-D-Fructopyranose
73%

β-D-Fructopyranose
21%

β-D-Fructopyranose
6%

Figure 22.1 Cyclic tautomers of fructose. (From Hyvonen, L., Varo, P., and Koivistonen, P. Tautomeric equilibria of D-glucose and D-fructose. *J Food Sci* 42:657. With permission.)

Temperature, pH, concentration of the solution and presence of other sweeteners are factors that most influence sweetness intensity. Of these, only temperature exerts a significant effect on the mutarotational behavior of fructose in solution and on the transformation from the sweetest β-D-fructopyranose form, to an equilibrium state in which less sweet tautomeric forms are present (Shallenberger 1978; Hyvonen 1980). NMR measurements at different temperatures demonstrate that the change in relative sweetness is directly related to the shift in tautomeric equilibrium as temperature is increased, as illustrated in Table 22.4 (Scott 1980). Based on the correlation between fructose sweetness and mutarotational behavior, Shallenberger (1978) deduced that the furanoses are nearly devoid of sweet taste. While this deduction is based primarily on conformational considerations, it is evident that an increase in the furanose anomer at the expense of the pyranose form will cause a reduction in perceived sweetness.

In practice, the degree of sweetness loss caused by this partial change to other, less sweet furanose tautomers, can be minimized through the use of cold solutions and slightly acid conditions. Experience has shown that citrus-flavored beverage bases, sweetened with pure crystalline fructose and containing the usual amounts of acidulents, can realize a reduction of up to 50% of usual sweetener calories. Conversely, one of the least efficient uses of fructose is in hot coffee, in which mutarotation to furanose forms diminishes sweetness to the point that fructose is isosweet with sucrose. One

Table 22.4   The Tautomeric Equilibrium of Fructose at Different Temperatures

Temperature (°C)	α-D-Fructo-Furanose (%)	β-D-Fructo-Furanose (%)	β-D-Fructo-Pyranose (%)
20	7	24	69
40	7	31	62
60	9	33	58
80	11	38	51

further note in regard to fructose sweeteners, is that the temperature at the time of consumption determines the equilibrium state of the anomers. Thus, cakes made with pure crystalline fructose will taste sweeter after they have been allowed to cool than they will if tasted just out of the oven.

Sweetness Intensity Profile and Flavor Enhancement

The sweetness of fructose is perceived more quickly, peaks more sharply and with greater intensity, and dissipates sooner from the palate than either sucrose or dextrose. It is this early sweetness intensity profile that accounts for the flavor enhancement so often observed in fructose formulations. Many fruit, spice, and acid flavors come through with greater clarity and identity after the sweetness of fructose has dissipated; they are not masked by the lingering sweetness of sucrose. The use of fructose thus makes possible the formulation of a more flavorful product, or alternatively, offers an opportunity for cost savings through lower flavor use.

Sweetness Synergy

Fructose exhibits sweetness synergy when used in combination with other caloric or high-intensity sweeteners. The relative sweetness of fructose blended with sucrose, aspartame, saccharin or sucralose is perceived to be greater than the sweetness calculated from individual components in the blend (Batterman et al. 1988; Batterman and Lambert 1988; Van Tournout et al. 1985; Beyts 1989).

When aspartame received clearance for use in food products in 1981, it became the task of food scientists to determine precisely how to use this sugar substitute. Contrary to oft-expressed opinion, aspartame does not exhibit a sweetness profile exactly identical to that of sucrose. Some subtle organoleptic differences can be perceived, with the significance of these differences being dependent on the specific conditions under which the aspartame is employed. Hyvonen (1981) in Finland and Johnson (1982) in the United States discovered that combinations of aspartame and crystalline fructose could be used to achieve a synergistic sweetening effect and to minimize any lingering, nonsweet flavors of aspartame.

Thus, the sweetness synergy between fructose and other sweeteners offers formulators the choice of accepting finished products with greater sweetness, or reducing sweetener levels and accepting ingredient cost savings. Blending fructose with reduced levels of high intensity sweeteners has the added benefit of eliminating their bitter, metallic or lingering aftertastes that are disagreeable to many consumers.

Colligative Properties

Fructose is a monosaccharide molecule with very different colligative properties than dextrose, another monosaccharide, or sucrose, a disaccharide. Colligative properties are those physical properties

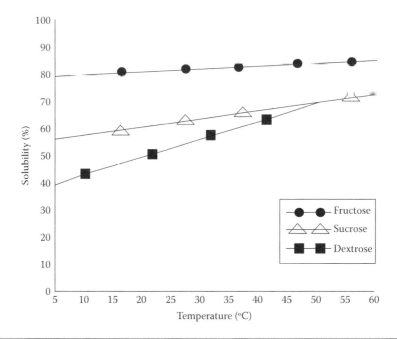

Figure 22.2 **Solubility of sucrose, dextrose (glucose) and fructose with varying temperatures.**

dependent solely on the concentration of particles in the specific system of interest. The concentration of particles is, in turn, dependent on the solubility and molecular weight of the particles.

The solubility of fructose, sucrose and dextrose versus temperature is illustrated in Figure 22.2. Fructose is more soluble at all temperatures than either sucrose or dextrose. Since it is half the molecular weight of sucrose and possesses greater solubility, it is easy to understand how the concentration of particles and concomitant colligative properties are accentuated with fructose.

Osmotic pressure, water activity and freezing point depression are three colligative properties that are accentuated with fructose. Fructose creates higher osmotic pressure and lower water activity than sucrose, dextrose or higher saccharides, resulting in greater product microbial stability. Fructose depresses the freezing point more than sucrose. When used in dairy desserts (both soft-serve and hard-frozen), the tendency of fructose to depress the freezing point can be countered with the addition of higher molecular weight corn sweeteners, gums or stabilizers. The depressed freezing point is actually an asset when fructose is substituted for sucrose in frozen juice concentrates.

Hygroscopicity and Humectancy

Fructose is quicker to absorb moisture (hygroscopicity) and slower to release it to the environment (humectancy) than sucrose, dextrose or other nutritive sweeteners. At approximately 55% relative humidity (RH), fructose begins absorbing moisture from the environment; sucrose does not absorb appreciable moisture until the RH exceeds 65%.

While the "yin" of fructose hygroscopicity can present challenges in ingredient handling and storage (see below), the "yang" of fructose humectancy is a valuable functional attribute. Fructose is useful in sustaining product moistness at low relative humidity, retarding sweetener recrystallization at high sweetener solids, delaying product staling, improving product eating qualities and prolonging product shelf life.

Browning and Flavor Development

Fructose is a reducing sugar whereas sucrose is not. If conditions of temperature and pH are favorable, reducing sugars can undergo a series of chemical condensation and degradation reactions with proteins and amino acids that produce flavored compounds and are responsible for product browning. In baked goods, the golden crust and delectable flavors and aromas of breads and cakes are the highly anticipated fruits of reducing sugars like fructose. If taken to extreme, however, this valuable attribute can result in surface burning and the development of undesirable off-flavors.

It should be noted that sucrose, a non-reducing sugar, may only significantly contribute to browning and flavor development after inversion to its constituent reducing sugars, fructose and dextrose.

Starch-Fructose Synergy

It has been observed that the gelatinization of starch in heated foods is altered in the presence of carbohydrates. Carbohydrate sweeteners delay gelatinization of the starch or cause the starch to gelatinize at a higher temperature. Bean et al. (1978) tabulated differences in the loss of birefringence in three different sugars over a range of concentrations and temperatures (Table 22.5). In a comparison of the effects of fructose versus sucrose, White and Lauer (1990) used Differential Scanning Calorimetry to demonstrate that fructose causes starch to gelatinize at a lower temperature than sucrose. The delay in starch gelatinization appears to be due to the change in water mobility induced by the starch-fructose combination.

Table 22.5 Temperature Range for Loss of Birefringence by Starch in Bleached Commercial Cake Flour in Several Sugar Solutions and % Loss Birefringence

Sugar	Sucrose			Glucose			Fructose		
Solutions (% w/w)	2% (°C)	50% (°C)	98% (°C)	2% (°C)	50% (°C)	98% (°C)	2% (°C)	50% (°C)	98% (°C)
None	55	58.5	62.5	55	58.5	62.5	55	58.5	62.5
10	58	61	65	58.5	61	64.5	—	—	—
20	59	63.5	69.5	61	64	67.5	59	63	67
30	66	70	74	66	68	72	—	—	—
40	73	76	79	68.5	73	76	67	71	74
50	82	84	87	77	79	83	—	—	—
57	90	91.5	94	82	84	86.5	78	80	83
60	93.5	94.5	96.5	85	86.5	90.0	81	83	85
62	—	—	—	88	89.5	91.5	84	86	88.5
65	98	101	104	90	91.5	94	85	87	90
70	104.5	106	Boiled	95	97.5	100.5	89	91	94.5
73	—	—	—	—	—	—	91.5	94.5	97
80	—	—	—	—	—	—	99	103	105

Metabolism of Fructose

Clouds of suspicion descended on fructose briefly in the late 1980s, just as United States crystalline fructose production was taking off, and returned for an extended period in the mid 2000s. The work of Reiser, Hallfrisch and others challenged the safety of fructose, citing potential adverse effects on glucose tolerance, uric acid production, hyperlipidemia and copper status (Blakely et al. 1981; Reiser et al. 1985, 1989; Hallfrisch et al. 1986). A highly regarded monograph published in 1993 and edited by Forbes and Bowman (1993) refuted many of these challenges through a comprehensive examination of fructose intake and metabolism.

The clouds returned with a vengeance in 2004, this time focused specifically on high fructose corn syrup. Bray et al. (2004) published a hypothesis that high fructose corn syrup (HFCS) was uniquely responsible for obesity (i.e., the only nutritive sweetener), based on a mathematical association between increased use of HFCS and increased rates of obesity in the United States between 1970 and 2000. Bray's hypothesis resonated with a research community and general populace desperate for a quick fix to the growing obesity epidemic. For six years, HFCS was thoroughly maligned by scientists, health professionals, legislators, journalists and the public.

By 2010, consensus was reached among knowledgeable scientists that HFCS was not nutritionally unique and that it was, in fact, metabolically equivalent to sucrose (Soenen and Westerterp-Plantenga 2007; Melanson et al. 2007; Angelopoulos et al. 2009; Stanhope et al. 2008), and a sweetener with similar composition, sweetness, caloric value and metabolism. This consensus was supported by the American Medical Association (American Medical Association 2008) and the American Dietetic Association (American Dietetic Association 2009); expert panels convened by the Center for Food, Nutrition, and Agriculture Policy (University of Maryland) (Forshee et al. 2007), Experimental Biology (Fulgoni 2008) and the International Life Sciences Institute (ILSI)-USDA Agricultural Research Service (Murphy 2009); and both noted nutritionists and outspoken nutrition critics. Despite this endorsement, a number of food and beverage manufacturers driven by negative public perception, began to replace HFCS with sucrose. This move was decried by one nutritionist as "100% marketing and 0% science," since exchange of one sweetener for another with the same metabolic profile produced no improvement in nutritional quality (Ludwig 2009).

After conceding HFCS-sucrose nutritional equivalence, Bray shifted critical attention to the fructose moiety common to both (Bray 2010a; Bray 2010b). Support for a unique fructose health risk relies on ecological, epidemiologic and randomized controlled studies. Since ecological studies cannot establish cause-effect relationships and epidemiologic studies provide weak evidence linking suspected risk factors with health outcomes, the strongest evidence for a fructose effect comes from randomized, controlled trials comparing fructose to glucose against various metabolic endpoints. Papers demonstrating untoward health effects proliferated, with claims that fructose constituted an increased risk for a litany of health disorders, including pancreatic cancer cell proliferation (Liu et al. 2010), cardiovascular disease (Nguyen et al. 2009), dementia (Stephan et al. 2010), diabetes (Montonen et al. 2007), excessive food intake (Lane and Cha 2009), gout (Doherty 2009), hypertension (Jalal et al. 2010), kidney disease (Cirillo et al. 2009), liver disease (Abdelmalek et al. 2010), metabolic syndrome (Perez-Pozo et al. 2010) and myocardial infarction (Gul et al. 2009).

Whether fructose truly constitutes a health risk has not been established and continues to be spiritedly debated. Critics of fructose versus glucose experimentation argue that these experimental diets are divorced from real-world human fructose experience:

- People don't consume a diet sweetened exclusively with fructose or glucose; they consume complex diets containing both sugars in roughly equal amounts, whether from sweeteners or fruits and vegetables.
- People don't consume 25%–40% or 60%+ of calories as fructose, as is typically fed to human or animal test subjects, respectively.
- Such extreme diets are in reality probing biochemical pathways under excess and duress, and should not be used to guide public health policy.

Fructose appears to behave very differently when glucose is present. Coss-bu et al. (2009) reported that when ingested alone, fructose contributes little to glycogen synthesis; after co-ingestion of fructose and glucose with the resultant insulin response from the glucose, however, fructose becomes a significant contributor to glycogen synthesis. Marriott et al. (2009) reported that human fructose intakes range from 5% to 9.4% to less than 17% of calories in the 25th, 50th, and 95th percentile population groups. Finally, two recent reviews contend that not only does fructose *not* cause relevant changes in biologically significant measures like triglyceride levels or bodyweight (Dolan et al. 2010), but that there are benefits to be gained from moderate levels of fructose in the diet (Livesey 2009).

Absorption

Fructose in the diet occurs either bound covalently to glucose in the disaccharide, sucrose, or as the free monosaccharide. Free monosaccharide fructose can originate from dietary fruits and berries, or from sweeteners like honey, high fructose corn syrup, and crystalline or liquid fructose. Ingested sucrose is hydrolyzed to fructose and glucose by sucrase enzymes associated with the brush border of the intestinal epithelium. The resulting monosaccharides are immediately transported through the brush border membrane by the disaccharidase-related transport system, without being released to the lumen (Ugolev et al. 1986).

Free fructose absorption is still incompletely understood. One prominent theory is that monosaccharides are absorbed from the intestinal lumen across the brush-border membrane into cells by the Na(+)-dependent glucose transporter, SGLT1, and the facilitated fructose transporter, GLUT5 (Ferraris 2001).

Riby et al. (1993) reported that the capacity for fructose absorption is small compared to sucrose and glucose. They found that the simultaneous ingestion of glucose could prevent fructose malabsorption by increasing intestinal absorptive capacity for fructose. This suggested an alternate theory, that the pair of monosaccharides might be absorbed by the disaccharidase-related transport system as if they were products of the enzymatic hydrolysis of sucrose. The human intestinal capacity for fructose absorption in the absence of glucose appears to vary significantly from one individual to another. It is important to note US dietary consumption data, which indicate that typical glucose-to-fructose ratios are more than adequate to support fructose absorption in the general population (Park and Yetley 1993). Individuals who experience symptoms of malabsorption are advised to avoid consumption of products in which fructose is the sole carbohydrate.

Metabolism

All dietary fructose is absorbed and transported by the intestinal epithelium into the hepatic portal vein. The active hepatic enzyme system for metabolizing fructose efficiently extracts this sugar into the liver, leaving relatively low fructose concentrations in systemic blood vessels (Mayes 1993). Significantly, its entry into liver cells and subsequent phosphorylation by fructokinase is insulin independent (Mehnert 1971). Following cleavage by liver aldolase, the resulting trioses can be utilized for gluconeogenesis and glycogenesis or the synthesis of triglycerides, or they can enter the glycolytic pathway. There is a widespread belief that fructose and its metabolites are quantitatively and absolutely converted to fat in the liver. However, the ultimate fate of these trioses is dependent

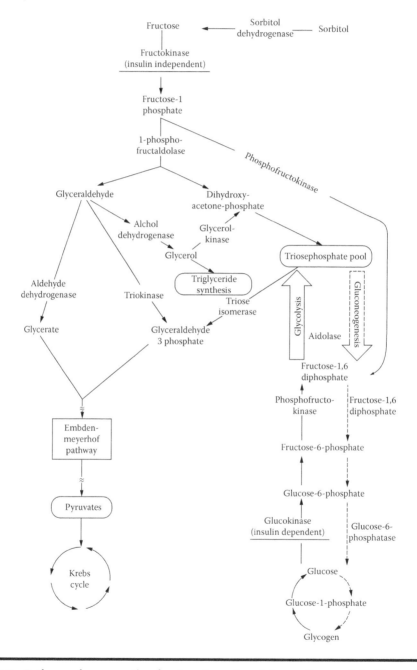

Figure 22.3 Pathways for converting fructose to various intermediary metabolites and energy.

on the metabolic state of the individual and can be significantly altered by the presence of glucose, as mentioned earlier. Figure 22.3 illustrates the pathways involved in fructose metabolism.

Fructose and Diabetes

Studies by Crapo et al. (1980) in the early 1980s showed that acute administration of fructose results in lower glycemic and insulin responses in normal subjects, individuals with impaired glucose tolerance and patients with non-insulin-dependent diabetes mellitus. Figures 22.4 and 22.5 illustrate these comparative effects.

Increased awareness of the harmful side effects of diabetes and the necessity for keeping blood sugar levels of diabetics close to normal without inducing severe hypoglycemia led diabetologists and food technologists to renew investigations into the use of fructose as a preferred sweetener for diabetics (Talbot and Fisher 1978). Though short-term studies showed that substitution of fructose

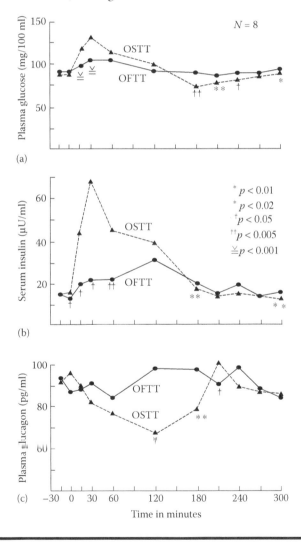

Figure 22.4   Comparison of glucose (a), insulin (b), and glucagon (c) responses to oral sucrose versus oral fructose (OSTT = oral sucrose tolerance test; OFTT = oral fructose tolerance test).

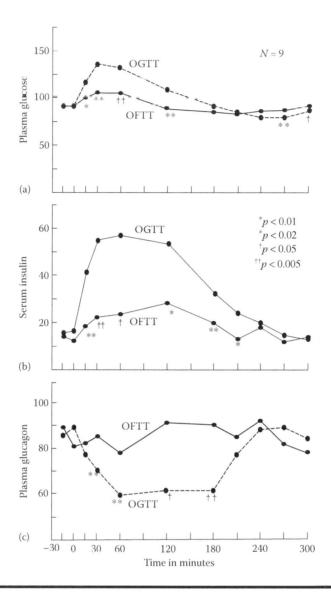

Figure 22.5 Comparison of glucose (a), insulin (b), and glucagon (c) responses to oral glucose versus oral fructose (OGTT = oral glucose tolerance test; OFTT = oral fructose tolerance test).

for sucrose in the diets of individuals with diabetes improves glycemic control, long-term effects are still inconclusive (Gerrits and Tsalikian 1993). The American Diabetes Association has been reluctant to endorse fructose use for diabetics; however, their reluctance appears to be based on results from extreme fructose experimentation described above.

Food Intake

The influence of fructose on appetite and satiety is another controversial and apparently unresolved issue. The result obtained appears highly dependent on the experimental model tested: sweetener type and amount used, type of food tested, solid versus liquid, and so forth. There is evidence that fructose at exaggerated levels (30% of calories as fructose) and in the absence of glucose influences appetite hormones (reduces leptin and increases ghrelin (Teff et al. 2004)) but not under physiologic conditions. Melanson and Soenen demonstrated that high fructose corn syrup and sucrose are equally satiating at more moderate intakes (Melanson et al. 2007; Soenen and Westerterp-Plantenga 2007).

Physical Performance

Athletes have long used dietary supplements in an effort to sustain peak physical performance. While much of the data concerning fructose supplementation is contradictory, fructose feeding before or during exercise can enhance performance under certain conditions. Fructose intake prior to exercise appears to spare muscle glycogen by elevating liver glycogen, thereby prolonging activity. In addition, there is good evidence to suggest that the addition of fructose supplementation during ultra-endurance events can improve performance by 126% (Craig 1993). As indicated earlier, gastrointestinal discomfort, created by intake of large amounts of fructose in the absence of glucose, can hinder athletic performance. Since absorption capacity varies widely in the general population, the benefits of fructose will also vary from one athlete to another. Ongoing research continues to refine the role fructose can play in enhancing athletic performance.

Glycemic Effect

Jenkins et al. determined the effect of 62 commonly eaten foods and sugars fed to groups of human volunteer subjects. They constructed a Glycemic Index, defined as "the area under the blood glucose response curve for each food expressed as a percentage of the area after taking the same amount of carbohydrate as glucose" (Jenkins et al. 1981, 363). Table 22.6 reports the glycemic index for several sweeteners and foods relative to glucose, which is assigned a value of 100.

These glycemic values suggest that simple diabetic carbohydrate exchanges based on carbohydrate content may not, in fact, be an accurate predictor of physiological response. A clinical comparison of orally administered fructose, sucrose, and glucose in subjects with reactive hypoglycemia was reported in *Diabetes Care* (Crapo et al. 1982, 512). The investigators determined that the use of 100 g loads of pure fructose as the sweetener in cakes and

Table 22.6 Glycemic Index

Glucose	100
Sucrose	59 ± 10
Fructose	20 ± 5
Maltose	105 ± 12
Apples	39 ± 3
Raisins	64 ± 11
White bread	69 ± 5

Source: From Jenkins et al., *Am J Clin Nutr* 34(3):362–366, 1981.

beverages, or by itself, resulted in a significantly reduced glycemic effect, as indicated by markedly less severe glucose and insulin responses. The authors concluded, "fructose may thus prove useful as a sweetening agent in the dietary treatment of selected patients with reactive hypoglycemia."

Practical Applications of Crystalline Fructose

Crystalline fructose was first promoted as a nutritionally beneficial sweetener by virtue of its unique metabolic disposition in the body. Early product formulations targeting diet and health conscious consumers included powdered diet and sports beverages, nutritional candy bars, and specialty diabetic and dietetic food items. Breakthrough technology to crystallize fructose from HFCS feedstock was implemented in the mid-1980s, enabling large scale and low cost fructose production. Only then did formulation scientists attempt to integrate crystalline fructose into high volume, mainstream food products.

Providing incentive to alert food scientists looking for applications for crystalline fructose was a new class of dietetic foods defined by the US Food and Drug Administration (US FDA). Products in this class could be labeled "reduced-calorie" if they contained at least one-third fewer calories than their full-calorie counterpart (Gardner 1978). US FDA has since determined that products with 25% fewer calories that their full calorie counterparts may be labeled "reduced-calorie" (US CFR 2010). Crystalline fructose was immediately recognized as a critical ingredient in achieving the necessary reduction in sweetener calories (Osberger and Linn 1978). Substituting crystalline fructose for other sweeteners in existing formulas is not simply a case of plugging in a factor to determine the quantity of fructose required. Considerable food technology expertise is required to formulate reduced calorie products because unwanted texture and body changes arise, in many instances, from the reduction in sweetener solids.

The availability of a new dry sweetener alternative to sucrose that combined high relative sweetness, dry physical form, and unique functional properties coincided with emerging opportunities to formulate food products for new regulatory classes. This collision of tool with opportunity provided all the impetus food scientists needed to formulate crystalline fructose into many food and beverage categories, with the following benefits:

Dry Mix Beverages, Puddings, and Gelatins

- Reduced sweetener content and concomitant lower calories are possible due to the intense sweetness and dry form of crystalline fructose.
- Fruit flavors are beneficially enhanced, permitting reduced levels of these expensive ingredients.
- Fructose-starch synergy sets puddings in about half the time required for all-sucrose products, permitting reduced starch use, cleaner flavor and improved product performance.

The formulas in Tables 22.7 through 22.9 illustrate these concepts.

Lite Pancake Syrups and Carbonated Beverages

- Reduced total sweetener use and "Lite" or reduced calorie label claims are possible, due to the sweetness synergy between fructose and sucrose, saccharin or aspartame.

Table 22.7   Reduced-Calorie Lemonade Mix[a]

Krystar® crystalline fructose	93.0 lb
Citric acid (anhydrous fine granular)	5.45 lb
Ascorbic acid	0.42 lb
Carrageenan (Viscarin #402)	0.10 lb
Riboflavin	0.0016 lb
Syloid #244	0.50 lb
Permaseal clouding agent FD-9208-B	0.20 lb
Naturalseal lemon flavor FD-8949-D	0.33 lb

[a] Usage: Mix 8 oz of above product with sufficient water to make 1 gal of reduced-calorie lemonade. Chill and serve.

■ Fruit flavors are enhanced and high intensity sweetener off-flavors may be eliminated as lower levels are used.
■ Fructose provides lasting sweetness when paired with a sweetener like aspartame, which has a limited shelf life due to thermal decomposition.

Breakfast Cereals

■ Flavor enhancement and sweetness synergies improve product performance or allow cost reductions (e.g., cinnamon sugar coating).
■ Simple sugars may be moved down the list of ingredients on the nutrition label, a strategy aimed at health-conscious consumers (White and Parke 1989).

Baked Goods

■ A reducing sugar like fructose used in place of a non-reducing sugar like sucrose improves product flavor and browning development in microwave and conventional ovens.
■ The increased solubility of fructose makes it resistant to recrystallization. The development of soft, moist cookies was made possible by the partial substitution of fructose for sucrose in these formulations.

Table 22.8   Low-Calorie Fructose Lemonade Mix[a]

Fructose + saccharin blend (98:1)	89.1 lb
Citric acid (anhydrous fine granular)	9.2 lb
Ascorbic acid	0.7 lb
Carrageenan (Viscarin #402)	0.16 lb
Riboflavin	0.0025 lb
Permaseal clouding agent FD-9208-D	0.32 lb
Naturalseal lemon flavor FD-8949-D	0.53 lb

[a] Usage: Mix 5 oz of above product with sufficient water to make 1 gal of low-calorie lemonade. Chill and serve.

Table 22.9   Reduced-Calorie Gelatin Dessert

Crystalline fructose	79 lb, 0 oz
300 Bloom gelatin	14 lb
Fumaric acid	3 lb, 8 oz
Sodium citrate	2 lb
Syloid No. 244	1 lb
Spray-dried color and flavor	As required

- Humectant properties of fructose can replace glycerin to improve product moisture and shelf life in baked goods and bakery fillings. Multiple benefits gained include eliminating the off-flavor of glycerin, removing glycerin from the product label, and ingredient cost savings (Anonymous 1995).
- It is possible to exploit the fructose-starch synergy to control the starch gelatinization temperature in cakes formulated with fat replacers. Reduced calorie cake formulas like those illustrated in Table 22.10 allow bakers to produce cakes comparable in height and volume to full-fat products (Horton et al. 1990).
- Reduced fructose-starch cook temperatures may allow reduced product heat damage, faster line speeds and lower energy costs.

Dairy Products

- Fructose may sweeten both the base and fruit in all-nutritively-sweetened yogurt for enhanced fruit flavor and ingredient image, and ingredient cost savings.
- Enhanced chocolate flavor allows reduced costs and fewer calories in chocolate milk.

Table 22.10   Reduced Calorie Cake Mixes

Ingredient	White Cake (%)	Yellow Cake (%)	Chocolate Cake (%)
Krystar® crystalline fructose	47.28	47.28	44.90
Hi-ratio bleached cake flour	38.75	38.75	37.72
N-Flate (National Starch)	7.71	7.71	7.52
Henningsen type P-20 egg white solids	3.44	3.44	3.35
Baking powder	2.24	2.24	1.49
Givaudan FD-9993 vanilla powder	0.58	0.58	0.57
De Zaan 11SB cocoa powder	—	—	4.17
Baking soda	—	—	0.28
Roche 1% CWS dry β-carotene	—	0.15	—

Note: Directions for use: Blend 16 oz (454 g) of reduced-calorie cake mix with 10 oz of water for 30 s at low speed, followed by 5 min whipping at high speed. Transfer the heavy batter into two 9-inch round cake pans (or one 9½ inch × 13½ inch rectangular pan) that have been sprayed with a low-caloric, nonstick spray. Bake at 325°F 30–32 min.

Confections

- Fructose reduces the water activity of chocolate bar caramel fillings, resulting in greater microbial stability.
- It enhances hard candy flavors and improves product color.
- Fructose is more resistant to re-crystallization than sucrose or dextrose. This quality can obviate the need for invertase when fructose replaces sucrose in liquid center confections like chocolate covered cherries.
- Its lower relative viscosity offers faster production speeds.
- Use of a reducing sugar like fructose in place of a non-reducing sugar like sucrose allows shorter cook time to develop color and flavor.
- The fructose-starch synergy increases line efficiency by reducing starch levels and the viscosity of the deposit hot mix.
- Fructose comprising 35% of the premium chocolate or carob coatings to enrobe energy bars permitted a thinner carob coating to be applied without reducing total sweetness (Zimmermann 1974). The authors warned, however, that temperatures must not exceed 120°F or agglomeration of crystals and severe glazing of pumps and pipelines will result.
- Most attempts at making hard candies with fructose have been unsuccessful, primarily because boiled candy made from fructose will not set properly. The addition of maltodextrin (44%), however, reportedly improved the set in a hard candy marketed in Italy (Xyrofin 1982).

Frozen Dairy Products and Novelties

- Calorie reduction and flavor enhancement are possible, but care must be taken to balance the freezing point depression with higher molecular weight compounds like corn syrups.

Frozen Fruit Packs and Juice Concentrates

- Greater sweetener solubility and lower osmotic pressure speed fruit infusion, preserve fruit integrity and prolong storage stability.
- Improved solubility, resistance to recrystallization and depressed freezing point hasten frozen juice concentrate reconstitution in water.
- Fructose enhances the natural flavors in fruit packs and juices.

Sports Drinks

- The increased solubility of fructose relative to sucrose and dextrose allows formulation to greater caloric density, particularly in chilled beverages.
- Flavor enhancement and compatible sweetness intensity profile help mask unpalatable mineral and vitamin supplements apparent in sucrose-sweetened beverages.

Fructose Marketing

Crystalline fructose has been available as an industrial food ingredient in the United States since 1975. Fructose sales were initially strong as a result of intense marketing to educate the medical community, the health food industry and health-conscious consumers about the metabolic and sweetening advantages of crystalline fructose. Sales reached a plateau after several years, however, due chiefly to three contributing factors:

- Higher sales price relative to competing caloric and high intensity sweeteners
- Sales efforts targeting the relatively small medical and health food markets
- The arrival of significant quantities from Europe and Asia at prices sharply below domestic prices

US crystalline fructose manufacture would likely have failed early on had not A.E. Staley Manufacturing Company (now Tate & Lyle PLC) and Archer Daniels Midland, two leaders in the corn wet milling industry, decided in the mid 1980s that the time was ripe to begin large-scale production of crystalline fructose. While long overdue in the view of many industry observers, this decision to enter the crystalline fructose market made a great deal of sense: corn wet millers had access to low cost raw materials and were expert in refining the high fructose corn syrup feed stock to crystalline fructose. Furthermore, sales and distribution channels already in place to supply the food industry with corn sweeteners and starches enabled them to market crystalline fructose to this enormous industry far more aggressively than had been attempted to date.

Economies of scale, improved manufacturing processes, and increased awareness among food scientists of its unique physical and functional properties, have spurred the growth of crystalline fructose manufacturing and formulation into finished foods and beverages. Though manufactured today in many countries throughout the world, crystalline fructose production is clearly dominated by two US companies, Tate & Lyle PLC and Archer Daniels Midland. Actual production volumes are closely guarded and difficult to come by, however, it is roughly estimated that crystalline fructose manufacturing grew from 12 million pounds in 1986 to approximately 185 million pounds in 1998 (Haley 1999). Over the same time period, the selling price of crystalline fructose dropped in the United States from over $1.00/lb to approximately $0.35/lb. At this price, crystalline fructose was fully able to compete with sucrose and dextrose on a sweetness and functionality basis. In 1993, Vuilleumier estimated the potential crystalline fructose market to be 200–300 million pounds (Vuilleumier 1993).

Regulatory Status

Crystalline fructose, HFCS and sucrose are Generally Recognized as Safe (GRAS) in the United States. The *Food Chemicals Codex* and *United States Pharmacopoeia* define fructose as containing not less than 98% or more than 102% fructose, and not more than 0.5% dextrose. Only crystalline fructose and crystalline fructose syrup meet this definition; HFCS and sucrose do not.

The Codex Alimentarius Commission defines fructose as "purified and crystallized D-α-fructose." These requirements, as well, are met only by crystalline fructose (Hanover and White 1993).

Storage and Handling

Relative humidity (RH) is the most important factor in crystalline fructose storage and handling. Table 22.11 illustrates the effect of RH on the moisture absorption of crystalline fructose (Finnish State Technical Research Center 1973). Product lumping is caused by:

Table 22.11 The Effect of Relative Humidity on the Moisture Absorption of Crystalline Fructose at 20°C[a]

Relative Humidity (%)	Days Exposed								
	1	2	5	7	9	12	16	20	26
33	0.02	0.01	0.02	0.02	0.01	0.01	0.01	0.01	0.01
48	—	0.03	0.03	0.03	0.03	0.03	—	—	—
55	—	0.06	0.06	0.06	0.06	0.05	—	—	—
60	0.10	0.12	0.13	0.13	0.13	0.13	0.13	0.13	0.13[b]
66	0.32	0.52	1.12	1.51	1.87	2.45	3.16	3.84	4.89
72	0.68	1.33	3.34	4.66	6.01	7.91	10.3	12.2	14.7
76	0.91	1.78	4.40	6.14	7.86	10.3	12.8	14.9	17.7
82	1.27	2.49	6.14	8.67	11.0	13.7	16.9	19.7	23.8

[a] The amount of moisture absorbed is indicated as a percent of dry weight.
[b] When moisture content exceeds 0.7%, fructose becomes increasingly wet and sticky. Processing conditions should be controlled to achieve values above this line.

- An initial increase in RH, causing surface water absorption and localized dissolving of fructose at crystal surfaces
- Subsequent lowering of RH, resulting in release of moisture, recrystallization and sticking of adjacent crystals to one another

A conditioned air system with RH of 55% or less and a maximum temperature of 24°C is recommended for bulk handling crystalline fructose. Crystalline fructose is typically packaged and shipped to customers in 25 kg multiwall (foil-lined) bags, 600 and 1000 kg tote bags, and bulk railcar. Recommended storage conditions are RH less than 50% and 21°C.

Fructose syrups made by dissolving crystalline fructose in water can be stored at 21°C–29°C. These pure fructose syrups may be held indefinitely with little likelihood of crystallization and are microbially stable (Hanover 1992; Hanover and White 1993).

References

Abdelmalek, M.F., Suzuki, A., Guy, C., Unalp-Arida, A., Colvin, R., Johnson, R.J., and Diehl, A.M. 2010. Increased fructose consumption is associated with fibrosis severity in patients with nonalcoholic fatty liver disease. *Hepatology* 51(6):1961–1971.

American Dietetic Association. 2008. Hot Topics: High fructose corn syrup and weight status. Available from http://www.eatright.org. Accessed December 2008.

American Medical Association. 2008. *AMA finds high fructose syrup unlikely to be more harmful to health than other caloric sweeteners.* Available from http://www.ama-assn.org/ama/pub/category/print/18691.html. Accessed June 20, 2008.

Angelopoulos, T.J., Lowndes, J., Zukley, L., Melanson, K.J., Nguyen, V., Huffman, A., and Rippe, J.M. 2009. The effect of high-fructose corn syrup consumption on triglycerides and uric acid. *J Nutr* 139(6):1242S–1245S.

Anonymous. 1995. Fructose replaces glycerin in bakery fillings. *Food Prod Design* 4(11):93.

Batterman, C.K., Augustine, M.E., and Dial, J.R. 1988. Sweetener Composition. US 4,737,368: A.E. Staley Manufacturing Company.

Batterman, C.K., and Lambert, J. 1988. Synergistic sweetening composition. International Patent Publication WO88/08674. A.E. Staley Manufacturing Company.

Bean, M.M., Yamazak, W.T., and Donelson, D.H. 1978. Wheat starch gleatinization in sugar solutions II: Fructose, glucose and sucrose – cake performance. *Cereal Chem* 55 (6):945

Beyts, P.K. 1989. Sweetening compositions. U.K. Patent Application GB2210545A.

Blakely, S.R., Hallfrisch, J., Reiser, S., and Prather, E.S. 1981. Long-term effects of moderate fructose feeding on glucose tolerance parameters in rats. *J Nutr* 111(2):307–314.

Bray, G.A. 2010a. Fructose: pure, white, and deadly? Fructose, by any other name, is a health hazard. *J Diabetes Sci Technol* 4(4):1003–1007.

Bray, G.A. 2010b. Soft drink consumption and obesity: it is all about fructose. *Curr Opin Lipidol* 21(1):51–57.

Bray, G.A., Nielsen, S.J., and Popkin, B.M. 2004. Consumption of high-fructose corn syrup in beverages may play a role in the epidemic of obesity. *Am J Clin Nutr* 79 (4):537–543.

Cirillo, P., Sautin, Y.Y., Kanellis, J., Kang, D.H., Gesualdo, L., Nakagawa, T., and Johnson, R.J. 2009. Systemic inflammation, metabolic syndrome and progressive renal disease. *Nephrol Dial Transplant* 24(5):1384–1387.

Coss-Bu, J.A., Sunehag, A.L., and Haymond, M.W. 2009. Contribution of galactose and fructose to glucose homeostasis. *Metabolism* 58(8):1050–1058.

Craig, B.W. 1993. The influence of fructose feeding on physical performance. *Am J Clin Nutr* 58(Suppl. 5):815S–819S.

Crapo, P.A., Kolterman, O.G., and Olefsky, J.M. 1980. Effects of oral fructose in normal, diabetic, and impaired glucose tolerance subjects. *Diabetes Care* 3(5):575–582.

Crapo, P.A., Scarlett, J.A., Kolterman, O.G., Sanders, L.R., Hofeldt, F.D., and Olefsky, J.M. 1982. The effects of oral fructose, sucrose, and glucose in subjects with reactive hypoglycemia. *Diabetes Care* 5(5):512–517.

Doherty, M. 2009. New insights into the epidemiology of gout. *Rheumatology (Oxford)* 48:ii2–ii8.

Dolan, L.C., Potter, S.M., and Burdock, G.A. 2010. Evidence-based review on the effect of normal dietary consumption of fructose on development of hyperlipidemia and obesity in healthy, normal weight individuals. *Crit Rev Food Sci Nutr* 50(1):53–84.

Ferraris, R.P. 2001. Dietary and developmental regulation of intestinal sugar transport. *Biochem J* 360(Pt 2):265–276.

Finnish State Technical Research Center. 1973. Research Report #A-6759-73. Helsinki.

Forbes, A.L., and Bowman, B.A. 1993. Health effects of dietary fructose. *Am J Clin Nutr* 58:721S.

Forsberg, K.H., Hamalainen, L., Melaja, A.J., and Virtanen, J.J. 1975. US 3,883,365: Suomen Sakeri Osakeyhtio (Finnish Sugar Company).

Forshee, R.A., Storey, M.L., Allison, D.B., Glinsmann, W.H., Hein, G.L., Lineback, D.R., Miller, S.A., Nicklas, T.A., Weaver, G.A. and White, J.S. 2007. A critical examination of the evidence relating high fructose corn syrup and weight gain. *Crit Rev Food Sci Nutr* 47(6):561–582.

Fulgoni, V., 3rd. 2008. High-fructose corn syrup: Everything you wanted to know, but were afraid to ask. *Am J Clin Nutr* 88(6):1715S.

Gardner, S. 1978. Special dietary foods. *Fed Register* 43:185.

Gerrits, P.M., and Tsalikian, E. 1993. Diabetes and fructose metabolism. *Am J Clin Nutr* 58(Suppl):796S–799S.

Gul, A., Rahman, M.A., and Hasnain, S.N. 2009. Influence of fructose concentration on myocardial infarction in senile diabetic and non-diabetic patients. *Exp Clin Endocrinol Diabetes* 117(10):605–609.

Haley, S. 1999. *Personal Communication*. Washington, DC: US Economic Research Service.

Hallfrisch, J., Ellwood, K., Michaelis, O.E., Reiser, S., and Prather, E.S. 1986. Plasma fructose, uric acid, and inorganic phosphorus responses of hyperinsulinemic men fed fructose. *J Am Coll Nutr* 5(1):61–68.

Hanover, L.M. 1992. Crystalline fructose: Production, properties, and applications. In *Starch Hydrolysis Productions: Worldwide Technology, Production, and Application*, F.W. Schenck, and R.E. Hebeda (Eds.). New York: VCH Publishers, Inc.

Hanover, L.M., and White, J.S. 1993. Manufacturing, composition, and applications of fructose. *Am J Clin Nutr* 58(Suppl. 5):724S–732S.

Horton, S.D., Lauer, G.N., and White, J.S. 1990. Predicting gelatinization temperatures of starch/sweetener systems for cake formulation by differential scanning calorimetry II. *Eval Appl Mod* 35(8):734, 737, 739.

Hyvonen, L. 1980. *Varying Relative Sweetness*. University of Helsinki EKT-546.

Hyvonen, L. 1981. Research report: Synergism between sweeteners. Helsinki, Finland: University of Helsinki.

Hyvonen, L., Varo, P., and Koivistonen, P. 1977. Tautomeric equilibria of D-glucose and D-fructose. *J Food Sci* 42:657.

Jalal, D.I., Smits, G., Johnson, R.J., and Chonchol, M. 2010. Increased fructose associates with elevated blood pressure. *J Am Soc Nephrol* 21(9):1543–1549.

Jenkins, D.J.A., Wolever, T.M., Taylor, R.H., Barker, H., Fielden, H., Baldwin, J.M., Bowling, A.C., Newman, H.C., Jenkins, A.L., and Goff, D.V. 1981. Glycemic index of foods: A physiological basis for carbohydrate exchange. *Am J Clin Nutr* 34(3):362–366.

Johnson, L. 1982. Nutley, NJ: Hoffmann-La Roche (Unpublished report).

Lane, M.D., and Cha, S.H. 2009. Effect of glucose and fructose on food intake via malonyl-CoA signaling in the brain. *Biochem Biophys Res Commun* 382(1):1–5.

Liu, H., Huang, D., McArthur, D.L., Boros, L.G., Nissen, N., and Heaney, A.P. 2010. Fructose induces transketolase flux to promote pancreatic cancer growth. *Cancer Res* 70(15):6368–6376.

Livesey, G. 2009. Fructose ingestion: Dose-dependent responses in health research. *J Nutr* 139(6):1246S–1252S.

Ludwig, D.S. 2009. *Crain's Chicago Business*. Available from http://www.chicagobusiness.com/article/20090905/ISSUE01/100032337. Accessed September 7, 2009.

Marriott, B.P., Cole, N., and Lee, E. 2009. National estimates of dietary fructose intake increased from 1977 to 2004 in the United States. *J Nutr* 139(6):1228S–1235S.

Mayes, P.A. 1993. Intermediary metabolism of fructose. *Am J Clin Nutr* 58(Suppl. 5):754S–765S.

Mehnert, H. 1971. In *Handbuch des Diabetes Mellitus*. Munich: J.F. Lehmanns Verlag.

Melanson, K.J., Zukley, L., Lowndes, J., Nguyen, V., Angelopoulos, T.J., and Rippe, J.M. 2007. Effects of high-fructose corn syrup and sucrose consumption on circulating glucose, insulin, leptin, and ghrelin and on appetite in normal-weight women. *Nutrition* 23(2):103–112.

Montonen, J., Jarvinen, R., Knekt, P., Heliovaara, M., and Reunanen, A. 2007. Consumption of sweetened beverages and intakes of fructose and glucose predict type 2 diabetes occurrence. *J Nutr* 137(6):1447–1454.

Moskowitz, H. 1974. In *Sugars in Nutrition*, H. Sipple, and K. McNutt (Eds.). New York: Academic Press, Inc.

Murphy, S.P. 2009. The state of the science on dietary sweeteners containing fructose: Summary and issues to be resolved. *J Nutr* 139(6):1269S–1270S.

Nguyen, S., Choi, H.K., Lustig, R.H., and Hsu, C.Y. 2009. Sugar-sweetened beverages, serum uric acid, and blood pressure in adolescents. *J Pediatr* 154(6):807–813.

Osberger, T., and Linn, H. 1978. In *Low Calorie and Specialty Diet Foods*. West Palm Beach, FL: CRC Press, Inc.

Park, Y.K., and Yetley, E.A. 1993. Intakes and food sources of fructose in the United States. *Am J Clin Nutr* 58(Suppl. 5):737S–747S.

Perez-Pozo, S.E., Schold, J., Nakagawa, T., Sanchez-Lozada, L.G., Johnson, R.J. and Lillo, J.L. 2010. Excessive fructose intake induces the features of metabolic syndrome in healthy adult men: Role of uric acid in the hypertensive response. *Int J Obes (Lond)* 34(3):454–461.

Reiser, S., Powell, A.S., Scholfield, D.J., Panda, P., Ellwood, C., and Canary, J.J. 1989. Blood lipids, lipoproteins, apoproteins, and uric acid in men fed diets containing fructose or high-amylose cornstarch. *Am J Clin Nutr* 49(5):832–839.

Reiser, S., Smith, Jr., J.C., Mertz, W., Holbrook, J.T., Scholfield, D.J., Powell, A.S., Canfield, W.K., and Canary, J.J. 1985. Indices of copper status in humans consuming a typical American diet containing either fructose or starch. *Am J Clin Nutr* 42(2):242–251.

Riby, J.E., Fujisawa, T., and Kretchmer, N. 1993. Fructose absorption. *Am J Clin Nutr* 58(Suppl. 5):748S–753S.

Scott, R. 1980. Nutley, NJ: Hoffmann-La Roche (Unpublished).

Shallenberger, R.S. 1978. *Pure and Applied Chemistry*. London: Pergamon Press Ltd.

Soenen, S., and Westerterp-Plantenga, M.S. 2007. No differences in satiety or energy intake after high-fructose corn syrup, sucrose, or milk preloads. *Am J Clin Nutr* 86(6):1586–1594.

Stanhope, K.L., Griffen, S.C., Bair, B.R., Swarbrick, M.M., Keim, N.L., and Havel, P.J. 2008. Twenty-four-hour endocrine and metabolic profiles following consumption of high-fructose corn syrup-, sucrose-, fructose-, and glucose-sweetened beverages with meals. *Am J Clin Nutr* 87(5):1194–1203.

Stephan, B.C., Wells, J.C., Brayne, C., Albanese, E., and Siervo, M. 2010. Increased fructose intake as a risk factor for dementia. *J Gerontol A Biol Sci Med Sci* 65(8):809–814.

Talbot, J.M., and Fisher, K.D. 1978. The need for special foods and sugar substitutes by individuals with diabetes mellitus. *Diabetes Care* 1(4):231–240.

Tannahill, R. 1973. *Food in History*. New York: Stein and Day.

Teff, K.L., Elliott, S.S., Tschop, M., Kieffer, T.J., Rader, D., Heiman, M., Townsend, R.R., Keim, N.L., D'Alessio, D., and Havel, P.J. 2004. Dietary fructose reduces circulating insulin and leptin, attenuates postprandial suppression of ghrelin, and increases triglycerides in women. *J Clin Endocrinol Metab* 89(6):2963–2972.

Ugolev, A.M., Zaripov, B.Z., Iezuitova, N.N., Gruzdkov, A.A., Rybin, I.S., Voloshenovich, M.I., Nikitina, A.A., Punin, M.Y., and Tokgaev, N.T. 1986. A revision of current data and views on membrane hydrolysis and transport in the mammalian small intestine based on a comparison of techniques of chronic and acute experiments: Experimental re-investigation and critical review. *Comp Biochem Physiol A Comp Physiol* 85(4):593–612.

US CFR (United States Code of Federal Regulations). 2010. Nutrient Content Claims for the Calorie Content of Foods. 21CFR 101.60(4)(i). http://www.accessdata.fda.gov/scripts/cdrh/cfdocs/cfcfr/CFRSearch.cfm?fr=101.60&SearchTerm=reduced%20calorie. Accessed May 25, 2010.

Van Tournout, P., Pelgroms, J., and Van der Meerer, J. 1985. Sweetness evaluation of mixtures of fructose with saccharin, aspartame or acesulfame K. *J Food Sci* 50:469.

Vuilleumier, S. 1993. Worldwide production of high-fructose syrup and crystalline fructose. *Am J Clin Nutr* 58(Suppl. 5):733S–736S.

White, D.C., and Lauer, G.N. 1990. Predicting gelatinization temperature of starch/sweetener systems for cake formulation by differential scanning calorimetry. I. Development of a model. *Cereal Foods World* 35(8):728.

White, J.S., and Parke, D.W. 1989. Fructose adds variety to breakfast. *Cereal Foods World* 34(5):392–398.

Xyrofin. 1982. Sweet Topic No. 35. Baar, Switzerland.

Zimmermann, M. 1974. Fructose in the manufacture of dietetic chocolate. *Manuf Confect* 39.

Chapter 23

High Fructose Corn Syrup

Allan W. Buck

Contents

Introduction

More than 40 years have elapsed since the first commercial shipment of modern high fructose corn syrup (HFCS) in the United States. By the end of the 20th century, production peaked at more than 18 billion pounds dry weight (USDA 2010) with about 42% of the current US sweetener consumption coming from HFCS and 45% from sucrose products (i.e., table sugar, medium invert sugar, invert sugar, etc.) (USDA 2010). HFCS was developed initially as a liquid alternative to sucrose. It demonstrated improvements over liquid sucrose products and was successful as a result of stable pricing, superior and consistent quality, widespread availability and ease of use. Coinciding with the introduction and use of HFCS were tremendous advancements in programmable logic controller (PLC) and computer technology that were implemented in food processing facilities for process control, quality control, and manufacturing efficiencies. This high quality, microbial stable syrup was ideal for automation, improving areas such as batch metering, temperature control, yields and consistency, and is therefore of benefit to all facets of operations including procurement, quality, engineering, and overall plant operations.

Initially, one product was supplied with a concentration of 42% fructose on a dry basis. Because of its quality, sweetening properties, and availability as a bulk liquid, economical sweetener, the product, 42% HFCS, rapidly penetrated large markets formally held exclusively by liquid sucrose products.

Table 23.1   Market Division between 42% and 55% HFCS

Year	% 42% HFCS[a]	% 55% HFCS[a]
1978	88	12
1982	50	50
1987	35	65
1992	42	58
1993	41	59
1994	40	60
1995	39	61
1996	38	62
1997	37	63
1998	36	64
1999	36	64
2000	38	62
2001	38	62
2002	39	61
2003	40	60
2004	40	60
2005	40	60
2006	40	60
2007	39	61
2008	38	62
2009	37	63

Source: From US Department of Agriculture, Economic Research Service (USDA). Sugar and Sweetener Yearbook Tables. http://www.ers.usda.gov/Briefing/Sugar/data.htm yearbook, 2010.

[a] USDA/ERS Table 23.30—US high fructose corn syrup (HFCS) supply and use, by calendar year

Not being quite as sweet as medium invert sugar used by the soft drink industry, technical demands from this industry led to the development of a 55% fructose product with equivalent sweetness levels. The development was brought about by the commercialization of a fractionation process that allows molecular separation of fructose from dextrose and concentration of the sweeter fructose fraction. This product, typically produced at levels of 90% fructose, was made available as a commercial product, but was usually blended back with 42% HFCS to make 55% HFCS. The transition from sucrose to 100% HFCS in major soft drink brands occurred over about a five year period (1980–1985), as the corn wet milling industry increased production capacity to match needs and proved to the soft drink industry that quality and consistency in HFCS products were available. In the end, 55% HFCS had become the standard sweetener for carbonated beverages in the United States and the ratio between 42% and 55% HFCS has remained fairly constant since approval was given for the major soft drink brands. Table 23.1 shows the market breakdown for the two major HFCS products. HFCS matured to the point that growth was a function of increased population consumption of food products and not on penetration of new markets. Currently, the total consumption of nutritive sweeteners is declining in the United States (USDA 2010). Table 23.2 shows major events leading to the maturity of HFCS in the United States and Figure 23.1 depicts the growth in pounds.

Although corn is an economical and abundant source of starch for the manufacture of HFCS, the technology for the manufacture of high fructose syrups (HFS) is applicable to starch sources other than corn. For all practical purposes, properties are identical and the remainder of this chapter will refer to that product made from cornstarch which is such an important part of the US food chain.

Most of this chapter will deal with the common commercial products 42% and 55% HFCS. Commercial products with higher levels of fructose (around 90% HFCS) are available but in limited commercial supply compared with the widespread use of 42% and 55% HFCS. Because of the high level of fructose, many of the applications, functional and nutritional properties of the 90% products are similar to pure crystalline fructose.

Table 23.2   Brief History of HFCS

1957	Patent by Marshall and Kooi disclosed a microbial enzyme capable of isomerizing dextrose to fructose
1964	Clinton Corn Processing Company started work to commercialize a process
1965	Clinton Corn initiated work with Japanese Agency of Industrial Science and Technology to commercialize process
1967	First commercial shipment of HFCS by Clinton Corn was 15% fructose and was made by batch system using soluble glucose isomerase
1968	First commercial shipment of 42% HFCS by Clinton Corn-batch process and with both soluble and insoluble glucose isomerase
1972	Clinton Corn brought on stream first continuous plant using immobilized enzyme
May 1978	First large-scale 90% HFCS plant by Archer Daniels Midland. Permitted volume production of 55% HFCS
Jan. 1980	HFCS approved as 50% of the sweetener in brand Coca-Cola
Apr. 1983	HFCS approved as 50% of the sweetener in brand Pepsi-Cola
Nov. 1984	100% level of 55% HFCS approved for both brands

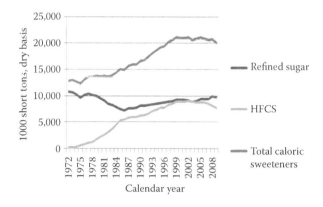

Figure 23.1 US Shipments of Caloric Sweeteners. (From US Department of Agriculture, Economic Research Service (USDA). 2010. Sugar and Sweetener Yearbook Tables. http://www. ers.usda.gov/Briefing/Sugar/data/table49.xls.Table 49—U.S. total estimated deliveries of caloric sweeteners for domestic food and beverage use, by calendar year.)

Manufacture

Manufacture of HFCS is a highly sophisticated, automated operation starting with a continuous flow of corn and ending with a continuous output of finished sweetener. Manufacturing plants are relatively new and use the latest technology in separations, purification and computer automated control, resulting in the industry providing the highest quality food ingredient. The production of HFCS involves four major processing steps: (a) wet milling of corn to obtain starch, (b) hydrolysis of the starch to obtain dextrose, (c) conversion of a portion of the dextrose to fructose, and (d) enrichment of the dextrose-fructose stream to increase the fructose concentration.

Corn Wet Milling

Referring to Figure 23.2, the objective of the corn wet milling process is to separate the starch from the other parts of the corn kernel (hull, gluten and germ). Cleaned, shelled corn is soaked or steeped in a battery of tanks (steeps) in warm water containing 0.1% to 0.2% sulfur dioxide. The steepwater, which swells and softens the kernel to facilitate separation of the various components, flows countercurrently and is ultimately drawn off and replaced by fresh water. The steeped corn is degerminated in a water slurry by passing it first through a shearing mill that releases the germ and subsequently to a continuous liquid cyclone, which separates the germ for oil extraction. After fine grinding, the separated endosperm and hull are screened and the resulting slurry is passed to a continuous centrifuge for starch and gluten separation. The separated starch fraction is filtered and redispersed in water repeatedly to reduce solubles (CRA 1986) and to provide a pure starch slurry available for further processing into starch-based products such as HFCS, corn syrups, and ethanol.

42% High Fructose Corn Syrup

The process for making 42% HFCS is shown in Figure 23.3. The starch slurry made from the corn wet milling process is treated with alpha amylase enzyme at high temperature to liquefy the

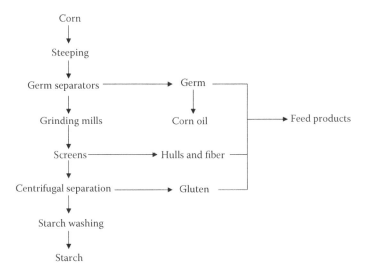

Figure 23.2   Corn wet milling process.

Figure 23.3   Production of 42% HFCS.

starch to a dextrose equivalent (DE) of 10–20. This material is then treated with a saccharifying enzyme, glucoamylase, which hydrolyses the liquefied starch to a dextrose content of about 95%. This high dextrose syrup is filtered, decolorized with carbon, and ion exchange refined to remove salts and other ionic compounds. It is then passed through a fixed-bed column of immobilized glucose isomerase enzyme. This isomerizing enzyme converts glucose to fructose (Figure 23.4) and reaches equilibrium at about 45% fructose. This material is again filtered and refined by carbon and ion exchange treatment and evaporated to become the commercial 42% HFCS.

55% *High Fructose Corn Syrup*

The refined glucose isomerase-treated liquor may be passed through fractionation units to enrich the fructose content to around 90%. This material is filtered and refined by carbon and ion exchange treatment. It may be evaporated and provided as a commercial product with properties similar to crystalline fructose in liquid applications, however, more typically, this high fructose-containing syrup is blended with the 42% HFCS to a concentration of about 55% fructose and evaporated to become the commercial 55% HFCS. The dextrose from the fractionation unit is fed back to the isomerization columns in the 42% HFCS process. This process is shown in Figure 23.5 and a comparison of the composition of these commercially available products is shown in Table 23.3.

For a thorough description of the corn wet milling process, HFS manufacture, and special unit operations, see White 1992, May 1994, and CRA 1986.

Characteristics

Typical chemical and physical characteristics are shown in Table 23.4. As the fructose level increases, so does sweetness. This is the major characteristic provided by HFCS.

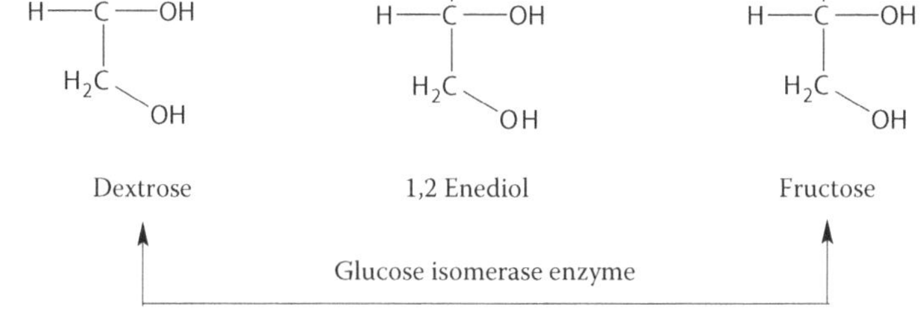

Figure 23.4 Isomerization of dextrose to fructose.

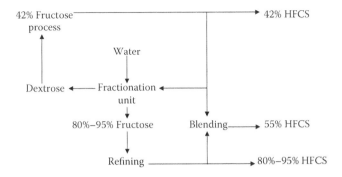

Figure 23.5   Production of 55% HFCS.

Conventional corn syrups are characterized by their DE. DE is defined as the measurement of the total reducing sugars in the syrup calculated as dextrose and expressed as a percentage of the total dry substance. HFCS is composed of a high amount of reducing sugars. Because the DE of HFCS is so high, HFCS is not characterized by DE but rather by the percent of fructose on a dry basis. Measurement of reducing sugars is based on the reducing action of aldose sugars on certain metallic salts (e.g., copper sulfate in Fehling's Solution). Fructose, being a ketose sugar, is converted to an aldose under the alkaline conditions of the tests and behaves as dextrose. The reducing characteristic is one that particularly distinguishes HFCS from sucrose. Sucrose is not a reducing sugar and therefore does not participate in reducing reactions such as nonenzymatic browning until it has inverted into its constituent monosaccharides of fructose and dextrose.

pH for all products is low. Because these products are ion exchange refined, they have little buffering capacity and the pH tends to drift down with time. Because of the low buffering capacity and low acidity, HFCS has little effect on the pH of foods. Lower pH favors color stability for reducing sugars.

The treatment with carbon in the production of HFCS yields products with low color and flavor. Ion exchange treatment provides a product with extremely low ash, which helps to maintain the color and flavor stability of HFCS by removing color precursors and catalysts. Both treatments serve to reduce trace components that may promote or otherwise catalyze color and flavor development. The refining treatments and resulting lack of contaminant result in extremely bland products with few off-flavor notes.

Table 23.3   Typical Composition of High Fructose Corn Syrup[a]

	42% HFCS	*55% HFCS*
Fructose, %	42	55
Dextrose, %	52	41
Higher Saccharides, %	6	4
Ash, %	0.03	0.03
Dry Solids, %	71	77

[a] Dry solids basis.

Table 23.4   Typical Characteristics of HFCS

	42% HFCS	55% HFCS	90% HFCS
Saccharides			
Fructose, %	42	55	90
Dextrose, %	52	41	8
Highers, %	6	4	2
Sweetness	1.0	1.0+	1.2
Dry Solids, %	71	77	77
Moisture, %	29	23	23
Color, RBU	<25	<25	<25
Ash, %	0.03	0.03	0.03
Viscosity (cps)			
80°F	150	700	575
90°F	100	400	360
100°F	70	250	220
pH	3.5	3.5	3.5
Lbs./gal. (20°C)	11.23	11.55	11.56
Fermentable solids, %	96	98	99
Flavor	Sweet, bland	Sweet, bland	Sweet, bland

HFCS products are extremely high in fermentable extract. This would be expected because of the high amount of monosaccharides and disaccharides (94%–100%) and negligible oligosaccharide composition.

HFCS is microbiologically stable and meets the most stringent specifications requested by the food industry. The stability is a function of the high osmotic pressure created by the predominance of the monosaccharides in their composition and the high solids levels. Forty-two percent HFCS is supplied at lower solids than the others because its greater dextrose content renders it more susceptible to crystallization.

HFCS has lower viscosity compared to other nutritive sweeteners such as sucrose or conventional corn syrups but is equivalent to dextrose solutions and invert sugar because of the similar monosaccharide composition. At equal solids levels, HFCS products have comparable viscosity; however, as supplied, 42% HFCS at lower solids has a lower viscosity.

Availability, Transport, Storage, Handling, and Shelf Life

The finished products are typically shipped from the manufacturing plants to the customer in bulk jumbo railcars equipped with food-grade epoxy lining or in bulk stainless steel tank trucks. Tank trucks normally have self-contained positive displacement pumps for transferring the product to customer storage tanks. In addition to direct plant shipments, customers may be supplied the product by way of distributors or sweetener transfer facilities. Hundreds of strategically located facilities around the United States exist for the purpose of transferring the product from railcar to storage tanks, trucks, drums, and so forth, for distribution to the customer.

Railcars are equipped with steam coils in the event the product cools sufficiently to promote crystallization of the more insoluble dextrose component. When this occurs, the syrup can be reliquefied with gentle heat. All HFCS products are microbially stable and, under recommended storage conditions, the shelf life is indefinite and depends on the sensitivity of the final application to color development. HFCS as manufactured, is a clear, colorless liquid. It tends to develop a light straw color (color value of about 25 reference basis units [RBU]) after prolonged time and/or higher temperatures. The product remains usable for many months under recommended storage conditions, maintaining sensory, physical, and chemical specifications. In general, storage temperatures of around 80°F to 90°F will allow the product to be pumpable, prevent crystallization, and minimize color development. Cooler storage temperatures will extend shelf life based on color development. Forty-two percent will crystallize less than around 68°F, 55% at slightly cooler temperatures and 90% will not crystallize at reasonable storage temperatures.

Stainless steel or mild steel tanks lined with food-grade epoxy lining are recommended for storage. Tanks should be equipped with sterile ultraviolet conditioning systems. These units are used to prevent condensation in the tank head-space that may drip and dilute the surface of syrup and make it less microbially stable. The conditioning systems consist of blowers that pass filtered air past ultraviolet lights for sterilization before being introduced into the tank headspace.

Most users will take advantage of the convenience provided by bulk handling systems; however, these products may also be available in intermediate bulk containers, drums, or pails. Users of these containers should be equipped to warm the contents if necessary to facilitate pumping or reliquefy the contents.

HFCS is sold by the pound. It is commonly used by metered volume. The change in density with temperature may be sufficient for very large volume users to effect inventory and yield calculations, so the effect of temperature must be taken into account when measured by volume. Table 23.5 demonstrates the change in weight on the basis of temperature.

Table 23.5 Effect of Temperature on Density

Temperature (°F)	Pounds/Gallon	
	42% HFCS	55% HFCS
60	11.258	11.573
70	11.230	11.542
80	11.201	11.511
90	11.172	11.479
100	11.142	11.448

Specifications and Analysis

HFCS is a consistent and high-quality bulk sweetener in the US market. Because of refining to remove nonsweetener components and extensive microbiological controls in the process, it is able to meet very stringent specifications. Composition, identification, and testing requirements for HFCS have been published in the *Food Chemicals Codex* (*FCC*) with the latest update being the Seventh Edition in 2010 (Table 23.6). Because the beverage manufacturers are the single largest users of these products, stringent guidelines and test procedures have been published by the International Society of Beverage Technologists (ISBT 2010). Those guidelines are shown in Table 23.7 and were the result of extensive collaboration between the soft drink manufacturers and the corn wet millers. Many of the test procedures incorporated were adapted from standard analytical methods developed and published by the Corn Refiners Association (CRA 2009).

Table 23.6   Food Chemical Codex 7—High-Fructose Corn Syrup Specification

Description
High-Fructose Corn Syrup (HFCS) occurs as a water white to light yellow, somewhat viscous liquid that darkens at high temperature. It is a saccharide mixture prepared as a clear, aqueous solution from high-dextrose-equivalent corn starch hydrolysate by the partial enzymatic conversion of glucose (dextrose) to fructose, using an insoluble glucose isomerase preparation that complies with 21 CFR 184.1372 and that has been obtained from a pure culture fermentation that produces no antibiotics. It is miscible in all proportions with water
Functional Use in Foods   Nutritive sweetener
Requirements
Identification   Add a few drops of the sample solution to 5 ml of hot alkaline cupric tartrate TS. A copious red precipitate of cuprous oxide forms
Assay   42% *HFCS*: Not less than 97.0% total saccharides, expressed as a percent of solids, of which not less than 42.0% consists of fructose, not less than 92.0% consists of monosaccharides, and not more than 8.0% consists of other saccharides   55% *HFCS*: Not less than 95.0% total saccharides, expressed as a percent of solids, of which not less than 55.0% consists of fructose, not less than 95.0% consists of monosaccharides, and not more than 5.0% consists of other saccharides
Arsenic (as As)   Not more than 1 mg/kg
Color   Within the range specified by the vendor
Heavy Metals (as Pb)   Not more than 5 mg/kg
Lead   Not more than 0.1 mg/kg
Residue on Ignition   Not more than 0.05%
Sulfur Dioxide   Not more than 0.003%
Total Solids   42% *HFCS*: Not less than 70.5%; 55% *HFCS*: Not less than 76.5%

Table 23.7 International Society of Beverage Technologists—Analytical Guidelines

42% HFCS	Parameter	55% HFCS
71.0% +/− 0.5%	Solids	77.0% +/− 0.5%
70 μmhos/cm Maximum	Conductivity	70 μmhos/cm Maximum
42% Minimum	Fructose	55% Minimum
92% Minimum	Dextrose & Fructose	95% Minimum
1.15 CRA (×100) Max. 20 ICUMSA Units Maximum	Color	1.15 CRA (×100) Max. 20 ICUMSA Units Maximum
6.0 mg/kg Maximum	Sediment	6.0 mg/kg Maximum
3.3–4.5	pH	3.3–4.5
Typical with no Objectionable Taste, Odor, or Appearance	Taste, Odor, & Appearance	Typical with no Objectionable Taste, Odor, or Appearance
50.0 mg/kg Maximum	Chlorides	50.0 mg/kg Maximum
3 mg/kg Maximum	Sulfur dioxide (SO_2)	3 mg/kg Maximum
80 μg/kg Maximum at 11%	Acetaldehyde	80 μg/kg Maximum at 11%
5 μg/kg Maximum at 11% Solids	Isovaleraldehyde	5 μg/kg Maximum at 11% Solids
75 mg/kg Maximum	5-(Hydroxymethyl)-2-furfural	75 mg/kg Maximum
2.0 mg/kg Maximum	Furfural	2.0 mg/kg Maximum
0.5 μg/kg Maximum	2-Aminoacetophenone	0.5 μg/kg Maximum
Passes Test	Floc Potential	Passes Test
1.0 mg/kg Maximum	Iron	1.0 mg/kg Maximum
1.0 mg/kg Maximum	Copper	1.0 mg/kg Maximum
1 mg/kg Maximum	Arsenic	1 mg/kg Maximum
0.1 mg/kg Maximum	Lead	0.1 mg/kg Maximum
200 CFU per 10 g Maximum	Total Mesophilic Bacteria	200 CFU per 10 g Maximum
10 CFU per 10 g Maximum	Yeast	10 CFU per 10 g Maximum
10 CFU per 10 g Maximum	Mold	10 CFU per 10 g Maximum

Because HFCS products are of high and consistent purity, useful properties such as solids and specific gravity can be determined by accurately measuring the refractive index. The CRA-sponsored studies by the Augustana Research Foundation in 1983 resulted in a regression model being generated showing the refractive index-dry solids-density relationship of specific corn sweeteners. This information is used in the physical constants shown in Table 23.8. Complete tables covering all solids ranges are available from HFCS manufacturers and CRA.

Table 23.8 HFCS Physical Constants

Solids%	Refractive Index 20°C	Refractive Index 45°C	Refractometer Brix 1966 ICUMSA	Specific Gravity (in air) 20°C	Lbs. per US Gal. at 20°C	Hydrometer Brix 20°C
42% HFCS						
11.0	1.3493	1.3457	10.92	1.04414	8.689	10.99
70.5	1.4632	1.4577	69.10	1.34717	11.211	69.54
71.0	1.4643	1.4589	69.55	1.35022	11.236	70.02
71.5	1.4655	1.4600	70.04	1.35327	11.261	70.50
72.0	1.4667	1.4612	70.54	1.35633	11.287	70.99
55% HFCS						
11.0	1.3492	1.3456	10.90	1.04418	8.689	11.00
76.5	1.4774	1.4716	74.84	1.38454	11.522	75.38
77.0	1.4786	1.4728	75.31	1.38769	11.548	75.87
77.5	1.4798	1.4740	75.78	1.39084	11.574	76.35

Source: Complied from CRA, ISBT and ICUMSA.

The most critical characteristics of HFCS include the saccharide profile, dry solids, and color. Saccharide profile is best determined by the use of high pressure liquid chromatography analysis. Polarimeters are often used to confirm acceptance of the correct product; however, because of the complex saccharide composition, they are not suitable for specification confirmation.

Color of HFCS is from virtually colorless to light straw. Because light straw is a subjective visual observation, spectrophotometric tests have been developed to measure this parameter. For years, soft drink bottlers used tests measuring RBUs on liquid sugars and set a specification of 35 RBU maximum. With the advent of HFCS with lower typical color, that specification was lowered to 20 RBU maximum. Since that time, the ISBT has adopted the CRA color test. A 20-RBU syrup is equivalent to a CRA color (times 100) of 1.15 maximum. Color values can be mathematically converted between the two units and, for the convenience of those more familiar with RBU color, that conversion equation is as follows:

$$RBU = \frac{CRA \times 100 + 0.0417}{0.0479}$$

The most commonly tested attributes are dry solids and pH, probably because many quality control laboratories are equipped to perform these tests. Dry solids are determined by refractometer. Many refractometers are calibrated with the Brix scale rather than refractive index. For sucrose solutions, Brix corresponds to percent solids in solution. The term Brix cannot be used when designating solids concentrations in other nutritive sweetener solutions such as invert sugars, HFCS, and corn syrups (Junk and Pancoast 1980). From Table 23.8 it can be seen that for HFCS products, the refractive index-Brix relationship is not the same as the refractive index-solids relationship. For 42% HFCS, the Brix value is roughly 1.4 less than the actual delivered solids, and for 55% HFCS, the difference is about 1.7. The difference is less at lower solids levels and at 10% solids the difference is only 0.1 degrees Brix. Also, it is important to realize the effect of temperature on the converted dry solids value. There is a 0.1% solids error in the dry solids reading for every 1°C difference from 20°C. A refractive index or Brix reading taken at 25°C, for example, will result in a 77.0% solids value being reported as 76.5%.

Because of the high purity of HFCS, it has low ionic strength and buffering capacity. For this reason it is imperative that pure dilution water be used and that residual buffer solutions be removed from the probes to prevent erroneous pH results. Although pH is the most commonly used test, titratable acidity is a better indicator of the effect HFCS may have on the pH of a food system.

Properties

Functionally, HFCS provides sweetness, flavor enhancement, viscosity, humectancy, nutritive solids, fermentability, and reducing characteristics such as flavor and color development. HFCS can be used to control crystallization, modify freezing point, raise osmotic pressure, and lower water activity.

By far, sweetness is the most important functional property of HFCS. Table 23.9 shows the relative sweetness of several sweeteners compared with sucrose. Although sweetness levels may be comparable, the actual sweetness may vary in a food system depending on such factors as the temperature, pH, and solids. Sweetness is also affected by the flavor profile (Shallenberger 1998). HFCS has a rapid sweetness release. Compared to sucrose, HFCS is considered to enhance fruit flavors, spicy notes, and tartness. This is because the sweetness of HFCS is quickly perceived and does not linger. Sucrose on the other hand has a slower onset of sweetness but a longer lasting action. This can have a flavor-masking effect. The combination of the two can result in a synergism that results in a system perceived sweeter than the individual ingredients are on their own.

Over the shelf life of a product, acidic food systems (such as most beverages) sweetened with sucrose will invert (i.e., convert the disaccharide sucrose into its respective free monosaccharides, fructose and dextrose), changing the sugar profile and consequently the flavor and sweetness of the product. HFCS has an advantage over sucrose in acidic food systems because it does not invert. Table 23.10 illustrates the effect of sucrose inversion.

Although HFCS as supplied has a significantly lower viscosity than sucrose at the same level, the difference is minimal at lower solids levels. Viscosities of sweetener systems as delivered are shown in Table 23.11.

Table 23.9 Relative Sweetness of Nutritive Sweeteners

Sweetener	Sweetness Relative to Sucrose
Sucrose	1.0
Crystalline Fructose	1.2–1.8
90% HFCS	1.2–1.6
42% HFCS	1.0–
55% HFCS	1.0+
Sucrose Medium Invert	1.0+
Dextrose	0.7

Table 23.10   Chemical Comparison between Sucrose and HFCS

100 g Sucrose total inversion 52.6 g dextrose + 52.6 g fructose
100 g HFCS (dry basis) yields 94 to 100 grams dextrose and fructose

HFCS is an economical humectant and widely used for that purpose in the food industry. It has greater sweetening power than other common humectants such as glycerine and sorbitol.

Having a DE of around 97, HFCS is composed predominately of reducing sugars. Reducing sugars react with proteins in food systems to generate brown pigments through the Maillard or nonenzymatic browning reaction. The reaction is greatly enhanced at elevated temperatures such as in baking, giving rise to the brown crust color of items such as breads and rolls. At room temperature and below, the reaction is greatly reduced.

The monosaccharide makeup of HFCS results in changes in colligative properties versus the disaccharide sucrose. HFCS depresses the freezing point of food systems compared to sucrose, making a lower freezing, softer product. This property can be an advantage in some applications but a disadvantage in others. When identical freezing points are required, conventional corn syrups are often used to balance the freezing point.

The high osmotic pressure of HFCS is also a result of the monosaccharide composition. This property makes HFCS as supplied microbially stable and has been suggested as one reason monosaccharides are better for pickling and preserving processes such as in the manufacture of pickles and processed fruit.

The smaller molecules also result in lower water activity compared with the disaccharide sucrose. Low water activities allow greater microbial stability at higher moisture content, which often improves texture and shelf life.

Applications

HFCS replaces sucrose on a solids basis in most foods, except where a dry sweetener is required, such as dry mixes and when hygroscopicity is a concern as in some confections. Some of the diverse product categories that find HFCS as an ingredient are bakery products, beverages of all types, processed fruits, condiments, frozen desserts, jams, jellies and preserves, pickles, wines, and liqueurs.

Beverages

By far the largest market segment is in beverages. Most 55% HFCS is used in this market. The major cola brands use that product because of its higher sweetness level. The availability of HFCS

Table 23.11   Viscosity of Nutritive Sweeteners at 100°F

	Viscosity (centipoises)
42% HFCS	70
55% HFCS	250
67.5° Brix Liquid Sucrose	115
42 DE/43 Bé Corn Syrup	21,000

as a bulk liquid sweetener along with its consistent carbohydrate composition, narrow dry solids specification and low ash level, allows for highly automated bottling lines that use online instrumentation, which continuously meters and blends beverages to exacting beverage specifications at rates exceeding 1700 12-ounce cans per minute.

Typically beverages use 10%–12.5% HFCS solids for desired sweetness. More acidic, stronger flavored beverages are at the high end. Sports drinks with isotonic requirements encouraging quick gastric emptying are typically at lower solids levels of about 2%–9% (Coleman 1999). HFCS is also used as a fermentation adjunct in wines and as a sweetener for wine coolers and liqueurs.

Baking

The second single largest market for HFCS is the baking industry. Again, HFCS as a bulk liquid sweetener is ideal for large, automated bread and bun manufacturing lines. It is ideal in yeast raised goods such as breads, buns, rolls, and yeast raised donuts because it is directly fermentable by yeast without the inverting action required before sucrose is used. In general, at equal solids levels, proof times are slightly shorter using HFCS compared with sucrose and volume, crust color, flavor, and grain are the same (Strickler 1982).

Besides its use as a sweetener and fermentable sugar, it is widely used in the baking industry as a humectant. It is incorporated into items such as cakes, cookies, and brownies as a humectant and, compared with other commonly used humectants such as glycerine and sorbitol, it has increased sweetness. In this regard it is commonly used in bakery fillings as well.

High substitution levels for sucrose in chemically leavened products such as cakes and cookies, result in different characteristics in the finished product. Cookies are plumper and softer, with less spread (Dubois 1986) but with lower water activity (Dubois 1987). Cakes are denser with less volume and flatter tops (Ohr 1999), which is reported to be due to starch gelatinization at lower temperature (White and Lauer 1990). Under equal conditions, chemically leavened bakery products are browner. In general, cakes of equal quality can be achieved with a partial replacement (10%–60%) of sugar with HFCS; however; changes in formulation can be made to achieve quality products at higher levels. Color may be controlled or minimized by lowering baking temperatures and lowering the pH slightly. The effect on leavening action may be modified by using higher protein flour and/or a highly emulsified shortening.

Processed Foods

The fruit canning industry uses a tremendous amount of HFCS, usually in combination with corn syrup and sucrose. HFCS is also being used in vegetable canning such as peas, beets and corn.

Reducing properties are important in maintaining the bright red color of tomato catsup and strawberry preserves (Palmer 1982), and HFCS as a sweetener is used in combination with conventional corn syrups in these items.

A large amount of HFCS is used in the production of jams, jellies and preserves, usually in combination with conventional corn syrups to provide solids and sweetness. The HFCS/corn syrup formulation is an economical sweetener system that enhances fruit flavors, prevents crystallization, and preserves color.

Ice Cream and Dairy Products

HFCS is used in ice cream and other frozen desserts at levels from 2%–8%. Removing the disaccharide sucrose from frozen desserts and replacing it with HFCS has an effect on the freezing point and texture of the product. In some cases, this is a benefit and in other cases both sucrose and HFCS are the preferred combination in these products. Formulators adjust the mixture with corn syrup, maltodextrin or other bulking agents to fine-tune the sweetness, freezing point, and texture of the system.

Dairies incorporating HFCS in frozen desserts are likely to manufacture chocolate milk as well. HFCS works well in this beverage and has an ingredient sparing effect on cocoa powder, allowing a 5%–15% reduction (Strickler 1982).

Nutritional, Safety, and Regulatory Status

HFCS containing 42% or 55% fructose is affirmed Generally Recognized as Safe (GRAS) with no limitations other than current good manufacturing practices per Title 21 of the US Code of Federal Regulations part 184.1866.

The nutritional value of HFCS is essentially that of 100% available carbohydrate contributing 4 kcal/gram energy. Of the carbohydrate fraction, roughly 98% of HFCS is classified as sugars (monosaccharides and disaccharide) for nutritional labeling. Most experts agree that nutritionally HFCS and sucrose are the same (American Medical Association 2008; American Dietetic Association 2004). Once the combination of glucose and fructose found in HFCS and sugar is absorbed into the blood stream, the two sweeteners appear to be metabolized similarly in the body (Coulston and Johnson 2002; Sigman-Grant and Jorita 2003; Melanson et al. 2007; Zukley et al. 2007; Lowndes et al. 2007; Akhavan and Anderson 2007; Almiron-Roig and Drewnowski 2003).

In 1983, the US FDA formally listed HFCS as safe for use in foods and, as mentioned previously, affirmed that decision in 1996. In the GRAS affirmation review by the US FDA, the agency concluded that HFCS "contains approximately the same glucose to fructose ratio as honey, invert sugar, and the disaccharide sucrose" (Federal Register 1996, 43447). Table 23.12 shows the changes in the per capita sweetener consumption since 1970.

Glycemic index is a measure of the rate at which a food substance raises blood sugar after ingestion and is often used by persons with diabetes as an indicator of the suitability of a particular food or ingredient in their meal plan. Pure fructose has a very low glycemic index of about 19, compared with glucose at 100. Sugar and honey, both with similar compositions to HFCS, have moderate GI values that average 55 and 68 respectively, (Foster-Powell 2002). Although it has not yet been specifically measured, HFCS would be expected to have a moderate GI because of its similarity in composition to honey and sugar.

The name High Fructose Corn Syrup, coined in the 1970s to differentiate corn based sweeteners containing the sweet compound fructose versus those that contain none, has been a source of confusion to consumers, the media and even some scientists. HFCS, despite its name, is not high in fructose and is simply a type of corn sugar. It is in the same class of nutritional sweeteners as cane sugar, beet sugar, and honey that consist of fructose and glucose in approximately equal amounts. As of this writing a citizen petition to the US FDA has been filed by the Corn Refiners Association (CRA) that requests the name "corn sugar" be allowed as a common or usual name for listing on food ingredient labels. Among the data provided in support of the petition, are survey results showing consumer misperceptions of the ingredient. As a result of the name, many consumers wrongly believe that HFCS contains more fructose,

Table 23.12 US per Capita Sweetener Consumption[a]

Crop Year	Refined Sugar	HFCS[c]	Glucose Syrups	Dextrose	Total Corn Sweeteners	Honey and Edible Syrups	Total Caloric Sweeteners
1970	101.8	0.5	10.7	4.6	15.9	1.5	119.1
1971	102.1	0.8	11.2	4.6	16.7	1.4	120.2
1972	102.3	1.2	12.0	4.6	17.8	1.5	121.5
1973	100.8	2.1	13.1	4.6	19.7	1.4	122.0
1974	95.7	2.8	13.8	4.5	21.2	1.1	117.9
1975	89.2	4.9	14.0	4.4	23.3	1.4	113.8
1976	93.4	7.2	13.9	4.1	25.2	1.3	119.9
1977	94.2	9.6	13.8	3.9	27.3	1.3	122.8
1978	91.4	10.8	13.9	3.7	28.4	1.5	121.3
1979	89.3	14.8	13.5	3.5	31.8	1.4	122.6
1980	83.6	19.0	12.9	3.5	35.3	1.2	120.2
1981	79.4	22.8	12.9	3.4	39.1	1.2	119.8
1982	73.7	26.6	12.7	3.4	42.7	1.3	117.7
1983	70.3	31.2	13.0	3.4	47.6	1.4	119.3
1984	66.7	37.2	13.1	3.5	53.8	1.3	121.8
1985	62.7	45.2	13.5	3.5	62.2	1.3	126.2
1986	60.0	45.7	13.6	3.6	62.8	1.4	124.3
1987	62.4	47.7	13.8	3.6	65.2	1.3	128.8
1988	62.1	49.0	14.3	3.7	66.9	1.2	130.2
1989	62.8	48.2	12.8	3.5	64.6	1.2	128.5
1990	64.4	49.6	13.6	3.6	66.8	1.2	132.4
1991	63.6	50.3	14.0	3.7	68.0	1.3	132.9
1992	64.2	51.8	15.1	3.6	70.5	1.4	136.1
1993	63.8	54.5	15.8	3.7	73.9	1.4	139.2
1994	64.4	56.2	15.9	3.8	75.9	1.4	141.6
1995	64.9	57.6	16.3	4.0	77.9	1.3	144.1
1996	65.1	57.8	16.4	4.0	78.2	1.4	144.7

Table 23.12 (Continued) US per Capita Sweetener Consumption[a]

Crop Year	Refined Sugar	HFCSC	Glucose Syrups	Dextrose	Total Corn Sweeteners	Honey and Edible Syrups	Total Caloric Sweeteners
1997	64.9	60.4	17.3	3.7	81.4	1.3	147.7
1998	64.9	61.9	17.1	3.6	82.7	1.3	149.0
1999	66.3	63.7	16.3	3.5	83.5	1.5	151.4
2000	65.5	62.7	15.8	3.4	81.8	1.5	148.9
2001	64.5	62.6	15.5	3.3	81.4	1.3	147.3
2002	63.3	62.9	15.5	3.3	81.6	1.5	146.5
2003	61.0	61.0	15.2	3.1	79.3	1.4	141.7
2004	61.7	59.9	15.6	3.3	78.9	1.3	141.9
2005	63.2	59.2	15.3	3.3	77.8	1.5	142.5
2006	62.6	59.3	13.8	3.1	75.2	1.6	139.4
2007	62.3	56.3	13.7	3.0	73.0	1.3	136.7
2008	65.4	53.2	13.4	2.8	69.3	1.5	136.2
2009[b]	63.6	50.1	13.0	2.7	65.8	1.4	130.7

Source: From US Department of Agriculture, Economic Research Service (USDA). 2010. Sugar and Sweetener Yearbook Tables. http://www.ers.usda.gov/Briefing/Sugar/data.htm yearbook.

[a] Dry basis.
[b] Preliminary.
[c] USDA/ERS Table 23.49—US total estimated deliveries of caloric sweeteners for domestic food and beverage use, by calendar year

is higher in calories and is sweeter than other typical sweeteners such as sugar, honey and fruit juice concentrates.

Economics/Cost

The cost of the raw material, corn, has a major influence on industry economics; however, other factors can seriously offset that parameter. Net corn costs are determined after subtracting the value of corn wet milling by-products of corn oil, corn gluten feed, and corn gluten meal. Strong demand and pricing for these by-products has the net effect of reducing overall corn costs, whereas weak markets drive the effective price of the raw material up. Major additional costs are for energy (steam and electricity), manpower (salaries, wages, and overhead), and supplies (chemicals, enzymes, packaging, etc.). These are in addition to water and waste treatment costs, depreciation, maintenance, and so forth, required to operate the process.

Historical prices of HFCS and sucrose are shown in Table 23.13 (USDA 2010).

Table 23.13 Pricing Sugar, HFCS 42 and 55

Calendar Year	Sugar[a]	HFCS 42[b]	HFCS 55[b]
2000	42.41	12.76	15.72
2001	43.42	13.66	17.05
2002	43.10	15.76	18.50
2003	42.68	16.38	19.07
2004	42.64	17.98	20.54
2005	43.54	18.62	21.12
2006	49.58	19.35	21.77
2007	51.48	23.41	25.49
2008	52.91	27.57	29.32
2009	57.03	31.51	32.95

[a] USDA 2010, Table 23.6—US retail refined sugar price, cents per pound
[b] USDA 2010, Table 23.9—US prices for high fructose corn syrup, wholesale list price, cents per pound dry weight

References

Akhavan, T., and Anderson, G.H. 2007. Effects of glucose-to-fructose ratios in solutions on subjective satiety, food intake, and satiety hormones in young men. *Am J Clin Nut* 86(5):1354–1363.

Almiron-Roig, E., and Drewnowski, A. 2003. Hunger, thirst, and energy intakes following consumption of caloric beverages. *Physiol Behav* 79:767–774.

American Dietetic Association. 2004. Use of nutritive and nonnutritive sweeteners. *J Am Diet Assoc* 104:255–275.

American Medical Association. 2008. *Report 3 of the Council on Science and Public Health (A-08): The Health Effects of High Fructose Corn Syrup*. Chicago: American Medical Association.

Coleman, E. 1999. Sports Drink Research. Food Technology. March, pp. 104–108.

Corn Refiners Association (CRA). 1986. Corn Refiners Association Symposium on Food Products of the Wet Milling Industry, 25 September. Washington, DC: CRA.

Corn Refiners Association (CRA). 2009. Analytical Methods of the Member Companies of the Corn Refiners Association. October.

Coulston, A.M., and Johnson, R.K. 2002. Sugar and sugars: Myths and realities. *J Am Diet Assoc* 102(3):351–353.

Dubois, D. (Ed.). July 1986. AIB Technical Bulletin.

Dubois, D. (Ed.). July 1987. AIB Technical Bulletin.

Federal Register. 1996. Rules and Regulations. Vol. 61, No. 165, Friday, August 23, 1996.

Foster-Powell, K., Holt, S.H.A., and Brand-Miller, J.C. 2002. International table of glycemic index and glycemic load values. *Am J Clin Nutr* 76:5–56.

International Society of Beverage Technologists (ISBT). 2010. High Fructose Syrups 42&55 Quality Guidelines and Analytical Procedures Revision 5. April.

Junk, W.R., and Pancoast, H.M. 1980. *Handbook of sugars*. 2nd edition, Westport, CT: AVI Publishing Company.

Lowndes, J., Zukley, L.M., Nguyen, V., Angelopoulos, T.J., and Rippe, J.M. 2007. The Effect of High-Fructose Corn Syrup on Uric Acid Levels in Normal Weight Women. Presented at the June 2007 meeting of The Endocrine Society. Program Abstract #P2-45. June.

May, J.B. 1994. Wet milling: Process and products. In *Corn: Chemistry and technology*, S.A. Watson and P.E. Ramstad (Eds.), 377–397. St. Paul: American Association of Cereal Chemists, Inc.

Melanson, K.J., Zukley, L., Lowndes, J., Nguyen, V., Angelopoulos, T.J., and Rippe, J.M. 2007. Effects of high-fructose corn syrup and sucrose consumption on circulating glucose, insulin, leptin, and ghrelin and on appetite in normal weight women. *Nutrition* 23(2):103–112.

Milling and Baking News. 12 March 1996. p. 23.

Ohr, L.M. 1999. Prepared Foods. March.

Palmer, T.J. 1982. Nutritive Sweeteners from starch. In *Nutritive sweeteners*, G.G. Birch and K.J. Parker (Eds.), 83–108. London: Applied Science Publishers.

Park, Y.K., and Yetley, E.A. 1993. Intakes and food sources of fructose in the US. *Am J Clin Nutri.* 58:737S–747S.

Shallenberger, R.S. 1998. Sweetness theory and its application in the food industry. *Food Tech* 52(7):72–76.

Sigman-Grant, M., and Jorita, J. 2003. Defining and interpreting intakes of sugars. *Am J Clin Nutr* 78(4):815S–826S.

Strickler, A.J. 1982. Corn syrup selections in food applications. In *Food carbohydrates*, D.R. Lineback and G.E. Inglett (Eds.), 12–24. Westport, CT: AVI Publishing Company.

US Department of Agriculture, Economic Research Service (USDA). 2010. Sugar and Sweetener Yearbook Tables. http://www.ers.usda.gov/Briefing/Sugar/data.htm yearbook.

White, J.S. 1992. Fructose syrup: Production, properties, and applications. In *Starch hydrolysis products*, F.W. Schenck and R.E. Hebeda (Eds.),177–199. New York: VCH Publishers, Inc.

White, D.C., and Lauer, D.N. 1990. Predicting gelatinization temperature of starch/sweetener systems for cake formulation by differential scanning calorimetry. *Cereal Foods World* 35:728–731.

Zukley, M., Lowndes, J., Melanson, K.J., Nguyen, V., Angelopoulos, T.J., and Rippe, J.M. 2007. The effect of high fructose corn syrup on post-prandial lipemia in normal weight females. Presented at the June 2007 Meeting of The Endocrine Society. Program Abstract #P2-46. June.

Chapter 24

Isomaltulose

Anke Sentko and Jörg Bernard

Contents

Introduction

Isomaltulose is a slow release carbohydrate. Although a disaccharide, its physiological properties differ significantly from traditional sugars. Isomaltulose does not promote dental caries and has a low glycemic blood glucose response. Isomaltulose is hydrolyzed and absorbed slowly and completely in the small intestine. Due to this digestion process, the energy is provided in a balanced and sustained way and without gastrointestinal distress. These key physiological properties make isomaltulose unique.

Isomaltulose is an isomer of sucrose. Its chemical name is 6-O-α-D-glucopyranosyl-D-fructofuranose. The glucose and fructose moieties are linked α-1,6-glycosidic in isomaltulose instead of α-1,2-glycosidic as in sucrose. Due to this, digestion slows down and physiology differs.

Production

Isomaltulose is a natural constituent of honey (Siddiqua and Furgala 1967), occurring at levels up to 1%, and has also been found in sugar cane extract (Egglestone and Grisham 2003; Gómez Bárez et al. 2000; Takazoe 1985).

Commercial isomaltulose is derived from sucrose by enzymatic rearrangement (isomerization). The isomerization is catalyzed by an immobilized enzyme preparation of non-viable cells of Protaminobacter rubrum followed by crystallization. The enzyme and its production-organism were discovered by Südzucker in the 1950s (Schiweck 1980; Weidenhagen and Lorenz 1957). The criteria for characterization and purity and the corresponding analytical methods for isomaltulose are laid down in the *Food Chemicals Codex* (*FCC*) (Figure 24.1).

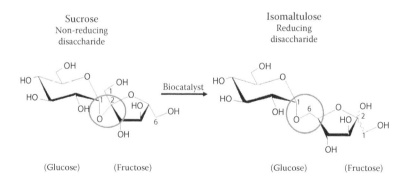

Figure 24.1 Enzymatic rearrangement of sucrose to isomaltulose (structural formula).

Sensory Properties

The sweetness quality of isomaltulose is similar to sucrose. It is perceived quickly, it is refreshing and heavy without any aftertaste. Its sweetening power in a 10% solution is 0.48 related to a sucrose solution.

Isomaltulose is reported to mask the off-flavors of some intense sweeteners (Anonymous 1985).

Physical and Chemical Properties

Physical Properties

The appearance of isomaltulose, a white crystalline material, is similar to that of sucrose. Isomaltulose crystallizes in the form of a monohydrate (Dreissig and Luger 1973); no anhydrous crystalline phase has been found (Gehrich 2002). The crystal water is so tightly bound, that during milling or other strong mechanical treatment, no release of crystal water is observed. Furthermore, it is practically non-hygroscopic. Thus, crystalline isomaltulose hydrate can be used in applications sensitive to moisture, for example, chocolate. Like sorbitol hydrate, isomaltulose hydrate melts within its crystal water. Its melting point (123°C–124°C) is lower than that of sucrose (186°C). Properties of isomaltulose crystals and amorphous isomaltulose are summarized in Table 24.1.

The solubility of isomaltulose, as seen in Figure 24.2, although lower than that of sucrose, is adequate for most applications (Schiweck 1980). The viscosity of isomaltulose solutions is very similar to that of sucrose solutions (Schiweck 1980). Likewise, the vapor pressure (water activity) and the freezing point depression of aqueous isomaltulose solutions are similar to those of sucrose solutions.

Table 24.1   Physio-chemical Properties of Isomaltulose

	Isomaltulose	*Sucrose*
Molecular mass in g/mol	360,318 (hydrate)	342,303
Crystal group	Rhombic P2₁P2₁P2₁[a]	Monoclinic[c]
Specific rotation	103 … 104°	186 ± 4[c]
Melting temperature in °C	123 … 124	46.41[c]
Heat of fusion in kJ/mol	61[d]	65[c]
Heat of solution in kJ/mol	23.46 crystal[b]	6.22 crystal[b]
	−17.25 amorphous[b]	−13,79 amorphous[b]
Glass transition temperature in °C	62[b]	65[c]

[a] Dreissig and Luger 1973.
[b] Gehrich 2002.
[c] Bubnik et al. 1995.
[d] Südzucker CRDS, internal data.

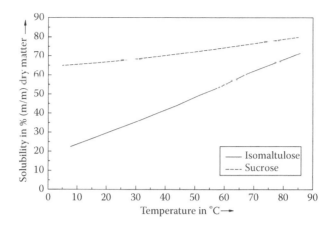

Figure 24.2   Solubility of isomaltulose.

Chemical Properties

Isomaltulose is very acid stable compared with sucrose. Figure 24.3 illustrates that even at pH 1 with HCl, a 10% isomaltulose is stable for more than 30 minutes when held at 95°C, whereas a 10% sucrose solution is nearly completely hydrolyzed under these conditions. Likewise, it is proven that an 8% (m/m) isomaltulose solution at pH 3 (typical value for beverages) is stable for at least nine months at room temperature (see the section "Beverage Applications"). Due to the lower melting temperature and the higher stability of the glycosidic linkage, isomaltulose shows less caramelization during heat treatment than sucrose. Isomaltulose as a reducing sugar is subject to Maillard reactions. The color formation during the reaction of isomaltulose and lysine is plotted in Figure 24.4. The microbial stability of products prepared with isomaltulose is very good.

Isomaltulose is one of the most active saccharides in terms of antioxidative behavior. This behavior can be demonstrated with various methods like trolox equivalent antioxidative capacity (TEAC), ferric reducing antioxidant power (FRAP), Chapon, or mitteleuropäische brautechnische analysenkommission (MEBAK). Figure 24.5 shows the difference of the reductive potential

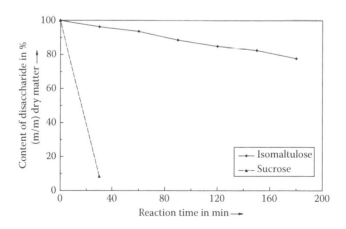

Figure 24.3   Acid stability of isomaltulose at elevated temperature (pH 1 with HCl, 10% solution). (From Südzucker AG Mannheim/Ochsenfurt CRDS n.d.)

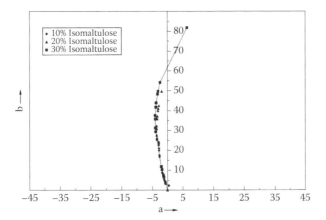

Figure 24.4   Color formation during Maillard reaction.

of isomaltulose versus fructose with respect to ferric salts at 60°C. Although this property is by far less developed compared to classical antioxidative agents, a significant gradation can be observed to other saccharides. Nevertheless, the concentration of saccharides in food products is high compared to classical antioxidative compounds and thus the antioxidative capacity could be sufficient to increase the shelf life of a final product containing isomaltulose.

Physiological Properties

With respect to its physiological properties, isomaltulose is a fully, yet slowly available carbohydrate. It has a low effect on blood glucose and insulin levels; it allows a more balanced, longer lasting supply of energy and supports fat oxidation. In addition, isomaltulose is kind to teeth.

The physiological properties of isomaltulose are largely related to its slower hydrolysis by intestinal enzymes. Isomaltulose has been shown to be hydrolyzed to glucose and fructose by intestinal α-glycosidasis (Dahlqvist et al. 1963). In studies utilizing human small intestinal mucosa homogenates as an enzyme source, isomaltulose was hydrolyzed with Vmax of 26%–45% compared to

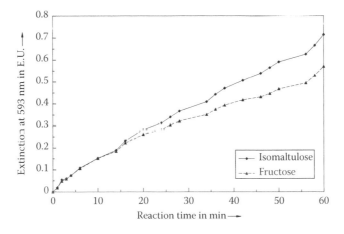

Figure 24.5   Antioxidative potential of isomaltulose and fructose in Fe(II)/Fe(III) system.

sucrose (Lina et al. 2002). The slower hydrolysis and absorption of isomaltulose, respectively of its constituents, results in a slower rise and lower maximum increase of blood glucose and corresponding insulin responses compared to sucrose (Lina et al. 2002).

Although the release of isomaltulose is slower, it is nevertheless completely available and no significant amounts reach the large intestine. This was demonstrated in a study with healthy subjects with an end ileostomy. Fifty grams of isomaltulose were incorporated in a breakfast meal and in another trial taken as a drink, in both cases after overnight fasting. The determined apparent digestibility was 95.5% and the apparent absorption was estimated to be 93.6% when taken as a drink. Intake in a meal resulted in an apparent digestibility of 98.8% and an apparent absorption of 96.1%, demonstrating virtually complete digestion and absorption in the small intestine, independent of the mode of intake (Holub et al. 2010).

While isomaltulose is slowly released, it is nevertheless completely available, resulting in an energy conversion factor for food labeling purposes of 4 kcal/g, the general value for carbohydrates.

Dental Health

Isomaltulose is virtually not used by the oral flora as a substrate for fermentation. No significant amounts of acids are produced and therefore the demineralization process known from fermentable, cariogenic carbohydrates does not occur after eating isomaltulose. The plaque pH telemetry, an *in vivo* method developed at the University of Zürich, enables the observation of plaque pH changes on the tooth surface during and after the ingestion of a carbohydrate-based food. This has been established as the basic method for dental claims in several countries. The corresponding pH telemetry curve of isomaltulose in Figure 24.6 shows that the ingestion of isomaltulose, unlike that of sucrose, is not followed by a decrease in plaque pH below the value of 5.7, where

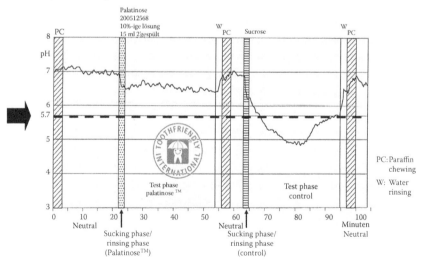

pH-Telemetry test of a 10% isomaltulose drink versus a 10% sucrose drink

Figure 24.6 pH Telemetry curve of isomaltulose demonstrating toothfriendliness as established by Imfeld, T. (University of Zürich). (From Van Loveren, C. 2009. Oral and dental health—prevention of dental caries, erosion, gingivitis and periodon-titis. ILSI Europe Concise Monograph Series, ILSI Europe; Prof. Imfeld, zahnmedizinisches zentrum der universität zürich, 07.03.2006, pH telemetrietest.)

demineralization processes occur (Van Loveren 2009). Therefore, isomaltulose is toothfriendly, and enables the production of toothfriendly foods.

Isomaltulose is included in the list of non-cariogenic carbohydrate sweeteners in the context of the US FDA health claim approval on carbohydrate sweeteners and dental caries (US FDA 2008).

In Japan, isomaltulose has Food for Specified Health Uses (FOSHU) status due to its dental properties (Japanese MHLW).

Effect on Blood Glucose and Insulin

The slow release of isomaltulose, due to the four to five times slower hydrolysis and absorption, is reflected in its blood glucose response curve and insulin response. Compared to sucrose, the blood glucose response of isomaltulose is lower, more balanced and prolonged (i.e., extended over a longer period of time), still delivering energy while the blood glucose response curve after sugar intake is already below baseline. These basic characteristics were demonstrated in a number of blood glucose response studies including, for instance, the glycemic index (GI) study at Sydney University (SUGiRs 2002) and the study by Holub et al. (2010).

The glycemic index and insulinemic index of isomaltulose were determined at Sydney University, based on an intake of 50 g isomaltulose and using the same amount of glucose as the reference, resulting in a glycemic index of 32 and an insulinemic index of 30 (SUGiRS 2002). Accordingly, isomaltulose is a disaccharide carbohydrate with a very low glycemic index, and enables the production of carbohydrate-based foods with a low or reduced blood glucose response.

On a long-term perspective, in a diet primarily based on slowly available carbohydrates, leading to lower blood glucose and insulin levels throughout the day, isomaltulose is thought to be supportive in metabolic control. Findings from a four week intervention study by Holub et al. (2010) show that a daily intake of 50 g isomaltulose with various foods as part of a normal diet is well tolerated and may even contribute to improved carbohydrate metabolism parameters over a longer period of time as a consequence of its lower and prolonged glycemic response. In this study, adults with impaired fat metabolism (hyperlipidemia) consumed 50 g isomaltulose or sucrose daily with various foods, as part of a controlled diet. The regular isomaltulose consumption was well tolerated with no adverse effects on blood lipids (cholesterol, LDL, HDL, TAG, apo) or cardiovascular risk markers (oxidized LDL, NEFA). Moreover, carbohydrate metabolism parameters, namely fasting blood glucose levels and insulin resistance, were significantly lower after four weeks of isomaltulose intake compared to the beginning, while no such significant differences were seen with sucrose.

Effect on Fat Oxidation

As a result of its more steady blood glucose supply at a low glycemic level and the corresponding insulin profile, the ingestion of isomaltulose is associated with less suppression of fat oxidation in comparison to readily available carbohydrates. Higher levels of fat oxidation with isomaltulose in comparison to conventional high glycemic carbohydrates have been observed in physically active people, as well as in overweight persons under mostly sedentary conditions.

Findings with physically active people showed that the intake of isomaltulose-containing sports drinks before and during an intense endurance-type activity on a bicycle ergometer, led to a higher contribution of fat utilization on total energy expenditure than the consumption of respective drinks with the high glycemic carbohydrate maltodextrin (Koenig et al. 2007). With isomaltulose, the postprandial increase in blood glucose and insulin levels was lower and concentrations of free fatty acids were higher than with maltodextrin. In parallel, the Respiratory

Quotient (RQ), determined from gas exchange in breath using indirect calorimetry, was significantly lower throughout the entire test period, indicating a higher contribution of fat oxidation in total energy expenditure. Also, Achten et al. (2007) reported a higher contribution of fat oxidation with isomaltulose in comparison to sucrose during a 150 minute exercise of moderate intensity by trained men. An increased contribution of fat oxidation is particularly relevant in endurance-type physical activity because it can potentially spare carbohydrate sources in glycogen stores in favor of enhanced endurance performance.

In addition, overweight persons may profit from a higher contribution of fat oxidation in energy metabolism. In a study at the University of Freiburg with overweight people with insulin resistance, the intake of meals containing isomaltulose in comparison with conventional readily available carbohydrates (i.e., a sucrose-glucose syrup combination) resulted in lower rises in blood glucose and lower daily insulin levels as well as in an increase in fat oxidation by up to 28% (Kozianowski 2007).

The effect of isomaltulose on fat oxidation in overweight persons under mostly sedentary conditions has been investigated in a study by van Can et al. (2009), which was conducted with ten healthy, overweight persons in a randomized, single-blind, cross-over design. In this study, the ingestion of an isomaltulose-containing drink (75 g, 400 ml) with a standardized mixed meal at lunch time caused lower glucose and insulin responses in comparison to sucrose and subsequently less inhibition of postprandial fat oxidation. Fat oxidation rates were higher by about 14%.

Gastrointestinal Tolerance

The gastrointestinal tolerance of isomaltulose is similar to sucrose (Holub et al. 2010; Lina et al. 2002). As isomaltulose is virtually completely digested and absorbed in the small intestine, no gastrointestinal distress occurs, as this sign of physiological overload is related to nutrients that are nondigestible or low digestible. Even when taken in high amounts during physical activity, no gastrointestinal discomfort was reported. Having in mind that other nutritive sweeteners or other carbohydrates with a low glycemic response have their characteristic property because they are low digestible or nondigestible, thus are related to gastrointestinal distress at some point, this positive aspect of isomaltulose needs to be emphasized.

Toxicological Evaluations

The chemical nature of isomaltulose and its known physiology clearly demonstrate that isomaltulose is safe. Biochemical studies and toxicological studies were evaluated (Lina et al. 2002) resulting in the conclusion that the use of isomaltulose is of no health concern. This is reflected as well in several assessments of isomaltulose safety in regulatory processes for approval from various major authorities, resulting in the confirmation of the food status of isomaltulose and its use in food in general.

Applications and Product Development

On the basis of its physical, chemical, organoleptic and excellent physiological properties, isomaltulose is used in a wide range of applications. It is a disaccharide carbohydrate which is fully digestible, slowly released, low glycemic, and kind to teeth. It has a clearly defined and stable chemical structure. Beverages or firm formulations in confectionery are the main applications. Different types of isomaltulose are available under the brand name Palatinose, produced by Beneo-Palatinit. Isomaltulose has most recently been used in sports drinks, ice tea, chocolates, chewy and gummy candies, chewing gum and coated confectionery. Existing processing equipment from sugar treating processes can be used for all applications without requiring major changes (Südzucker AG Mannheim/Ochsenfurt CRDS n.d.). Only formula and process parameter modifications are recommended to optimize processes and products.

The spectrum of isomaltulose applications is listed in Table 24.2.

Beverage Applications

Ready-to-drink Products, Soft Drinks, Sports Beverages, Dairy Drinks, Toothfriendly Children's Drinks

The most important application of isomaltulose can be found in beverages (Cegelski et al. 2009b). Due to its chemical stability, also in an acidic environment, isomaltulose exhibits no hydrolysis in low pH beverages. Figure 24.7 demonstrates the stability of an isomaltulose ready-to-drink beverage compared to a sucrose based formulation for at least one year.

Many sports drinks with a pH value of ~3 currently on the market contain sucrose, which hydrolyzes into glucose and fructose in this acidic environment. In this case the number of osmoactive particles increases and the isotonic balance is destroyed. In contrast, isomaltulose stays stable under these conditions and can act as an ingredient for isotonic, hypotonic and hypertonic beverage concepts, as it helps to maintain the osmolality of the product (Cegelski et al. 2009a). This behavior provides a solution for applications in water, fruit or dairy based products.

The use of isomaltulose in soft drinks results in a sucrose-like, natural, sweet perception without any aftertaste. As its sweetening power is about 50% of sucrose formulation, the sweetness may be adjusted with the use of a combination with sugars like fructose or intense sweeteners. Isomaltulose based soft drinks are characterized by the same natural sweetness profile as those with sugar and have a fully rounded mouth feeling.

Liquid Dairy Applications Containing Proteins

Isomaltulose, as a reducing disaccharide, is subject to Maillard reaction but less so than lactose. Discoloration during heat treatment of milk based products like yogurt can be modulated with the addition of isomaltulose. With short time treatment under ultra high temperature (UHT) conditions, isomaltulose containing protein solutions are stable without colorization during pasteurization for 10 minutes at 90°C.

Because of its higher bacterial and chemical stability compared to sucrose, isomaltulose is used as a sweetener in dairy products containing active cultures with acidophilus and bifidus bacteria. These bacteria cannot split isomaltulose and the sweetness level remains constant.

Alcohol-reduced Beers, Alcohol-free Beers, Beer Mixes, for example, Shandy and Malt Based Beverages

Isomaltulose displays a high stability against fermentation by most yeasts and bacteria. This can effectively be used in the brewery production of alcohol-reduced beers, alcohol-free beers, beer mixes, for example, shandy and malt based beverages with a non-fermentable bulk sweetener. The application of isomaltulose increases the final extract in these products, resulting in increased

Table 24.2 Spectrum of Isomaltulose Applications

Application	Palatinose™ PAP-N	Palatinose™ PAP-PF/PA	Palatinose™ DC
	Crystalline	Powder	Agglomerated
Soft drinks	X	X	—
Sports beverages	X	X	—
Milk and yogurt products	X	X	—
Beer	X	X	—
Instant products	X	X	X
Low boiled candies	X	X	—
Gummies/Jellies	X	X	—
Coated products	X	X	—
Chewing gum			
• coating	X	X	—
• center	X	X	—
Compressed tablets	X	—	X
Baked goods	X	X	—
Icings	X	X	—
Fillings (fat based)	X	X	—
Chocolate	X	X	—
Cereal bars	X	X	—
Ice cream	X	X	—
Jam/fruit spread	X	X	—

palate fullness, body and an optimized true-to-type sensorial profile (Pahl et al. 2008). In addition, diacetyl formation, which normally develops during the fermentation process and disappears during the storage process, can be reduced by the addition of isomaltulose. In the finished product, isomaltulose has a positive influence on the mouthfeel and gives body to these kinds of beverages. In beer specialty products, such as beer/lemonade mix drinks, isomaltulose cannot be

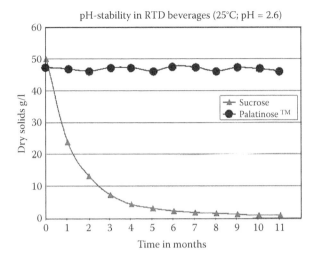

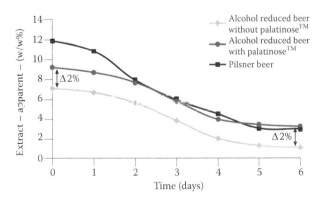

Figure 24.7 **Isomaltulose and sucrose sports drinks during a storage test over 11 months at 25°C and pH 2.5–2.7. (From Pahl et al.,** *Brewing Sci* **62:49–55, 2008.)**

fermented by a wide range of beer contaminants including *Lactobacillus brevis* and *Saccaromyces diastaticus*. If isomaltulose is the predominant carbohydrate in the beer-mix product, an increase of microbiological stability can be achieved. Parameters like foam stability and stability against occurring turbidity are not influenced. Isomaltulose survives fermentation during the beer brewing process, resulting in a final extract similar to traditional pilsner type beer (Figure 24.8) (Pahl et al. 2010).

The key properties of isomaltulose in beer and beer specialty products are listed below:

■ Enhances microbiological stability of beer mix products
■ Cannot be converted by most yeasts and bacteria
■ Reduces the diacetyl formation (diacetyl is a substance which develops during the fermentation process and disappears during the storage process)
■ Optimizes sensorial profile and taste
■ Increases body and improves mouthfeel

Figure 24.8 **Use of isomaltulose in beer-type beverage providing body.**

■ Enhances the long-term stability
■ Reduces the development of fermentation by-products

In addition, isomaltulose can be used to enhance the stability of food products which are sensitive to oxygen. This complies with the findings in different solid and liquid matrixes, for example, beer, where isomaltulose decreases the formation of aging products like E-2-Nonenal significantly.

Instant Drinks

Isomaltulose can be used to improve the disintegration of instant granules in liquids. In these formulations, isomaltulose, being a very low hygroscopic and free-flowing powder (Palatinose PAP-PF) or agglomerate (Palatinose DC), hardly forms lumps and significantly reduces the water absorption in blends with, for example, fructose which is reflected in the sorption curves of powder blends stored at 25°C/65% relative humidity (Figure 24.9) (Südzucker AG Mannheim/Ochsenfurt CRDS n.d.).

The glass transition temperature of isomaltulose is, at 62°C, considerably higher compared to fructose (10°C) or glucose (35°C). This, therefore, favors isomaltulose during spraying or mechanical exposure compared to fructose and glucose, as treatment of carbohydrates may often develop amorphous parts during this process. The amorphous moiety leads to an increased hygroscopicity affecting the product stability, especially if the glass transition temperature is low compared to room temperature. Furthermore, isomaltulose develops a very low portion of amorphous material during these treatments and the agglomerated isomaltulose combines a very low hygroscopicity with an excellent flowability. The water activity of isomaltulose and sucrose is equal.

Confectionery Applications

Compressed Tablets/Lozenges

Isomaltulose in direct compressible (DC) quality is particularly suitable for the production of chewable tablets and lozenges by direct compression. It possesses both an excellent flowability and compressibility as can be seen in Figure 24.10 (Südzucker AG Mannheim/Ochsenfurt

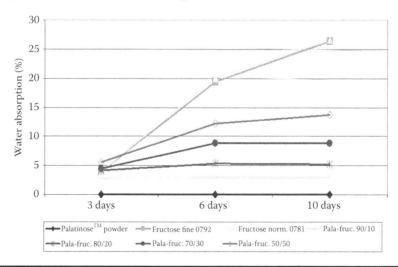

Figure 24.9   Water uptake of isomaltulose powder compared to fructose or powder blends at 25°C/65% rh.

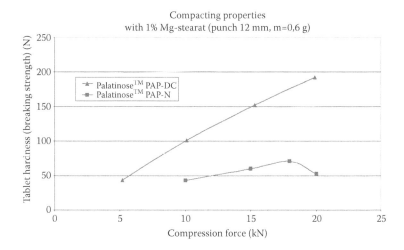

Figure 24.10   Strength force — compression force — profile of Palatinose™ DC and Palatinose™ PAP-N. (From Südzucker AG Mannheim/Ochsenfurt CRDS n.d.)

CRDS n.d.). Because of the low hygroscopicity, isomaltulose tablets have good storage stability, helping to protect moisture-sensitive ingredients. Isomaltulose tablets exhibit sucrose-similar organoleptic properties (Beneo Palatinit GmbH 2010). Isomaltulose DC can be used in fruit or mint-tasting tablets due to its mild sweetness and low cooling effect.

Chocolate

Isomaltulose is suitable for chocolate applications. Due to its stable bound crystal water, isomaltulose can be applied to the traditional conching process. The production process is mainly the same as for conventional chocolate. With isomaltulose, toothfriendly chocolate can be produced without a laxative potential, as the use of polyols is not necessary. Other fermentable carbohydrate sources need to be considered and excluded from the matrix as well, for example, in the production of toothfriendly milk chocolate, milk powder should be replaced by casein, as even in lactose-reduced milk powder the residual content of fermentable carbohydrates is still too high to produce toothfriendly chocolate.

Isomaltulose-containing chocolates are characterized by a satin sheen and a subtle glaze. Isomaltulose provides a full-bodied mouthfeel and a crispy snap which are comparable to conventional chocolate.

Chewing Gum

Isomaltulose, with its stable crystal structure, can be applied as dry matter in chewing gum formulation by replacing a conventional sugar, for example, sucrose on a one to one basis. Chewing gum sticks, pellets, or gum balls, and also coatings can be manufactured with isomaltulose (Südzucker AG Mannheim/Ochsenfurt CRDS n.d.), as it combines low hygroscopicity with sufficient solubility for cost effective coating processes. Its crystals remain stable in the chewing gum mass, which leads to a soft and smooth texture. The coating of isomaltulose is a comparatively short process.

The stickiness of the gum mass during the production process is low. As the water uptake and the drying out of the chewing gums during storage is at a low level, chewing gums with isomaltulose show good stability with a long shelf life (Südzucker AG Mannheim/Ochsenfurt CRDS n.d.). The slow dissolution of isomaltulose supports a long lasting flavor release and sweetness.

Coated Products

Isomaltulose can be used as a coating bulk agent also at low process temperatures. The coating procedure itself depends on the type of coating equipment and on the type of center. Moreover, isomaltulose can replace sucrose in all sorts of pan-coating (Südzucker AG Mannheim/Ochsenfurt CRDS n.d.). Coated products with isomaltulose have an excellent crunch and very good color stability.

Coating Possibilities

- Hard coating with isomaltulose suspension or solution, possibly in combination with isomalt powder
- Soft coating with isomaltulose in combination with another non-recrystallizing sugar substitute
- Chocolate coating with isomaltulose chocolate

The low hygroscopicity of isomaltulose leads to products with a good shelf life. Simple carton packaging of isomaltulose coated products is sufficient.

Isomaltulose coatings are stable against abrasion forces experienced during the coating process. This reduces damage and results in well-defined corners and edges of the coated products.

Other Applications

Jellies, Gummies, Marshmallows

Jellies, gummies or marshmallows can be produced with isomaltulose. Common production equipment can be used with isomaltulose. For the development of toothfriendly and non-laxative products, syrup based on polydextrose is used in combination with isomaltulose, the latter in an amount between 13% and 15%. The shelf life of isomaltulose gummies and jellies is good because of the low hygroscopicity of isomaltulose (Beneo Palatinit GmbH 2010).

Chewy Candies

Non-cariogenic chewy candies with a good flavor profile and stability can be produced with isomaltulose at levels between 30% and 35%. For the production of toothfriendly chewy candies, a combination of isomaltulose with polydextrose can be used. Also within this product group, the low hygroscopicity of isomaltulose results in improved storage stability which could decrease the packaging costs.

Baked Goods

Isomaltulose can be used as a one-to-one replacer for sucrose, for example, in cakes and other baked goods, without adversely affecting the product quality. It can be added to sweet and to salted products in order to improve the texture. The use of isomaltulose results in low water activity, reduced caramelization and good water binding capacity. The reduced sweetness intensity can be increased by the addition of further sugars, for example, fructose or intense sweeteners.

Fondant, Icings, Glazings

Isomaltulose can be used at levels up to 80% in combination with liquid components to control the crystallization level in pasty products. Fondant for fillings or icings and glazings for decoration purposes as well as the protection of bars and baked goods against drying are known applications. A very short crystallization/drying time in glazings can be obtained with isomaltulose. The low hygroscopicity is the reason for the excellent shelf life of isomaltulose-based fondants.

Regulatory Status

Isomaltulose is a food like sugar or flour and is used in many countries all over the world. In Japan, isomaltulose has been available since 1985. Isomaltulose is approved as FOSHU in Japan. In the United States, a generally recognized as safe (GRAS) file was submitted by Südzucker/Beneo-Palatinit and accepted by the US FDA. Further, in 2008 the US FDA added isomaltulose to the list of nutritive sweeteners that do not promote tooth decay and a related health claim was approved. Some countries have approval processes for food and food ingredients that were not previously eaten in that country, for example, the European Union and its member states (2005) and Australia/New Zealand (2007). Approvals were achieved for both regions. Many countries on all continents confirmed the food status and the possible use in food for the benefit of the consumer having the physiological properties of isomaltulose in mind. Specifications are laid down in the EU(2005) and in the *FCC* (*Food Chemicals Codex* 2010–2011).

References

Achten, J., Jentjens, R.L., Brouns, F., and Jeukendrup, A.E. 2007. Exogenous oxidation of isomaltulose is lower than that of sucrose during exercise in men. *J Nutr* 137(5):1143–1148.

Anonymous. 1985. Palatinose. Low cariogenic sugar with a bright future. *Food Engineering* 57(5):75.

Australia New Zealand Food Standards Code, Amendment No. 92, Gazette of the Commmonwealth of Australia No. FSC 34. 2007. http://www.foodstandards.gov.au/foodstandards/changingthecode/gazettenotices/amendment922august203626.cfm. Accessed June 6, 2011.

Beneo Palatinit GmbH. 2010. Information Brochure: A technical view on Palatinose™ - the next generation sugar.

Bubnik, Z., Kadlec, P., Urban, D., and Bruhns, M. 1995. *Sugar technologists manual*. 8th edition. Berlin, Germany: Bartens Verlag.

Cegelski, M., Dörr, T., Radowski, A., and Schneider, J. 2009a. Kohlenhydrate in Sortgetränken, sensorische Eigenschaften und Osmolalitätsstabilität. *Brauwelt* 6:134–139.

Cegelski, M., Dörr, T., Radowski, A., and Schneider, J. 2009b. Kohlenhydrate in Sprortgetränken, Untersuchungen von im Markt befindlichen Sprotgetränken. *Brauwelt* 7:176–180.

Dahlqvist, A., Auricchio, S., Semenza, G., and Prader, A. 1963. Human intestinal disaccharidases and hereditary disaccharide intolerance. The hydrolysis of sucrose, isomaltose, Palatinose (isomaltulose) and a 1,6-alpha-oligosaccharide (iso-malto-oligosaccharide) preparation. *J Clin Invest* 42:556.

Dreissig, W., and Luger, P. 1973. Die Strukturbestimmung der Isomaltulose. *Acta Cryst* B29:514–521,

Egglestone, G., and Grisham, M. 30 April 2003. Oligosaccharides in cane and their formation on cane deterioration. *Oligosaccharides in Food and Agriculture* 849:211–232.

European Commission Decision 2005/581/EC of 25 July 2005 authorising the placing on the market of isomaltulose as a novel food or novel food ingredient under Regulation (EC) No 258/97 of the European Parliament and of the Council (notified under document number C(2005) 2776). http://eur-lex.europa.eu/LexUriServ/LexUriServ.do?uri=CELEX:32005D0581:EN:NOT

Food Chemicals Codex (*FCC*). 2010–2011, 7th edition, pp. 546–548. Rockville, IN: United States Pharmacopeial Convention.

Gehrich, K. 2002. Phasenverhalten einiger Zucker und Zuckeraustauschstoffe. PhD Thesis, TU Braunschweig.

Gómez Bárez, J.A., Garcia-Villanova, R.J., Elvira Garcia, S., Rivas Palá, T., Gonzálas Paramás, A.M., and Sánchez Sánchez, J. 2000. Geographical discrimination of honeys through the employment of sugar patterns and common chemical quality parameters. *Eur Food Res Technol* 210:427–444.

Holub, I., Gostner, A., Theis, S., Nosek, L., Kudlich, T., Melcher, R., and Scheppach, W. 2010. Novel findings on the metabolic effects of the low glycaemic carbohydrate isomaltulose (Palatinose). *Br J Nutr* 103(12):1730–1737.

Japanese Ministry of Health, Labour and Welfare (MHLW). Food for Specified Health Uses (FOSHU): http://www.mhlw.go.jp/english/topics/foodsafety/fhc/02.html. Accessed June 6, 2011.

Koenig, D., Luther, W., Polland, V., and Berg, A. 2007. Carbohydrates in Sports Nutrition - Impact of the Glycemic Index. *AgroFood* 18(5):9–10.

Kozianowski, G. 2007. Physiological functionalities of the novel low glycemic carbohydrate isomaltulose (Palatinose™). *Ann Nutr Metab* 51(suppl 1):157.

Lina, B.A., Jonker, D., and Kozianowski, G. 2002. Isomaltulose (Palatinose™): A review of biological and toxicological studies. *Food Chem Toxicol* 40(10):1375–1381.

Pahl, R., Dörr, T., Radowski, A., and Hausmanns, S. 2010. Isomaltulose: A new, non fermentable sugar in beer and beer specialities. *Beverage Tech* 2:1–4.

Pahl, R., Methner, F.J., Schneider, J., Kowalczyk, J., Radowski, A., and Hausmanns, S. 2008. Study on the applicability of isomaltulose (Palatinose™) in beer and beer specialities, and its remarkable results. *Brewing Sci* 62:49–55.

Schiweck, H. 1980. Palatinit® - Herstellung, technologische Eigenschaften und Analytik palatinithaltiger Lebensmittel. *Alimenta* 19:5–16.

Siddiqua, I.R., and Furgala, B. 1967. Isolation and characterization of oligosaccharides from honey. *J Apicultural Res* 6(3):139–145.

Südzucker AG Mannheim/Ochsenfurt CRDS. Genuine data. In press.

Sydney University's Glycaemic Research Service (SUGiRS). 2002. Glycaemic Index Report - Isomaltulose.

Takazoe, I. 1985. New trends on sweeteners in Japan. *Int Dent J* 35:58–65.

US Food and Drug Administration (US FDA). 27 May 2008. 21CFR§101.80 Food Labeling: Health Claims: Dietary non-cariogenic carbohydrate sweeteners and dental caries. *Fed Reg* 73(102):30299–30301.

US Food and Drug Administration (US FDA): Inventory of GRAS Notice 000184: Isomaltulose. http://www.accessdata.fda.gov/scripts/fcn/gras_notices/grn000184.pdf. Accessed June 6, 2011.

Van Can, J.G., Ijzerman, T.H., van Loon, L.J., Brouns, F., and Blaak, E.E. 2009. Reduced glycaemic and insulinaemic responses following isomaltulose ingestion: Implications for postprandial substrate use. *Br J Nutr* 102(10):1408–1413.

Van Loveren, C. 2009. Oral and dental health—prevention of dental caries, erosion, gingivitis and periodontitis. ILSI Europe Concise Monograph Series, ILSI Europe.

Weidenhagen, R., and Lorenz, S. 1957. Palatinose (6-O alpha-D-glucopyranosyl-D-fructofuranose), ein neues bakterielles Umwandlungsprodukt der Saccharose. *Zeitschr Zuckerind* 7:533–534.

Chapter 25

Trehalose

Alan B. Richards and Lee B. Dexter*

Contents

---

* Deceased.

Introduction

Trehalose is a unique disaccharide with important functional properties. Although these properties have been recognized for many years, it is only recently that the use of trehalose in a wide variety of products has been considered economically feasible. This change resulted from the development of a novel system of manufacture that dramatically reduces production costs, making it possible to integrate trehalose into a wide range of cost-sensitive products. The various physical properties as they relate to food applications are now being explored. This chapter reviews various aspects of trehalose and provides preliminary information on its functional properties, potential use, regulatory status, and safety profile.

Historical Background

Trehalose is believed to have been first isolated in 1832 from the ergot of rye by Wiggers (Birch 1963). The name "trehalose" was coined years later when Berthelot extracted the same disaccharide as the primary sugar contained in a cocoon-like secretion of a beetle found in the Iraqi desert. The insect cocoon or shell was known as "trehala manna" (Leibowitz 1943). Other insects make similar structures using different sugars (Leibowitz 1943, 1944). Trehala manna was described as a sticky raw material, and Leibowitz reported that about 30%–45% of the dry weight of the manna was trehalose (Leibowitz 1943, 1944). Bedouins have been known to gather manna, found on leaves of various plants, for use as a sweetening agent. Some believe that it may be similar to the manna mentioned in the Old Testament of the Bible, recorded as being eaten by the Israelites in the wilderness (Bergoz 1971).

Historical Consumption of Trehalose and Natural Occurrence

Trehalose appears to have been a part of the human diet since the beginning of mankind. It occurs in a wide range of fungi, insects, and other invertebrates such as lobster, crab, prawns, and brine shrimp (Elbein 1974; Miyake 1997; Sugimoto 1995). In addition, it can be isolated from certain plants such as sunflowers, particularly the seeds. Other common foods in which trehalose occurs are honey, wines, sherries, invert sugars, breads, and mirin (a sweet saké used for cooking) (Elbein 1974; Miyake 1997). Both brewer's and baker's yeast accumulate trehalose, the latter to concentrations of 15%–20%, on a dry weight basis (Van Dijick et al. 1995). Certain bacteria, including *Escherichia coli*, a common enteric bacteria of the human digestive tract, can synthesize trehalose (Leslie et al. 1995; Nicolaus et al. 1988).

Early man likely had a much greater exposure to trehalose than modern man because of greater dietary dependence on fungi, insects, and aquatic invertebrates. Although many of these substances are not common in our modern diet, several items are still consumed by a large portion of the population and provide a constant exposure to trehalose in the course of a normal diet. Modern food sources shown to contain significant quantities of trehalose are honey (0.1%–1.9%), mirin (1.3%–2.2%), brewer's yeast (0.01%–5.0%), and baker's yeast (15%–20%) (Aso et al. 1962; Elbein 1974; Sugimoto 1995; Van Dijick et al. 1995). Commercially grown mushrooms can contain 8%–17% (w/w) trehalose. Reported values of trehalose from natural sources may vary substantially because of experimental conditions, analytical methods, and the life-cycle stage of the organism assayed.

Elbein provides an extensive review of the reported general occurrence of trehalose and the number of derivatives of trehalose that are found in nature (Elbein 1974). Further work will likely show the presence of trehalose in many additional organisms.

Since 1995, trehalose has been incorporated into hundreds of food products in Japan. Total consumption of trehalose added to food products in Japan was approximately 30,000 metric tons in 2010. It is also sold in many other countries including the United States and those in the EU for use in processed foods.

Chemical Composition

α,α-Trehalose (α-D-glucopyranosyl α-D-glucopyranoside) has commonly been called mushroom sugar or mycose. It is a conformationally stable, nonreducing sugar consisting of two glucose molecules bound by an α,α 1,1 glycosidic linkage (Figure 25.1). After ethanol extraction or enzymatic synthesis, trehalose crystallizes to a stable dihydrate form (Birch 1963; Sugimoto 1995). Trehalose can also be found in nature as α,β (neotrehalose) and β,β (isotrehalose) isomers; however, these are uncommon, and they possess chemical and physical properties that are distinct from α,α-trehalose. In this review "trehalose" will be used to designate the α,α-trehalose form. The chemical formula and molecular weight of trehalose dihydrate are $C_{12}H_{22}O_{11} \cdot 2H_2O$ and 378.33, respectively. The Chemical Abstracts Service Registry Number is 6138-23-4.

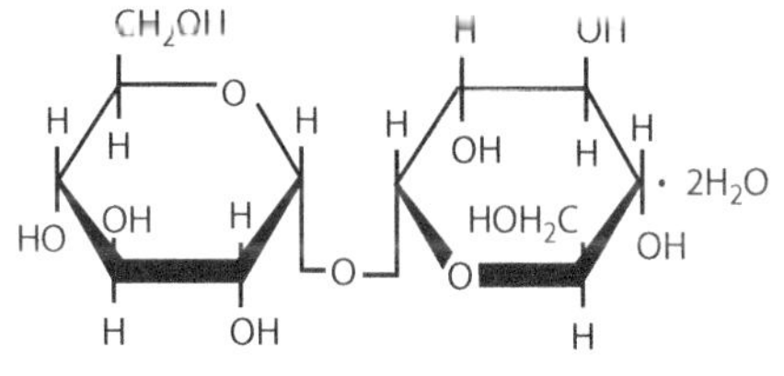

Figure 25.1 Structure of α,α-trehalose dihydrate.

Physical Properties

Physical properties reported in this section are, in a large part, obtained from assays of trehalose produced enzymatically from starch by Hayashibara Biochemical Laboratories (HBC) of Okayama, Japan; however, data from other sources are included (Sugimoto 1995)

Trehalose is available as a white crystalline powder that is colorless and of prismatic rhomboid form. X-ray diffraction reveals cell parameters of a = 12.33, b = 17.89, and c = 7.66 (Brown et al. 1972; Taga et al. 1972). The orthorhombic cell formula is given as $P2_12_12_1$, and the units per cell are 4. The theoretical density of 1.511 g/cm^3 agrees closely with the actual value of 1.512 g/cm^3. Reports of the melting point of trehalose dihydrate range from 94°C–100°C (Birch 1963). Continued heating above 100°C vaporizes the water of hydration, resulting in resolidification; a second melting point is reached at 205°C (Birch 1963). The melting point of trehalose dihydrate produced and assayed by HBC (97°C) equals the value listed in reference texts (Budavari 1989; HBC 1997). The melting point of HBC is 210.5°C, approximately 7.5°C higher than reported in the literature (Budavari 1989; HBC 1997).

Birch reports data from several studies in which the optical rotation of various preparations of trehalose ranged from +177° to +197.14°, but gives a value of +199° for his own anhydrous preparation (Birch 1963, 1965). Unpublished data from assays performed by HBC (1997) give a specific rotation equal to $[\alpha]_D^{20}$ + 199° (c = 5).

Production of Trehalose

Historical Processes for Trehalose Production

The isolation of trehalose from natural sources has been ongoing for several decades. Historically, the production of trehalose has been limited to small quantities obtained by extraction from microorganisms such as baker's yeast (Birch 1963; Koch and Koch 1925; Steiner and Cori 1935; Stewart et al. 1950). One complex purification method produced maximum yields of 0.7%, whereas later procedures resulted in yields of up to 16% of active dried yeast (92% solids) (Koch and Koch 1925; Stewart et al. 1950).

In the 1980s and 1990s investigators focused on the use of bacterial synthesis, transgenic technology, or enzymatic conversion for the production of trehalose (Sugimoto 1995). Trehalose has been isolated during the fermentation of n-alkanes by *Arthrobacter* sp. in quantities of 5–6 g/L (Suzuki et al. 1969). Three Japanese patents, issued between 1975 and 1993, described the production of trehalose by bacteria using common carbon sources such as glucose and sucrose. Genetic recombination has been attempted by insertion of a gene that converts glucose into trehalose in a sugar-producing crop (Sugimoto 1995). Murao et al. reported on an enzymatic method for producing trehalose from maltose using trehalose- and maltose-phosphorylases, which provided a yield of 60% (Murao et al. 1985). Trehalose has also been produced enzymatically by a combination of acid reversion of D-glucose and trehalose (Schick et al. 1991; Sugimoto 1995).

Although these newer methods have reduced the cost of trehalose, they have not been able to provide trehalose at a cost low enough for use in most food applications. This has effectively restricted its use to research, pharmaceuticals, and high-value cosmetics.

Hayashibara Manufacturing Process

HBC screened soil samples for bacteria that could produce trehalose from relatively inexpensive carbon sources such as starch or maltose.

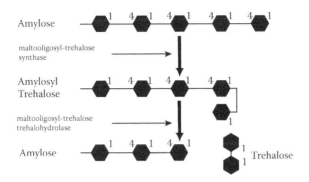

Figure 25.2 Enzymatic production of Hayashibara trehalose from starch. A debranching enzyme is used to produce the amylose substrate.

One group of bacteria was shown to convert maltose to trehalose in a manner that was more efficient than previously reported enzyme systems (Murao et al. 1985; Nishimoto et al. 1995). A second bacterial group was found to use starch as the substrate for trehalose production (Nakada et al. 1995a, 1995b). HBC scientists isolated and characterized the two different trehalose-producing enzyme systems from these organisms (Nakada et al. 1995a, 1995b; Nishimoto et al. 1995). Those species that produced trehalose directly from maltose required only one novel enzyme, trehalose-synthase (Nishimoto et al. 1995). The bacteria that produced trehalose from starch possessed two novel enzymes, maltooligosyl-trehalose synthase and maltooligosyl-trehalose trehalohydrolase (Nakada et al. 1995a, 1995b).

HBC has commercialized the latter process using starch to manufacture trehalose. This new process has reduced the price of trehalose approximately 100-fold. Before this discovery, the cost of trehalose in Japan ranged from 20,000–30,000 Yen (¥) per kilogram ($170–250/kg). Food-grade trehalose is now sold in Japan for about 300 ¥/kg ($3.00/kg).

The direct conversion of starch to trehalose on a commercial scale depends on a system in which several enzymes co-react in the same medium (Nakada et al. 1995a, 1995b; Sugimoto 1995; Tsusaki et al. 1995). Isoamylase debranches the starch, producing amylose (Figure 25.2). Amylose is then converted to amylosyl-trehalose by maltooligosyl-trehalose synthase, which identifies the reducing end glucose units and catalyzes the conversion of α-1,4 bonds into α,α 1,1 bonds. In the next step, maltooligosyl-trehalose trehalohydrolase interacts with amylosyl-trehalose molecules, hydrolyzing the α-1,4 bond between the amylosyl-moiety and the trehalose, releasing the trehalose into the media. This reaction terminates when the degree of polymerization of amylose becomes two or less; that is, when only glucose or maltose remains. This enzymatic system, followed by conventional methods of saccharide production, results in high-purity crystalline trehalose (Nakada et al. 1995a, 1995b; Sugimoto 1995; Tsusaki et al. 1995).

Technical Qualities

In nature, trehalose functions as one of the primary molecules responsible for maintaining the bioactivity of cell membranes during stress caused by freezing, desiccation, or high temperatures (Roser 1991a). It is present in relatively high concentrations in organisms that can be desiccated or frozen and on rehydration or thawing return to essentially full function. Examples include brine shrimp, baker's yeast, brewer's yeast, insects, and certain plants (Roser 1991a). In nature, trehalose is thought to function through a number of mechanisms to assist such organisms as baker's yeast in resuming normal metabolic activity after freezing or drying. These mechanisms include the following:

1. Water replacement—Trehalose has been shown to replace structural water molecules that are hydrogen bonded to the end groups of proteins. By substituting for water, trehalose maintains the three-dimensional structure of protein molecules, and hence their ability to remain active after drying or freezing (Crowe et al. 1990; Kawai et al. 1992; Roser 1991a, 1991b).
2. Formation of a glassy solid—Scientists have found that trehalose can form an amorphous glassy matrix around the vital tertiary structures of proteins and phospholipids, thus protecting them during drying and freezing (Colaço and Roser 1994; Crowe et al. 1990; Rudolph and Crowe 1985).
3. Chemical inactivity—The chemical inertness of trehalose may be important in protecting biomembranes from destructive molecular reactions (Bolin and Steele 1987).
4. Formation of a tight-fitting complex with water—Trehalose has the ability to form a hydrogen-bonded complex with water. This complex retards ice crystal formation and allows trehalose to penetrate more deeply than other sugars into certain membrane structures (Donnamaria et al. 1994; Portmann and Birch 1995).

Because of these and other characteristics, trehalose possesses a number of technical qualities that indicate that it may provide significant beneficial properties to foods, cosmetics, and pharmaceuticals. Other properties of interest include mild sweetness, low hygroscopicity, good solubility in water, lack of color in solution, a high glass transition temperature, stability over a wide range of pH and temperatures, and chemical stability (Colaço and Roser 1994; Green and Angell 1989; Roser 1991a, 1991b; HBC 1997). Because of its stability, trehalose resists caramelization and doesn't participate in Maillard reactions in food systems (Bolin and Steele 1987; Colaço and Roser 1994; Karmus and Karel 1993; HBC 1997). Trehalose is also known to stabilize proteins and starches (Colaço and Roser 1994; Colaço et al. 1992; Crowe et al. 1990; Kawai et al. 1992; Roser 1991a, 1991b; HBC 1997).

Sweetness and Organoleptic Properties

Intensity and Persistence

Portmann and Birch compared the intensity and persistence of various concentrations of trehalose to other sugars (Portmann and Birch 1995). Using a trained taste panel, the authors studied the perceived sweetness of trehalose versus other sugars at concentrations of 2.3%, 4.6%, 6.9%, and 9.2%. The report showed that the perceived sweetness of trehalose increased in a nonlinear fashion as the concentration increased. The ratios of the intensity of sweetness at the various concentrations between sucrose and trehalose were 6.5:1, 5.0:1, 3.5:1, and 2.5:1, respectively. Comparisons of sucrose to glucose and fructose showed that the sweetness intensity ratio did not change with increasing concentration.

Increasing the concentration of trehalose also leads to a disproportionate increase in the perception of the persistence of sweetness as compared with sucrose (Portmann and Birch 1995). The ratio of the persistence of sweetness of sucrose to trehalose was 1.5:1 (2.3%), 0.8:1 (4.6%),

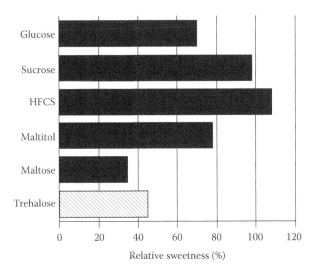

Figure 25.3   The relative sweetness of various sugar solutions compared with a 22.2% (w/w) solution of trehalose.

0.6:1 (6.9%), and 0.5:1 (9.2%), whereas glucose and fructose showed little change. The ratio of the intensity and persistence of sucrose to maltose had a similar pattern as trehalose but was less dramatic.

HBC performed a comparative sweetness test using a 22.2% solution (w/w) of trehalose and various concentrations of sucrose (HBC 1997). The study showed that trehalose at 22.2% is about 45% as sweet as sucrose (Figure 25.3). The studies suggest that although the sweetness of trehalose is less intense, it can be more persistent than sucrose (Portmann and Birch 1995).

Taste Quality

Trehalose was also tested for taste quality by a Japanese taste panel. A 22.2% trehalose solution was compared with a 10% sucrose solution (HBC 1997). Results showed that 17 of 20 subjects preferred the taste of trehalose to that of sucrose. Anecdotal responses to the quality of the taste of trehalose have described it as mild with a clean aftertaste that seems to clear the palate easily.

Taste Chemistry of Trehalose

Trehalose, a nonreducing disaccharide, is similar to other reducing disaccharides in that it does not produce as great a sensation of sweetness as does sucrose. Disaccharides are also generally accepted to be less sweet than the monosaccharides from which they are made, with the notable exception of sucrose. Trehalose is believed to have only one glucose molecule occupying the binding site on the sweet taste receptor (Lee and Birch 1975). This is thought to result from steric hindrance around the C-1 site on the glucopyranose ring.

Physicochemical studies suggest that the shape and charge of trehalose are such that water molecules form a closer (tighter) association with trehalose than with other saccharides. Theoretically, these small molecular clusters can migrate deeper into the taste epithelium (Green and Angell

1989; Shamil et al. 1987). Trehalose and the associated water may concentrate in these areas more efficiently than other sugars, which would result in a more persistent sweet taste (Portmann and Birch 1995).

In contrast, the perception of sweetness intensity has been hypothesized to be enhanced by the ability of water molecules to dissociate or move in relation to an associated sugar (Mathlouthi 1984). It has been suggested by Mathlouthi that free water molecules are required to allow the transport of Na^+/K^+ ions across the membrane of the sweet taste receptor. The more free water, the higher the possible membrane potential (Na^+/K^+ transport), resulting in the sensation of more intense sweetness (Mathlouthi 1984). Trehalose holds the water of hydration tightly so that water is not available to facilitate the membrane potential. Both macroscopic and physiochemical results suggest that there is an optimal molecular volume for binding to sweet taste receptors, which corresponds to a particular packing of the water molecules around dissolved sugar molecules (Mathlouthi et al. 1993). The tight interaction of the trehalose/water complex could inhibit the bonding of trehalose to the receptor, further reducing the sweetness compared with other sugars (Portmann and Birch 1995).

Hygroscopicity

Low hygroscopicity is one of the hallmark qualities of trehalose. Figure 25.4 shows the moisture content of trehalose after storage at 25°C for seven days at various humidity levels. The initial water content of trehalose dihydrate is 9.54%, whereas anhydrous trehalose is 0.65%. The dihydrate form remained stable up to a relative humidity (RH) of approximately 92%. The moisture content of anhydrous trehalose increased as RH increased from 35%–75% and then paralleled the moisture content of the dihydrate until 92% RH was reached. The maximum moisture content was nearly 14% (HBC 1997). Table 25.1 provides comparative data on the water absorption (%) of several sugars stored for 10 days at 90% RH and 25°C. It appears that trehalose could be of benefit compared with other sugars in dry blending operations and other applications in which low hygroscopicity is desired.

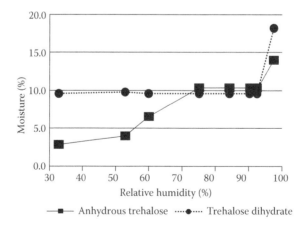

Figure 25.4 Hygroscopicity of anhydrous and dihydrous trehalose. Trehalose was stored at various relative humidity concentrations for seven days at 25°C and then measured for the percent moisture.

Table 25.1 Water Absorption Ratio in Percent of Various Sugars Stored at 90% Relative Humidity and 25°C for 10 Days

Sugar	Storage Time in Days			
	1.0	**3.0**	**6.0**	**10.0**
Trehalose	0.2	0.0	0.1	0.2
Lactose	0.2	0.3	0.3	0.7
Maltose	3.5	6.6	8.4	8.4
Glucose	4.0	7.7	10.8	11.3
Sucrose	5.1	15.0	33.4	50.0

Solubility and Osmolarity

Trehalose is soluble in water and insoluble in absolute alcohol. Aqueous ethanol has been used to crystallize trehalose, which produces the dihydrate form (Birch 1965). The solubility of trehalose in 70% aqueous ethanol has been reported as 1.8 g/100 ml (Birch 1965). Table 25.2 provides data on the solubility of trehalose in water at various temperatures (HBC 1997). Trehalose has a solubility similar to maltose up to about 80°C at which temperature the solubility increases at a rate greater than the other sweeteners. The osmolarity of trehalose is also essentially the same as maltose (Table 25.3).

Glass Transition Properties

Like most sugars, trehalose is capable of forming amorphous gels called glasses, which are characterized by high viscosity and low molecular mobility (Crowe et al. 1996; Green and Angell 1989). Glasses form as water is removed from sugar solutions during thermal- or freeze-drying (Ding et al. 1996). These vitreous solids can become sticky (or rubbery in the case of polymers) when their glass transition temperature is exceeded or there is an increase in moisture. In this viscous liquid state, many commonly used sugars, such as sucrose, become unstable and either crystallize or decompose chemically. Crystallization will usually result in an anhydrous sugar with a consequent increase in water activity in the residual phase. This, in turn, causes an increase in the rate of browning and other degradative reactions in foods (Karmus and Karel 1993).

Trehalose solutions, on the other hand, solidify to form heat-stable glasses that are capable of binding water and thus controlling water activity (Crowe et al. 1996; Ding et al. 1996; Roser 1991a, 1991b). The glass transition temperature of these glasses is higher than for any other disaccharide

Table 25.2 Water Solubility at Various Temperatures of Trehalose Produced by the Hayashibara Method on an Anhydrous Basis

Temperature (°C)	10	20	30	40	50	60	70	80	90
g/100 g H_2O % Saturation	55.3	68.9	86.3	109.1	140.1	184.1	251.4	365.9	602.9
Concentration (w/w)	35.6	40.8	46.3	52.2	58.3	64.8	71.5	78.5	85.8

Table 25.3 Comparison of the Osmotic Pressure of Concentrations of Trehalose and Maltose

	Concentration of Sugar (% w/w)			
Sugar	5	10	20	30
Trehalose	193[a]	298	690	1229
Maltose	195	299	676	1221

[a] Osmotic pressure measured in milliosmoles.

(Green and Angell 1989). If crystallization of trehalose glasses does occur, the dihydrate forms, rather than the anhydrous crystal, resulting in a reduction of the water activity (Aldous et al. 1995).

Because of its special glass-forming properties, trehalose can be used to stabilize proteins and lipids against the denaturation caused by desiccation and freezing (Colaço et al. 1992; Crowe et al. 1990, 1998; Ding et al. 1996). The glass transition temperature for pure trehalose is the highest of any disaccharide (115°C) (Ding et al. 1996). This temperature is approximately 43°C higher than that of sucrose. The significance of a high glass transition temperature and the unique characteristics of trehalose glasses are discussed later.

Several studies have shown that proteins and other cell membrane components become embedded in the vitreous matrix formed by trehalose during drying (Colaço and Roser 1994; Colaço et al. 1992; Crowe et al. 1985, 1990, 1998). Proteins dried in the presence of trehalose are physically constrained from undergoing degradative reactions (Crowe et al. 1996). Several reports in the literature have cited instances in which enzymes preserved in trehalose have remained active, even after exposure and extended storage at high temperatures (Colaço et al. 1992).

A review of the literature suggests that there may be several important reasons for the superiority of trehalose in preventing the destructive chemical and physical reactions that accompany drying and freezing. Trehalose glasses are purported to function by a number of mechanisms including the following:

1. Reduction in molecular mobility—The formation of a stable glass reduces molecular mobility and prevents destructive physical forces, like fusion, from taking place during freezing or drying.

2. Reduction in water activity—Water activity is necessary for degradative interactions such as the Maillard reaction, which occurs between proteins and other biocomponents (Karmus and Karel 1993). Trehalose glasses are hydrophilic and bind water tightly if excess water is introduced into the system (Ding et al. 1996). Trehalose glasses are almost unique in their ability to form a dihydrate compound, when excess water is present, effectively lowering the water activity (Aldous et al. 1995). Ding et al. stated that the kinetics of most chemical reactions can be essentially shut down by the high viscosity of vitreous trehalose glasses, even at temperatures well above the glass transition temperature (Ding et al. 1996). They demonstrated that trehalose glasses can absorb up to 0.15 moles of water per mole of trehalose without loss of rigidity at ambient temperature (Ding et al. 1996). Others have reported that the freeze-drying of a 0.25 molar (9.45%) trehalose solution did not remove all the water. After storing the preparation for 48 hours under vacuum, the sample retained 0.02 g of water/g of

trehalose (Crowe et al. 1996). The glass transition temperature determined for this preparation was 92°C. When the sample was stored at ambient temperature and high humidity for 66 hours, it absorbed water vapor, and 70% of the sample reverted to trehalose dihydrate (Crowe et al. 1996).

3. Immobilization of active end-groups—Proteins are reported to contain between 0.25–0.75 g of water/g of protein (Roser 1991b). Phospholipids also reside in an aqueous environment. As water is removed during drying, the active end-groups usually associated with hydrogen bonded water are exposed and can bind to other molecules. This denaturation of both proteins and phospholipids, which causes distortion of the three-dimensional structure of the molecules, results in the loss of function. Trehalose can substitute for water by hydrogen bonding to the end-groups, thus minimizing changes and preserving the natural structure and function of the protein or phospholipid.

4. Nonhygroscopicity—Trehalose glasses are hydrophilic but not hygroscopic. It is reported that they are dry to the touch (Roser 1991b). This means that vapor pressure of water in equilibrium with trehalose glasses is high relative to other sugars. Trehalose glasses, therefore, are able to lose water to the environment, rather than absorbing it (Roser 1991a, 1991b). This becomes important in maintaining a level of water activity below the threshold at which destructive interactions occur (Karmus and Karel 1993).

By contrast, the glass structure of other sugars commonly used in the food industry are hygroscopic, thus becoming tacky in the presence of water vapor. When moisture from the atmosphere is absorbed, the surface layers of the sugar dissolves, and crystallization can occur, because this is the most stable state for most sugars. The onset of crystallization allows weeping, particularly with sucrose-based confections (Roser 1991b). Water, because of its plasticizing effect, creates a more mobile structure in these types of glasses, inducing degradative chemical reactions. These reactions can denature proteins and other food components (Karmus and Karel 1993).

The mobile state will also occur if the storage temperature exceeds the glass transition temperature of a particular sugar-containing matrix. If food products are stored above their glass transition temperature, crystallization is likely, with an accompanying increase in browning reactions (Karmus and Karel 1993). Because trehalose has a very high glass transition temperature and can control water vapor by forming the dihydrate, it can remain in the glass state at temperatures higher than other sugars. Glasses formed from trehalose, therefore, have a greater capability for preserving or stabilizing proteins, lipids, or carbohydrates embedded in them (Ding et al. 1996).

Other interesting properties of trehalose glasses have been reported. Crowe et al. found that because trehalose glasses are permeable to water, products suspended in trehalose solutions can be dried to very low residual water contents without resorting to extreme dehydration methods (Crowe et al. 1996). In another study, cytochrome c was dried with 27% trehalose (w/w) to a 6% moisture content under vacuum at ambient temperature. The compound was found to resist denaturation at temperatures up to 115°C, whereas the authors reported that the normal temperature at which cytochrome c denatures is 65°C (Crowe et al. 1996). In general, proteins protected by trehalose retain full activity after exposure to high temperature (Colaço et al. 1992).

It has been reported that the hydrophilic nature of trehalose glasses makes them impenetrable to the volatile lipid molecules responsible for food aromas and flavors. These molecules are trapped within the trehalose glass and cannot dissipate to the atmosphere. Therefore, the aromas and flavors are still present when the preparation is reconstituted (Crowe et al. 1996; Roser 1991a). It is speculated that trehalose might be particularly effective in stabilizing foods and biomaterials that are subject to high humidity and temperatures (Crowe et al. 1996). Roser has reported that trehalose glasses do not attract significant quantities of water at a relative humidity of 90% or less (Roser 1991a).

Although a great deal of applications research remains to be done, HBC has demonstrated that trehalose can be added to foods that undergo phase transitions in preparation or storage (HBC 1997). Such foods include frozen bakery products, frozen desserts, dried fruits and vegetables, glazes, and spray-dried products.

Storage Stability

The storage stability of a 10% solution of trehalose was tested by dividing aliquots of the solution and placing them in the dark at 25°C and 37°C for 12 months (HBC 1997). Samples were examined for pH, color development, turbidity, and residual sugar content monthly for the first six months and thereafter at 9 and 12 months. After 12 months, there was no development of color or turbidity, nor was there any degradation of trehalose at either temperature. Samples stored at 25°C and 37°C for 12 months, showed a drop in pH from an initial value of 6.80 to 5.27 and 5.15, respectively.

Four percent solutions of trehalose were prepared at nine different pH concentrations. The samples were heated to 100°C for 8 and 24 hours (HBC 1997). The buffer systems (0.02 mM) used were acetate (pH 2–5), phosphate (pH 6–8), and ammonium (pH 9–10). After incubation, the samples were evaluated for pH and residual sugar content. The results of pH changes are presented in Table 25.4. Samples appeared stable for the first eight hours and were relatively stable for the 24-hour incubation period. High-performance liquid chromatography analysis of the samples showed that even after 24 hours of incubation at 100°C, all samples retained greater than 99% of the original concentration of trehalose.

Table 25.4 The pH of 4% Trehalose Solutions Prepared in 0.02 mM Buffers and Incubated for 8 and 24 Hours at 100°C

Incubation Time (hr)		
0	8	24
3.46	3.41	3.83
3.71	3.64	4.02
4.25	4.22	4.50
5.14	5.10	5.25
6.69	6.67	6.52
7.46	7.44	7.13
8.25	8.14	7.64
8.88	8.91	8.42
9.86	9.85	9.16
Mean change ± SD	0.04 ± 0.03	0.37 ± 0.19

Table 25.5   The Absorbance of 10% Trehalose and 10% Maltose Buffered Solutions (0.033 M Sodium Phosphate in Water) Incubated at 120°C for Up to 90 Minutes

Incubation Time (min)	Absorbance at 480 nm	
	Trehalose	*Maltose*
0	0.005	0.000
30	0.013	0.051
60	0.010	0.121
90	0.012	0.184

Heat Stability and Caramelization

The heat stability of trehalose was examined in a test system in which a 10% trehalose solution was made in a pH 6.0 buffer (HBC 1997). The buffered solution was incubated in an oil bath at 120°C for up to 90 minutes. Maltose control samples were also included. Solutions were analyzed for absorbance at 480 nm (Table 25.5). The trehalose solution remained stable after a slight initial increase in absorbance, whereas the maltose solution increased at each sampling time. At 90 minutes, the absorbance of the maltose sample was 15-fold greater than that of trehalose. The increase in absorbance is related to the production of colors associated with the caramelization process. Trehalose resisted the production of such color and may be an effective inhibitor of undesired caramelization.

Maillard Reaction

The stability of 10% solutions of trehalose or maltose were tested in 0.05 M buffered solutions (pH 4.0, 5.0, 6.0, 6.5, and 7.0) containing 1% glycine (HBC 1997). The solutions were placed in an oil bath at 100°C, and samples were taken at 0, 30, 60, and 90 minutes. After cooling, the samples were analyzed for changes in absorbance at 480 nm. After 90 minutes the absorbance values of maltose were 0.003 (pH 4.0), 0.014 (pH 5.0), 0.324 (pH 6.0), 0.610 (pH 6.5), and 0.926 (pH 7.0), whereas trehalose values were 0.000, 0.002, 0.010, 0.010, and 0.012, respectively. In another experiment, 10% solutions of trehalose and maltose were combined with a 5% (w/w) solution of polypeptone. Samples were placed in an oil bath at 120°C for up to 90 minutes. Color change was measured at 480 nm. Figure 25.5 displays the data from the experiment. These data suggest that trehalose may be an excellent sugar to use when inhibition or control of the Maillard reaction is desired.

Viscosity

Trehalose has a relatively low viscosity, even at higher concentrations (HBC 1997). Figure 25.6 summarizes the viscosity of trehalose solutions from 5%–40% (w/w) at 25°C and 37°C. Even at 40%, the viscosity did not rise above 5.7 cP.

Nutritional Profile

Trehalose produced by HBC is the only product currently available in commercial quantities to the food industry. A nutritional analysis of several lots of this product showed that it is composed of more than 90% carbohydrate. The remaining components consisted of water at about 9.7%

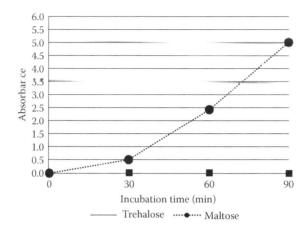

Figure 25.5   The change in absorbance at 480 nm of 10% solutions of trehalose or maltose containing 5% (w/w) of a polypeptone solution. Solutions were incubated at 120°C for 90 minutes.

(dihydrate), trace quantities of lipid with an average content of 0.025%, and low concentrations of protein (0.007%–0.008%) and ash (0.002%–0.010%). Sodium concentrations ranged from nondetectable to 0.065%, with residual starch being undetectable. These values compared favorably with those of other saccharides listed in the *Food Chemicals Codex*.

Energy values based on the preceding data calculate to 3.62 kcal/g. This value correlates well with the analytical results, 3.46 kcal/g (HBC 1997).

Shelf Life and Transport

To ensure that its product continued to meet food grade specifications over an extended shelf life, HBC conducted analyses on commercial grade samples of trehalose stored for up to 24 months. Twenty kilograms of trehalose was stored at 25°C in three-layer Kraft paper bags in which one of

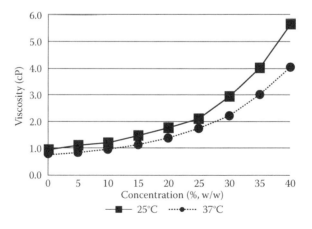

Figure 25.6   The viscosity of trehalose solutions from 5% to 40% (w/w) at 25 and 37°C measured in centipoise (cP).

the layers was polyethylene. One hundred gram samples were taken from the same bag at 0, 1, 3, 6, 9, 12, 18, and 24 months and tested for eight variables. The results confirmed that the material maintained integrity over the storage period, with little variation in the product.

Since becoming commercially available in Japan in November 1995 until December 2010, a cumulative total of more than 200,000 metric tons has been sold. Approximately 30,000 metric tons were sold in 2010. There has not been a recall of trehalose because of degradation of the product.

No data on the stability of trehalose in specific products is available. Because trehalose is chemically, thermally, and pH stable, however, it suggests that trehalose may aid in the stabilization of the products in which it is used.

Transport of trehalose is not restricted in any of the other countries in which it is approved for use.

Cost and Availability

Trehalose is currently being sold in Japan for approximately 300 Yen (¥) per kilogram ($3.00/kg @ 100 ¥ to $1.00). Prices vary slightly depending on the exchange rate of the yen. To the authors' knowledge, HBC is the only generally commercially available source of a purified trehalose product. It is commercially available through marketing partners throughout the world. Other groups are undoubtedly investigating alternative methods for the production of trehalose, but none are known to be commercially available at this time.

Regulatory Status

Quadrant Holdings (Cambridge, United Kingdom) received approval for trehalose to be used as a novel food from the Ministry of Agriculture, Fisheries and Food in the United Kingdom. The approval, granted in 1991, is for the use of trehalose as a cryoprotectant for freeze-dried foods. Use levels were limited to 5% for each formulation. Trehalose produced by HBC was approved as a food additive in Japan in 1995 and as a food ingredient in Korea and Taiwan in 1998. Hayashibara International Inc. self-affirmed trehalose as Generally Recognized as Safe (GRAS) in May, 2000, and received a letter of no objection from the US Food and Drug Administration (US FDA) in October, 2000. In the US, the use of HBC is limited only by current Good Manufacturing Practices. JECFA reviewed the safety profile of trehalose in June 2000, and assigned an acceptable daily intake as "not specified." It was approved as a Novel Food in the EU in 2001, and is also approved as a Novel Food in Australia, Canada, and Brazil. Trehalose is approved in most all countries throughout the world.

Applications

Numerous published and unpublished studies have shown that relatively low concentrations of trehalose in food formulations can reduce sweetness; stabilize protein matrices, flavors, colors, and fatty acids; reduce starch retrogradation; maintain the texture of coatings; and prevent weeping (Colaço and Roser 1994; MacDonald and Lanier 1991; Roser 1991a, 1991b; Roser and Colaço 1993; Rudolph and Crowe 1985; HBC 1997).

Protein Stabilization

Trehalose appears more effective in stabilizing proteins against damage caused by drying or freezing than other sugars tested (Colaço and Roser 1994; Crowe et al. 1990; Rudolph and Crowe 1985). Trehalose has also been shown to help maintain delicate protein structures after thawing and to stabilize disulfide bonds, thereby inhibiting the formation of odors and off-flavors.

HBC found that a 5% addition of trehalose to an egg white preparation, subsequently frozen for five days and then thawed, resulted in almost no protein denaturation compared with a control preparation (Figure 25.7). Protein denaturation was measured by the change in turbidity before and after freezing. Relative denaturation of preparations containing other sugars ranged from 14%–58% (HBC 1997).

In a second example, pulverized carrots were mixed with 10% (w/w) of various sugars. Samples were dried for 64 hours at 40°C and stored for seven days. The superoxide dismutase (SOD)–like activity was measured (Figure 25.8). Trehalose appeared to maintain three times as much SOD-like activity as sucrose (HBC 1997). Studies using seven other vegetables dried with and without trehalose showed a similar SOD-like protective activity. Preservation of enzymes under various stress conditions has been reported by others (Colaço and Roser 1994; Crowe et al. 1996).

The effect of trehalose on protein stability was also demonstrated in a coffee/milk drink. A 3% addition of trehalose (25% of added sugar) reduced the coagulation of casein during sterilization at 121°C for five minutes and suppressed color, taste, and pH changes during subsequent storage (HBC 1997).

Stabilization of Starch

Trehalose is currently being used in Japan to retard starch retrogradation in such products as Udon noodles (0.2% of flour), clam chowder (0.4% of product), and traditional Japanese confectioneries (10%–50% of sugars). Although the mechanism is not yet fully understood, applications research has shown that trehalose can be effective in stabilizing starch. Basic tests were performed

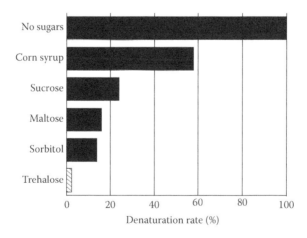

Figure 25.7 Egg white (95 g) was mixed with 5 g of various sugars and stored for five days at −20°C. The samples were thawed and the turbidity was assayed. Egg white without added sugar served as control and was given the relative denaturation value of 100%.

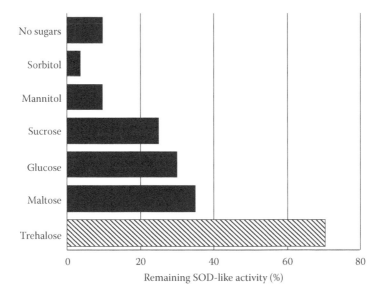

Figure 25.8 **Pulverized carrots were mixed with various sugars and dried at 40°C for 64 hours. The samples were rehydrated and the SOD-like activity was measured. The results presented are the percent of SOD-like activity obtained as compared with the activity before the samples were dried.**

to assess the ability of trehalose to inhibit starch retrogradation. A 1% cornstarch solution was mixed with 6% of various sugars. The mixture was gelatinized and cooled. The turbidity of the solution was tested before and after 12 hours of storage. The percent change was regarded as the amount of retrogradation of the starch solution (Figure 25.9) (HBC 1998). A similar experiment was performed using equal volumes of a 2% starch solution mixed with a 12% sugar solution.

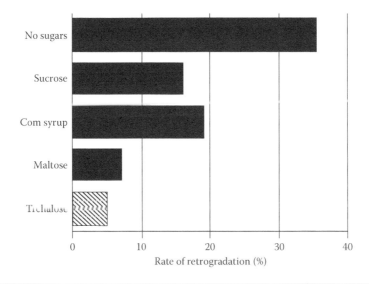

Figure 25.9 **A 1% corn starch solution was mixed with various sugars (6%). The samples were then gelatinized, cooled to 4°C and stored for 12 hours. The percent change in the turbidity before and after storage is reported as the amount of starch retrogradation.**

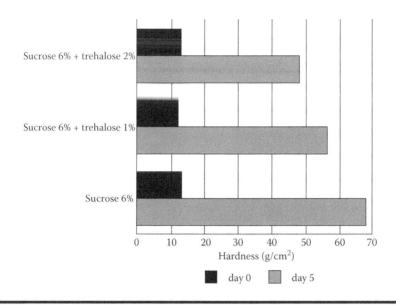

Figure 25.10 Bread was baked using 6% sucrose and 1%, 2%, or no trehalose. The hardness of the bread was assayed after cooling the bread (4°C) and again after five days of storage at 4°C.

After gelatinization, the solutions were stored at 4°C for 12 hours. Results were similar to those obtained in the first experiment (Figure 25.9). The percentage of retrogradation with trehalose was 13%, whereas that for sucrose was 50% (HBC 1998).

Applications of this effect were tested in several Japanese food systems. The hardness of bread made with and without trehalose was measured on day 0 and after five days storage at 4°C (Figure 25.10). The addition of 2% trehalose to the mix provided a reduction in hardness after five days of about 32%. The effects of trehalose in frozen sponge cakes and dinner rolls were also investigated. Hardness measurements on thawed sponge cakes showed that trehalose (1.7%) reduced starch retrogradation and produced a product that was 28% softer than controls after thawing. A 1% trehalose addition to dinner rolls containing sucrose resulted in less heat shock when the rolls were thawed using an electric range, than when sucrose was used alone. In addition, trehalose-produced rolls maintained a softer texture for up to 16 hours after thawing (Figure 25.11) (HBC 1997).

Prevention of Fat Decomposition

The ability of trehalose to protect membrane phospholipid layers subjected to heat stress or freeze/thaw cycles has been examined (Crowe et al. 1985, 1990, 1998). Results showed that trehalose was more effective in maintaining the integrity of vesicle membranes than standard protectants such as glycerol or dimethylsulfoxide. The ability of trehalose to stabilize free fatty acids has also been studied (HBC 1998). One hundred grams of four different fatty acid solutions were combined with 1 ml of 5% solutions of sucrose, sorbitol, or trehalose. The mixtures were heated for one hour at 100°C. Fatty acid concentrations before and after heating were measured by gas chromatography. Results showed that trehalose markedly reduced the decomposition of the fatty acids tested compared with control samples or those with sucrose or sorbitol (Figure 25.12).

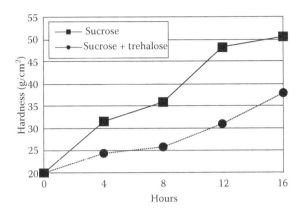

Figure 25.11 **Dinner rolls with or without 1% trehalose (dry weight) were baked and frozen at –18°C for seven days. The rolls were thawed in an electric range and the hardness was measured every four hours for 16 hours.**

Texture, Flavor, and Color Stabilization

Colaço and Roser studied the use of trehalose to preserve the fresh qualities of a number of foods (Colaço and Roser 1994). They maintained that, because trehalose was not particularly sweet, it would not significantly change the flavor of foods to which it was added. The authors blended fresh eggs with trehalose and dried the mixture at 30°C–50°C. An odorless, yellow-orange powder that could be stored at room temperature was produced. When this powder was rehydrated, the product was reported indistinguishable from an equivalent fresh egg mixture.

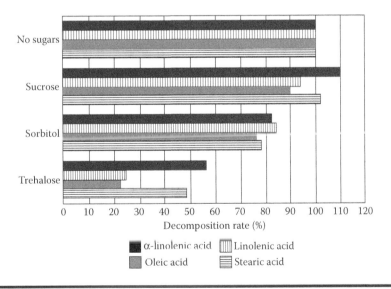

Figure 25.12 **One of four fatty acid solutions (100 g) was mixed with 1 ml of a 5% solution of sucrose, sorbitol, or trehalose or without adding a sugar solution (control). The solutions were heated for one hour at 100°C, and the amount of remaining fatty acids was compared with preincubation measurements. The results are recorded as percent decomposition of the control sample. Fatty acids were assayed using gas chromatography.**

The ability of trehalose to preserve the fresh aroma and texture of herbs, fresh fruit slices, and purees was also studied (Colaço and Roser 1994; Nazzaro et al. 1997). Fruit purees dried with trehalose resulted in shelf-stable powders with little aroma. On rehydration, these powders recovered the texture and smell of fresh fruit. Control samples dried without trehalose were more difficult to reconstitute, and a taste panel described the aroma and flavor as denatured or "cooked." The authors used gas chromatography to quantitate the preservation of fruit volatiles from bananas, which had been vacuum-dried at 37°C–40°C with or without 10% (w/w) trehalose. After several months storage at room temperature, the banana purees dried without trehalose exhibited a loss of flavor volatiles, whereas those dried with trehalose showed a minimal loss. It was concluded that the purees dried with trehalose retained most of the volatiles typical of fresh fruit (Colaço and Roser 1994; Nazzaro et al. 1997).

Similar results were noticed for fresh herbs dried with trehalose. Control products were reported to have lost flavor, aroma, and color during storage, whereas rehydrated herbs dried with trehalose maintained their fresh color and organoleptic properties (Colaço and Roser 1994).

Spinach was heated at 90°C for 30 seconds to inactivate enzymes. The spinach was shredded and 10% (w/w) of various sugars were added. The spinach/sugar preparations were dried at 55°C for 16 hours and heated in an oven at 100°C for three hours to simulate maximum commercial heat stress (HBC 1998). Color measurements using an L*a*b* colorimeter were taken after three hours in the oven. Figure 25.13 shows the differences in green color (a-value) compared with a control sample where no sugar was added. In this study, trehalose provided the most beneficial effect for color preservation.

Ten grams of fish paste was mixed with 5 ml of 5% solutions of various sugars. The prepared fish paste was boiled for 15 minutes. Percent release of trimethylamines (a primary component of fishy odors) was measured for the various sugar-containing preparations and a control sample (Figure 25.14).

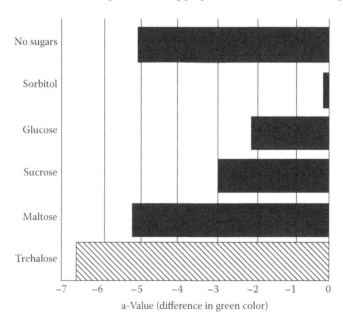

Figure 25.13 Various sugars (10% w/w) were mixed with fresh spinach. Samples were dried at 55°C for 16 hours, after which they were heated in an oven at 100°C for three hours. Color measurements were taken after heating and reported as a-values.

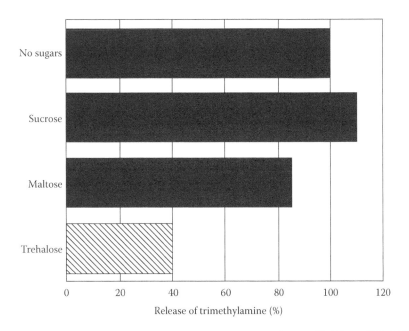

Release of trimethylamine (%)

Figure 25.14 **Ten grams of fish paste was mixed with 5 ml of a 5% solution of sucrose, maltose, or trehalose. A fourth sample without the addition of a sugar was used as a control. The samples were boiled for 15 minutes, and the relative release of trimethylamine was measured, with the control representing 100%.**

A relative release value of 100% trimethylamine was used as the control when the fish paste was boiled in the absence of any of the sugars. In comparison, fish paste boiled after mixing with trehalose solution released 40%, whereas the fish paste using sucrose released approximately 110% (HBC 1998). In a similar experiment, trehalose or sorbitol was combined with fish paste at 10% and 20% to test the release of aldehyde, ethylmercaptan, and trimethylamine (fish-related smells) after boiling for 15 minutes. Ten percent trehalose reduced the release of these chemicals compared with sorbitol by 3.1-, 2.9-, and 4.7-fold and in samples containing 20% by 2.8-, 2.4-, and 5.2-fold, respectively.

In food formulations, trehalose has been shown to preserve the color and texture of cold or frozen foods. A 4% gelatin raspberry mousse was prepared using sucrose or with a partial substitution of sucrose by trehalose. The mousse was frozen ($-18°C$) for one week and then thawed at $4°C$. The addition of trehalose helped maintain 97.9% of the hardness and integrity of the dessert compared with the mousse containing only sucrose (81%) (HBC 1997). We have not been able to duplicate this data in the United States.

Japanese commercial food processors have used trehalose in a variety of food categories since 1995. They report that trehalose masks bitterness and enhances flavor in beverages, reduces sweetness, retards starch retrogradation, increases moisture-holding capacity in bakery products, prevents Maillard reactions in candies and light-colored soups, enhances reconstitution of dried noodles, and prevents hygroscopicity in jellies and in various toppings (HBC 1997). Table 25.6 lists the multifunctional benefits of trehalose reported for foods and provides the approximate amounts required for each effect. These concentrations are based on estimated values used in Japanese commercial products discussed previously and on laboratory trials conducted by HBC.

Table 25.6   Technical Effects of Trehalose

Food Category	Technical Effect	Approximate Trehalose Addition
Bakery products	Moisture retention	2% flour
	Shelf life extension	
	Crumb softener	
	Reduced sweetness	
	Reduced hygroscopicity	
Frozen bakery products	Protein preservation freeze-thaw stabilization	13%–18%
	Shelf life extension	
	Crumb softener	
Frozen desserts	Freeze-thaw stabilization	13%–18%
	Texture stabilization	
Dairy-based foods and toppings	Texture stabilization	2%–12.5%
	Texture stabilization	
	Flavor profile improvement	
Dried, frozen, or processed fruits and vegetables	Color stabilization	5% of carrier solution
	Texture stabilization	
	Flavor profile improvement	
	Masks bitterness	
Beverages	Flavor profile improvement	0.4% of product to 50% of sugars
	Color stabilization	
	Reduced sweetness	
	pH stabilization	
	Masks bitterness	
Jellies and gelatins	Moisture retention	15%–30% of sugars
	Reduced sweetness	
	Reduced hygroscopicity	
	Color stabilization	
	Flavor profile improvement	

Table 25.6 (Continued)   Technical Effects of Trehalose

Food Category	Technical Effect	Approximate Trehalose Addition
Confectionery	Reduced syneresis	5%–40% of product
	Shelf life extension	5%–80% of sugars
	Moisture retention	
	Reduced sweetness	
	Reduced hygroscopicity	
	Texture improvement	
	Flavor profile improvement	
Meats/fish/eggs	Protein preservation	2%–10%
	Moisture retention	
	Texture improvement	
	Masks cooking odors	

Cariogenicity

The cause of dental caries and the method for intraoral plaque–pH telemetry have been well documented and discussed in Chapters 13, 19, and 21 of this book. Trehalose was compressed into a lozenge with mint flavor. The change in pH during consumption of the mint was studied under a standard protocol in which a plaque-covered pH sensor is integrated into a removable, mandibular, restorative device (Imfeld 1983; Imfeld and Muhlemann 1978). The pH did not drop below the critical value of 5.7 in any of the four subjects tested. With regard to nonfermentability, the mints would, therefore, qualify for the "toothfriendly" claim.

A similar study was performed in Japan where trehalose was incorporated into a chocolate candy (HBC 1997). Four subjects dissolved the chocolate (5.1 g) in their mouths, and the plaque pH was measured by an indwelling electrode over a 30 minute period. A 10% sucrose solution was used as a positive control. The pH did not drop below 5.7 in any of the four subjects consuming the chocolate. This suggests that trehalose taken under these conditions does not promote dental caries.

In vitro studies on the fermentability of trehalose using *Streptococcus mutans* have shown that trehalose can be fermented (British Sugar 1998); however, the fermentation rate was lower than that of sucrose. *In vivo* plaque-pH assays using the two different trehalose-containing products indicate that the time during which trehalose is in contact with dental plaque is insufficient to result in critical plaque acidification. Furthermore, the amount of acid formed during the period of trehalose exposure was either too small to reduce the pH below 5.7 or may have been neutralized sufficiently by increased saliva production during consumption of the mint or chocolate.

Metabolism

Humans have long consumed trehalose in various foods, primarily young mushrooms and baker's yeast (Sugimoto 1995; Miyake 1997). At present, trehalose is being used by the Japanese food industry at a rate of more than 2,000 metric tons per month. This will likely increase on an international scale as more functional applications are found. The mechanism by which dietary trehalose is metabolized in mammals has been studied and appears straightforward. Trehalose is not assimilated intact into the body and has not been detected in blood. Like other disaccharides, it is hydrolyzed on the brush border of the intestinal enterocytes. The enzyme that hydrolyzes trehalose into its two glucose units is trehalase (Birch 1963; Elbein 1974). Trehalase is tightly bound to the surface of the external side of the membrane microvilli in the small intestine (Galand 1989; Maestracci 1976). It is highly specific for trehalose, appears to have the highest concentration in the proximal and middle jejunum, and declines toward the distal ileum (Asp et al. 1975). Trehalase hydrolyzes trehalose in close proximity to the enterocytes where the two glucose molecules are transported into the body by a well-known active transport system (Ravich and Bayless 1983). Glucose uptake in the small intestine is efficient, and studies have shown that each 30 cm of jejunum (10% of the total length of the small intestine) can absorb 20 g of glucose per hour (Holdsworth and Dawson 1964). In addition, during the first hour after digestion, the stomach releases only about 50 g of glucose into the duodenum (Holdsworth and Besser 1968). Disaccharides are known to retard gastric emptying, and although trehalose has not been specifically tested, it is likely that it would have a similar effect (Elias et al. 1968). Thus all ingested trehalose is hydrolyzed to glucose and absorbed in the small intestine.

Only a few studies have examined the level of trehalase activity in the gut. These have not always used the same endpoints or experimental conditions, so it is not possible to make direct comparisons. It can be inferred from these studies that, except in a few specific instances, the concentration of trehalase in the small intestine of humans is sufficient to handle substantial amounts (at least 50 g in a single ingestion) of trehalose.

Gudmand-Høyer et al. (1988) noted that trehalase and lactase activities in the gut are similar. In the experience of these authors, a lactase activity of 6.0 IU/g protein or less results in malabsorption of lactose, whereas activities from 6.0 to 8.0 IU/g protein were considered intermediate and may or may not result in malabsorption. The authors suggested that it may be appropriate to use a similar standard for trehalase. They referred to a study of trehalase activity of intestinal biopsies from 248 Danish patients in which the lowest level of activity in the group was 8.3 IU/g. Furthermore, no trehalase deficiencies were identified in more than 500 biopsies specimens of Danish subjects (Gudmand-Høyer et al. 1988).

Welsh described intestinal trehalase activity in 123 Caucasian subjects ranging in ages from one month to 93 years from the southwestern United States (Welsh et al. 1978). No significant differences in trehalase activity for all age groups or sex-based differences were found. The lowest recorded values were 7 and 8 IU/g in two infants 0 to 2 years of age (n = 70). No statistically significantly differences in trehalase activity were found for any age group or sex. Importantly, trehalase activity did not appear to wane with age.

In a study of 100 consecutive normal biopsy samples from adults (72 men, 28 women), two subjects with low trehalase activity (2.7 and 1.5 IU/g) were identified (Bergoz et al. 1982). It was suggested by these authors that a trehalase value of less than 5 IU/g might result in intolerance of trehalose ingestion, which is similar to the conclusion of Gudmand-Høyer et al. (1988). In a second study, Bergoz tested 16 control subjects for their ability to assimilate trehalose (Bergoz 1971). Each subject drank 50 g of glucose in water, followed within two days by a similar preparation of trehalose. Blood glucose values were assayed after ingesting glucose or trehalose, and the ratios of glucose absorbed were calculated. All control subjects tolerated both test solutions and assimilated the glucose hydrolyzed from trehalose (ratio = 0.70, range, 0.31–1.42); however, the time to peak blood glucose concentrations was slower after trehalose ingestion. A third study by Bergoz et al. (1973) reported on 50 hospitalized control subjects. Patients were given glucose and trehalose oral tolerance tests as described previously. Control subjects had a median absorption ratio of 0.70 (range, 0.31–1.52) and tolerated the trehalose exposure. These values were essentially the same as in the second study.

Twenty patients in Czechoslovakia with no bowel symptoms or disease were examined for trehalase activity (Mardzarovová-Nohejlova 1973). None of these subjects were considered to be trehalase-deficient (Welsh et al. 1978).

Murray et al. (2000) reviewed several reports on trehalase activity in various groups of patients, and presented data on 369 subjects from the United Kingdom with healthy intestinal histology. Of the 369 subjects, only one was shown to be slightly deficient in trehalase activity. The authors used a different standard to classify trehalase deficiency, and although not identical to the other methods it appears that the data can be compared.

In summary, no indication of trehalase deficiency was reported in 500 biopsies specimens in a Danish population (Gudmand-Høyer et al. 1988). Welsh's group studied 123 subjects from the Southwestern United States. The lowest value of trehalase activity was 7 IU/g in an infant(s) less than one year of age; however, there were no statistical differences in means between age groups. In addition, no adults had low enzymatic levels (Welsh et al. 1978). Bergoz et al. assayed biopsy specimens from 100 controls and showed substantial trehalase values in 98 samples (Bergoz et al. 1982). Bergoz found no abnormalities in trehalose absorption in 16 control patients (Bergoz 1971). Fifty hospitalized patients showed normal trehalose absorption (Bergoz et al. 1973). Twenty patients sampled in Czechoslovakia did not display a trehalase deficiency, and only one of the 369 was found to have low activity in the United Kingdom (Mardzarovová-Nohejlova 1973; Murray et al. 2000). Taken together, it appears that when trehalase activity assays or trehalose absorption tests are performed on hundreds of control subjects, only a few individuals can be identified with low trehalase activity. Importantly, this percentage (less than 1%) appears substantially less than for lactase and possibly other disaccharidase deficiencies (Dahlqvist 1974).

Trehalase is also found in human peripheral blood, urine, kidney and liver, and urine (Birch 1972; Stork et al. 1977). The serum concentrations of trehalase appear to be fairly constant in an individual, although they can vary widely within the population. Significant sex or age differences have not been reported. The presence of trehalase in these areas of the body do not appear to be related to the absorption of trehalose from the gut.

Malabsorption and Intolerance

Malabsorption and/or intolerance to trehalose has been reported (Bergoz et al. 1973, 1982; Dahlqvist 1974; Gudmand-Høyer et al. 1988; Mardzarovová-Nohejlova 1973; Murray et al. 2000; Welsh et al. 1978). The clinical symptoms and the pathophysiology appear to be identical to those seen in other disaccharide malabsorption syndromes and are therefore self-limiting (Gudmand-Høyer et al. 1988; Dahlqvist 1974). Few specific studies have been performed to identify the prevalence of trehalose intolerance or have specifically related intestinal trehalase activity with trehalose intolerance. It is believed that trehalose intolerance is substantially less frequent than lactose intolerance with similar self-limiting consequences (Dahlqvist 1974).

As reported previously, 100 consecutive normal biopsy specimens were evaluated for trehalase activity (Bergoz et al. 1982). Two subjects had low trehalase activity. One subject with the lowest trehalase activity was tested for trehalose malabsorption and intolerance. No elevation in blood glucose was observed after consumption of the trehalose; intestinal symptoms developed within the first hour. The authors suggested that trehalase values less than 5 IU/g would result in symptoms. Bergoz also identified a 71-year-old woman who reported abdominal pain when she ingested mushrooms (Bergoz 1971). The patient and 16 control subjects were tested for glucose and trehalose absorption and tolerance by consuming 50 g of both sugars in water. The woman experienced intestinal distress starting about 20 minutes after ingesting trehalose, but not glucose. The mean absorption ratio of trehalose to glucose was 0.02 for the patient compared with 0.7 for controls.

A 24-year-old man complained of intestinal symptoms after eating mushrooms (Mardzarovová-Nohejlova 1973). The patient's father, uncle, and two cousins also reported experiencing similar symptoms when they ate mushrooms. Jejunal biopsies were taken from immediate family members and 20 control patients (Mardzarovová-Nohejlova 1973). Trehalose and glucose tolerance was tested with 50 g of each disaccharide. Trehalose ingestion caused intestinal symptoms in the father and son; no trehalase activity was detected in their biopsy tissue.

Murray et al. reported that only one of 369 subjects with normal histology was deficient in trehalase; however, the subject was not tested for malabsorption or intolerance (Murray et al. 2000). Gudmand-Høyer et al. studied intestinal biopsies from 97 adult (50 women, 47 men) residents of Greenland (Gudmand-Høyer et al. 1988). Eight subjects had trehalase activity values of less than 6 IU/g protein and another six had values less than 8 IU/g. Three subjects with low trehalase activity (less than 6 IU/g) were given 50 g of trehalose dissolved in water. No glucose was assimilated into the blood. Lactase deficiency is found in approximately 60% of Greenlanders, and sucrase deficiency is not uncommon. The percentage of trehalase deficiency in a fairly closed genetic group and in a single family suggests a hereditary basis, as is seen in other disaccharide deficiencies (Bergoz et al. 1982; Gudmand-Høyer et al. 1988).

At-risk Populations

It appears from studies that the consumption of all disaccharides, including trehalose, can be of concern in patients with intestinal malabsorption disorders (Berg et al. 1973; Bergoz et al. 1973; Murray et al. 2000; Redeck and Dominick 1983). Patients with juvenile diabetes were tested for trehalase activity. It does not appear that this condition results in a significant depression of trehalase activity, although the mean enzymatic activity of these patients in one study was lower than controls (Ruppin et al. 1974).

Cerda and coworkers reported that in non-insulin-dependent diabetic patients with chronic pancreatic insufficiency, there was approximately a two-fold increase in trehalase and other disaccharidases (Cerda et al. 1972). Patients with chronic renal failure appear to have reduced levels of trehalase, but the reduction appears to be less than for other disaccharidases (Dennenberg et al. 1974). The more common condition of lactase deficiency (lactose intolerance) has not been shown to be correlated with a deficiency of trehalase in the general population (Asp et al. 1971). From this information, it appears that the consumption of trehalose under various conditions presents no more risk than any other disaccharide in a normal diet.

Safety Studies

Trehalose is considered a natural sugar for which humans have evolved and maintained an intestinal disaccharidase. Because trehalose is now only a minor component of the human diet, little interest has been shown in studying possible effects of higher consumption. In the early 1990s, Quadrant Holdings of Cambridge, England, began investigating the possible use of trehalose for the health and food industry and commissioned seven toxicological studies. With the advent of the enzymatic production method of HBC and the subsequent reduction in cost, several additional safety studies were initiated. These studies have been reported in a review publication (Richards et al. 2002).

Mouse Micronucleus, Chromosome Aberration, and Bacterial Mutation Assays

All studies were performed by standard methods using trehalose (Sugimoto 1995; Richards et al. 2002). Peak dosing concentrations for each assay were 5000 mg/kg, 5000 µg/ml and 5000 µg/plate, respectively. On the basis of the criteria established by the protocol, it was concluded that trehalose did not show a positive response (negative treatment effect) in any of the test systems.

Single Dose Oral Toxicity

Four studies were performed on mice, rats, and beagle dogs. The first study was performed on rats using trehalose produced by HBC (Richards et al. 2002). The remaining three studies were performed using pharmaceutical grade trehalose produced by Pfanstiehl Laboratories, Inc. (Quadrant Holdings 1994). The results are provided in Table 25.7.

Table 25.7 Four Acute Toxicological Studies Performed on Mice, Rats, and Beagle Dogs Using Two Different Preparations of Trehalose

Study	Dose	Conclusions
Trehalose crystals: acute oral toxicity to the rat	Oral gavage 16 g/kg	The acute lethal dose is greater than 16 g/kg body weight. Piloerection in all rats within five minutes of dosing lasting through day 1. Occurred at later intervals, but stopped by day 4.
An acute toxicity study of trehalose in the albino mouse	Intravenous (IV) 1 g/kg Oral gavage (OG) 5 g/kg	No treatment-related mortality, body weight changes, or systemic toxicity was evident.
An acute toxicity study of trehalose in the albino rat	Intravenous (IV) 1 g/kg Oral gavage (OG) 5 g/kg	No acute signs of toxicity with either IV or OG dose.
An acute toxicity study of trehalose in the beagle dog	Intravenous (IV) 1 g/kg six day washout Oral capsules 5 g/kg	No acute signs of toxicity with either IV or OG dose.

Acute Eye Irritation Study

Six male New Zealand White rabbits received a single 0.1 ml dose of a 10% trehalose solution in the right eye (Quadrant Holdings 1994). The left eye was used as a control. Eyes were examined at 1, 24, 48, and 72 hours after application. No evidence of irritation was observed.

Seven-Day Continuous Infusion Study

Three groups of five male Sprague-Dawley rats were dosed by continuous infusion with a 10% solution of trehalose delivered at a rate of 400 mg/kg/hr, a 5% solution of dextrose at 200 mg/kg/hr, or a 0.9% NaCl solution at 4 ml/kg/hr (Quadrant Holdings 1997). The solutions were administered for seven consecutive days through tail vein catheters. No clinical signs, changes in body weights, or treatment-related findings at necropsy were observed. The only untoward effect reported was osmotic nephrosis in the animals given trehalose. This is thought to result from a lack of trehalase in the kidney of rats.

Fourteen-Day Toxicity Studies

A 14-day toxicity study in CD-1 albino mice with pharmaceutical grade trehalose was performed (Quadrant Holdings 1994). Doses of 1 g/kg/day by intravenous (IV) injection, 2.5 g/kg/day by subcutaneous injection, or 5 g/kg/day by oral gavage were administered for 14 consecutive days. No gross or microscopic lesions attributable to trehalose were observed, and trehalose appeared to be innocuous to the mice treated in these studies. In addition, there were no effects on body weight or food consumption.

Beagle dogs were given doses of 1 g/kg/day by IV injection, 0.25 g/kg/day subcutaneous injection, or 5 g/kg/day orally for 14 consecutive days (Quadrant Holdings 1994). No gross or microscopic lesions attributable to trehalose were observed, indicating that trehalose appeared to be nontoxic. The only condition observed was self-limiting loose stools in dogs given trehalose orally. The loose stools are thought to be caused by osmotic pressure bringing water into the gut, similar to that seen in human and other animals with lactase or other disaccharidase deficiencies.

Thirteen-week Toxicity Study

A subchronic 13-week study was conducted to assess the effects of up to 50,000 ppm of trehalose in mice (Richards et al. 2002). Trehalose was well tolerated and there was no evidence of toxicity. There was a slight reduction in food consumption, with a concomitant reduction in weight gain in male mice. It is believed that palatability was the cause of reduced food intake. Slight increases in plasma glucose and reduction in plasma bilirubin and potassium were observed. The results indicate a no-toxic-effect-level of 50,000 ppm.

Embryotoxicity, Teratogenicity, and Two-generation Reproductive Studies

Trehalose was fed in the diet to mated female rats from 0 to 21 days, and to female rabbits from 0 to 29 days of gestation (Richards et al. 2002). The concentration of trehalose was 2.5%, 5%, and 10% of the diet for both species, which equated to an intake of 1.4–2.0, 2.8–3.9, and 5.5–7.8 g/kg/day for rats, and 0.21–0.77, 0.48–1.34, and 1.04–2.82 g/kg/day for rabbits, respectively. Analysis of the results from both studies indicated that trehalose did not induce maternal or developmental toxicity, even at the highest concentration.

Two generations of male and female rats were fed trehalose in the diet at concentrations of 2.5%, 5%, and 10% (Richards et al. 2002). The consumption by males of the two generations ranged from 1.24–2.90, 2.38–5.65, and 4.89–12.43 g/kg/day, respectively. The range of consumption by female rats (including premating, during gestation, and during lactation) was 1.06–5.46, 2.22–11.49, and 4.40–23.2 g/kg/day, respectively. No treatment effects were observed on any variables of adult males and females, offspring, or reproduction.

Conclusions

Trehalose is a relatively new entry to the food industry, but has had a long history as a natural part of the human diet. Although trehalose consumption in particular foods is not a large part of our modern diet, the total amount of trehalose consumed can be substantial when the contribution of these trehalose-containing foods is considered as a whole.

The ability to digest trehalose depends on the presence of the enzyme trehalase. From available information, it appears that only a relatively small percentage of people in a western population lack this enzyme. Intolerance to trehalose appears to be far less common than intolerance to lactose, and both are believed to have a genetic basis.

Approximately 30,000 metric tons of trehalose added to food products were consumed in the Japanese market during 2010 without any known reports of intolerance. In addition, symptoms of intolerance to trehalose are identical to those observed in individuals with intolerances to other disaccharides. Safety studies have demonstrated that there are no consistent untoward effects associated with the consumption of trehalose.

From information presented in this chapter, it would appear that trehalose has functional properties that may be of great interest to the food, cosmetic, and pharmaceutical industries. Two decades ago, the cost of trehalose precluded its use in all but the highest value products. With the advent of the new enzymatic production process invented by HBC, trehalose is now priced where its use in cost-sensitive applications can by justified.

Trehalose is approved as a food additive in Japan, as a GRAS notified substance in the United States, as a novel food in the EU, Canada, Australia and Brazil, and as a food (food ingredient) in most other countries of the world. The Joint Food and Agriculture/World Health Organization Expert Committee on Food Additives (JECFA) has determined trehalose is a food with no specified acceptable daily intake (ADI).

References

Aldous, B.J., Auffret, A.D., and Frank, F. 1995. The crystallization of hydrates from amorphous carbohydrates. *Cryo-Letter* 16:181–186.

Aso, K., Watanabe, T., and Hayasaka, A. 1962. Studies on the MIRIN. II. Fractionation and determination of sugars in Mirin (sweet saké) carbon column chromatography. *Tohoku J Agric Res* 13:251–256.

Asp, N.G., Berg, N.O., Dahlqvist, A., Jussila, J., and Salmi, H. 1971. The activity of three different small-intestinal β-galactosidases in adults with and without lactase deficiency. *Scand J Gastroenterol* 6:755–762.

Asp, N.G., Gudman-Høyer, E., Andersen, B., Berg, N.O., and Dahlqvist, A. 1975. Distribution of disaccharides, alkaline phosphatase, and some intracellular enzymes along the human small intestine. *Scand J Gastroenterol* 10:647–651.

Berg, N.O., Dahlqvist, A., Lindberg, T., and Nordén, A. 1973. Correlation between morphological alterations and enzyme activities in the mucosa of the small intestine. *Scand J Gastroenterol* 8:703–712.

Bergoz, R. 1971. Trehalose malabsorption causing intolerance to mushrooms. *Gastroenterol* 60:909–912.

Bergoz, R., Bolte, J.P., and Meyer zum Bueschenfelde, K.H. 1973. Trehalose tolerance test. *Scand J Gastroenterol* 8:657–663.

Bergoz, R., Vallotton, M.C., and Loizeau, E. 1982. Trehalose deficiency. *Ann Nutr Metabol* 26:291–295.

Birch, G.G. 1963. Trehalose. In *Advances in carbohydrate chemistry*, M.L. Wolfrom and R.S. Tyson (Eds.), vol. 18, pp. 201–225. New York: Academic Press.

Birch, G.G. 1965. A method of obtaining crystalline anhydrous α,α-trehalose. *J Chem Soc* 3:3486–3490.

Birch, G.G. 1972. Mushroom sugar. In *Health and food*, G.G. Birch, L.F. Green, and L.G. Plaskett (Eds.), pp. 49–53. New York: Elsevier Sciences.

Bolin, H.R., and Steele, H.J. 1987. Nonenzymatic browning in dried apples during storage. *J Food Sci* 52:1654–1657.

British Sugar plc. 1997. pH telemetry expert opinion. Reference 85 Trehalose. *pH Laboratory Lucerne* 1–11.

Brown, G.M., Rohrer, D.C., Berking, B., Beevers, C.A., Gould, R.O., and Simpson, R. 1972. The crystal structure of α,α-trehalose dihydrate from three independent x-ray determinations. *Acta Cryst* B28:3145–3158.

Budavari, S. 1989. Trehalose dihydrate. In *The Merck Index*, 11th edition, p. 1508. New Jersey: Merck and Co.

Cerda, J.J., Preiser, H., and Crane, R.K. 1972. Brush border enzymes and malabsorption: Elevated disaccharides in chronic pancreatic insufficiency with diabetes mellitus. *Gastroenterol* 62:841.

Colaço, C.A.L.S., and Roser, B. 1994. Trehalose, a multifunctional additive for food preservation. In *Food packaging and preservation*, Chapter 7, M. Mathlouthi (Ed.), pp. 123–140. Maryland: Aspen.

Colaço, C.A.L.S., Sen, S., Thangavelu, M., Pinder, S., and Roser, B. 1992. Extraordinary stability of enzymes dried in trehalose: Simplified molecular biology. *Bio Tech* 10:1007–1010.

Crowe, J.H., Carpenter, J.F., Crowe, L.M., and Anchordoguy, T.J. 1990. Are freezing and dehydration similar stress vectors? A comparison of modes of interaction of stabilizing solutes with biomolecules. *Cryobiol* 27:219–231.

Crowe, J.H., Crowe, L.M., Carpenter, J.F., Rudolph, A.S., Wistrom, C.A., Spargo, B.J., and Anchordoguy, T.J. 1998. Interactions of sugars with membranes. *Biochem Biophys Acta* 947:367–384.

Crowe, L.M., Crowe, J.H., Rudolph, A., Womersley, C., and Appel, L. 1985. Preservation of freeze-dried liposomes by trehalose. *Arch Biochem Biophys* 242:240–247.

Crowe, L.M., Reid, D.S., and Crowe, J.H. 1996. Is trehalose special for preserving dry biomaterials? *Biophys J* 71:2087–2093.

Dahlqvist, A. 1974. Enzyme deficiency and malabsorption of carbohydrates. In *Sugars in nutrition*, Chapter 13, H. Sipple (Ed.), pp. 187–217. New York: Academic Press.

Dennenberg, T., Lindberg, T., Berg, N.O., and Dahlqvist, A. 1974. Morphology, dipeptidases and disaccharidases of small intestinal mucosa in chronic renal failure. *Acta Med Scand* 195:465–470.

Ding, S.P., Fan, J., Green, J.L., Lu, Q., Sanchez, E., and Angell, C.A. 1996. Vitrification of trehalose by water loss from its crystalline dihydrate. *J Thermal Anal* 47:1391–1405.

Donnamaria, M.C., Howard, E.I., and Grigera, J.R. 1994. Interaction of water with α,α-trehalose in solution: Molecular dynamics simulation approach. *J Chem Soc Faraday Trans* 90:2731–2735.

Elbein, A.D. 1974. The metabolism of α,α-trehalose. In *Advances in carbohydrate Chemistry and Biochemistry*, R.S. Tipson and D. Horton (Eds.), vol. 30, pp. 227–256. New York: Academic Press.

Elias, E., Gibson, G.J., Greenwood, L.F., Hunt, J.N., and Tripp, J.H. 1968. The slowing of gastric emptying by monosaccharides and disaccharides in test meals. *J Physiol* 194:317–326.

Galand, G. 1989. Brush border membrane sucrase-isomaltase maltase-glucoamylase and trehalase in mammals. *Comp Biochem Physiol* 94B:1–11.

Green, J.L., and Angell, C.A. 1989. Phase relations and vitrification in saccharide-water solutions and the trehalose anomaly. *J Phys Chem* 93:2880–2882.

Gudmand-Høyer, E., Fenger, H.J., Skovbjerg, H., Kern-Hansen, P., and Madsen, P.R. 1988. Trehalase deficiency in Greenland. *Scand J Gastroenterol* 23:775–778.

Hayashibara Company Ltd (HBC). 1997. Trehalose Technical Data (Fifth edition).

Hayashibara Company Ltd. (HBC). 1998. Unpublished data.

Holdsworth, C.D., and Besser, G.M. 1968. Influence of gastric emptying-rate and of insulin response on oral glucose tolerance in thyroid disease. *Lancet* 2:700–703.

Holdsworth, C.D., and Dawson, A.M. 1964. The absorption of monosaccharides in man. *Clin Sci* 27:371–379.

Imfeld, T. 1983. Identification of low caries risk dietary components. In *Monographs in oral science*, H.M. Myers (Ed.), vol. 11, p. 117. Basel, Switzerland: Karger.

Imfeld, T., and Muhlemann, H.R. 1978. Cariogenicity and acidogenicity of food, confectionery and beverages. *Pharm Ther Dent* 3:53.

Karmus, R., and Karel, M. 1993. The effect of glass transition on Maillard browning in food models. In *Maillard reactions in chemistry, food and health*, T.P. Labuza and G.A. Reieccius (Eds.), pp. 182–187. London: Royal Society of Chemistry.

Kawai, H., Sakurai, M., Inoue, Y., Chûjô, R., and Kobayashi, S. 1992. Hydration of oligosaccharide: Anomalous hydration ability of trehalose. *Cryobiol* 29:599–606.

Koch, E.M., and Koch, F.C. 1925. The presence of trehalose in yeast. *Science* 61(1587):570–572.

Lee, C.K., and Birch, G.G. 1975. Structural functions of taste in the sugar series: Binding characteristics of disaccharides. *J Sci Food Agri* 26:1513–1521.

Leibowitz, J. 1943. A new source of trehalose. *Nature* 152:414.

Leibowitz, J. 1944. Isolation of trehalose from desert manna. *J Biochem* 38:205–206.

Leslie, S., Israeli, E., Lighthart, B., Crowe, J.H., and Crowe, L.M. 1995. Trehalose and sucrose protect both membranes and proteins in intact bacteria during drying. *Appl Environ Microbiol* 61:3592–3597.

MacDonald, G.A., and Lanier, T. 1991. Carbohydrates and cryoprotectants for meats and surimi. *Food Technol* 3:150–159.

Maestracci, D. 1976. Enzymatic solubilization of the human intestinal brush border membrane enzymes. *Biochim Biophys Acta* 433:469–481.

Mardzarovová-Nohejlova, J. 1973. Trehalase deficiency in a family. *Gastroenterol* 65:130–133.

Mathlouthi, M. 1984. Relationships between the structure and the properties of carbohydrates in aqueous solution: Solute-solvent interactions and the sweet taste of D-fructose, D-glucose and sucrose in solution. *Food Chem* 13:1–13.

Mathlouthi, M., Bressan, C., and Portmann, M.O. 1993. Role of water structure in the sweet taste chemoreception. In *Sweet-Taste Chemoreception*, M. Mathlouthi (Ed.), pp. 141–174. London: Elsevier Applied Science.

Miyake, S. 1997. 4.01 Effects of trehalose in food applications. HBC unpublished company report 1–8.

Murao, S., Nagano, H., Ogura, S., and Nishino, T. 1985. Enzymatic synthesis of trehalose from maltose. *Agric Biol Chem* 49:2113–2118.

Murray, I.A., Coupland, K., Smith, J.A., Ansell, I.S., and Long, R.G. 2000. Intestinal trehalase activity in a UK population: Establishing a normal range and the effects of disease. *Br J Nutri* 83:241–245.

Nakada, T., Maruta, K., Mitzuzumi, H., Kubota, M., Chaen, H., Sugimoto, T., Kurimoto, M., and Tsujisaka, Y. 1995a. Purification and characterization of a novel enzyme, maltooligosyl trehalose trehalohydrolase, from Arthrobacter sp. Q36. *Biosci Biotech Biochem* 59:2215–2218.

Nakada, T., Maruta, K., Mitsuzumi, H., Tsukaki, K., Kubota, M., Chaen, H., Sugimoto, T., Kurimoto, M., and Tsujisaka, Y. 1995b. Putrification and properties of a novel enzyme, maltooligosyl trehalose synthase, from Arthrobacter sp. Q36. *Biosci Biotech Biochem* 59:2210–2214.

Nazzaro, F., Malanga, P.A., Immacolata, F., Zappia, V., Donsi, G., Ferrari, G., Nigro, R., and De Rosa, M. 1997. Trehalose as Innovative Agent for food osmo-drying processes. Food, Science and Technology Congress, Como, Italy.

Nicolaus, B., Gambacorta, A., Basso, A.L., Riccio, R., De Rosa, M., and Grant, W.D. 1988. Trehalose in archaebacteria. *System Appl Microbiol* 10:215–217.

Nishimoto, T., Nakano, M., Ikegami, S., Chaen, H., Fukuda, S., Sugimoto, T., Kurimoto, M., and Tsujisaka, Y. 1995. Existence of a novel enzyme converting maltose into trehalose. *Biosci Biotech Biochem* 59:2189–2190.

Portmann, M.O., and Birch, G.G. 1995. Sweet taste and solution properties of α,α-trehalose. *J Sci Food Agri* 69:275–281.

Quadrant Holdings. 1994. Unpublished data.

Ravich, W.J., and Bayless, T.M. 1983. Carbohydrate absorption and malabsorption. *Clin Gastroenterol* 12:335–356.

Redeck, U., and Dominick, H.C. 1983. Trehalose-load-test in gastroenterology. *Monatsschr Kinderheilkd* 131:19–22.

Richards, A.B., Krakowka, S., Dexter, L.B., Schmid, H., Wolterbeek, A.P.M., Waalkens-Berendsen, D.H., Shigoyuki, A., and Kurimoto, M. 2002. Trehalose: A review of properties, history of use and human tolerance, and results from multiple safety studies. *Food Chem Toxicol* 40:871–898.

Roser, B. 1991a. Trehalose, a new approach to premium dried foods. *Trends Food Sci Technol* 2(July):166–169.

Roser, B. 1991b. Trehalose drying: A novel replacement for freeze drying. *Bio Pharm* 4(September):47–53.

Roser, B., and Colaço, C.A.L.S. 1993. A sweeter way to fresher food. *New Scientist* 106(15 May):25–28.

Rudolph, A.S., and Crowe, J.H. 1985. Membrane stabilization during freezing: The role of two natural cryo-protectants, trehalose and proline. *Cryobiol* 22:367–377.

Ruppin, H., Domschke, W., Domschke, S., and Classen, M. 1974. Intestinale disaccharideasen bei juvenilem diabetes mellitus. *Klin Wochr* 52:568–570.

Schick, I., Fleckenstein, J., Weber, H., and Kulbe, K.D. 1991. In *Proceedings of the 2nd International Symposium Biochemical Engineering*, 126–129. Stuttgart: Fischer.

Shamil, S., Birch, G.G., Mathlouthi, M., and Clifford, M.N. 1987. Apparent molar volumes and tastes of molecules with more than one sapophore. *Chem Senses* 12:397–409.

Steiner, A., and Cori, C.F. 1935. The preparation and determination of trehalose in yeast. *Science* 82(2131):422–423.

Stewart, L.C., Richtmyer, N.K., and Hudson, C.S. 1950. The preparation of trehalose from yeast. *J Natl Inst Health* 72:2059–2061.

Stork, A., Fabian, E., Kozakova, B., and Fabianova, J. 1977. Trehalase activity in diabetes mellitus and in cirrhosis of the liver. In *Enzymology and its Clinical Use*, J. Horejsi (Ed.), pp. 37–40. Praha: Univerzita Kavlova.

Sugimoto, T. 1995. Production of trehalose by enzymatic starch saccharification and its applications. *Shokuhin Kogyo* 38:34–39.

Suzuki, T., Katsunobu, T., and Kinoshita, S. 1969. The extracellular accumulation of trehalose and glucose by bacteria grown on n-alkanes. *Agric Biol Chem* 33:190–195.

Taga, T., Senma, M., and Osaki, K. 1972. The crystal and molecular structure of trehalose dihydrate. *Acta Cryst* B28:3258–3263.

Tsusaki, K., Nishimoto, T., Nakada, T., Kubota, M., Chaen, H., Nakada, T., Maruta, K., et al. 1995. Purification and properties of a novel enzyme, maltooligosyl trehalose synthase, from Arthrobacter sp. Q36. *Biosci Biotech Biochem* 59:2210–2214.

Van Dijick, P., Colavizza, D., Smet, P., and Thevelein, J.M. 1995. Differential importance of trehalose in stress resistance in fermenting and nonfermenting Saccharomyces cerevisiae cells. *Appl Environ Micro* 61:109–115.

Welsh, J.D., Poley, J.R., Bhatia, M., and Stevenson, D.E. 1978. Intestinal disaccharide activities in relation to age, race and mucosal damage. *Gastroenterol* 75:847–855.

MULTIPLE INGREDIENT APPROACH

Chapter 26

Mixed Sweetener Functionality

Abraham I. Bakal

Contents

Introduction

Previous chapters have dealt with the properties and functionality of several currently approved sweeteners or compounds under development or consideration. The requirements for the ideal sweetener or sugar substitute, as usually defined (Bakal 1983) are to:

1. Have the taste and functional properties of sugar
2. Have low-calorie density on a sweetness equivalency basis
3. Be physiologically inert
4. Be nontoxic
5. Be noncariogenic
6. Compete economically with other sweeteners

An analysis of the organoleptic or functional properties of each single sweetener clearly shows that none of the currently known sugar substitutes comes close to the taste and functional properties of sucrose. Most exhibit one or more of these differences:

1. Taste properties such as sweetness lag, undesirable and lingering aftertaste, narrow taste profile, or bitterness. Saccharin, for example, is generally reported to have a bitter aftertaste (Salant 1972), stevioside to have a menthol or licorice aftertaste (Inglett 1970), aspartame to have a delayed sweetness (Beck 1974), and sucralose to have a lingering aftertaste.
2. Lack of bulking properties.
3. Stability problems during processing and storage. Aspartame, for example, loses its sweetness in aqueous solutions and is not stable at high temperatures (Beck 1978); thaumatin reacts with tannins and loses its sweetness (Higginbotham 1980).
4. Competitive prices. Saccharin and cyclamate reportedly cost, for example, on a sweetness equivalency basis, less than sucrose and other nutritive sweeteners. On the other hand, in the United States, some of the other sweeteners cost more (Personal communications 1999) than saccharin although the cost on a sweetness equivalency basis is trending down.

The food industry has partially overcome the bulking limitations in some selected applications. With the expansion of polydextrose use and the availability of polyols such as isomalt, maltitol and sorbitol, other reduced-calorie products based on combinations of polydextrose and/or on polyols and intense sweeteners have been introduced in the US market. Fibers and fibers with polydextrose have also recently found expanded use as bulking agents. However, this problem is universal for all intense sweeteners; thus it will not be discussed in this chapter. Except for the bulk issue, by far the most important limitations to consumer acceptability, are taste properties and to some extent cost.

The use of more than one sweetener provides the food technologist with a tool for overcoming some of the taste limitations. The advantages of combining sweeteners are many and some of the aims and goals are to:

1. Formulate products that closely imitate the taste and stability of their sugar-sweetened counterparts
2. Create totally new taste experiences by utilizing sweeteners in the same manner the food industry utilizes flavors (Porter 1983)
3. Meet cost restrains

Two interesting observations made by Paul in 1921 led to the subsequent practice of combining sweeteners. Paul, as cited by Mitchell and Pearson (1991), observed that the relative sweetness of an intense sweetener decreases with increasing concentration and that when combining two sweeteners, each sweetener will at least contribute the relative sweetness on the sweetness concentration curve. In other words, less of each sweetener is required when two or more sweeteners are used to achieve the same final sweetness achieved by the use of a single sweetener. These observations formed the basis for extensive work on the effects and benefits of using a combination of sweeteners.

Indeed, the use of combined sweeteners in food applications has been, and is currently being practiced by the food industry. Before the cyclamate ban in the United States, the food industry used a combination of saccharin and cyclamate in the formulation of products. This mixture showed synergistic properties and improved taste profile as shown in Figure 26.1. For comparison, the taste profile of saccharin against sucrose is shown in Figure 26.2. With the approval of aspartame, acesulfame potassium, and sucralose and their expanded use worldwide, several intense sweetener mixtures are being used in many foods and beverages. Figure 26.3 shows the taste profile of an aqueous solution sweetened with a saccharin/aspartame/cyclamate mixture in the ratio of

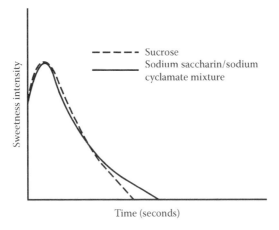

Figure 26.1 **Sweetness profile of aqueous solutions of sucrose versus a sodium saccharin/ sodium cyclamate combination (1:10).**

1:5:8, respectively. These data clearly demonstrate that combined sweeteners result in an improved sweetness profile when compared with each of the single sweeteners. In addition to the sweetness profile, the quality of the taste is significantly improved, as shown in organoleptic tests conducted in our laboratories. Verdi and Hood (1993) provide a good summary of some of the advantages of intense sweetener blends. Bakal and Cash (2006) discuss some of the applications and considerations that need to be taken into account when a sweetener combination is selected for use in a food product. Another review article by Saulo (2005) discusses the various sweeteners and their uses. A more recent study by Daniells (2009) provides a thorough discussion of sweetener blends for various food applications.

Neotame was approved in the United States in 2002 (Calorie Control Council) and is between 7,000 and 13,000 times sweeter than sugar. In addition to its sweetening qualities, it acts as a flavor enhancer. Since its approval, neotame has found uses in combination with other sweeteners, both intense and nutritive sweeteners (Torres 2005). One blend described by

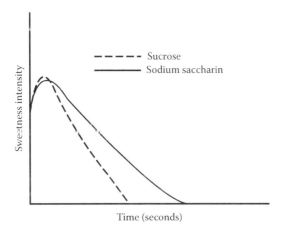

Figure 26.2 **Sweetness profile of aqueous solution of sucrose versus sodium saccharin.**

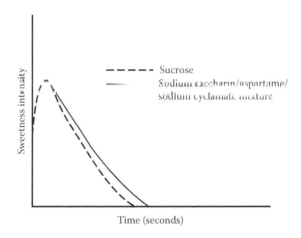

Figure 26.3 **Sweetness profile of aqueous solutions of sucrose versus a sodium saccharin/ sodium aspartame/sodium cyclamate combination (1:5:8).**

Torres, for example, consists of maltitol, neotame and acesulfame potassium that is claimed to be 600 times sweeter than sugar.

The recent Generally Recognized As Safe (GRAS) notification received regarding a stevia sweetener (US FDA 2008) has created renewed interest in natural sweetener blends. These include blends of stevia sweeteners with sugar and other natural sweeteners such as Lo Han Guo, fructose, erythritol, and so forth. Some of the recent developments vis-a-vis the regulatory situation for stevia sweeteners are discussed in Chapter 11 in this book. It is, however, important to note that while the first GRAS notification in the United States addressed a stevia sweetener rich in rebaudioside A (US FDA 2008), additional GRAS notifications have been published for stevia products that meet Joint FAO/WHO Expert Committee on Food Additives (JECFA) specifications and contain at least 95% steviol glycosides (US FDA 2010). These stevia products are finding use in mixed natural sweeteners.

GRAS notification for Lo Han Guo, also dubbed as monk fruit, is expected to increase interest in this natural sweetener and mixtures of monk fruit and stevia are expected to enter the market in the near future.

Applications of combined sweeteners in various products are discussed in the remaining section of this chapter.

Applications

Tabletop

Tabletop sweeteners are a major product category used by consumers for a variety of applications. They are commonly available as powders packed in bulk or sachets, as tablets, or as liquids. The effective amount of sweetener or sweetener combinations required to provide the sweetness equivalency of two teaspoons (10 g) of sugar to one cup (240 ml) of coffee is given in Table 26.1. This table clearly illustrates the reduction in the level of sweeteners ingested when a combination of sweeteners is used. When a mixture of saccharin/cyclamate is used, the amount of saccharin is reduced by a factor of 5 (i.e., 40 mg alone to 8 mg in combination), and the amount of cyclamate by a factor of 2.5 (i.e., 200 mg alone to 80 mg in combination), as compared to the use of any one

Table 26.1 Typical Effective Amounts of Sweeteners or Sweetener Combinations in Tabletop Applications

Sweetener	Amount (mg)[a]
Sodium saccharin	30–40
Sodium cyclamate	150–200
Aspartame	35–45
Acesulfame potassium	50–60
Sucralose	12–15
Rebaudioside A	25–40
Sodium cyclamate/sodium saccharin	80/8
Aspartame/sodium saccharin	5/15
Aspartame/Acesulfame potassium	15/15
Acesulfame potassium/aspartame	30/3
Aspartame/sodium saccharin/sodium cyclamate	10/4/30

Source: From Bakal, A.I. 1983. Functionality of combined sweeteners in several food applications, *Chem. Inc.*, September 19; Scott, D. (G.D. Searle) 1979. Use of dipeptides in beverages, *German Patent* 2,002,499. September 20.

[a] Effective amount required to provide the sweetness equivalency of two teaspoons (10 g) of sugar to one cup (240 ml) of coffee.

single sweetener. Similarly, in the case of an aspartame/saccharin mixture, the level of aspartame may be reduced by a factor of 2, and saccharin by a factor of 10.

Several patents and articles describe the synergism among aspartame, saccharin, cyclamate, acesulfame potassium, and other sweeteners. Scott described the use of aspartame in combination with saccharin, cyclamate, or both to provide beverages with improved taste and consumer acceptability (Scott 1979, 1971). A patent authored by Scott described the use of aspartame with saccharin in a ratio range of 15:1 to 1:15 (Scott 1971). A patent issued to General Foods discloses a sweetening composition containing aspartame, saccharin, and cyclamate (Finucase 1978). The inventor states that the sweetness is intense and lacks the lingering or bitter aftertaste associated with these sweeteners singly. A patent issued to E. R. Squibb proposes the use of "dipeptides" and saccharin, with the dipeptides masking the bitter aftertaste of saccharin (Lavia and Hill 1977, 1977; Hill and Lavia 1978). The saccharin to dipeptide ratio is 48:1.

Taste panels conducted in our laboratories on tabletop preparations confirm the superiority of a combination of aspartame with saccharin, cyclamate, or acesulfame potassium over any of these sweeteners singly. Panelists judged the taste quality of coffee sweetened with a combination of sweeteners, closely resembling the taste quality of sugar-sweetened coffee. Studies conducted in our laboratories suggest that aspartame/saccharin, aspartame/cyclamate, and aspartame/acesulfame potassium combinations have improved stability, even in hot aqueous solutions, and also exhibit superior taste profiles. The superior taste may be associated with flavor-enhancing properties of aspartame.

A commercial US tabletop product uses a blend of erythritol and rebaudioside A (rebiana). This tabletop product is packed in sachets each weighing 3.5 g and is reported to be as sweet as two teaspoons of sugar. Since erythritol is about 0.6–0.7 times as sweet as sugar, it contributes a significant amount of sweetness to the product in addition to the sweetness derived from the stevia.

Spoonable sweeteners have recently found increased consumer acceptance. These products usually consist of a spray dried mix of maltodextrin and sweetener with a bulk density of about 0.1 g/cc so that one teaspoon of this very light weight product is as sweet as one teaspoon of sugar. The process for preparing spoonable tabletop product involves making a solution of the sweetener or sweetener blend with a low degree of polymerization (DE) maltodextrin and spray drying this solution using special nozzles, and in most cases carbon dioxide injection, to control the bulk density of the spray dried product. These tabletop preparations are available in the United States and other markets and are typically packed in the United States in bags and in other countries in jars. This form is convenient to use in table use as well as in cooking and baking since it measures like sugar. In the United States, most of these products available in the market use a single sweetener, whereas combined sweeteners, especially aspartame and acesulfame potassium, are widely used in other parts of the world.

From 20% to 30% of tabletop sweeteners are sold in bulk (not in spoonable form) in the United States, a significant portion of which is used in cooking and baking. Consumers who cook with low-calorie sweeteners are usually those who must restrict their sugar intake for medical reasons. Saccharin, aspartame, acesulfame potassium, sucralose and stevia are approved for sale in bulk in the United States.

The liquid tabletop sweetener market represents approximately 10% of the US sales. Aspartame-based products may not be feasible for this application because of their limited stability in aqueous solutions. Other sweeteners in liquid form, however, are available in various markets and are popular in Japan among other markets.

In Europe, most tabletop products are formulated with a combination of sweeteners. The commercial products include traditional combinations of saccharin and cyclamate and a combination of acesulfame potassium and aspartame.

In summary, tabletop formulations employing a single sweetener meet the current needs of the consuming public; however, each preparation has one or more shortcomings, such as taste or stability. The consensus of experts and consumers is that tabletop preparations combining two or more sweeteners are superior in taste and stability, and more closely imitate the sweetness of sucrose.

Carbonated Beverages

Low-calorie soft drinks represent a significant segment of the market. Until the approval of aspartame for use in carbonated beverages in the United States, only saccharin was available after the ban of cyclamates. The level of saccharin used to sweeten one fluid ounce varies between 8 and 11 mg, the actual level depending on the soft drink flavor and the product brand. To achieve similar sweetness levels with aspartame as the single source sweetener, significantly higher amounts of aspartame are required. Table 26.2 summarizes the levels of saccharin and aspartame used in selected soft drinks and syrups, showing that, on the average, the amount of aspartame used is about 1.5 times the amount of sodium saccharin when each of the sweeteners is used alone.

Once regulatory approval was granted to aspartame, it became the most widely used sweetener in soft drinks.

Aspartame-based beverages lose sweetness as a function of storage time, temperature, and pH. Data submitted by G.D. Searle to the US Food and Drug Administration (US FDA) (US FDA 1982) indicate that about 50% of the initial aspartame remains in cola beverages stored at 30°C

Table 26.2 Typical Concentrations of Saccharin and Aspartame in Selected Soft Drinks and Syrups

Flavor	Sodium Saccharin[a] (mg/100 ml)	Aspartame[b] (mg/100 ml)
Soft Drinks		
Cola	31–42	57.7
Orange	37–38	92.6
Lemon-lime	26–42	50.1
Root beer	27–37	60.5
Syrups		
Cola	—	347.6
Orange	—	401.0
Lemon-lime	—	234.2
Root beer	—	355.2

[a] From Scott 1979.

[b] From Finucase 1978.

for 24 weeks (Figure 26.4). In cola syrups (pH 2.4), 75% of initial aspartame remains after 2 weeks storage at 30°C (Figure 26.5).

Combination sweeteners have been used by the soft drink industry for many years. Superior products were available when cyclamate and saccharin were both approved by the US FDA. Most products used 10:1 cyclamate/saccharin combinations because this ratio provided acceptable

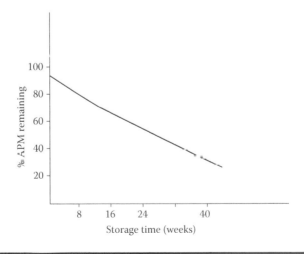

Figure 26.4 Aspartame (APM) stability in carbonated cola beverages stored at 30°C. (From US Food and Drug Administration (US FDA). Aspartame, 21CFR172, 804. March 19, 1982.)

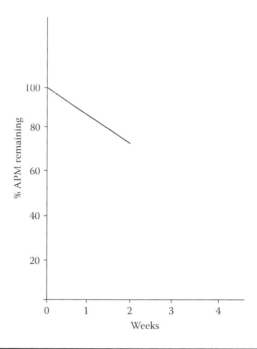

Figure 26.5 Aspartame (APM) stability in cola syrup at 30°C. (From US Food and Drug Administration (US FDA). Aspartame, 21CFR172, 804. March 19, 1982.)

products. The ban of cyclamate in the United States resulted in significant deterioration in the taste profile of soft drinks, which employed saccharin as the single sweetener. The use of aspartame in combination with saccharin is described by G.D. Searle (1982). Cola beverages sweetened with aspartame and saccharin at a ratio of 1:1 and stored at 20°C show significantly better sweetness stability than cola beverages sweetened with aspartame alone and stored under the same conditions. The performance of this combination of sweeteners is exceptional. The pH of the aspartame/saccharin product is lower than the pH of the aspartame-sweetened product (pH 2.8 versus 3.05, respectively). The data are shown in Figure 26.6.

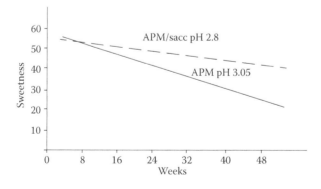

Figure 26.6 Average scores for sweetness of cola carbonated berverages stored at 30°C over a period of 52 weeks. (From US Food and Drug Administration (US FDA). Aspartame, 21CFR172, 804. March 19, 1982.)

Following the approval of aspartame in the United States, beverage manufacturers announced their plans to introduce soft drinks containing a combination of aspartame and saccharin. One cola beverage marketed at that time contained approximately 18 mg of saccharin and 8 mg of aspartame per 100 ml.

Marketing and other considerations have led major soft drink manufacturers to convert to 100% aspartame in the United States and in the United Kingdom in bottled and canned drinks. In other markets, however, products containing two or more intense sweeteners were introduced. A major soft drink manufacturer, for example, markets products that contain saccharin, aspartame, and acesulfame potassium.

Experiments conducted in our laboratory indicate that the addition of very small amounts of aspartame, in accordance with the specifications of the Lavia and Hill (1972) patent, improve the acceptance of saccharin-sweetened soft drinks. The cola beverages evaluated were sweetened with 33 mg saccharin per 100 ml. The addition of 0.7 mg aspartame to 100 ml of the saccharin-sweetened beverage (0.0007% concentration) resulted in a significant consumer preference for the beverage. It is clear that the use of aspartame at this concentration falls below the sweetness threshold of aspartame, which, according to Beck, is in the range of 0.007%–0.001% (Beck 1974).

The expiration of the US patent on aspartame and the approval of acesulfame potassium for use in soft drinks, resulted in marketing changes and renewed interest in using a combination of sweeteners. The most important of these developments is the introduction of a new brand of cola beverage, which is sweetened with a combination acesulfame potassium and aspartame. This product contains 70 mg of aspartame and 22.7 mg of acesulfame potassium per 8 oz another cola-based beverage uses a combination of saccharin and aspartame. This product contains 64 mg of saccharin and 19 mg of aspartame per 8 oz for comparison, the 100% sweetened cola contains 125 mg of aspartame per 8 oz.

Acesulfame potassium and aspartame show a sweetness enhancement of about 35% when used in combination. In addition to sweetness enhancement, aspartame broadens the taste profile of acesulfame potassium and brings the taste of the mixture closer to the sweetness profile of sucrose.

Aspartame and cyclamate combinations may also prove highly beneficial if cyclamates are approved in soft drinks. Acesulfame potassium and cyclamate is another combination that is of interest because it yields excellent taste quality and exceptional storage stability.

The approval of sucralose in the United States resulted in the introduction of soft drinks based on a combination of sucralose and acesulfame potassium. A major cola brand, for example, uses a combination of 30 mg acesulfame potassium and 40 mg sucralose per 8 oz of cola. Another major soft drink company uses 19 mg acesulfame potassium, 57 mg aspartame and 18 mg sucralose in their fruit flavored beverage.

Experiments conducted in our laboratories indicate that neohesperidin dihydrochalcone (NHDC), when used as a single sweetener, is inadequate for the preparation of soft drinks. When used in combination with saccharin, however, NHDC has a synergistic effect and gives improved taste perceptions. This finding is in accordance with observations described in the patent literature (Ishii et al. 1972).

The recent GRAS notifications of stevia sweeteners has resulted in the introduction of some all-natural soft drinks. This market, however, is still under development.

Moskowitz addressed the issue of sweetness optimization in cola-flavored beverages using combination sweeteners (Moskowitz et al. 1978) and presented a quantitative model for developing products acceptable to the consumer. Hoppe discussed the effect of various mixtures of sucrose, saccharin, and cyclamate on sweetness perception in aqueous solutions (Hoppe 1981). Van Tornout et al. (1985) evaluated the taste characteristics of mixtures of fructose with saccharin, aspartame, and acesulfame potassium in soft drinks. Their data indicate that combinations of small amounts of fructose with these intense sweeteners result in soft drinks that cannot be distinguished from sucrose-sweetened beverages. All of these data indicate the benefits to the consumer that can be derived from the use of combination sweeteners in soft drinks.

Dry Mixes

The dry mix category encompasses a variety of food products that are sold in dry form and are reconstituted by the consumer before use. They include beverage mixes, presweetened cereals, puddings, and desserts. In the United States, until the approval of aspartame, saccharin was the only intense sweetener used in this application. The approval of aspartame resulted in significant proliferation of these products in the market positioned as reduced-calorie and sugar-free. These aspartame-sweetened products have enjoyed wide consumer acceptance. To date, no dry mix products are available in the United States using combinations of saccharin and aspartame, although in the past, cyclamate/saccharin combinations were used. Table 26.3 provides a comparison of typical concentrations of saccharin and aspartame in selected products. Table 26.3 clearly shows that, in these applications, the effective aspartame level is about twice the effective sodium saccharin level. The approval of acesulfame potassium for use in these products and the expiration of the US patent on aspartame, resulted in the introduction of several products sweetened with a combination of aspartame and acesulfame potassium.

A combination of saccharin/aspartame, with a concentration of 15–20 mg sodium saccharin and 6–10 mg aspartame, in 100 g of finished product, yields a highly acceptable product and is significantly less expensive. In addition, such a combination represents a major reduction in the consumption of each of the sweeteners: almost a 50% reduction in sodium saccharin and about a 75% reduction in aspartame. Most importantly, results of consumer tests conducted in our laboratory clearly indicate the superiority of the products sweetened with a combination of the two sweeteners. Similarly, combinations of acesulfame potassium and aspartame yield products with a taste profile much more closely resembling that of the sucrose-sweetened products than those sweetened with either intense sweetener singly.

Recently, most commercial products use combinations of acesulfame potassium and aspartame as well as sucralose with other sweeteners in these dry mix applications.

Chewing Gums

Sugarless chewing gums represent a significant segment of the chewing gum market. These gums typically contain sorbitol, mannitol, and/or xylitol as the bulking agents to replace sugar. However,

Table 26.3 Typical Concentrations of Saccharin and Aspartame in Selected Dry-Mix Products

Product	Sodium saccharin[a]	Aspartame[b]
Beverage mix	27–34	40–55[b]
Gelatin dessert	21–28	19–44
Puddings	20–25	50–60

[a] Concentration in mg/100 g of reconstituted product.
[b] Beck 1978 suggests using 50–65 mg/100 g.

the sweetness intensity of sorbitol and mannitol is about half that of sucrose. These gums receive lower consumer acceptance because of lack of sweetness and flavor; therefore, intense sweeteners have been and are being used to improve the taste qualities of these products.

Saccharin is used in sugarless gums at concentrations between 0.1% and 0.2%. A typical chewing gum stick, weighing approximately 3 g, contains from 3 to 6 mg of saccharin. Several patents indicate the use of insoluble saccharin to produce chewing gums with long-lasting flavor (Bakal et al. 1978; Mackay et al. 1978). Taste tests comparing the sweetness quality of chewing gums with and without saccharin establish the superiority of the saccharin-sweetened products. Saccharin taste does not seem to be a limiting factor in this application because of the low saccharin concentration.

Figure 26.7 illustrates the sweetness-duration properties of chewing gums containing saccharin acid compared with chewing gums containing the same concentration of sodium saccharin. Figure 26.8, which shows the overall quality rating of the same chewing gums on a scale of 0 (dislike extremely), to 4 (neither like nor dislike), to 8 (like extremely), clearly shows consumer preference for the chewing gum containing the insoluble saccharin. Because of its relatively low sweetness intensity, cyclamate is not used as a single sweetener by the chewing gum industry. Before the US ban on cyclamates, cyclamate/saccharin combinations were commonly used at ratios between 2:1 and 10:1. Chewing gums using this combination of sweeteners were acceptable and had taste profiles similar to sugar-based products. US FDA regulations allow the use of aspartame in chewing gum, both as a sweetener and flavor enhancer (US FDA 1982). The literature

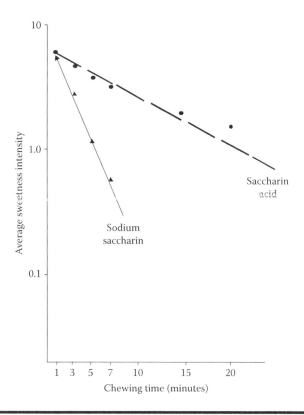

Figure 26.7 Sweetness intensity of chewing gums with sodium saccharin and saccharin acid.

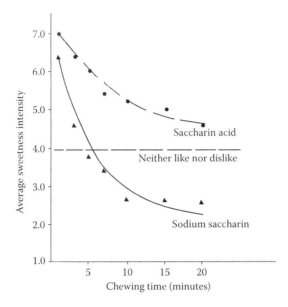

Figure 26.8   Overall acceptability rating of chewing gums with sodium saccharin and saccharin acid.

describes the use of aspartame in chewing gum at levels of 0.3% or higher (Bahoshy et al. 1976). These gums are reported to have longer-lasting flavors.

One of the problems encountered by chewing gum manufacturers in their use of aspartame, is the relative instability of aspartame in chewing gums. A patent issued to General Foods describes a method for stabilizing aspartame-containing gums by removing the calcium carbonate filler from the gum base formulation and replacing it with talc (Klose et al. 1981). This procedure apparently alters the pH of the gum base and results in a slower rate of aspartame decomposition.

A method investigated in the industry is the application of aspartame only on the surface of the gum through dusting or glazing (Cea and Glass 1981). This technique, although technically successful, presents consumer acceptance problems. Recently, encapsulation techniques have been investigated and are reportedly employed by some chewing gum manufacturers to overcome sweetness loss and to extend the sweetness and flavor of both aspartame- and saccharin-sweetened products.

In a 1978 paper, Beck describes the use of a combination of aspartame and saccharin in chewing gum. The ratio used was 2:1, aspartame to saccharin. This gum was preferred to gum containing either of the single sweeteners. Furthermore, although no stability data are available on this combination of sweeteners, it is believed that stability will be significantly improved. Several products containing aspartame and mixtures of aspartame and saccharin were evaluated. Figure 26.9 provides a sweetness profile of chewing gums containing a combination of saccharin and aspartame. The data clearly show that this product exhibits longer-lasting sweetness properties.

The approval of acesulfame potassium in the United States resulted in the introduction of chewing gums sweetened with this sweetener alone and with a combination of acesulfame potassium and aspartame. Such products are also available in Europe. Preliminary experiments conducted in our laboratory indicate that the use of a combination of sweeteners in chewing gums

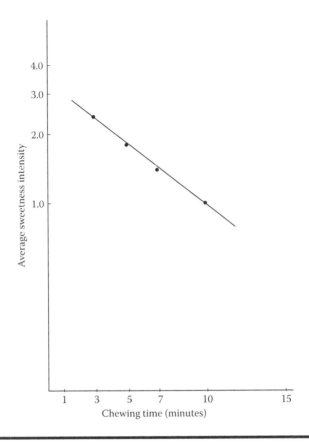

Figure 26.9 **Sweetness intensity of chewing gums containing a mixture of saccharin and aspartame.**

presents many advantages to the manufacturer and the consumer. Major US chewing gum manufacturers have introduced acesulfame potassium/aspartame-sweetened products and sucralose containing combinations of sweeteners.

Several other sweeteners have been investigated. Experiments conducted with NHDC indicate that this sweetener is good in chewing gum applications. It is stable and provides an acceptable sweetness profile and flavor-enhancing properties (Westall et al. 1972). Stevioside is currently used in sugarless gums in Japan. Although the products lack the sensory qualities of saccharin- or aspartame-sweetened products, they enjoy high consumer acceptance in Japan (Bakal 2007).

The recent GRAS acceptance of stevia sweeteners in the United States is expected to result in the introduction of new chewing gums and confectioneries sweetened with natural sweetener combinations such as stevia and sugar or stevia and erythritol.

Processed Foods

This application category includes a variety of food products. Heat-processed foods, which include baked goods, as well as canned dried foods, represent a variety of low-calorie products. Refrigerated and frozen products also employ low-calorie sweeteners.

Saccharin and sucralose are extensively used in baked products, especially by people with diabetes. Recipes employing saccharin tabletop sweeteners are available (Becker 1981) as well as recipes employing all other available and approved sweeteners.

The availability of a number of bulking agents such as polydextrose and the polyols (e.g., sorbitol, isomalt, and maltitol) have made it possible to introduce several reduced calorie and sugar-free cookies, cakes and cake mixes. In the United States, these products utilize acesulfame potassium, sucralose, a combination of acesulfame potassium and aspartame as the intense sweetener and other combinations of sweeteners including sucralose. Other baking products include combinations of high intensity sweeteners and sucrose.

The instability of aspartame at high temperatures limits its use in this application. Recipes are available, however, for preparing baked products to which aspartame is added after baking and cooling (Gibbons 1982). Aspartame added to baked goods at low levels improves their aroma (Chibata et al. 1976). Lim et al. (1989) reported on the use of combinations of intense sweeteners in cookies.

The use of aspartame in frozen desserts and other liquid products in the United States has proliferated since approval in these applications was granted by the FDA. Beck has reported that aspartame-sweetened frozen desserts were preferred by a taste panel to ice milk (Beck 1974, 1978). While this information is old, it illustrates the properties of aspartame as a sweetener in this application. Aspartame also performs well in frozen beverages (Finucane 1979), no-bake cheesecake, yogurt, and other products. No sugar added ice cream and frozen desserts have become available in the United States. These products generally use polyols and polydextrose and a combination of aspartame and acesulfame potassium. Other products labeled as no sugar added use sucralose and several sweetener combinations. Rebaudioside A sweetened frozen desserts are being looked at by the US industry as of this writing.

Cyclamate/saccharin combinations, when available, were extensively used in baked goods because of the heat stability of this combination. Cyclamate to saccharin ratios range from 2:1 to 10:1 in most uses, with 10:1 being the preferred ratio when organoleptic considerations were taken into account. Similarly, this combination was used in low-calorie frozen desserts, salad dressings, jams and jellies. Blends of acesulfame potassium and aspartame are used in Europe in this category of product applications.

The US market is actively seeking applications using all natural sweetener combinations to reduce the sugar content and the calories of several products. Fifty percent reduced sugar and calories orange juice is an example of this development. These products utilize stevia sweeteners. This interest has lead to numerous research activities seeking to identify sugar enhancers to reduce the natural, caloric sweetener content by 25% to 50%. Redpoint Bio announced in January 2010 that it has found that compound RP44, later identified as rebaudioside C, a component of stevia, enhances the sweetness of sucrose, fructose, and other caloric sweeteners by up to 50% (Redpoint Bio 2010).

Conclusions

The search for the ideal noncaloric sugar substitute will continue. The ideal sweetener must not only meet the taste and functional properties of sugar, but can also undoubtedly create opportunities for new products that are not yet possible using currently available ingredients. On the basis of available evidence, the likelihood of identifying a single compound capable of meeting all requirements of all food applications is extremely small. Thus, the use of multiple sweeteners, specifically designed to meet desired taste and functional characteristics in a specific food application, is the most logical and promising route.

The availability of natural intense sweeteners with acceptable taste profiles, creates opportunities for the development and introduction of low and reduced calorie and sugar free products that are sweetened with all natural sweeteners, including combinations of caloric and noncaloric natural sweeteners.

Market developments in the last few years clearly suggest that combinations of sweeteners are being well received by consumers.

References

Bahoshy, B.J., Klose, R.E., Nordstrom, H.A. (General Foods) 1976. Chewing gums of longer lasting sweetness and flavor, US Patent 3,943,258. March 9.

Bakal, A. 2007. Personal observations in Japan.

Bakal, A., and Cash, P. 2006. Sweetening the Pot. Prepared Foods, July 1.

Bakal, A.I. 1983. Functionality of combined sweeteners in several food applications, Chem. Inc. September 19.

Bakal, A.I., Witzel, F., Mackay, D.A.M., and Schoenholz, D. (Life Savers) 1978. Long-lasting flavored chewing gum including sweetener dispersed in ester gums and method, US Patent 4,087,557. May 2.

Beck, C.I. 1974. Sweetness, character, and applications of aspartic acid-based sweeteners, in *ACS Sweetener Symposium*, G.E. Inglett (Ed.), AVI Publishing, Westport, Conn., pp. 164–181.

Beck, C.I. 1978. Application potential for aspartame in low calorie and dietetic food, in *Low Calorie and Special Dietary Foods*, B.K. Dwivedi (Ed.), CRC Press, West Palm Beach, Fla., pp. 59–114.

Becker, G.L. 1981. *Cooking for the Health of It*, Benjamin, Elmsford, N.Y.

Calorie Control Council. Neotame at http://www.caloriecontrol.org/sweeteners-and-lite/sugar-substitutes/neotame. Accessed May 16, 2011.

Cea, T., and Glass, M. (Warner-Lambert) 1981. Aspartame sweetened chewing gum of improved sweetness stability, *S. African Patent* ZA 80 06,065. September 30.

Chibata, I., Miyoshi, M., Ito, H., Fujii, T., and Kawashima, K. (Tanabe Seiyaku) 1976. Aspartyl amide sweetening agents, US Patent 3,971,822. July 27.

Daniells, S. 2009. The science of sweetener blends. *Beverage Daily*, September 23.

Finucane, T.P. (General Foods) 1979. Methyl ester of L-aspartyl-L-phenylalanine in storage-stable beverages, *Canadian Patent* 1,050,812. March 20.

Finucase, T.P. (General Foods) 1978. Sweetening compositions containing aspartame and other adjuncts, *Canadian Patent* 1,043,158. November 28.

G. D. Searle and Company. 1982. Aspartame in carbonated beverages, Vol. 1, US Food Additive Petition No. 2A3661. August 13.

Gibbons, B. 1982. *Slim Gourmet Sweets and Treats*, Harper and Row, New York.

Higginbotham, J.D. 1980. Talin, a New Natural Sweetener from Tate and Lyle, Tate and Lyle, Reading, UK.

Hill, J.A., and Lavia, A.F. (E. R. Squibb) 1978. Artificial sweeteners, US Patent 3,695,898.

Hoppe, K. 1981. Taste interaction of citric acid with sucrose and synthetic sweeteners, *Nahrung* 25:11–14.

Inglett, G.E. 1970. Natural and synthetic sweeteners, *Hortscience* 5:139–141.

Ishii, K., Toda, J., Aoki, H., and Wakabayashi, H. (Takeda Chemical Industries) 1972. Sweetening composition, US Patent 3,653,923. April 4.

Klose, R.E., Bahoshy, B.J., Sjonvall, R.E., and Yeransian, J.A. (General Foods) 1981. Chewing gums of improved sweetness retention, US Patent 4,246,286. January 20.

Lavia, A.F., and Hill, J.A. (E.R. Squibb) 1972. Sweeteners with masked saccharin aftertaste, *French Patent* 2,087,843. February 4.

Lavia, A.F., and Hill, J.A. (E.R. Squibb) 1977. Sweetening compositions which mask the aftertaste of saccharin and potentiate its sweet taste contain saccharin together with an amount of a sweet tasting dipeptide in a quantity effective to mask the aftertaste of saccharin. US Patent 4,001,455. January 4.

Lim, H., Setser, C.S., and Kim, S.S. 1989. Sensory studies of high, potency multiple sweetener systems for shortbread cookies with and without polydextrose, *J. Food Science*, 54:624–628.

Mackay, D.A.M., Witzel, F., Dwivedi, B.K., and Schoenholz, D. (Life Savers) 1978. Long-lasting flavored chewing gum, US Patent 4,085,227. April 18.

Mitchell, M.L., and Pearson, R.L. 1991. Saccharin, in *Alternative Sweeteners*, 2nd edition, Nabors and Gelardi, Editors, Marcel Dekker, Inc. p. 132.

Moskowitz, H.R., Wolfe, K., and Beck, C. 1978. Sweetness and acceptance optimisation in cola flavored beverages using combinations of artificial sweeteners—a psycho-physical approach, *J. Food Quality* 2:17–26.

Paul, T. 1921. *Chem. Zeit.* 45:38–39.

Personal communications with sweetener suppliers. 1999.

Porter, A.B. 1983. Effectiveness of multiple sweeteners and other ingredients in food formulation, Chem. Ind., September 19.

Redpoint Bio. Press release re RP44. http://www.redpointbio.com. Accessed January 12, 2010.

Salant, A. 1972. Non-nutritive sweeteners, in *Handbook of Food Additives*, 2nd edition. T.E. Furia (Ed.), CRC Press, Cleveland, pp. 523–585.

Saulo, A. 2005. Sugars and sweeteners in foods, College of Tropical Agriculture and Human Resources. U. of Hawaii, March.

Scott, D. (G.D. Searle) 1971. Saccharin-dipeptide sweetening compositions, *British Patent* 1,256,995.

Scott, D. (G.D. Searle) 1979. Use of dipeptides in beverages, *German Patent* 2,002,499. September 20.

Torres, J.G. 2005. "Neotame-ing" the beast. Prepared Foods, May 1.

US Food and Drug Administration (US FDA). 1982. Aspartame, 21*CFR*172, 804. March 19.

US Food and Drug Administration (US FDA). 2008. Agency Response letter to Cargill, GRAS No. GRN 00253, December 17.

US Food and Drug Administration (US FDA). 2010. Agency Response letter to Pure Circle, GRAS notice No. GRN 000323. July 9.

Van Tornout, P., Pelgroms, J., and Van Der Meeren, J. 1985. Sweetness evaluation of mixtures of fructose with saccharin, aspartame or acesulfame-K. *J. Food Science* 50:469–472.

Verdi, R.J., and Hood, L.L. 1993. *Food Tech.* 47:94–102.

Westall, E.B., Scanlan, J., and Sahaydak, M. (Nutrilite Products and Warner-Lambert) 1972. Preservation and stability of flavors in chewing gum and confectioneries, *German Patent Offen.* 2,155,321 May 10; *British Patent* 1,310,329; US Patent 3,821,417.

Chapter 27

Polydextrose

Michael H. Auerbach, Helen Mitchell, and Frances K. Moppett

Contents

Introduction

Polydextrose is an ingredient designed to be the ultimate companion to high-intensity sweeteners. Polydextrose gives the bulk, texture, mouthfeel, and functional attributes of caloric sweeteners. These are attributes that are often lost in the formulation of calorie-reduced products. The key to the performance of polydextrose is its caloric value of one calorie per gram (Auerbach et al. 2007) and its metabolism as a dietary fiber (Craig et al. 1998, 2000). These properties, in combination with its excellent water solubility, make polydextrose unique in its application in reduced- and low-calorie and fiber fortified foods. When used to replace sugars and fats, polydextrose contributes only 25% of the calories of sugars and 11% of the calories of fats.

Unlike the serendipitous discovery of many unique products, polydextrose is the result of a targeted research program to discover an ingredient that would replace the bulk of sugar when high-intensity sweeteners are used to replace the sweetness of sugar. The goal of the program was to fill the need for a reduced-calorie bulking agent for the reduced-calorie foods market. For such food products to be widely accepted, they must be functionally comparable to their fully caloric counterparts. Achieving desirable body, mouthfeel, and texture is critical to gaining commercial success.

Polydextrose received US Food and Drug Administration (US FDA) approval for use in select food categories in 1981 (FAP 46 FR 30080) and for foods in general in 2007 (FAP 72 FR 46562). It was not until the mid-1980s, however, that significant commercial success was realized for this product. Proven successful formulations and a mushrooming of approved high-intensity sweetener applications have resulted in a secure place for polydextrose in reduced-calorie food products. The ingredient is approved in more than 75 countries and is used widely throughout the world.

Commercial Production

Polydextrose is a randomly bonded melt condensation polymer of glucose. This unique product is a patented material invented in Pfizer Central Research Laboratories by Dr. Hans Rennhard (Rennhard 1975). The patent describes the process for manufacture and applications in food of this novel carbohydrate substitute.

Polydextrose is manufactured by the bulk polycondensation *in vacuo* of a molten mixture of glucose, sorbitol, and either citric acid or phosphoric acid, in approximately an 89:10:0.1–1 mixture. The final product of this reaction is a weakly acidic, water-soluble polymer that contains minor amounts of free sorbitol and glucose and bound citric or phosphoric acid. As produced, polydextrose conforms to the typical composition described in Table 27.1. Polydextrose is supplied to the food industry in compliance with the *Food Chemicals Codex* (*FCC* 2010) and JECFA compendial specifications (JECFA 2006).

Table 27.1 Characteristics of Polydextrose

Compendial designation	*Food Chemicals Codex*
Appearance	White to light tan powder
Odor	None
Polymer	>90%[a]
Glucose + sorbitol	<6%[a]
Levoglucosan	<4%[a]
Water	<4%[a]
Citric acid (free)	<0.1%
pH (10% w/v solution)	2.5–6.5
Solubility in water (25°C)	80%
Optical rotation	+60
Viscosity (cps, 50% solution)	35

[a] Anhydrous, ash-free basis

The theoretical chemical structure of polydextrose is illustrated in Figure 27.1. This structure represents the types of bonding that can occur during polymerization. The R group may be hydrogen, glucose, or a continuation of the polydextrose polymer. As evidenced by this representative structure, polydextrose is a very complex molecule, being highly branched with varied glucose linkages. In fact, all possible glycosidic linkages with the anomeric carbon of glucose are present, including alpha- and beta-1–2, 1–3, 1–4, and 1–6, with some branching; the 1–6 linkage

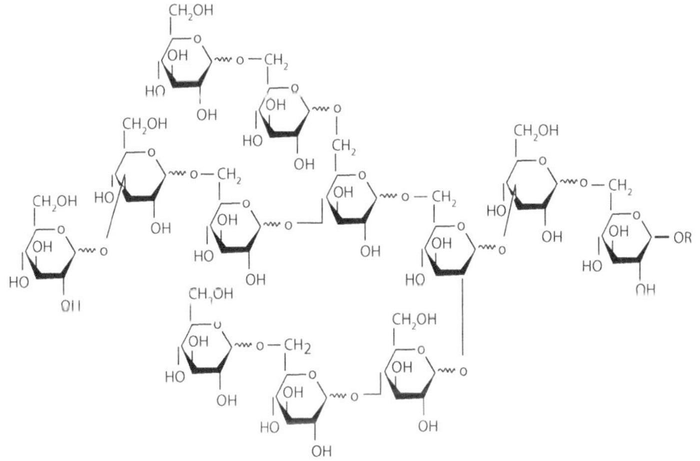

Figure 27.1 Representative structure for polydextrose. R = H, sorbitol, sorbitol bridge, or more polydextrose.

Table 27.2 Approximate Molecular Weight Distribution of Polydextrose

Molecular Weight Range	Percent
162–5,000	88.7
5,000–10,000	10.0
10,000–16,000	1.2
16,000–18,000	0.1

predominates. It is these chemical parameters that result in the water-soluble, reduced-calorie nature of polydextrose.

The typical molecular weight distribution of the polymer is shown in Table 27.2. The average molecular weight is 2,000–2,500 with an average degree of polymerization (DP) of 12–15. During the manufacturing process, the size of the polymer is controlled to restrict the formation of large molecular weight molecules and prevent the formation of insoluble material. Further discussion on this topic is presented by Allingham (1982) and Beereboom (1981).

Product Forms

Litesse® is the brand name for improved forms of polydextrose. Litesse is produced from polydextrose using additional processing to reduce acidity and bitterness, thereby improving the flavor profile. Litesse II and Litesse Ultra are further refinements. The Litesse family of products is unique in its ability to vary from a bland, neutral powder through a colorless, mildly sweet liquid. Table 27.3 summarizes the different attributes of the Litesse grades.

Table 27.3 Polydextrose and Litesse Product Forms

	Polydextrose	Litesse	Litesse II	Litesse Ultra[a]
Taste	Tart	Bland	Clean	Very clean
	Acid	Neutral	Mildly sweet	Mildly sweet
Color	Cream	Cream	Cream	White
Acidity (mEq/g)	0.1	0.03	0.003	0.002
pH range (10% w/v aqueous)	2.5–3.5	3.0–4.5	3.5–5.0	4.5–6.5
Maillard reaction	Yes	Yes	Yes	No

[a] Litesse® Ultra is a reduced version of Litesse® prepared by catalytic hydrogenation of polydextrose. It does not contain any reducing groups and will not take part in Maillard reactions. Litesse® Ultra produces a very stable, clear, water-white solution and is sugar free.

Properties

Caloric Content

The key property that gives polydextrose its important role in the formulation of reduced calorie foods is the caloric value of one calorie per gram (Coleman 1981; Auerbach et al. 2007). This value is significant when polydextrose is used to replace sucrose, which has a caloric value of 4 calories per gram. The degree of caloric reduction is even greater when polydextrose is used to replace fat, which has 9 calories per gram.

The reason polydextrose has a low caloric value is that it is a large, complex, randomly bonded polymer that is not broken down by mammalian digestive enzymes. The one calorie is a result of the metabolism of microorganisms in the intestinal tract. This subject is discussed more fully in the section "Metabolism."

Water Solubility

The excellent water solubility of polydextrose differentiates it from insoluble bulking agents such as cellulosic products. It is soluble to approximately 80% at 25°C. Polydextrose acts in a similar manner to sucrose in that it results in a clear solution without haze or turbidity. Good mechanical mixing is required to prepare concentrated solutions. The rate of solubility will depend on the efficiency of the mixing equipment as determined by shear speed and rate of addition. Use of moderate heat and slow rate of addition aid in making these solutions. A preblend with another water-soluble ingredient is another technique that will greatly speed the dissolution rate of polydextrose.

Viscosity

The viscosity that polydextrose contributes to a solution is greater than that contributed by an equal amount of sucrose. Figure 27.2 illustrates the viscosity of 70% solutions of polydextrose, sucrose, and sorbitol at varying temperatures. In a manner similar to that of sucrose solutions, the viscosity of polydextrose solutions decreases with an increase in temperature.

The viscosity-enhancing effect of polydextrose plays an important role in food uses such as reduced-calorie dressings and puddings. Figure 27.3 shows the viscosity of 20%–70% solutions of polydextrose, sucrose, and sorbitol. This illustrates the potential for using polydextrose to replace the viscosity typically provided by high sugar content.

Humectancy

Under conditions of high relative humidity, polydextrose is fairly hygroscopic. Figure 27.4 illustrates the water pickup of polydextrose in storage at 75% and 52% relative humidity. In food products, polydextrose functions as a humectant and can play an important role in product quality by controlling the rate of moisture gain or loss. An example of this property is found in the baked goods area. Polydextrose can retard the loss of moisture, which helps to protect against staling. In this way, polydextrose serves as an important ingredient in extending shelf life.

Another important characteristic of polydextrose in solution is the effect that it has on water activity (a_w). Water activity is a measure of the availability of the water to participate in chemical reactions, physical reactions, and support of microbial growth in a food product.

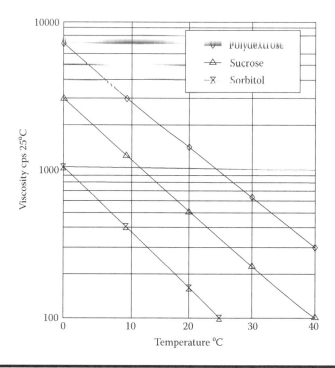

Figure 27.2 **Temperature/viscosity relationships for 70% aqueous solutions of polydextrose.**

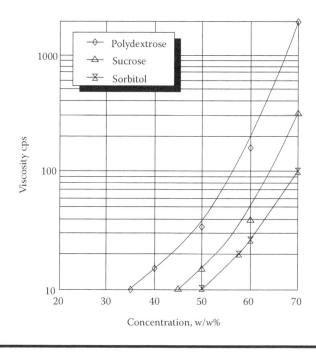

Figure 27.3 Concentration/viscosity relationship.

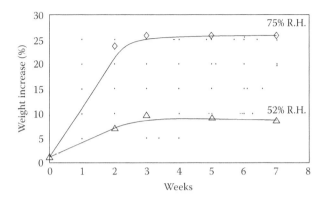

Figure 27.4   Hygroscopicity of polydextrose at 25°C.

The water activity of various polydextrose solutions is reviewed in Table 27.4. At equal concentrations, polydextrose has less effect on water activity than smaller molecular weight products such as sucrose and sorbitol.

Taste

Polydextrose is not sweet. It is used to replace the physical functionality of high-calorie ingredients such as sugar, but not the sweetness. In some food products, a less sweet taste may be desirable, and thus at least partial sucrose replacement can be achieved without compensating for the loss in sweetness. In most cases, it is necessary for calorie-modified foods to have the same quality and quantity of sweetness as the fully caloric counterpart. In such foods, sweetness can be provided by a high-intensity sweetener. It is in these applications that polydextrose is an ideal companion product for high-intensity sweeteners.

Noncariogenicity

Studies carried out at the University of Zurich (Muhlemann 1980) and Stoesser et al. demonstrated that polydextrose has a very low potential for promoting dental caries (Griffiths and Auerbach 2008). These tests showed polydextrose (a) to be inert to *in vitro* systems, (b) not to cause caries in rats, and (c) to pass the intra-oral proximal plaque pH telemetry tests in humans. These three tests are good indicators of the cariogenicity of foods. As a direct result of these studies, the Swiss government allows "Safe for Teeth" labeling associated with polydextrose.

Table 27.4   Water Activity of Various Polydextrose Solutions

	Solution Concentration (% w/w)	
Sweetener	*50%*	*60%*
Sorbitol	0.90	0.85
Sucrose	>0.95	0.91
Polydextrose	>0.95	0.92

Stability

Polydextrose is a very stable ingredient. The hygroscopic nature of the product requires good pack aging and reasonable storage under conditions of low humidity to avoid moisture pickup.

Polydextrose solution is also very stable. The high solids level does not support microbial growth. The solution can darken after prolonged storage at elevated temperatures, and, therefore, storage under cool temperatures is recommended.

Functions of Polydextrose

The unique combination of properties found in polydextrose results in a variety of functional attributes that are important in the formulation of reduced-calorie, low-calorie, and sugarless products. The principal change that is made in these food products is the removal of sucrose. High-potency sweeteners replace the sweetness lost when sucrose is removed. In most food products, however, a great deal more than sweetness is lost along with the sugar. Sucrose also provides the functional attributes of bulk, mouthfeel, humectancy, viscosity, freezing point depression and preservation (Freeman 1982; Smiles 1982; Liebrand et al. 1985). As illustrated in Table 27.5, polydextrose can at least partially fill all these roles except for sweetness. In a similar manner, when calorie reduction is achieved by reducing the fat level, there are functional attributes of the fat that polydextrose can replace.

Bulk

Polydextrose is most commonly referred to as a "bulking agent." Probably the greatest single challenge when formulating reduced-calorie products is replacing the bulk of sugar. In liquid products such as beverages, this is not such an important issue. In others, however, it is critical. Probably the most dramatic examples of the importance of a bulking agent are in baked goods and confections, which rely heavily on the bulk of sugar to give the products their character. Any significant reduction in the bulk and body of these products would greatly decrease their acceptability.

Other ingredients that could be used to replace the bulk of carbohydrates and fats include such diverse ingredients as crude dietary fiber, maltodextrin, and sorbitol. These products, however, are limited in their usefulness because none of these alternative bulking agents have the combined properties of one calorie per gram and water solubility.

Table 27.5   Functional Attributes of Sugar and Polydextrose

	Sugar	*Polydextrose*
Bulk	Yes	Yes
Mouthfeel/Texture	Yes	Yes
Humectancy	Yes	Yes
Viscosity	Yes	Yes
Freezing point depressant	Yes	Yes
Preservation	Yes	Yes
Sweetness	Yes	No
Caloric Content	4.0 kcal/g	1.0 kcal/g

Mouthfeel/Texture

A pleasant, satisfying mouthfeel and texture is another important contribution of sucrose and fat. Limited success has been achieved using gums to enhance the texture of reduced-calorie foods. At higher levels, a slimy mouthfeel and gelled characteristics become noticeable. The use of polydextrose gives a comparable mouthfeel and general textural eating quality to the fully caloric food product. The product categories that particularly illustrate this function are puddings, frozen desserts, and salad dressings. A rich, creamy mouthfeel is particularly important and expected in these products.

Freezing Point Depression

One of the less obvious functions of sugar is as a freezing point depressant. This function is very important in making creamy, palatable frozen desserts. If the freezing point of a product is too low, the texture as consumed will be too soft. If the freezing point is too high, an unacceptably hard product results. In conventional ice cream type products, sucrose is the primary freezing point depressant and alternative ingredients such as polydextrose are typically compared with sucrose.

The effect that an ingredient has on freezing point depression is a function of its molecular weight. Sucrose is a disaccharide. A larger molecule such as polydextrose is somewhat less effective as a freezing point depressant. A smaller molecular weight ingredient such as sorbitol would have a greater effect on freezing point depression than either sucrose or polydextrose. Ideally, a balance can be achieved with several ingredients to match the effect of sucrose.

The comparative effects of polydextrose, sucrose, and sorbitol on freezing point have been reported (Baer and Baldwin 1984, 1985). The freezing point of a 5% solution of these three ingredients was as follows: polydextrose, $-0.147°C$; sucrose, $-0.298°C$; and sorbitol, $-0.613°C$.

Preservation/Osmotic Activity

Another function that sucrose serves, particularly at higher use levels, is that of preservation. Sugar at high concentrations reduces water activity (a_w). Under conditions of reduced water activity, there is less water available for the growth of microorganisms and there is also greater osmotic pressure. When the sucrose level is reduced in a recipe, the new product may have shelf life problems with respect to bacteria, yeast, or mold growth that the original product did not have. There are several ways this problem can be addressed. One approach is to replace sucrose with ingredients that have a similar effect on water activity. Polydextrose can be used to help maintain the soluble solids level while limiting the calories. In this regard, polydextrose can be used to decrease the water activity of a product as described previously. An equal weight of polydextrose will not have as great an effect on a_w as sucrose.

It may be beneficial to make formulation changes in reduced-calorie products to maintain a comparable shelf life. One technique is to decrease the pH to inhibit microorganisms. Also, the use of chemical preservatives such as sodium benzoate or potassium sorbate may be particularly useful in calorie-modified products.

Cryoprotectant

Several applications have been suggested for the cryoprotectant capability of polydextrose. Cryoprotection involves stabilizing foods against the damaging physical effects of freezing. This is a function that is well recognized for many soluble ingredients, particularly sugars. This function has been described in fish products, including surimi (Park and Lanier 1987; Park et al. 1988), and meat products (Park 1986).

Fiber

Because it is not digested in the small intestine, polydextrose conforms to many definitions of dietary fiber. The fiber properties of polydextrose are reviewed in detail in the literature (Craig et al. 1998). In the United States, two prestigious bodies recently published definitions of fiber, both of which encompass polydextrose.

■ American Association of Cereal Chemists (AACC) (2000):
  Dietary fiber is the edible parts of plants, or analogous carbohydrates, that are resistant to digestion and absorption in the human small intestine with complete or partial fermentation in the large intestine. Dietary fiber includes polysaccharides, oligosaccharides, lignin and associated plant substances. Dietary fibers promote beneficial physiological effects including laxation, and/or blood cholesterol attenuation, and/or blood glucose attenuation.
■ National Academy of Sciences (NAS) Dietary Reference Intake (DRI) expert panel (2002):
  Dietary Fiber consists of non-digestible carbohydrates and lignin that are intrinsic and intact in plants. Functional Fiber consists of isolated non-digestible carbohydrates that have been shown to have beneficial physiological effects in humans. Total Fiber is the sum of Dietary Fiber and Functional Fiber.

A unique market has opened in Japan for beverage products fortified with polydextrose as a soluble fiber (Hamanaka 1987). The Japanese definition is "polysaccharides, related polymers and lignins, which are resistant to hydrolysis by the digestive enzymes of man" (FHPD, 1990).

The recently adopted Codex Alimentarius definition also covers polydextrose:
  Dietary fiber means carbohydrate polymers* with ten or more monomeric units[†] which are not hydrolysed by the endogenous enzymes in the small intestine of humans and belong to the following categories:
  – edible carbohydrate polymers naturally occurring in the food as consumed
  – carbohydrate polymers, which have been obtained from food raw material by physical, enzymatic or chemical means and which have been shown to have a physiological

---

[*] When derived from a plant origin, dietary fiber may include fractions of lignin and/or other compounds when associated with polysaccharides in the plant cell walls and if these compounds are quantified by the AOAC gravimetric analytical method for dietary fiber analysis: Fractions of lignin and the other compounds (proteic fractions, phenolic compounds, waxes, saponins, phytates, cutin, phytosterols, etc.) intimately "associated" with plant polysaccharides are often extracted with the polysaccharides in the AOAC 991.43 method. These substances are included in the definition of fiber insofar as they are actually associated with the poly- or oligosaccharidic fraction of fiber. However, when extracted or even re-introduced into a food containing non digestible polysaccharides, they cannot be defined as dietary fiber. When combined with polysaccharides, these associated substances may provide additional beneficial effects.
[†] Decision on whether to include carbohydrates with monomeric units from 3 to 9 should be left to national authorities.

> effect of benefit to health as demonstrated by generally accepted scientific evidence to competent authorities
> – synthetic carbohydrate polymers which have been shown to have a physiological effect of benefit to health as demonstrated by generally accepted scientific evidence to competent authorities

Polydextrose may be labeled as fiber in 31 countries. The official validated AOAC method for determination of polydextrose in food is AOAC 2000.11 (Craig et al. 2000, 2001). Currently, the fiber content of polydextrose is 90% when analyzed using AOAC 2000.11 and utilizing the minimum polymer content of 90% as per the *FCC* specification. There is a new method for dietary fiber (AOAC 2009.01) to support the above Codex fiber definition. Preliminary work suggests that the fiber content of polydextrose will be ~80% by AOAC 2009.01.

Cariogenicity

Based on plaque pH telemetry studies (Griffith and Auerbach 2004), Litesse Ultra (hydrogenated polydextrose; HPDX) is a non-fermentable carbohydrate and does not promote dental caries. Seven human studies utilizing intra-oral in situ pH monitoring devices demonstrated that soft and hard confections made with HPDX did not cause the oral microflora-driven decrease in pH associated with the development of dental caries (Muhlemann 1980; Stoesser et al. 2005). A noncariogenicity label claim has not yet been approved in the United States, but is permitted in the EU.

Applications

The original polydextrose food additive petition (FAP) covered specific applications in eight food categories (FAP 9A3441): baked goods and mixes, chewing gum, confections and frostings, dressings, frozen dairy desserts, gelatins, puddings and fillings, and hard and soft candy. Examples of use in these categories are given below. A more recently approved FAP extended this coverage to all foods except for meat, poultry, baby food and infant formula (see 21 CFR 172.841).

With the development of new forms of polydextrose and Litesse®, it is now possible to choose a bulking agent that best fits the end-use application (see Table 27.3 for a summary of polydextrose and Litesse attributes). Selection of the appropriate Litesse form offers greater flexibility in terms of color, taste, and product claims (Litesse Ultra is sugar free).

Polydextrose may be used to fulfill any of six main functions:

1. Bulking agent – to provide bulk or substance to a food without adding significant calories
2. Formulation aid – to promote or produce a desired physical state
3. Humectant – to promote retention of moisture
4. Texturizer – to affect the appearance or mouthfeel of the food
5. Fiber – to add soluble dietary fiber to the food
6. Noncariogenic – to reduce or prevent formation of dental caries

Baked Goods

This food category includes baked goods and baking mixes. A one-third calorie reduction in these products would require a significant fat and carbohydrate reduction, which would alter the physical and organoleptic acceptability of the baked good. Therefore, polydextrose plays a particularly important role in the formulation of reduced-calorie baked goods (FHPD, 1990; Kim et al. 1986). In many baked goods formulations, polydextrose can be used without a high-intensity sweetener.

Recent work reported on the good textural qualities of polydextrose shortbread cookies formulated with high-intensity sweeteners (Lim et al. 1989). For sweetness, the study evaluated synergistic combinations of aspartame, cyclamate, saccharin, and acesulfame potassium in the cookies.

The benefits of polydextrose in cake-type products include volume, tenderness, structure, and eating quality. Typical use levels in such products would be from 7% to 15% by weight.

Chewing Gum

Polydextrose can be used to make chewing gum with good shelf life and elasticity. A high-intensity sweetener would be required to provide sweetness to the gum. Several patents exist for low-calorie gum products that cite the use of polydextrose (Klose and Sjonvall 1987; Cherukuri et al. 1988).

Confections and Frostings

This area of application represents products in which the functional properties of carbohydrates and fats are critical to the character of the food. Polydextrose is not crystalline, and therefore the crystalline/sugar texture cannot be achieved. The bulk and mouthfeel of taffy-like products is more closely matched using polydextrose. Polydextrose use level would typically be in the range of 25%.

Salad Dressings

Many pourable salad dressings are surprisingly high in calories, containing up to 60–80 calories per tablespoon. Of their calories, approximately 80%–95% comes from fat (Goulart 1985). Therefore, making reduced-calorie dressing involves, primarily, reducing the oil level.

Polydextrose finds use in this application by replacing the functionality and mouthfeel that was contributed by the fat. Some dressings do contain a significant level of sugar. Russian, thousand island, and French-style dressings are highest in sugar. These dressings are good candidates for the replacement of sucrose with polydextrose and a high-intensity sweetener. A number of highly palatable reduced-calorie dressings have been formulated with polydextrose and enjoy considerable success in the marketplace.

Frozen Dairy Desserts

This area of use typically involves the reduction of both fat and sugar in the formulation of reduced-calorie products (Anonymous 1984). Sugar is a multipurpose ingredient providing bulk, mouthfeel, sweetness, and freezing point depression to the final product. The recipe in Table 27.6 is for a reduced-calorie, nonfat product.

The firmness of a frozen dessert can be manipulated by altering the freezing point. Freezing point depression is a function of molecular weight of key ingredients, and thus, a balance of polydextrose with a lower molecular weight product such as sorbitol, may best match the firmness desired.

Table 27.6 Nonfat Ice Cream

Ingredient	Percent
Sucrose	11.00
Nonfat milk solids	9.00
Corn syrup solids	7.40
Litesse®	4.00
Dairy-Lo®	3.00
Stabilizer/emulsifier	0.50
Flavor, vanilla, N&A	0.30
Flavor, art. mouthfeel type	0.03
Fat	<0.5%
Total solids	35%
Calories	127 calories per 100 g, 90 calories per 4 fl. oz serving at 90% overrun

Vitamin A should be added at a level ranging from 600–800 IU, depending on the reference ice-cream

Processing Procedure

Dairy-Lo is compatible with conventional ice cream manufacturing equipment and processes.

- Mix liquid ingredients
- Add Dairy-Lo and other dry ingredients with moderate agitation
- Pasteurize at 180°F (83°C) for 25 seconds
- Homogenize
- Cool and age
- Flavor the mix
- Freeze, package
- Harden

Very pleasant-tasting products are possible without the use of a high-intensity sweetener. A series of regulatory approvals in 1987 and 1988, however, extended the use of aspartame to include frozen dairy desserts, and this resulted in a surge of very successful reduced-calorie products (Goff and Jordan 1984; Reddy 1985). Typical use levels of polydextrose in these products range from 7%–15%.

Within this category, frozen yogurt has become very popular. Frozen yogurt is lower in fat than ice cream, and thus calorie reduction is achieved primarily from sugar replacement. Here again, polydextrose serves the same varied functions of bulk, texture, and freezing point depression.

Gelatins, Puddings, and Fillings

A 50% calorie reduction is readily achieved in an instant pudding-type product when polydextrose is used with a high-intensity sweetener (Anonymous 1983). Using polydextrose results in a creamier texture with a more uniform dispersion of cocoa and starch. It also functions as a formulation aid and helps prevent lumping and non-uniformity.

Hard Candy

A major portion of the sugar in hard candy can be replaced with polydextrose (Klacik 1989). Clarity and good bite are still achievable with a reduced sugar level. A high-intensity sweetener is needed to supply the sweetness expected in hard candy products.

Polydextrose is amorphous and does not crystallize at low temperatures or high concentrations so it can be used to control the crystallization of polyols and sugars and therefore the structure and texture of the final product. This is analogous to conventional sugar confectionery production in which glucose syrups are used to prevent or control sucrose crystallization.

Litesse Ultra is a particularly useful ingredient in the production of sugar-free and reduced-calorie hard candy and offers both product and processing improvements when used at low levels in polyol mixes.

While acknowledging that some polyols work very well in this application, there are still some technical issues associated with their use in hard candy systems. From production scale trials using Litesse Ultra in combination with polyols, the following general conclusions have been drawn:

1. Increasing the proportion of Litesse Ultra increases the molten mass viscosity, improving the handling characteristics of sugar-free candy. Cooling times are reduced and the products may be processed using conventional stamped and depositing technology.
2. Low-level additions of Litesse Ultra reduce the graining of isomalt candies.
3. The stability of Litesse Ultra combination may be equivalent or better than sugar/glucose candy.
4. Clear, transparent products with excellent flavor release are possible.
5. Calorie reductions greater than 50% are possible.

Soft Candy

The use of polydextrose as a partial sugar replacement in soft candy serves two key functions: providing bulk and humectancy (Liebrand and Smiles 1981; Cridland 1987).

Other Categories

Use of polydextrose in sweet sauces, toppings and syrups (FAP 9A4126), peanut spreads (FAP 7A3998; Anonymous. 1989), and fruit spreads (FAP 8A4068) demonstrate the flexibility of the ingredient across a range of applications where sugar is reduced or replaced.

Metabolism

The basis for the one calorie per gram caloric value of polydextrose lies in the difficulty that mammalian digestive enzymes have in attacking a large and randomly bonded polymer.

Extensive feeding studies with man and animals have confirmed the caloric contribution (Auerbach et al. 2007).

The definitive work on the metabolism of polydextrose studied the disposition of ^{14}C radio-labeled polydextrose in feeding studies. The caloric value of one calorie per gram and the testing methods were fully endorsed by numerous regulatory agencies, including the US FDA (Coleman 1981) and the Health Protection Branch of Canada (Lee 1993).

A portion of the polydextrose molecule is metabolized in the intestinal tract by microorganisms. As by-products of this metabolism, these microorganisms produce volatile fatty acids and carbon dioxide. The volatile fatty acids are absorbed in the large intestine and used as an energy source. Therefore, it is the microbial function that contributes the one calorie per gram of ingested polydextrose. This is a phenomenon common to essentially any complex carbohydrate in the diet.

Another area of metabolism that has been closely investigated is the effect of polydextrose on insulin demand and the implications of this on use by diabetic patients. Studies show that the use of polydextrose does not create an insulin demand using glucose tolerance techniques.

As described previously, tests show that polydextrose has a very low cariogenicity potential.

Gastrointestinal Toleration

Numerous clinical studies show that polydextrose is well tolerated when consumed in moderation as part of a normal daily diet. This has been confirmed by consumer results in the years since polydextrose was approved (Flood et al. 2004).

When consumed in excessive amounts, polydextrose can have a laxative effect. Studies of adults showed a mean laxative threshold of 90 g/day (Raphan 1979). This compares with a level of 70 g of sorbitol for a similar effect.

Regulatory Status

As noted above, polydextrose is approved for use in foods in 75 countries. It can be labeled as 1 calorie/gram in 74 countries (2 calories/gram in Germany) and as fiber in 31 countries. A brief summary of the regulatory approvals of polydextrose around the world is given below. The reader should also be aware that regulations are constantly being changed and updated.

FDA Approval

There were years of extensive research into the safety, properties, and applications of polydextrose before the approval of the polydextrose food additive petition in 1981 by the US FDA. The regulations allow for the safe use of polydextrose as a multipurpose food additive when used in accordance with good manufacturing practices as a bulking agent, formulation aid, humectant, and texturizer. Polydextrose is approved for use in the United States in all foods except for meat, poultry, infant formula and baby food under 21 CFR 172.841. A catalytically hydrogenated version of polydextrose, commercially designated Litesse Ultra, was approved by the FDA in 1998 (FAP 7A4556).

There is no maximum established use limit for polydextrose in the United States. Good manufacturing practices limit the quantity used to the minimum amount necessary to accomplish the intended purpose in the food.

International

Numerous national and supranational expert groups have assessed the safety of polydextrose. Without exception, they concluded that polydextrose is safe for human use. Both the Joint FAO/WHO Expert Committee on Food Additives (JECFA) and the EU Scientific Committee for Food (SCF) allocated an acceptable daily intake (ADI) of "not specified" in 1987 and 1990, respectively.

Polydextrose has been approved as a Miscellaneous Food Additive by the European Union and may be used at *quantum satis* levels. Polydextrose is listed as INS 1200 and E1200.

Japan's Ministry of Health and Welfare (MOHW) recognizes polydextrose as a food rather than a food additive. Polydextrose also conforms to the generally accepted Japanese definition of dietary fiber and many others. The energy value of 1 kcal/g is accepted in every country in which it is approved, except for Germany.

Labeling

When polydextrose is used in foods for special dietary purposes such as reduced-calorie foods in the United States, it must be labeled in accordance with 21 CFR Part 105. Should a single serving of food contain more than 15 g of polydextrose, the label must read: "Sensitive individuals may experience a laxative effect from excessive consumption of this product."

For nutritional labeling in the United States, polydextrose should be included in the carbohydrate section of the Nutrition Facts Panel. It may also optionally be included below "Total Carbohydrate" under "Other Carbohydrate." The calorie content of 1 kcal/g is used when determining total calories per serving. Use of polydextrose may also permit comparative nutrient content claims such as "light," "low-calorie," "reduced sugar," "sugar-free," and/or "no added sugar" under 21 CFR 101.60.

"Polydextrose" is the officially recognized name and should appear as such in the ingredients declaration.

References

Allingham, R.P. 1982. Polydextrose—A new food ingredient: Technical aspects. In *Chemistry of foods and beverages: Recent development*. New York: Academic Press, p. 293.

Anonymous. 1989. Polydextrose in peanut butter. US Patent 4,814,195 (March 21, 1989).

Anonymous. 1984. All the satisfaction of ice cream, but it's a diet bar. *Food Eng* 56(9):50–51.

Anonymous. 1983. Reduced calorie desserts made with aspartame and polydextrose (puddings, gelatins). *Food Processing*, pp. 21–22, July.

Auerbach, M.H., Craig, S.A.S., Howlett, J.F., and Hayes, K.C. 2007. Caloric Availability of Polydextrose. *Nut Rev* 65:544–549.

Baer, R.J., and Baldwin, K.A. 1984. Freezing points of bulking agents used in manufacture of low-calorie frozen desserts. *J Dairy Sci* 67:2860–2862.

Baer, R.J., and Baldwin, K.A. 1985. Bulking agents can alter freezing. Dairy Field, February 1985.

Beereboom, J.J. 1981. Technical aspects of polydextrose, Polydextrose Trade Press Briefing, May 28.

Cherukuri, S.R., Hriscisce, F., and Weis, Y.C. 1988. Low calorie chewing gum and method for its preparation, *European Patent* 0252874A.

Coleman, E.C. 1981. US Food and Drug Administration, letter dated April 21, to E.F. Bouchard (Pfizer).

Craig, S.A.S., Holden, J.F., Troup, J.P., Auerbach, M.H., and Frier, H.I. 1998. Polydextrose as soluble fiber: physiological and analytical aspects. *Cereal Foods World* 43:370.

Craig, S.A.S., Holden, J.F., and Khaled, M.J. 2000. Determination of polydextrose as dietary fiber in foods. *J AOAC Int* 83:1006–1012.

Craig, S.A.S., Holden, J.F., and Khaled, M.J. 2001. Determination of polydextrose in foods by ion chromatography: collaborative study. *J AOAC Int* 84, 472–478.

Cridland, A. 1987. Developments in diabetic chocolate. *Confectionery Manufacturing and Marketing* 24(10).

Food Chemicals Codex (FCC). 2010. 7th Edition, United States Pharmacopeia, pp. 811–817.

Flood, M.T., Auerbach, M.H., and Craig, S.A.S. 2004. A Review of the Clinical Toleration Studies of Polydextrose in Food *Food Chem Toxicol* 42:1531–1542.

Food Additive Petition (9A3441) 46 FR 30080 (June 5, 1981).

Food Additive Petition (6A4763) 72 FR 46562 (August 21, 2007).

Food Additive Petition (9A4126) 59 FR 37419 (July 22, 1994).

Food Additive Petition (7A3998) 59 FR 37419 (July 22, 1994).

Food Additive Petition (7A4556) 63 FR 57596 (October 28, 1998).

Food Additive Petition (8A4068) 59 FR 37419 (July 22, 1994).

Foundation for Health and Physical Development (FHPD). (1990). Ministry of Health and Welfare, Japan.

Freeman, T.M. 1982. Polydextrose for reduced calorie foods. *Cereal Foods World* 27(10):515–518.

Goff, D.H., and Jordan, W.K. 1984. Aspartame and polydextrose in a calorie-reduced frozen dairy dessert. *J Food Sci* 49:306–307.

Goulart, F.S. 1985. How to read a salad dressing label. *Nutr Dietary Consultant* 6(6):37–38, June.

Griffiths, J.C., and Auerbach, M.H. 2008. Lack of Dental Enamel Toxicity (Caries Formation) by Litesse Ultra (Hydrogenated Polydextrose). Poster presentation at 47th Society of Toxicology annual meeting, Seattle.

Hamanaka, M. 1987. Polydextrose as a water-soluble food fibre. *Food Industry* 30(17):73–80.

Joint FAO/WHO Expert Committee on Food Additives (JECFA). 2006. Compendium of Food Additive Specifications, Food and Nutrition Paper 52, Add. 6.

Kim, K., Hansen, L., and Setser, C. 1986. Phase transitions of wheat starch-water systems containing polydextrose. *J Food Sci* 51(4):1095–1097.

Klacik, K.J. 1989. Continuous production of sugar-free hard candies. Manufacturing Confectioner, pp. 61–67, August.

Klose, R.E., and Sjonvall, R.E. 1987. Low calorie, sugar-free chewing gum containing polydextrose. *European Patent* 0123742B1.

Lee, N. 1993. Health and Welfare Canada, letter dated September 23, to J. DiMarzo (Pfizer).

Liebrand, J.T., Smiles, R.E., and Freeman, T.M. 1985. Functions and applications for polydextrose in foods. *Food Technol Aust* 37(4):166–167.

Liebrand, J.T., and Smiles, R.E. 1981. Polydextrose for reduced calorie confections. *Manu Conf* 61(11):35–36.

Lim, H., Setser, C.S., and Kim, S.S. 1989. Sensory studies of high potency multiple sweetener systems for shortbread cookies with and without polydextrose. *J Food Sci* 54(3):625–628.

Muhlemann, H.R. 1980. Polydextrose—a low-calorie sugar substitute. *Swiss Dent* 2(3):29.

Park, J.W., and Lanier, T.C. 1987. Combined effects of phosphates and sugar or polyols on stabilization of fish myofibrils. *J Food Sci* 52(6):1509–1513.

Park, J.W., Lanier, T.C., and Green, D.P. 1988. Cryoprotective effects of sugar, polyols, and/or phosphates on Alaska pollack surimi. *J Food Sci* 53(1):1–3.

Park, J.W. 1986. Effects of cryoprotectants on properties of beef protein during frozen storage, dissertation. North Carolina State University.

Raphan, T. 1979. Study IV, Section E-C3 of FAP 9A3441, Pfizer communication.

Reddy, K.C. 1985. Effect of aspartame and polydextrose levels on organoleptic, physiochemical and economic values of ice cream and ice milk products. M.S. Thesis, Mississippi State University.

Rennhard, H.H. (Pfizer). 1975. Dietetic foods, US Patent 3,876,794.

Smiles, R.E. 1982. The functional applications of polydextrose. In *Chemistry of Food and Beverages: Recent Developments*. New York: Academic Press, pp. 305–322.

Stoesser, L., Tietze, W., Heinrich-Weltzien, R., Kruger, C., Griffiths, J.C., and Auerbach, M.H. 2005. Polydextrose—"ein zahn-freundlicher" Kohlenhydrat-Füllstoff, Oralprophylaxe & Kinderzahnheilkunde 27:144–149.

US Code of Federal Regulations 21 CFR 172.841.

Chapter 28

Other Low-Calorie Ingredients: Fat and Oil Replacers

Ronald C. Deis

Contents

Introduction

The interest in fats and oils and their presence in foods has gone through a number of developmental stages, each causing a change in priorities at various times in product development cycles. These stages include: the elimination of saturated fats, the "fat-free" phase, the fat and calorie reduction phase, the reduction of trans fats, and the search for healthy fats from natural and biosynthetic routes. All of this is gradually evolving into a more rational design of better-quality, more healthy products as we learn more about the effects of changes in diet. The need to eliminate saturated fats and to increase fiber became apparent in the 1980s. The number of companies marketing fiber products surged enormously, then ebbed as the fiber frenzy dwindled. As we entered the 1990s, the message that saturated fats were "bad" progressed into the need to eliminate fats entirely, that to be "fat free" was golden. This spawned a race to develop new fat replacers to fill the need, many claiming to be the one ingredient to replace fat in all food systems. As noted by M. Glicksman in 1991, "every food company, ingredient supplier, and biotechnology company is looking for a colorless, odorless liquid that looks, tastes, and functions like oil but has no calories and is less expensive than water," what Glicksman referred to as the "oily Grail" (Glicksman 1991). Some categories, such as dairy and salad dressing, were able to produce reasonably acceptable fat-free or reduced-fat alternatives over time. In other areas, such as baked goods (in which less water was available in the system and any added water caused noticeable differences), the quality of these products was poor, and consumer acceptance has dwindled over time. During this era of product development, the work on fat replacement led to better understanding of ingredient interactions, resulting in a reasonable lowering of fat and calories in foods using, for the most part, conventional food ingredients. Throughout the 2000s, product development became more pre-occupied with fad diets before settling into a concentration on sugar and calorie reduction. Reduction of trans fats is still an issue, and still a focus of concentration. "Fat free" may still be attractive in some categories, but is not the driving force it once was and, as a result, many of the "fat replacer" and "fat substitute" products have been replaced by a blend of more traditional ingredients. Cost and consumer awareness have played some role in this. It is interesting to note that many consumer studies now note that "*low* fat," "*low* sugar," and "*low* sodium" are the diet trends of interest in 2012 rather than the "free" trends of the 1990s and 2000s.

Fat is Essential and Functional

Reports concerning the negative effects of fat and oil consumption had caused many consumers to focus on fat elimination, but this has become more balanced in consideration of the body's fat requirements (Deis 1996b). Fats act as important energy sources, especially during growth or at times when food intake might be restricted. Protein and carbohydrates provide about 4 kcal/g, but fats provide more energy at about 9 kcal/g. Linoleic and linolenic acids are regarded as essential fatty acids that aid in the absorption of vital nutrients, regulation of smooth muscle contraction, regulation of blood pressure, and growth of healthy cells. On the negative side, the US Surgeon General has stated that consumption of high levels of fat is associated with obesity, certain cancers, and possibly gallbladder disease, and also noted that strong evidence exists for a relationship between saturated fat intake, high blood cholesterol, and coronary disease. Rather than viewing all fats as "bad," nutritionists urge consumers to control the percentage of calories as fat in their diets and to limit levels of saturated fat and polyunsaturates. Current government guidelines state that total fat intake should be between 20% and 35% of total calories, with most fats coming from sources of polyunsaturated and monounsaturated fatty acids. Concerted efforts from ingredient suppliers and product developers have reduced the use of saturated fats such as lard, beef tallow, butterfat, coconut oil, and palm oil and increased the use of vegetable oils with higher percentages of polyunsaturates and monounsaturates.

Most of the fat consumed in the United States comes from salad and cooking oils, followed by frying fats and bakery shortenings, then meat, poultry, fish, and dairy products (cheese, butter, margarine). Each of these applications is unique in its requirements for fat functionality (Table 28.1). In fried foods, oil acts as a heat-transfer medium, but also becomes a component of the food. Because of this dual function, the oil must meet a number of requirements. It must have good thermal and oxidative stability, good flavor, good shelf life, and acceptable cost.

Fats and oils provide important textural qualities to certain foods. Much of this is due to specific melting qualities and crystal structure, and qualities are provided by the "shortening"

Table 28.1 Functional Properties of Fats in Different Food Categories

Frozen desserts	Fried Foods	Meats	Baked Goods
Flavor	Flavor	Flavor	Flavor
Viscosity/body	Heat transfer	Mouthfeel	Viscosity/body
Creaminess	Crispness	Juiciness	Richness
Mouthfeel	Aroma	Tenderness	Texture
Opacity	Color	Texture	Shortness
Heat shock	Heat stability	Binding	Tenderness
Stability	Migration	Heat transfer	Flakiness
Overrun			Aeration
Melt			Elasticity
			Leavening
			Lubricity
			Moisture
			Retention
			Shelf life
			Water activity

Source: Adapted from Lucca, P.A., and Tepper, B.J., *Trends Food Sci Techol* 5:12–19, 1994.

effect of fats, primarily in baked goods. Fats provide baked goods with a characteristic rise, flakiness, tenderness, strength, "shortness," and cell structure that are not apparent in fat-free varieties (McWard 1995; Stockwell 1995). Fats and oils are also essential to lubrication of foods in two ways: as release agents during cooking and as lubricants during chewing, causing a cooling and coating sensation picked up as moistness in baked goods. Fats modify flavor release and affect mouthfeel by providing viscosity and coating effects and also possess their own characteristic flavors (animal fats, olive oil, and peanut oil are classic examples).

All the functional factors are impacted by the type of fats used. In turn, the type and composition have health consequences. Some of the issues surrounding functionality are better understood by looking at the nature of fat and the composition of the predominant fats and oils available for food use. Fat molecules consist of three fatty acids linked to a glycerol backbone. Native fats and oils are made up of mixtures of a wide range of fatty acids arranged in varying ratios and positions on these triglycerides. This determines the characteristics of a particular fat and creates a wide selection of properties. If a fatty acid is "saturated," it means that the maximum number of attachment sites (four) on the carbon atom are filled by another attached carbon or a hydrogen atom, and only single bonds exist. Unsaturated fatty acids contain one or more double bonds between carbon atoms in the chain. If an unsaturated fatty acid contains one double bond, it is referred to as monounsaturated (oleic, 18:1). If more than one double bond exists, the fatty acid is polyunsaturated (linoleic, 18:2; linolenic, 18:3). If an oil contains a predominance of saturated fatty acids, such as palm oil or coconut oil, it is commonly referred to as a saturated fat. Conversely, if the fat is predominantly unsaturated, it is referred to as an unsaturated fat.

Fats and oils are very complex substances from a number of origins, made up of a number of fatty acids in infinite combinations, and chemically modified in a number of ways. This chemical complexity results in a very complex mix of results in terms of appearance, texture, flavor, mouthfeel, and processing characteristics. Although it was the goal of many initial fat replacers to encompass all categories of food, fat functionality is category-specific. Performance in dairy products or salad dressing can be expected to be different than performance in baked goods or snacks. Secondly, fat functionality is product specific, for example, since the percent of moisture in a cheesecake is far more than that of a cookie, it can be expected that fat replacers will perform differently in these products. Finally, the need to consider adjustment of all other ingredients in the formulation means that fat replacement is formulation-specific. Fat replacement in any formulation requires attention to more than fat – support ingredients are required to address mouthfeel, texture, aeration or structure, color, flavor, handling characteristics, and shelf stability. Other factors to consider are cost, regulatory concerns, safety, packaging needs, label claims, and availability.

Fat Replacers – An Overview

Most fat replacement is accomplished by using water effectively, and air can be entrapped as a texture aid in many products. The first fat replacers were primarily air, water, and emulsifiers. Because early products were unsuccessful and the US Nutrition Labeling and Education Act (NLEA) eliminated the use of emulsifiers as nonfats, combinations of ingredients in a systems approach found more success. Finally, when the consumer discovered that "reduced fat" or "fat free" did not necessarily mean less calories, development became more focused toward caloric reduction and fat reduction.

The term fat replacer is a generic term for any bulking agent or ingredient that somehow replaces fat in a system. Fat extenders serve to extend the usefulness of a reduced amount of fat in a food. This could be an emulsifier or something coated with a fat, so that the fat is still a part of the system. A fat substitute actually has the characteristics of fat but is absorbed differently (or not absorbed at all) by the body, resulting in less caloric density. An example of this would be olestra, caprenin, or salatrim. Fat barriers reduce the amount of oil migration into a product (doughnuts, French fries). A number of film-formers such as starches and celluloses could be placed here. Fat mimetics are ingredients that somehow partially imitate fat, usually by binding water. Many of the carbohydrate and protein fat replacers would fall into this category.

Another way to look at fat replacers, is to place them into general application classes. The simplest classification is to classify them as (a) modified fats or (b) water binders. It is more fair and easier to discuss properties when they are classified by their chemical identity; carbohydrate, protein, or fat. Many fat replacers do function by binding water, but thinking of this in terms of their chemical class helps to explain how they accomplish this and whether more functional ingredients are available. A number of general reviews on fat replacers have been published (Deis 1996a; Giese 1996a,1996b; Haumann 1998; Swanson 1998; Thayer 1992; Calorie Control Council 2010a; Akoh 1998; Deis 1997) and can be consulted for more specific information. Because virtually every ingredient that participates in water binding or structure setting in foods could be considered a fat replacer, this will be a general overview of those potential ingredients.

Carbohydrates as Fat Replacers

At many ingredient trade shows, over 50% of the products promoted as "fat replacers" have been carbohydrates, although this is not as active an area after 2000 as it was in the 1990s. Carbohydrates generally mimic fat by binding water, thus providing lubrication, slipperiness, body, and mouthfeel. Carbohydrate adjustments can positively (or negatively) influence shelf life, freezing characteristics, and mouthfeel by affecting the physical state of the final product. Processing parameters such as pH, temperature, shear, and compatibility with other ingredients, as well as the rheological character of the carbohydrate, must be considered.

Guar, locust bean, and xanthan gums are effective thickeners across a number of food categories (Dziezak 1991; Sanderson 1996). Pectins can form soft to hard gels and are widely used in jams, jellies, and tomato-based products. Alginates and carrageenans are commonly used in ice-creams, puddings, fruit gels, and salad dressings. Most gums, depending on form and/or processing conditions, can be used as gelling agents or thickeners. Gellan gum, approved as a food additive in 1990, is produced by *Sphingomonas elodea* (known earlier as *Pseudomonas elodea*). Gellan exists in two forms. a native acylated form and a deacylated form. Both forms have a glucose, glucuronic acid, rhamnose backbone (2:1:1), forming a linear tetrasaccharide repeating unit. The acylated form provides elastic gels; the deacylated form provides a more brittle gel. Gellan is compatible with a number of other gums (xanthan, locust bean), starches, and gelatin, to manipulate the type of gel, elasticity, and stability and can form strong brittle films exhibiting oil and moisture barrier properties.

Fiber should always be considered for fat replacement and as a replacement for flour and other caloric ingredients (Gorton 1995). "Dietary fiber" as a classification encompasses a wide range of fiber sources that vary in their physical properties. Two subclasses are recognized, soluble and insoluble, which are very different in chemistry and physiological effects. One of the more acceptable definitions for dietary fiber (at this point, there is no globally accepted definition) is:

Dietary fiber means carbohydrate polymers[*] with ten or more monomeric units,[†] which are not hydrolysed by the endogenous enzymes in the small intestine of humans and belong to the following categories:

■ Edible carbohydrate polymers naturally occurring in the food as consumed
■ Carbohydrate polymers, which have been obtained from food raw material by physical, enzymatic or chemical means and which have been shown to have a physiological effect of benefit to health as demonstrated by generally accepted scientific evidence to competent authorities

---

[*] When derived from a plant origin, dietary fiber may include fractions of lignin and/or other compounds when associated with polysaccharides in the plant cell walls and if these compounds are quantified by the AOAC gravimetric analytical method for dietary fiber analysis: Fractions of lignin and the other compounds (proteic fractions, phenolic compounds, waxes, saponins, phytates, cutin, phytosterols, etc.) intimately "associated" with plant polysaccharides are often extracted with the polysaccharides in the AOAC 991.43 method. These substances are included in the definition of fiber insofar as they are actually associated with the poly- or oligo-saccharidic fraction of fiber. However, when extracted or even re-introduced into a food containing non digestible polysaccharides, they cannot be defined as dietary fiber. When combined with polysacchrides, these associated substances may provide additional beneficial effects.
[†] Decision on whether to include carbohydrates from 3 to 9 monomeric units should be left to national authorities" (CAC 2009).

■ Synthetic carbohydrate polymers which have been shown to have a physiological effect of benefit to health as demonstrated by generally accepted scientific evidence to competent authorities

There is general agreement that this is a very good, well thought out definition, with the exception of the passage concerning monomeric units. Codex has left the decision on whether to include carbohydrates from three to nine monomeric units to national authorities, which could lead to difficulties in any attempts to harmonize definitions. There is sufficient evidence and backing to include these carbohydrates, and "physiological effects of benefit to health" needs to be better defined (Calorie Control Council 2010b).

Cellulose is the most abundant source of insoluble dietary fiber, as well as the most abundant carbohydrate in nature, making up a significant part of the mass of a plant. Cellulose is a linear polymer of β-1,4-linked d-glucose which is undigestible by the human gastrointestinal tract (in contrast with the α-1,4-linked glucose in starch which is highly digestible). Sources are predominantly plant cell walls; foods high in insoluble fiber are whole grains, cereals, seeds, and skins from fruits and vegetables. Cellulose is available in a number of forms, from mechanically disintegrated forms to fermentation-derived to chemically substituted or hydrolyzed versions (Figure 28.1). The most common cellulose product found in foods is microcrystalline cellulose (MCC). Many forms are available: MCC as is, MCC with carboxymethylcellulose (CMC) added, or MCC in combination with other gums such as guar and sodium alginate, as well as a range of particle sizes. MCC and its derivatives have been used extensively to replace calories and fat. Other cellulose derivatives such as methylcellulose and hydroxypropyl methylcellulose (HPMC) have been used for years as multipurpose thickeners because of their ability to hydrate and build viscosity quickly to form clear gels of varying strengths.

Powdered cellulose, at 99% + total dietary fiber (TDF), is marketed in various fiber lengths to provide a range of water-holding capabilities. Powdered cellulose is derived from wood pulp, treated to remove lignin and other impurities, then milled to a range of fiber lengths from 22 to 120 mm in length. Chemically, it is 90% β-1,4 glucan plus approximately 10% hemicelluloses. Because it is almost pure TDF, powdered cellulose is considered to be noncaloric. Depending on fiber length, powdered cellulose can retain 3.5–10 times its weight in water (longer fiber lengths are able to retain more water, also increasing the viscosity of the food system).

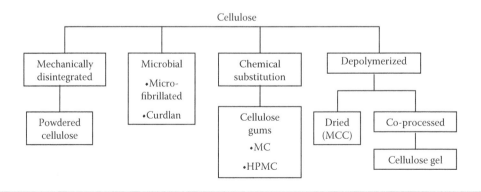

Figure 28.1 Forms of cellulose commercially available.

Two fermentation-derived cellulosic thickeners and stabilizers were introduced in the latter part of the 1990s. One was marketed as PrimaCel® by Monsanto, and was produced by the microbial fermentation of *Acetobacter xylinum* combined with sucrose and carboxymethylcellulose coagents to promote dispersion. It is still marketed by CP Kelco as "microfibrous cellulose," primarily for use for suspension in surfactant thickened systems and concentrated detergents. It was originally developed and patented by Weyerhaeuser Company. The product was previously known as Cellulon®. A Generally Recognized As Safe (GRAS) petition was accepted for filing by the US Food and Drug Administration (US FDA) in 1992. These fibers are extremely fine, with a 0.1 mm to 0.2 mm diameter, forming a strong, stable colloidal network.

Another fermentation-derived gum, curdlan, found some success in Japan and was approved by the US FDA for use as a formulation aid, processing aid, stabilizer, thickener, or texturizer in foods (21 CFR 172.809(b)). Curdlan is a unique polysaccharide with potential uses as a texture modifier and/or gelling agent in processed meats, noodles, surimi-based foods, and processed cooked foods (Yotsuzuka 2001). Discovered in 1966 at Osaka University, curdlan (common name) is a polysaccharide produced by *Alcaligenes faecalis* var. myxogenes. It is a linear β-1,3 glucan, insoluble in water, alcohol, and most inorganic solvents, and is indigestible; virtually 100% TDF. Once suspended, curdlan produces a weak low-set gel if heated to 60°C and then cooled to less than 40°C. Gel strength also increases with increased product concentration. If heated to greater than 80°C, a stronger, thermo-irreversible gel forms.

Another Asian product, konjac flour, is a centuries-old ingredient obtained by grinding the root of the Amorphophallus konjac plant (also known as elephant yam). Konjac is GRAS and has been listed (*FCC* monograph) as konjac, konjac flour, konjac gum, and konnyaku (Tye 1991). The average molecular weight of konjac is 200,000 to 2 million daltons (average, 1 million), with short side branches and acetyl groups positioned at C-6 every 6 to 20 sugar units. Konjac is able to form very strong, thermally reversible gels with carrageenan, xanthan gum, and locust bean gum. Adding a base (potassium hydroxide, sodium hydroxide, calcium hydroxide, or potassium carbonate or sodium carbonate) forms a thermally stable, nonmelting gel. Konjac also forms a heat-stable gel with starch when it is cold-set by raising the pH.

Insoluble fiber is an important tool for fat reduction and for caloric reduction. Powdered cellulose, cellulose gums, microcrystalline cellulose, and cellulose gels, as well as other plant fibers (oat fiber, soy fiber, wheat fiber, rice flour, hydrolyzed oat flour, etc.), present a wide variety of possibilities in terms of water-binding, viscosity, film-forming, gelling, and pulpiness. Most of these are regarded as "natural," if that is a consideration. When considering these for use, many factors should be considered, including water-holding capacity, texture, TDF, caloric density, prebiotic effects, ingredient legend compatibility, color, and cost to name a few. The US Department of Agriculture's Agricultural Research Service (USDA) has had an ongoing program for many years to develop usable products from agricultural by-products such as grain hulls (oat, corn, rice, soybean, peas) and brans (corn, wheat). This resulted in some products that have exhibited potential as fat-replacing ingredients, including oatrim, Z-Trim, and Nu-Trim (Inglett 2006). Developed as a fat replacer, oatrim was patented by USDA and licensed to ConAgra, Quaker Oats, and Rhône-Poulenc. Oatrim is enzymatically hydrolyzed oat flour containing 5% β-glucan soluble fiber. Starch in oat flour or bran is hydrolyzed by α-amylase to form a more soluble material (oat β-glucan-amylodextrins) labeled as "oatrim" or "hydrolyzed oat flour." Similar to other carbohydrate and protein fat replacers, oatrim can form a gel with water to mimic fat in a number of food applications. From a caloric standpoint, if the gel contains 25% oatrim at 4 kcal/g and the remaining 75% is water at 0 kcal/g, the gel is 1 kcal/g. USDA-conducted studies suggest oatrim might have some hypocholesterolemic benefits, but this product is not currently included in the list of oat products that can carry nutritional labeling to that effect. On October 9, 1996, the USDA announced the development of another product, dubbed "Z-Trim" (for zero calories). Whereas oatrim was developed from the inner, starch-containing part of the hull or bran, Z-Trim was developed from the more cellulosic, outer portion. In a process similar to the alkaline/hydrogen peroxide process, which led to a USDA-patented oat fiber, the hulls of oats were treated in a multistage process to remove the lignin. The resulting cellular fragments were purified, dried, and milled. This dried powder could later be rehydrated to form a gel or be incorporated directly into a food. Nu-Trim was introduced by the USDA in 1998 as a physically modified soluble fiber product with properties similar to oatrim.

Another fiber-containing ingredient that can take the place of fat in foods is resistant starch. Because of its reduced caloric content, it functions mainly as a bulking agent, although it can have other beneficial effects in the finished product (Mohr et al. 2010). The term "resistant starch" was coined in the early 1980s, but scientists had discussed its dietary effects years before that time. Resistant starches are starches and products of starch degradation that resist enzymatic digestion and act like dietary fiber. Resistant starch is present in many foods. It is naturally found in coarsely ground or chewed cereals, grains, or legumes as a physically inaccessible starch (RS1). It also can be found in bananas, high-amylose starch, and raw potato as naturally resistant or ungelatinized granules (RS2). A third type of resistant starch (RS3) is generated by retrograding starch during food processing. This variety can occur naturally in products such as bread, cereals, and cooked potatoes. Recently identified, another type is classified as RS4 (Brown 2004). This is representative of starch that has been rendered resistant by chemical modification. Currently, no commercial products of this type exist. As knowledge of TDF developed through the 1980s, resistant starches frequently were discussed, but no significant attempts were made to commercialize them until the 1990s. The first resistant starch released and marketed as such was an RS2 based on a high-amylose corn starch hybrid. The product, Hi-Maize®, was developed by Starch Australasia Ltd. and won the 1995 Australia Institute of Food Science and Technology Industry Innovation Award in Australia. Hi-Maize® contained approximately 20%–25% total dietary fiber and was introduced into several breads and extruded cereals as a functional fiber in Australia. National Starch and Chemical Company (now National Starch Food Innovation, recently acquired by Corn Products International), Bridgewater, NJ, was granted a US patent on a common process to produce RS3 starch. This resulted in the 1994 launch of Novelose®. National Starch continued to expand its Novelose line, offering products with more than 40% TDF, and also acquired the rights to Hi-Maize®. Resistant starch represents a functional fiber alternative. The RS3-resistant starch, for example, has been shown to significantly improve expansion and eating quality for extruded cereals and snacks SP. The RS2-resistant starch performs well in baked goods because of its small granule size and low water-holding capacity. The RS2 product can be labeled as "cornstarch," and the RS3 as "maltodextrin," both already recognizable terms on many ingredient legends. They compare well to several natural grain sources, contain little fat, are white in color, and neutral in flavor. A 1996 American Institute of Baking study (Ranhotra et al. 1996) determined that although nearly one third of the resistant RS3 starch consumed is fermented, it produces essentially no energy value.

Resistant maltodextrins are an indigestible dextrins produced by the acid and enzyme hydrolysis of starch (Brouns et al. 2007). These are claimed as soluble fiber and would be stated on an ingredient legend as "maltodextrin" or "resistant maltodextrin." Resistant maltodextrins typically contain α-1,4, α 1,6, and β-1,2, β-1,3, and β-1,6 glucosidic bonds, are very soluble in water, and have low viscosity. Another product with excellent solubility is partially hydrolyzed guar gum, which also offers the advantage of low viscosity (Yoon et al. 2008).

After starch, inulin is the most abundant nonstructural polysaccharide in nature, being the energy reserve in thousands of plants. As an oligosaccharide, inulin is extremely well known and widely used in Asia. Oligofructose is present naturally in onions, asparagus, leeks, garlic, artichokes, bananas, wheat, rye, and barley. For the purified form, manufacturers generally turn to the more concentrated sources, which include chicory (more than 70% inulin on dry solids) and Jerusalem artichoke (also more than 70%). Inulin from agave has also been characterized (Lopez et al. 2003) and has recently entered the market. One of the differences noted in agave fructans is the fact that they are more highly-branched, leading to greater solubility and clarity in beverages. Chemically, inulin is a $2 \rightarrow 1$ fructan with the general formula:

$$G\!fn$$

where G = glucosyl unit, f = fructosyl unit, and n = number of fructosyl units linked $n(2)$. The degree of polymerization (DP) ranges from 2–60.

Oligofructose, another product on the US market, contains a mixture of Gfn and independent fructosyl units, with an overall DP of 2–20. Inulin can be used for fat replacement in food products in dry and gel form because a 30%–40% solids gel has a fatty feel. The gel strength can be varied to result in a low-calorie fat replacement for specific uses. Several studies indicate the caloric value of inulin is approximately 1.0 to 1.5 kcal/g. Inulin is metabolized preferentially by bifidobacteria in the colon, thereby providing many benefits. However, the uses of inulin in foods are still poorly understood. Many Asian and European uses have focused on the health benefits, which are still relatively unknown to the US consumer.

In most cultures, plants or plant-derived ingredients (such as flour) have long been used as thickening agents. In plants, starch is a reserve carbohydrate, deposited as granules in the seeds, tubers, or roots. These starch granules differ in size and shape, depending on the plant source (Table 28.2). Granules of rice starch are small (3–8 µm), polygonal in shape, and tend to aggregate, thereby forming clusters. Cornstarch granules are slightly larger (approximately 15 µm) and round to polygonal. Tapioca granules are even larger (approximately 20 µm), with rounded shapes that are truncated at one end. Wheat starch tends to cluster in several size ranges: normal granules are approximately 18 µm; larger granules average about 24 µm; and smaller granules average approximately 7 to 8 µm, with round to elliptical shapes. Potato starches are oval and very large, averaging 30 to 50 µm. It is important to note these variations in granule size and shape because they

Table 28.2 Starches—Sources and Modifications

Source
Corn (common, waxy, high amylose) Rice Tapioca Potato Wheat
Chemical modification
Cross-linked Substitution Acid hydrolysis (particle gel)
Physical modification
Pregelatinization (instant)
Functional Native Starches

yield distinct differences in viscosity development, stability, mouthfeel, and rate of gelatinization in products. Starch is a carbohydrate polymer, consisting of anhydroglucose units linked together by λ-D-(1 → 4) glucosidic bonds arranged as two major types of polymers, known as amylose and amylopectin. Amylose is primarily linear, containing anywhere from 200 to 2000 anhydroglucose units. Amylopectin is a branched polymer, also connected by λ-1-D (1,4) glucosidic linkages, but with periodic branch points created by β-D-(l,6) glucosidic linkages. Amylopectin is typically much larger than amylose, with molecular weights in the millions. Amylose and amylopectin contain an abundance of hydroxyl groups, creating a highly hydrophillic (or water-loving) polymer that readily absorbs moisture and disperses well in water. Because amylose is linear, it has a tendency to align itself in a parallel nature with other amylose chains, leading to precipitation (in dilute solutions) or retrogradation (in high solids or gels). On the plus side, it also can lead to the formation of strong films, which are extremely useful in certain food applications. The negative side is that amylose can detract from the clearer food products by contributing opacity and also tends to mask delicate flavors (Deis 1998).

Because of its branching, amylopectin forms clearer gels, often favored in the food industry, that do not form strong films or gels. Retrogradation occurs less readily. In the native form, most starches contain 18%–28% amylose, with the remainder as amylopectin. Corn and wheat starches contain approximately 28% amylose, whereas potato, tapioca, and rice varieties are closer to 20%. Two genetic varieties of corn have become popular and well accepted in the industry: "waxy" starch, which contains amylopectin with practically no amylose, and high-amylose starches. Of the high amylose starches, two varieties exist: approximately 55% amylose and approximately 70% amylose. Obviously, waxy starches develop weak gels with excellent clarity and are poor film formers. On the other hand, high-amylose starches form very opaque, strong gels and are excellent film formers. Because of the number of hydroxyl groups available on the starch polymer, starches are, very fortunately, easily modified for a wide array of food-industry applications.

In addition to their corn-based products, several manufacturers have tapioca starches as part of their product mix, and tapioca and waxy corn are frequently featured as part of the same class or series of product. Tapioca, potato, and rice starches have been recognized to provide flavor advantages over corn, and these crops, plus wheat, are recognized as offering allergenicity advantages because of their low protein levels. Within the processed meats industry, modified corn, wheat, and potato starches compete with proteins as water binders in low-fat meat applications. Using starches as water binders in a fat-replacement system maintains consistent moisture, flavor, and texture throughout product shelf life.

In a discussion of starches, it should also be noted that syrups produced by enzyme and/or acid hydrolysis of starches from any of these sources can be used to control texture and water activity in food products. High fructose syrups (42% or 55% fructose) are extremely sweet and are low in molecular weight. Corn syrups are available in a wide range of dextrose equivalents, which can be used as a tool for adjusting the level of sweetness (sweetness decreases as the DE is decreased) and also controlling water activity (be aware of the degree of polymerization, or DP, range of the product). Maltodextrins and syrup solids can also be used in this way to manipulate the water activity of a formulation.

If calorie reduction is a consideration, monosaccharide and disaccharide polyols (sugar alcohols) should be considered (Deis 1993, 1994). Many of these are available in crystalline or syrup form, with caloric densities ranging from 0.2 to 3.0 kcal/g. These include sorbitol, erythritol, mannitol, isomalt, lactitol, xylitol, and maltitol.

Similar in function to the syrups, a hydrated polydextrose can contribute syrup-like qualities at 1 kcal/g. Polydextrose has been used in ice-creams, confectionery, and jams and jellies for

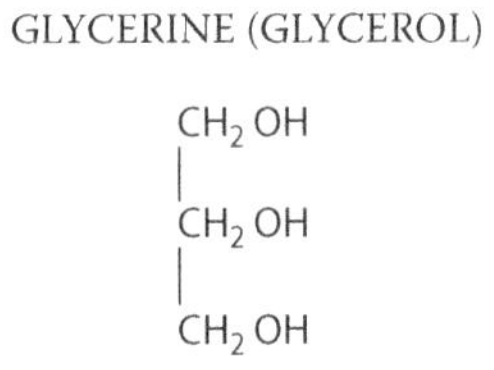

GLYCERINE (GLYCEROL)

CH$_2$ OH
|
CH$_2$ OH
|
CH$_2$ OH

Figure 28.2 Structure of glycerine.

caloric reduction. In high-moisture applications, it can provide a "slippery" mouthfeel that can somewhat mimic fat. In lower moisture applications, polydextrose can provide some hygroscopicity to the final product, which contributes to softening. The original polydextrose product was acidic and slightly bitter because of citrate residual in the product. Newer, less acidic versions are now available for lightly flavored foods.

Glycerin (Figure 28.2) is a carbohydrate well known for its use in fruits and candies as a control for drying and graining, as well as softness control (Croy and Dotson 1995). It is used in jelly candies, fudge, cake icings, cookie fillings, dried fruits, and citrus fruit peels. It prevents oil separation in peanut butter and acts as a softener and humectant in shredded coconut. Glycerin has been used extensively to influence texture and decrease water activity in low-moisture fat-free and fat-reduced products. It is extremely effective, but use is restricted by its sweet, astringent taste. A significant indirect use is the reaction of glycerin with fats and fatty acids to form monoglycerides and diglycerides, emulsifiers important to fat reduction.

Proteins as Fat Replacers

A protein's contribution to fat replacement is determined by the extent of denaturation, which affects flavor and the protein's solubility, gelling properties, and temperature stability (Table 28.3). Proteins are important as whipping agents, foam and emulsion stabilizers, and dough strengthened (Giese 1994). Gelatin and egg albumen have been used extensively in fat-reduced baked goods, frostings, and marshmallows. Soy proteins, egg albumen, wheat gluten, nonfat dry milk, caseinates, and whey protein concentrates (WPC) are often used to strengthen fat-reduced pasta, bread, and sweet goods.

Table 28.3 Functionality of Protein-based Fat Replacers

Water-binding
Emulsification
Viscosity
Film formation
Opacity
Whipping and foam stabilization
Binding

Whey proteins and whey protein concentrates (WPC) have been used extensively in dairy-based applications (Harrigan and Breene 1989). Some of these products are based on the "microparticulated" concept and consist of extremely small particles within certain size ranges, this is the concept initiated by Simplesse®, which was developed as a whey protein or egg protein-derived fat replacer (Corliss 1992). These particles mimic a fatlike sensation on the tongue. Other products, such as DairyLight® and DairyLo®, were based on controlled denaturation to provide a workable viscosity in processing, particularly in ice-creams. In processed meats, soy protein isolates and concentrates provide high protein quality, meat-like texture and appearance, excellent firmness, reduction in purge (or water loss), and brine retention in injected products. Soy proteins are used for a number of functions, but the top four functions are emulsification, fat absorption, hydration, and texture enhancement. Because they are concentrated proteins, soy proteins can also be used for film formation, adhesion, cohesion, elasticity, and aeration. Soy flour contains up to 50% protein and also retains carbohydrates, fiber, and fat (unless it is a defatted flour). Soy protein concentrates are made primarily by alcohol extraction of a portion of the carbohydrates from defatted, dehulled soybeans. As awareness of isoflavones has increased, manufacturers have developed other processes to preserve isoflavone content. Most of the fiber is retained in a soy protein concentrate and these must contain at least 65% protein on a moisture-free basis. The most concentrated soy protein source is isolated soy protein, required to have at least 90% protein on a moisture-free basis. These are generally extracted with water from defatted, dehulled soybeans. Because 90% of the product is protein, isolates contain very little fiber or other components.

Fat-based Fat Replacers

The fat-based fat replacers have the advantage of a closer relationship chemically with fats, and so physical appearance, thermal stability, and melting points may be a little closer (Table 28.4). Emulsifiers such as lecithin, monoglycerides and diglycerides, DATEM (diacetyl tartaric ester of monoglyceride), SSL (sodium stearoyl lactylate), polyglycerol esters, and sucrose esters have been used to extend the available fat in the system, allowing one to increase water content, more fully aerate if needed, improve processing characteristics, stabilize emulsions, and improve shelf life by complexing with starches and proteins (Table 28.5).

The concept of structured fats has been around for a number of years. Although medium-chain triglycerides have long been recognized for their nutraceutical potential, their major drawback may be education of the consumer and cost (Merolli et al. 1997). Medium-chain triglycerides (MCT, $C_6 - C_{12}$) are metabolized differently than long-chain triglycerides (LCT, $C_{14} - C_{24}$). LCTs are hydrolyzed, then reesterified to triglycerides, then imported into chylomicrons, which enter

Table 28.4   General Categories of Fat-based Fat Replacers

Emulsifiers
Molecular backbones to which fatty acids are attached in such a way that digestion is altered, but functional properties are retained (e.g., Olestra)
Glycerol backbones to which groups with poor digestibility are attached (e.g., Caprenin, Salatrim)

Table 28.5   Functional Properties of Emulsifiers

Increase water content
Aeration to reduce density
Use fats more efficiently
Improved processing
Stabilize emulsions
Release agent
Starch, protein interactions

the lymphatic system. MCTs bypass the lymphatic system. They are hydrolyzed to MC fatty acids, which are transported by way of the portal vein directly to the liver, where they are oxidized for energy. They are not likely to be stored in adipose tissue. For enteral and parenteral feeding, their advantage is already known. MCTs provide patients with an energy source similar to glucose, but with twice the caloric value.

Synthetically structured fats were actively researched for a number of years and, although they are not in the news as of 2010, they should be mentioned (Haumann 1997; Stauffer 1995). In any discussion of fat substitutes, these come to mind because they were designed to look and act like fats, but they contribute fewer calories and less fat. Two approaches to this were taken: (a) work from a glycerol backbone and attach planned ratios of long-chain (LC) saturated fatty acids with very low caloric density and shorter-chain (SC) fatty acids with slightly lower caloric density than LC fatty acids (caprenin, salatrim); or (b) attach fatty acids to a nonglycerol backbone in such a manner that the molecule is poorly absorbed in the body (olestra). Because the first method results in a triglyceride similar to what could be found in nature, the regulatory route is far shorter. Have it reviewed by an expert panel, file a GRAS petition, and commercialize. The second route is a little more complex. A full food additive petition is required and because of the amount of fat that could potentially be replaced in the diet, approval will be on a category-by-category basis until there is an adequate comfort level with any potential side effects.

Although the "ultimate fat substitute" (heat-stable, friable) seems to be in the olestra-type class, other developers of fats chose a more limited route. Several companies worked toward synthesizing carefully structured triglycerides with a glycerol backbone, viewing GRAS approval as a much faster route to regulatory acceptance. This is usually done through interesterification, a modification process that results in the rearrangement of the fatty acids of the triglyceride molecule. Through choices of starting materials (different oils or fats), catalysts and/or enzymes, and kinetics, this reaction can be more directed toward a relatively specific end product. This means that the choice of fatty acids involved, as well as their relative ratios, can be limited. Interesterification has been used for some time as a more randomized process to produce plastic fats from animal/vegetable fat blends for use in margarines. The first product commercialized under this grouping was caprenin, a reduced-calorie designer fat consisting of three fatty acids: capryllic (eight carbon atoms, no double bonds), capric (10 carbon atoms, no double bonds), and behenic acid (22 carbons, no double bonds). Behenic acid is only partially absorbed by the body, and the medium-chain fatty acids have lower caloric densities than longer-chain fatty acids, resulting in a total caloric density for caprenin of 5 kcal/g. Caprenin was commercialized as a cocoa butter replacer and was launched in two products. Unfortunately, the product had difficult tempering characteristics and appeared to increase serum cholesterol slightly, resulting in its withdrawal from the market.

As caprenin was being tested, another family of restructured fats was being developed by Nabisco Foods Group (now Kraft). Salatrim, which is an acronym for Short And Long Acyltriglyceride molecule, is a family of structured triglycerides based on the use of at least one SC fatty acid and at least one LC fatty acid (stearic, C-18). Salatrim triglycerides typically contain one or two stearic acids combined with specific ratios of SC fatty acids (acetic, C-2; propionic, C-3; and butyric, C-4). As with naturally occurring triglycerides, the properties of salatrim are dictated by the fatty acids used, as well as their position of the molecule. The first commercial product was developed to replace cocoa butter in confectionery applications. A GRAS petition was filed in December, 1993 and was accepted for filing by the US FDA in June, 1994. Safety studies have shown no effect on serum cholesterol, no effect on absorption of fat-soluble vitamins, and have verified the safety of the molecule. Salatrim is the generic name for this class of molecules, and it is the name used on an ingredient legend. Because of the lower caloric density of stearic acid and the SC fatty acids, salatrim contributes a total of 5 kcal/g. Because the US FDA had no regulation for food factors regarding fat reduction (only calories), these claims resulted in some controversy and discussion at the US FDA. In its 1994 petition, Nabisco claimed that because five-ninths of the fat available in salatrim was used by the body, the product would have a food factor of 5/9, related to both fat and calories. Danisco A/S marketed salatrims for a number of years and received European approval for its use.

The ideal fat replacer, or fat substitute, should have a physical appearance, thermal stability, and melting point close to that of the fat being replaced. The "synthetic oil" approach has been to devise a molecular backbone to which fatty acids can be attached such that digestion is altered. Olestra is a mixture of hexahepta-, and octa-fatty acid esters of sucrose (Brewer 1995). As opposed to the glycerol backbone of triglycerides, olestra has a sucrose backbone, to which six to eight long fatty acid chains have been added (70% of the molecules have eight long chains). Olestra is synthesized from sucrose and vegetable oil (cottonseed or soybean), and it has physical properties comparable to conventional fats used in savory snacks and crackers. The complexity of the molecule inhibits the activity of digestive enzymes required to break it down. Therefore, olestra passes through the body undigested, contributing no fat or calories to foods. Olestra is a thermally stable, fryable fat substitute that can substitute for all of the oil in a product, contributing essentially no fat and no calories (Yost 1994; Sullivan 1995a, 1995b). This has been commercialized in several lines of potato and tortilla chip products. On January 24, 1996, the US FDA approved olestra for use "in place of fats and oils in prepackaged ready-to-eat savory (i.e., salty or piquant, but not sweet) snacks. In such foods, the additive may be used in place of fats and oils for frying or baking, in dough conditioners, in sprays, in filling ingredients, or in flavors" (CPR 172.867c). Approval of olestra by the US FDA did not come without cost—nearly 30 years, 270 volumes of data, and more than 150 long-term and short-term studies. Olestra has found limited use, but can still be found in some snack foods.

The synthetic route to healthy fats is not easy. Other fat substitute projects have either stalled out or have been placed on hold as the olestra project has progressed. Pfizer Food Science Group (acquired by Cultor, which was acquired by Danisco A/S) developed a mixture of fatty acid esters of sorbitol (Sorbestrin), which was reported to be a thermally stable, fryable fat substitute with a caloric content of 1.5 kcal/g. This product is not available commercially, and its use would require a full food additive petition in the United States. Other potential fat substitutes based on the same idea, fatty acid esters of novel backbones, will face the same scrutiny (and expense), so we cannot expect any newcomers to this area in the foreseeable future. Projects (Table 28.6) reported at ARCO Chemical Company (EPG, or propoxylated glycerol esterified with fatty acids), Frito-Lay Inc. (DDM, or dialkyl dihexadecylmalonate), and Best Foods (TATCA, trialkoxycitrate and

Table 28.6   Other Acaloric or Low-calorie Fat Substitutes (No FAPs to Date)

EPG (esterified propoxylated glycerol) ARCO/CPC Patented 1989 Glycerin reacted with propylene oxide, esterified with fatty acids
DDM (dialkyl dihexabecylmalonate) Dicarboxylic acid esters of fatty alcohols Frito-Lay (1986)
TATCA (trialkoxytricarballylate, trialkoxycitrate) CPC/Best Foods (1985) Margarines

trioleyltricarballyates) have not received much press in recent years, but these also would require approval as food additives. Although successes have been achieved recently in what are termed "structured fats," these have not come without considerable time and cost. The laborious process of developing an olestra-like product will prevent most ingredient suppliers from exploring that route.

Blending Ingredients

Another strategy that has resulted in a number of useful products is the use of blends and co-processed ingredients. Several of these have found their way into fat replacement and can be useful if a cost/benefit exists. If these ingredients are targeted toward one category (such as baked goods or ice-cream), they are definitely worth assessing. If the product claims it "works in everything," one should question what else is needed in the formulation to support such a use. Fat does not have the same function in all foods, so it is unlikely that one blend could replace it across all categories. A number of blends have experienced some success, for example: stabilizer systems in ice-creams and baked goods; cellulose gums; many natural blends derived from fruit concentrates and pastes; and a number of specialty ingredients designed to act as fat replacers. Most of these follow the systems approach, and therefore are combinations of hydrocolloids and fiber, hydrocolloids and emulsifiers, hydrocolloids and opacifiers, and so forth. The advantage in using blends or co-processed ingredients depends on the applications. If it works, and the cost is right, why not? In the fast-track world of product development, this could mean the difference between success and failure. Blends also can be a way to better distribute small amounts of small-percentage ingredients on a functional carrier. Examples of this would include emulsifiers on a hydrocolloid carrier or enzyme-emulsifier co-processed ingredients. Flavors also are often incorporated into these ingredients. These can be flavors that help to supply missing notes that occur through the absence of a highly-flavored fat such as butter or may include other flavoring systems. If the supplier will work with the product developer to customize these ingredients, the effort is well worth it. Technical service from the supplier is essential to maximize the ingredient's performance. "Across the category board" blends and co-processed ingredients should be avoided unless the supplier is willing to customize.

Summary

Consumers expect more than just fat reduction. They have an increasing awareness that calories are just as important, if not more important, than fat levels, and that other elements of food are important in maintaining overall health. Many of these new ingredients also provide an added bonus, the presence of insoluble or soluble fiber for health benefits and calorie reduction. New fat-reduction strategies can be found in these three words: cost, benefit, and quality. Cost control is a constant consideration in product development, and the ingredients that have made inroads into the market have done so with a cost/benefit of the ingredient in mind. Quality improvement has caused product developers to maintain a watch for new and emerging ingredients that might provide a quality they need for certain formulations. Many say nothing new has emerged from ingredient technology during the last several years. This may seem the case on the surface, because regulatory approval is sometimes difficult, budgets are tight, and product development timelines do not always allow for evaluation of something new. Once developed, several years might transpire before a new ingredient "breaks through" in a market. New products, however, have emerged and new ideas incorporating old products have proven successful. In particular, fiber seems to be making a comeback because of its low calories and nutritional benefits. Many of these ingredients also deliver functional benefits that increase their value to designers trying to formulate lower fat, lower calorie foods.

References

Akoh, C.C. 1998. Fat replacers. *Food Technol* 52(3):47–53.

Brewer, M.S. 1995. Olestra and other fat substitutes. *FDA Backgrounder BG* 95–17.

Brown, I. 2004. Applications and uses of resistant starch. *J AOAC Intl* 87(3):727–732.

Brouns, F., Arrigoni E, Langkilde, A.M., Verkooijen, I, Fässler, C., Andersson, H., Kettlitz, B., Nieuwenhoven, M., Philipsson, H., and Amadò, R. 2007. Physiological and metabolic properties of a digestion-resistant maltodextrin, classified as type 3 retrograded resistant starch. *J Agric Food Chem* 55(4):1574–1581.

Calorie Control Council. 2010a. Fat replacers. http://www.caloriecontrolcouncil.org/sweeteners-and-lite/fat-replacers. Accessed June 7, 2011.

Calorie Control Council. 2010b. The Definition of Dietary Fiber. http://www.caloriecontrolcouncil.org/sweeteners-and-lite/fiber. Accessed June 7, 2011.

Codex Alimentarius Commission (CAC). 2009. Report of the Thirty-Second Session of the Codex Alimentarius Commission, FAO Headquarters, Rome, Italy, 29 June - 4 July.

Corliss, G.A. 1992. Protein-based fat substitutes in bakery foods. *AIB Tech Bulletin* XIV(10):1–8.

Croy, C., and Dotson, K. 1995. Glycerin. *INFORM* 6(10):1104–1114.

Deis, R.C. 1993. Low calorie and bulking agents. *Food Technol* Dec:94.

Deis, R.C. 1994. Adding bulk without adding sucrose. *Cereal Foods World* 39(2):93–97.

Deis, R.C. 1996a. Designing low calorie foods. *Food Prod Design* Sept:29–44.

Deis, R.C. 1996b. New age fats and oils. *Food Prod Design* Nov:27–41.

Deis, R.C. 1997. Reducing fat: A cutting edge strategy. *Food Prod Design* March:33–59.

Deis, R.C. 1998. The new starches. *Food Prod Design* Feb:34–58..

Dziezak, J.D. 1991. A focus on gums. *Food Technol* 45(3):115–132.

Giese, J. 1994. Proteins as ingredients: Types, functions, applications. *Food Technol* 48(10):50–60.

Giese, J. 1996a. Fats, oils, and fat replacers. *Food Technol* 50(4):78–84.

Giese, J. 1996b. Fats and fat replacers: balancing the health benefits. *Food Technol* 50(9):76–78.

Glicksman, M. 1991. Hydrocolloids and the search for the "oily grail." *Food Technol* 45(10):94–103.

Gorton, L. 1995. Fiber builders. *Baking & Snack* Feb:46–49.

Harrigan, H.A., and Breene, W.A. 1989. Fat substitutes: sucrose esters and Simplesse. *Cereal Foods World* 34(3):261–267.

Haumann, B.F. 1997. Structured lipids allow fat tailoring. *INFORM* 8(10):1004–1011.

Haumann, B.F. 1998. The benefits of dietary fats—the other side of the story. *INFORM* 9(5):366–367, 371.

Inglett, G.E. 2006. Developing functional hydrocolloids from small grains Paper presented at American Oil Chemists' Society Meeting, St. Louis, MS.

Lopez, M.G. Mancilla-Margalli, N.A., and Mendoza-Diaz, G. 2003. Molecular structures of fructans from *Agave tequilana* Weber var. *azul J Agric Food Chem* 51(27):7835–7840.

Lucca, P.A., and Tepper, B.J. 1994. Fat replacers and the functionality of fat in foods. *Trends Food Sci Techol* 5:12–19.

McWard, C. 1995. Form and function. *Baking & Snack* Feb:52–54.

Merolli, A., Lindemann, J., and Del Vecchio, A.J. 1997. Medium chain lipids: New sources, uses. *INFORM* 8(6):597–603.

Mohr P, Quinin, S., Morell, M., and Topping, D. 2010. Engagement with dietary fibre and receptiveness to resistant starch in Australia. *Pub Health Nutr* 21:1–8.

Ranhotra, G.S., Gelroth, J.A., and Glaser, B.K. 1996. Energy value of resistant starch. *J Food Sci* 61(2):453–455.

Sanderson, G.R. 1996. Gums and their use in food systems. *Food Technol* 50(3):81–84.

Stauffer, C.E. 1995. Fats that cut calories. *Baking & Snack* May:44–49.

Stockwell, A. 1995. The quest for low fat. *Baking & Snack* May:32–42.

Sullivan, J. 1995a. Flaky pie shells that maintain strength after filling. *US Patent* #5,382,440.

Sullivan, J. 1995b. Shortbread having a perceptible cooling sensation. *US Patent* #5,378.486,

Swanson, B.G. 1998. Fat replacers: Part of a bigger picture. *Food Technol* 52(3):16.

Thayer, A.M. 1992. Food additives. *Chem Eng* June 15:26–44.

Tye, R.J. 1991. Konjac flour: Properties and applications. *Food Technol* 45(3):82–92.

Yost, R.A. 1994. Quick-setting sandwich biscuit cream fillings. *US Patent* #5,374,438.

Yoon, S., Chu, D., and Juneja, L.R. 2008. Chemical and physical properties, safety and application of partially hydrolized guar gum as dietary fiber. *J Clin Biochem Nutr* 42(1):1–7.

Yotsuzuka, F. 2001. Curdlan. In *Handbook of Dietary Fiber* M.L. Dreher, and S.S. Cho (Eds.). Boca Raton, Fl: CRC Press.

Chapter 29

Benefits of Reduced Calorie Foods and Beverages in Weight Management

Joan Patton, Beth Hubrich, and Lyn O'Brien Nabors

Contents

Introduction

Obesity has reached epidemic proportions. According to the Centers for Disease Control and Prevention (CDC), 68% of US adults and 32% of children aged 2–19 years are now either overweight or obese (Flegal et al. 2010; Ogden et al. 2010). Between 1980 and 2008 the rate of obesity doubled among children aged 2–5 years and tripled among children aged 6–11 years. Significantly, during the same time period, the number of obese adolescents increased from 5% to 18%. The numerous problems attendant to obesity are often perpetuated into adulthood, as obese children and adolescents are more likely to become obese as adults.

Both obesity and the struggle to lose weight are extremely complex, involving many factors in addition to diet, such as hereditary, environmental and psychological influences. The psychological burden of obesity is perhaps as damaging as the physical complications of the disease (Brownell 1984; Allon 1973). Obesity is, too often, unfortunately associated with laziness, lack of self control and lack of concern about physical appearance (Brownell and Wadden 1992; Allon 1973).

Today there is a growing awareness of the serious health risks related to obesity, including high blood pressure, diabetes, coronary heart disease and certain types of cancer. As a result, sensible dieting and fitness are increasingly important to many.

It is well established that losing weight is the result of consuming fewer calories than expended. Health experts agree that weight loss is best achieved by a combination of reducing caloric intake and increasing exercise/activity, but can be accomplished more slowly by a reduction in caloric intake or exercise alone. Weight management results from balancing food intake with energy expenditure.

Low-calorie sweeteners can play an important role in reducing calorie intake. "Light" foods and beverages containing low-calorie sweeteners, reduced in sugar and/or calories, provide good tasting products that are significantly lower in calories than their full-calorie counterparts. A broad range of good tasting light products is widely available.

According to Gallup Consumer Surveys, almost half of adults believe they are getting too much sugar in their diets (The Gallup Organization Inc. 2010). Consumers are most likely to look at both the calorie and sugar content on the nutrition label. In addition, about two-thirds of mothers believe their children are consuming too much sugar—which represents the single ingredient that mothers are most concerned about monitoring.

There are currently six low-calorie sweeteners approved for use in the United States alone. All provide no calories to foods and beverages and have been determined to be safe for use by the general population. They are among the most thoroughly tested ingredients, both individually and as a group, in the food supply. All have been carefully reviewed and deemed safe by the US Food and Drug Administration (FDA) and numerous other regulatory bodies and expert committees around the world, including the Joint Expert Committee on Food Additives of the World Health Organization and Food and Agriculture Organization (JECFA). The sweeteners have been determined safe by health organizations such as the American Diabetes Association (2002) and the American Dietetic Association (2004) as well. In addition, the American Dental Association (1998) has approved a position statement acknowledging the "Role of Sugar-Free Foods and Medications in Maintaining Good Oral Health." The low-calorie sweeteners currently available for use in the United States are acesulfame potassium, aspartame, neotame, saccharin, stevia sweeteners and sucralose.

Dr. James Hill et al. (2003) of the University of Colorado has reported that people are gaining an extra two pounds per year or 14 to 16 pounds over an eight-year period. He notes that a simple approach to preventing this weight gain is to cut out just 100 calories per day. This 100 calories per day can be cut by using reduced-calorie products in place of their full-calorie counterparts. Simply substituting a packet of low-calorie tabletop sweetener for sugar in coffee, on cereal or in ice tea three times a day, for example, is a savings of about 100 calories. Table 29.1 shows the number of calories that may be saved by substituting reduced-calorie products sweetened with low-calorie sweeteners for their full calorie counterparts.

Desire for Sweet Taste

The desire for sweet foods is not a recent phenomenon. There is historical evidence that humans have always had a preference for sweets. Examples include a 20,000-year-old cave painting of

Table 29.1   Calories Saved

Use	In Place of	Calories Saved
Sugar-free pudding (1/2 cup)	Pudding	70
Tabletop sweetener (one packet)	Sugar (2 tsp)	32
Sugar-free syrup (1/4 cup)	Syrup	180
Sugar-free preserves (1 tbsp)	Preserves	40
No-sugar added ice cream (1/2 cup)	Ice cream	60
Sugar-free gelatin (1/2 cup)	Gelatin	70
Light lemonade powered drink mix (8 oz.)	Lemonade powdered drink mix	55
Fat-free, light yogurt (8 oz. container)	Fat free yogurt	140
Diet soda (12 oz.)	Regular soda	150
Light cheesecake (1 slice)	Cheesecake	100
Light cranberry juice cocktail (8 oz.)	Cranberry juice cocktail	90

a man robbing a wild bees' nest and drawings of honey production in ancient Egyptian tombs (Pfaffmann 1977). Also, research has shown that the desire for sweets is inborn. Newborn infants have long been observed to react positively to sweetness, and a fetus at five months has been reported to respond positively to sweet-tasting stimuli (Maller and Desor 1973). Studies with adults, as well as infants, have demonstrated that the pleasant response to sweet is an innate, reflex reaction rather than a learned response (Pfafmann 1977; Maller and Desor 1973; Beidler 1975; Steiner 1973).

Some have claimed that low-calorie sweeteners perpetuate a desire for sweet foods and may stimulate appetite—thereby leading to increased food intake and weight gain. Such claims are essentially speculation and not well founded scientifically. Importantly, scientific evidence points to the opposite conclusion, that is, that low-calorie sweeteners do not stimulate appetite or food intake, nor do they cause weight gain. Low-calorie sweeteners and the reduced-calorie products containing them, in fact, may assist in reducing calorie intake and consequently weight.

Clinical Evidence: Body Weight

Scientific evidence supports the effectiveness of low-caloric sweeteners in helping control caloric intake. Historically, research was limited because: 1) low-calorie sweeteners are not drugs where efficacy can be readily determined clinically, and 2) until the approval of aspartame, it was almost impossible to devise a double-blind, crossover study using low-calorie sweeteners—because the taste of saccharin, the only low-calorie sweetener available for most of the 20th century, is distinct from that of sucrose, subjects could immediately detect its presence. The availability of aspartame made it possible to investigate more thoroughly the value of low-calorie sweeteners in weight management. Research now documents that consumption of low-calorie sweeteners can be useful in weight-loss efforts.

A six-week clinical study (Berryman et al. 1968) evaluated the effects of covert calorie dilution on body weight. Saccharin and cyclamate were covertly substituted for sugar in canned fruit and beverages in 25 subjects over a 41-day period. Male subjects lost an average of 3.7 pounds and females an average of 2.1 pounds, compared to an insignificant gain by the control group.

Kanders et al. (1988, 1990) evaluated the effect of aspartame on control of body weight in obese subjects. Fifty-nine obese men and women were recruited to participate in a pilot study with a 12-week multidisciplinary diet program. Recruits were randomly assigned to consume a nutrient balanced deficit diet (1000 ± 200 kcal/day) with or without aspartame. Although not statistically significant, women in the aspartame group ($N = 24$) lost 3.7 pounds more than women in the no-aspartame group ($N = 21$). Men showed the opposite trend, with those in the no-aspartame group ($N = 7$) losing about 4 pounds more than those in the aspartame group ($N = 4$).

Forty-six of these subjects (11 males) participated in a one-year follow-up study. Increased levels of physical activity, increased consumption of aspartame, and decreased desire for sweets were associated with maintenance of weight loss at the one-year follow-up. Aspartame intake at the end of follow-up was associated with better weight maintenance in male subjects. Although the small sample size prevents definitive conclusions, aspartame consumption did not cause weight gain and may be beneficial in promoting weight loss and maintenance when used as part of a multidisciplinary weight control program.

Blackburn et al. (1997) conducted a randomized, controlled, prospective clinical follow-up study to investigate whether the addition of aspartame to a multidisciplinary weight control program would improve weight loss and long-term control of body weight in obese women. One hundred and sixty-eight obese women, aged 20 to 60 years, were placed on a nutrient-balanced deficit diet (1000 ± 200 kcal/day) for three weeks. At the end of this time, the subjects were instructed to continue the balanced deficit diet and were randomly assigned either to consume aspartame-sweetened foods and beverages during the remaining 16 weeks of the active weight loss phase of the study, or to avoid such products. During the one-year weight maintenance phase and two-year follow-up periods, participants were encouraged to continue to consume or avoid aspartame-containing products according to their original group assignment.

During the active weight loss period, all subjects attended weekly one-hour sessions with instruction on behavioral and lifestyle strategies to facilitate weight loss. During the 12-month maintenance and the 19-month follow-up, the groups met monthly. Throughout the study, regular exercise, mainly walking, was strongly encouraged. Body weight, aspartame intake, exercise level and subjective ratings of hunger, desire for sweets and eating control were evaluated at baseline, 19, 71, and 156 weeks.

One hundred and thirty-six subjects completed the active weight loss phase; 125 subjects completed the maintenance phase; and 86 subjects completed the follow-up phase. Subjects in both treatment groups lost a mean of approximately 10% of body weight (10 kg) during the 19 weeks of active weight loss. Among subjects in the aspartame group, aspartame consumption was positively associated with weight loss. The desire for sweets decreased significantly in the aspartame group, but not in the no-aspartame group; hunger did not differ significantly from baseline in either treatment group, but eating control increased significantly in both treatment groups. Hunger and desire for sweets remained unchanged within both treatment groups during the maintenance phase (weeks 19–71). Eating control decreased significantly in both maintenance and follow-up phases in both groups, suggesting more uncontrolled eating during maintenance and follow-up.

At the end of the maintenance phase (week 71), subjects in the aspartame group experienced a 3.1% mean weight regain, and those in the no-aspartame group regained a mean of 4.9%. By the end of the follow-up phase (week 156), subjects in the aspartame group had regained an additional 2.4%, with a net weight loss from baseline of 5.1%. In contrast, subjects in the no-aspartame group had a gain of 5.4%, with a net weight loss of 0.3% from baseline. Significant predictors of better weight control from baseline to week 156 included increased exercise, increased self-reported eating control, and initial treatment group assignment, where aspartame group subjects had an advantage over the no-aspartame group subjects. The researchers concluded that aspartame, as part of a multidisciplinary weight control program, may facilitate weight control.

Aspartame was associated with weight loss in two additional multi-week studies. Morris et al. (1989) investigated low-calorie sweetener consumption patterns of 35 overweight individuals before and after completing a 16-week weight loss program. The program consisted of either a low-fat diet or a low-fat diet combined with regular exercise. Each subject's use of saccharin and aspartame was self-reported via food diaries.

At the end of the 16-week period, women lost more than 15 pounds and men lost more than 20 pounds, while consuming aspartame and saccharin. Women increased their intake of low-calorie sweeteners by 34% (from 281 to 377 mg/day) by the end of the study. The researchers concluded: "These results suggest that consumption of artificial sweeteners is not a barrier to weight loss and that foods containing artificial sweeteners can be incorporated into a weight-loss program" (Morris et al. 1989, p. 94).

Tordoff and Alleva (1990a) conducted a long-term study monitoring the diet records and body weights of 30 normal-weight adults during three separate periods, each lasting three weeks. During each period, the subjects consumed 40 ounces daily of aspartame-sweetened soda or high fructose corn syrup (HFCS)-sweetened soda or no experimental drinks.

The researchers observed that drinking aspartame-sweetened soda decreased the sugar and calorie intake of both sexes significantly compared to the control period. Consumption of aspartame-sweetened soda also led to a non-significant decrease in body weight in both sexes combined, while consumption of HFCS-sweetened soda resulted in a significant weight gain in both men and women.

The researchers noted that although some studies have associated low-calorie sweeteners with an increase in appetite and short-term food intake, this study showed that "drinking large volumes of aspartame-sweetened soda . . . reduces sugar intake and thus may facilitate the control of calorie intake and body weight" (Tordoff and Alleva 1990a, p. 963).

Rabin et al. (2002) conducted a 10-week study in which overweight subjects were given either supplemental drinks, or food containing sucrose or similar drinks and foods containing artificial sweeteners. Those consuming sucrose-containing products experienced increases in total energy intake, body weight, fat mass, and blood pressure. This was not observed in subjects consuming artificial sweetener-containing products. The authors concluded that the most likely reason for the differences between the groups was the use of large amounts of beverages, resulting in over consumption of energy on the high-sucrose diet and suggest that those overweight may want to consider choosing beverages containing artificial sweeteners, rather than sucrose, to prevent weight gain.

Rolls (1991) published a review of 45 studies examining the effects of low-calorie sweeteners on hunger, appetite, and food intake, which provides compelling evidence that low-calorie sweeteners are helpful in controlling weight. From her review, which included a thorough discussion of the major allegations related to the benefits of low-calorie sweeteners, Rolls (1991) concluded: "Preliminary clinical trials suggested that aspartame may be a useful aid in a complete diet-and-exercise program or in weight maintenance. Intense sweeteners have never been found to cause weight gain in humans" (p. 872). She added, "If the individual uses the consumption of a low-calorie food as an excuse to eat a high-calorie food, or if the individual is not actively trying to restrict intake, daily energy intake may remain unchanged. However, if intense sweeteners are part of a weight control program, they could aid caloric control by providing palatable foods with reduced energy. It needs to be stressed that there are no data suggesting that consumption of foods and drinks with intense sweeteners promotes food intake and weight gain in dieters" (p. 872).

An additional review further supports the benefits of low-calorie sweeteners in weight control. de la Hunty et al. (2006) reviewed 16 studies that examined the effect of replacing sugar with aspartame alone or aspartame in combination with other intense sweeteners on caloric intake or bodyweight. The study assessed the amount of energy compensation and whether the use of aspartame-sweetened food and drinks is an effective way to lose weight. Significant reductions in energy intakes with aspartame compared with all types of controls except with aspartame as compared with non-sucrose controls such as water were found. de la Hunty and colleagues (2006) state "the meta-analyse demonstrate that using foods and drinks sweetened with aspartame instead of sucrose, results in a significant reduction in both energy intakes and body weight" (p. 116). Weight reduction of approximately 0.2 kg/wk was common for individuals using aspartame-containing foods and beverages.

Bellisle and Drewnowski (2007) reviewed evidence from laboratory, clinical and epidemiological studies in the context of current research on energy density, satiety, and the control of food intake. The authors state a long known fact—low-calorie sweeteners are not appetite suppressants. However, in randomized controlled trials, the gold standard of research, low-calorie sweeteners have been shown to be associated with some modest weight loss. The authors conclude that the ultimate effect of low-calorie sweeteners is dependent upon their integration into a reduced-calorie diet.

Magnuson (2007) conducted the largest comprehensive review of aspartame data to date. Experts in the fields of toxicology, epidemiology, biostatistics, metabolism, pathology, neurology, and pediatrics spent nine months examining over 500 aspartame studies. Studies claiming a link between aspartame consumption and weight gain were part of the review. Following their critique of aspartame data, the researchers concluded, "Aspartame is a well-characterized, thoroughly studied, high intensity sweetener that has a long history of safe use in the food supply and can help reduce the caloric content of a wide variety of foods" (Magnuson 2007, p. 702).

Mattes and Popkin (2008) reviewed 224 professional studies on the effects of non-nutritive sweeteners on appetite, food intake, and weight. The researchers concluded: "taken together, the evidence summarized by us and others suggests that if non-nutritive sweeteners are used as substitutes for higher-energy-yielding sweeteners, they have the potential to aid in weight management."

An additional study published by Phelan et al. (2009) compared dietary strategies, including the use of low-calorie sweeteners in diet beverages, among weight loss maintainers (WLM) and those who had always been a normal weight. Researchers analyzed the food intake and dietary restraint of more than 300 individuals over three separate 24-hour periods, and found that WLM use more dietary strategies to accomplish their weight loss maintenance, including increased consumption of artificially sweetened beverages. Phelan (2009) concluded: "Our findings…suggest that the use of artificially sweetened beverages may be an important weight control strategy among WLM (weight loss maintainers)" (p. 1188). The authors also noted: "The current study suggests that WLM use more dietary strategies to accomplish their WLM, including greater restriction of fat intake, use of fat and sugar modified foods, reduced consumption of caloric beverages, and increased consumption of artificially sweetened beverages" (p. 1189).

Brown et al. (2010) published a review study of the metabolic effects of low-calorie sweeteners in children, concluding the "jury is still out" regarding the effects of low-calorie sweeteners on obesity, diabetes, and glucose metabolism in children. The article is based on just 18 out of a possible

116 studies (from the initial literature search) and half of these were observational in nature. Only three of the studies evaluated were randomized control trials. In her conclusion Brown compares animal research using low-calorie sweeteners to that in humans stating: "Our growing understanding of the active metabolic role played by such chemicals in animal models should spur further research into the effects of these common food additives in humans" (Brown et al. 2010, p. 310).

The existing clinical evidence contradicts any suggestion that low-calorie sweeteners cause people to overcompensate for the calories saved in a given meal by eating more in a later meal. Additionally, extensive clinical research shows that low-calorie sweetener use does not result in increased calorie intake or lead to weight gain and may assist with weight loss and control.

Other Studies Assessing Food/Caloric Intake

Clinical evidence indicates that low-calorie sweeteners are effective in limiting calorie intake. In pioneering work in this area, Porikos and colleagues investigated the effect of covert caloric dilution on food intake in three separate clinical studies (Porikos et al. 1977, 1982; Porikos and Pi-Sunyer 1984; Porikos and Van Itallie 1984). In one study, the diet of six normal-weight male subjects was reduced calorically by 25% by the covert substitution of aspartame-sweetened analogs for all menu items containing sucrose (Porikos et al. 1982). Although the subjects compensated somewhat for the caloric dilution, their caloric intake stabilized at 15% below their normal intake. The subjects did not show a shift in either sweetened or unsweetened food choices while their diet was being diluted, which contradicts claims that low-calorie sweeteners may encourage a desire for sweets.

In another study conducted by Porikos et al. (1977), the covert substitution of aspartame for sucrose in the diets of obese adults resulted in a 25% reduction in caloric intake. In a third caloric dilution study of obese and normal weight subjects, Porikos reported that obese subjects reduced their caloric intake by 16% compared to a baseline diet, and normal weight subjects reduced caloric intake by 16% (Porikos and Pi-Sunyer 1984; Porikos and Van Itallie 1984). Dr. Porikos concluded from her research that foods and beverages containing low-calorie sweeteners can offer an effective approach to dieting, noting: "They allow for a reduction in energy intake without alteration in taste and only minor changes in volume of diet. A dietary regimen which includes low-calorie versions of people's favorite foods, particularly sweets, should encourage compliance" (Porikos and Van Itallie 1984, p. 278).

Rolls also has conducted research regarding the effects of low-calorie sweeteners on caloric intake and hunger. In a study of 42 normal-weight men, the subjects were given 8 to 16 ounces of lemonade, sweetened to equal intensity with either aspartame or sucrose, or the same volumes of water, or no drink (Rolls et al. 1990). Subjects were separated into three groups receiving the drinks at different times: with a self-selection lunch or 30 or 60 minutes before lunch. Researchers found that there was no instance in the three experiments in which hunger ratings or intake under the aspartame versus water conditions differed, and concluded: "Thus these data do not support the hypothesis that aspartame-sweetened drinks increase food intake" (Roll et al. 1990, p. 25).

In a study conducted by Drewnowski et al. (1994a), the effects of four breakfast preloads on hunger ratings, energy intakes and taste responsiveness profiles were examined in 24 normal-weight adults. The breakfasts consisted of 400 g of creamy white cheese ("fromage blanc") with maltodextrin or water, and differed in energy value (700 kcal vs. 300 kcal) and the nature of the sweetener. High-calorie breakfasts were sweetened with sucrose or with aspartame, while low-calorie ones contained aspartame or were not sweetened at all. Daily energy intakes following breakfast were the same for all four breakfasts. No calorie compensation was observed: subjects given 300 kcal breakfasts had lower total daily intakes than when given 700 kcal. The researchers concluded: "These data do not support the notion that intense sweeteners increase hunger or result in increased energy intakes in normal-weight subjects" (Drewnowski et al. 1994a, p. 344).

Rogers et al. (1988) have speculated that low-calorie sweeteners can affect appetite and caloric intake via postingestive effects. According to this theory, when low-calorie sweeteners bypass the sweet taste, i.e., when administered in capsule form, they affect certain hormones involved in appetite regulation.

In two separate clinical studies, these researchers investigated the effects of encapsulated aspartame on motivation to eat and caloric intake. In the first study, food intake was measured in 27 normal-weight individuals following preloads of 234 or 470 mg of encapsulated aspartame, 234 mg of aspartame dissolved in water, or a placebo (control) (Rogers et al. 1990).

The encapsulated aspartame reduced motivation to eat and significantly reduced calorie intake one hour later compared with both the aspartame solution and control preloads. The 234 mg dose reduced intake by between 9% and 14% (138 and 175 calories). The 470 mg dose had a similar effect, reducing intake by 150 calories. The aspartame solution and control preloads did not significantly alter motivational ratings or food intake. It should be noted that Blundell's data in the study regarding motivation to eat are contradicted by those of his previous two studies where he reported that aspartame in water stimulates appetite. The researchers concluded: "The results provided clear evidence of a predominant postingestive inhibitory action of aspartame on appetite . . ." (Rogers et al. 1990, p. 1241).

In a subsequent double-blind study, Blundell and his colleagues (Rogers et al. 1991) measured hunger and food intake in 16 adults following a preload of encapsulated aspartame or its breakdown components, L-aspartic acid or L-phenylalanine. On the same day for four consecutive weeks, the subjects were given either 200 mg L-aspartic acid, 200 mg L-phenylalanine, 400 mg aspartame or placebo one hour before a self-selected test meal. None of the treatments had a significant effect on hunger, either before or following the test meal. However, the aspartame treatment reduced food intake at the test meal by 15% compared with the placebo, while aspartame's components had no significant effect on intake. Aspartame also did not result in a rebound increase in hunger in the post meal interval (3.5 hours). Noting the absence of increased hunger despite the reduced intake, the researchers concluded: "This suggests that aspartame may act to intensify the satiating effects of ingested food" (Rogers et al. 1991, p. 739).

Black et al. (1993) compared the effects on appetite and food intake of different volumes of beverage, beverages with aspartame in solution, and beverages with aspartame in capsules. In contrast to Rogers et al. (1990), Black et al. (1993) reported that aspartame in capsules had no effect on appetite. Furthermore, the researchers concluded that appetite reduction after consumption of an aspartame-sweetened beverage is likely due to the volume of the drink and not the aspartame.

Numerous other studies, utilizing various methodologies, have evaluated the effect of aspartame on hunger, appetite and food intake. Replacing sucrose with aspartame in foods or beverages has not been shown to increase food intake or hunger in children (Anderson et al. 1989; Birch et al. 1989) and has not been shown to increase food intake in normal weight (Blundell and Hill 1987; Rolls et al. 1989, 1990; Black et al. 1991; Canty and Chan 1991; Drewnowski et al. 1994a, 1994b) or in overweight men and women (Rodin 1990; Drewnowski 1994b). Interestingly, all of these studies reported either unchanged or reduced motivation to eat regardless of whether the aspartame was delivered in a solid or liquid form.

Wilson (2000) compared the effect of plain milk, sucrose-sweetened milk, and aspartame-sweetened milk on mealtime caloric intake in young children. Children consumed more sweetened milk than plain milk. However, the researchers found that young children do not reduce caloric intake at a meal to compensate for the extra calories resulting from sucrose-sweetened milk, whereas aspartame increased milk consumption without providing the extra calories of sucrose-sweetened milk.

Ludwig et al. (2001) found no association between diet soft drink consumption and obesity incidence in children. Ludwig et al. enrolled 548 ethnically diverse children (aged 7–11 years) in a prospective study for 19 months. The researchers observed consumption patterns of diet soft drinks and other lifestyle variables. At the end of the study, Ludwig and colleagues concluded that diet soft drink consumption was not related to obesity incidence among the children followed in the study.

A study, conducted by Rodearmel et al. (2007), examined the effectiveness of moderate calorie reduction and exercise for preventing excessive weight gain in overweight children aged 7–14 years. More than 200 families with at least one overweight child participated in the six-month study. Half of the families were placed on a program that eliminated 100 calories per day using reduced-calorie products containing sucralose (Splenda®) and added 2,000 steps a day. After six months, two out of three of the children following the program lost or maintained weight, while half of the children in the control group increased their body mass index (BMI). The researchers concluded that mild calorie reduction (using low-calorie or reduced calorie products), combined with a moderate increase in physical activity, was effective in preventing weight gain in overweight adolescents.

Anton et al. (2010) compared the effects of preloads containing aspartame, stevia or sucrose on food intake, hunger, blood glucose, and insulin levels in both healthy and overweight individuals. Participants completed three separate food test days (with intake directly measured) during which they received preloads containing one of the sweeteners before lunch and dinner. Those who received the stevia or aspartame consumed fewer calories overall, did not overeat, and did not report increased feelings of hunger. Both aspartame and stevia resulted in reduced post meal glucose levels compared to sucrose. Post meal insulin levels were significantly reduced following the stevia preload compared to both sucrose and aspartame. The researchers noted: "In conclusion, participants did not compensate by eating more at either their lunch and dinner meal and reported similar levels of satiety when they consumed lower calorie preloads containing stevia and aspartame than when they consumed higher calorie preloads containing sucrose" (Anton et al. 2010, p. 42).

Studies on aspartame, appetite, and food intake have been reviewed in detail by Rolls (1991), Renwick (1994), Drewnowski (1995), and Rolls and Shide (1996). As Rolls and Shide (1996) concluded: "From evaluation of the available data, there is no consistent nor compelling evidence that the intense sweetener aspartame increases food intake or body weight" (p. 285).

Research on Hunger and Appetite

Research also has been conducted investigating the effect of low-calorie sweeteners on hunger and appetite. A human study conducted by Blundell and Hill (1986), reported in a letter to the editor of *The Lancet*, investigated the effects of aqueous solutions of sugar or aspartame or plain water on feelings of hunger versus fullness. An orally administered aspartame solution (162 mg/200 ml) reportedly was found to increase ratings of motivation to eat and decrease ratings of fullness. The researchers only measured subjective feelings of hunger and not whether food intake actually increased following the subjects' reports of "residual hunger" after aspartame ingestion; no food was offered to test actual behavior. The researchers do not state whether their study was performed under double-blind, controlled conditions. Blundell and colleagues (Rogers et al. 1988) completed further research suggesting that solutions of saccharin and acesulfame potassium also increase hunger ratings compared with water, but none of these sweeteners was found to increase food intake one hour later. It should be noted that the same researchers (Rogers et al. 1990) failed to replicate their findings of an appetite-stimulating effect of an aqueous aspartame solution in a later study.

Rogers and Blundell (1989) extended their investigations to examine food in a clinical study, which measured the effects of saccharin on hunger and food intake. The normal-weight subjects consumed a fixed amount of yogurt sweetened either with saccharin or glucose, with no sweetener, or with starch and saccharin, followed by lunch one hour later. Food intake was significantly greater throughout the day following consumption of the saccharin-sweetened yogurt (without starch) compared with the plain, unsweetened yogurt. The researchers theorized that the increase in intake may have been related to the sweetness of the yogurt and not to the saccharin itself, since the yogurt containing saccharin plus starch did not lead to greater intake than the equicalorie glucose-sweetened yogurt.

Appleton and Blundell (2007) further reviewed the affect of low-calorie sweetened beverages on short-term appetite. The researchers gave 10 women (habitual users of low-calorie sweetened beverages) and 10 women (non-habitual users of low-calorie sweetened beverages) various preloads (non-sweet/low-energy, sweet/low-energy and sweet/high-energy). The researchers then measured the test meal intake, total intake and subjective perceptions of appetite. Non-habitual users of low-calorie sweeteners had an increased appetite compared with habitual users, however there was no difference in the effect of energy on appetite between the two groups. The researchers noted: "The lack of response to sweet taste in high consumers of artificially-sweetened beverages can be explained as a result of the repeated experience of sweetness without energy by these consumers. This lack of response suggests an adaptation to sweet taste as a result of the habitual dietary pattern" (Appleton and Blundell 2007, p. 479).

Tordoff and Alleva (1990a) reported in their study of the effects of sweetness without calories on hunger that chewing an unflavored gum base with varying concentrations of aspartame increased ratings of hunger, but not proportionately with increases in aspartame concentration. Instead, subjects rated themselves hungrier after chewing moderate (0.3% or 0.5%) rather than high (1.0%) concentrations of aspartame. As the researchers did not measure food intake, one cannot infer from this study that aspartame's sweet taste can affect subsequent eating behavior. Noting several variables in this study, which affected ratings of hunger, the researchers concluded: "The results of this study point out the fragility of the influence of sweetness on appetite" (Tordoff and Alleva 1990a, p. 558).

Other research argues against the claim that low-calorie sweeteners may increase appetite. Rolls (1987) investigated the effect of aspartame versus sucrose on hunger and satiety. In a study of normal weight adults who were offered gelatin sweetened covertly with either sucrose or aspartame, subjects ate a constant weight of food despite the difference in calories. Hunger was suppressed equally by both desserts over the past hour, but there was no compensation for the caloric difference when additional food was offered an hour later. Rolls concluded: "It is clear that low energy sweeteners can be as satisfying as sugars during a meal, particularly when the subjects are unaware of the caloric manipulation" (Rolls 1987, p. 171).

In a subsequent study, Rolls et al. (1989) covertly substituted aspartame for sugar in gelatin or pudding preloads served to normal-weight adults two hours before a self-selection meal. Half of the 32 subjects were aware of the caloric manipulation, half were not. Both informed and uninformed subjects consumed significantly fewer calories from the aspartame preloads compared with the sucrose preloads. In the two-hour period between the preload and the test meal, both the low and high-calorie versions of the test foods equally suppressed ratings of hunger and the desire to eat.

Total intake of the preload and subsequent meal did not differ significantly between the sucrose and aspartame conditions. Awareness of the low caloric content of the aspartame-containing preloads did not cause these subjects to eat more at the subsequent meal. The researchers concluded: "Aspartame-sweetened foods can be of benefit in reducing hunger and increasing satiety" (Rolls et al. 1989, 126).

Using a different study method, Mattes (1990) evaluated the effects of sucrose and aspartame on hunger and food intake by keeping the calories and rated pleasantness of a test meal constant. The study's 24 normal-weight subjects consumed equicaloric breakfasts of either unsweetened cereal (control) or cereal sweetened with sucrose or aspartame. Half the subjects were aware of the cereal composition, half were uninformed. There were no significant differences in hunger ratings up to three hours following the meal and no significant difference in food intake throughout the day. The authors concluded. "The present data indicate that the use of aspartame in foods does not stimulate energy intake … " (Mattes 1990, p. 1043).

A study of Leon et al. (1989) investigated the effects of aspartame in 108 adults who received either 75 mg/kg of encapsulated aspartame (equivalent to approximately 10 liters of aspartame-sweetened beverage) or a placebo daily for 24 weeks. The study revealed no significant change in body weight of participants consuming aspartame. Citing numerous studies which failed to show any correlation between low-calorie sweetener intake and hunger or weight gain, the researchers refuted Blundell and Hill's claim that aspartame may increase appetite and concluded: "Taken together, these studies argue against a 'paradoxical' effect of aspartame on appetite" (Leon et al. 1989, p. 2323).

The effects of low-calorie sweeteners consumed in beverages on hunger and calorie intake were investigated in two studies. In a study by Black et al. (1991), the aspartame-sweetened preload was given three hours after a standard breakfast. The researchers noted the importance of controlling food intake prior to a test load as this may mask the effects of a low-calorie sweetener on subsequent food intake. The 20 normal-weight subjects consumed either 12-ounces or 24-ounces of an aspartame-sweetened soft drink, or water, one hour before a test lunch. Consumption of the 24-ounce preload significantly reduced hunger in the one-hour period following the preload, compared with mineral water and the 12-ounce preloads. None of the preloads had any effect on calorie intake. The researchers concluded: "(these) results add support to the growing body of evidence indicating that the consumption of foods and beverages containing nonnutritive sweeteners do not increase hunger and food intake" (Black et al. 1991, p. 810).

In a similar study by Canty and Chan (1991), the effects of saccharin as well as aspartame were evaluated. Twenty normal-weight subjects consumed approximately seven ounces of water or soft drink sweetened with saccharin, aspartame or sucrose, or water three hours following a standard breakfast and one hour before ad libitum consumption of a standard lunch. Overall, consumption of saccharin and aspartame did not increase hunger or food consumption compared with water. In fact, hunger ratings in the hour between preload consumption and lunch were generally highest for water, followed by aspartame, saccharin and sucrose. The researchers also found no significant differences in calorie intake associated with any of the preloads.

Research on the possible neurochemical effects of aspartame and phenylalanine (an amino acid component of aspartame) on hunger and food intake (specifically the balance of protein, carbohydrate and fat) was reported by Ryan-Harshman et al. (1987). This research involved two separate double-blind tests in normal weight men. In the first test, subjects were given varying levels of phenylalanine in capsule form one hour before lunch. In the second test, doses of both phenylalanine and aspartame were administered 105 minutes before eating. Tests using visual analog scales were administered at intervals before and after eating to assess subjective feelings of hunger and mood. Blood samples were taken from some subjects for plasma amino acid analysis.

The researchers found that plasma phenylalanine levels and ratios to competing amino acids rose significantly with all treatments except the lowest phenylalanine doses. However, in all doses studied, including a dose of 10 g (10,000 mg), neither aspartame nor phenylalanine affected calorie intake nor did they significantly affect intake ratios of protein, carbohydrate or fat. Also, no consistent relationship existed between plasma amino acid levels and food intake among the individuals who participated in the biochemical analysis. The researchers concluded that, in the doses studied, aspartame and phenylalanine "do not affect short-term energy and macronutrient intakes, and subjective feelings of hunger . . . ." (Ryan-Harshman et al. 1987, p. 247).

Insulin and Hunger

Some researchers have suggested that sweetness, including that provided by low-calorie sweeteners, increases hunger and attribute this to a proposed effect on insulin response (Rodin 1984a, 1984b; Remington 1985). Claims have been made that the body responds to low-calorie sweeteners as it does to sugar by triggering insulin release, which lowers blood sugar and stimulates hunger (Rodin 1984a, 1984b; Remington 1985). The claim has little basis in science.

Human studies have demonstrated that low-calorie sweeteners do not significantly affect insulin levels (Goldfine et al. 1969; Ambrus et al. 1976; Okuno et al. 1986; Shigeta et al. 1985; Carlson and Shah 1989; Colagiuri and Shah 1989; Horwitz et al. 1988). Although some studies have indicated an initial insulin response in animals and humans given orally administered saccharin solutions (Von Borstel 1985; Louis-Sylvestre 1976; Berthoud et al. 1981), two phases of insulin release normally occur when carbohydrates are consumed. The first, cephalic phase is marked by an immediate small rise in plasma insulin and accounts for only a small portion of the total insulin released. Von Borstel (1985) reported that this initial insulin release could be triggered by the mere taste, smell or other sensory qualities of food. The second phase begins several minutes later and lasts until blood sugar is normalized (Ganong 1973).

Early research indicated that only the cephalic phase of insulin secretion might be observed in animals and humans consuming low-calorie sweeteners, and then only some of the time (Von Borstel 1985). Bruce et al. (1987) failed to observe a cephalic phase insulin release in human subjects given aspartame-sweetened water.

Härtel et al. (1993) examined five test solutions (one each containing aspartame, acesulfame potassium, cyclamate, saccharin or sucrose) and water in a multiple crossover study. Sucrose resulted in a significant increase in plasma insulin and blood glucose concentrations compared to the low-calorie sweeteners and water. Generally, glucose concentrations were within the normal range, and there were no significant differences in plasma insulin or blood glucose concentrations with the four low-calorie sweeteners compared to water. The authors concluded that there is no cephalic insulin secretion with low-calorie sweeteners.

Teff et al. (1995) conducted a randomized study using one and three minute oral exposures in random order to solutions of aspartame, saccharin, and sucrose and water, which were expectorated. Apple pie was used as a modified sham-feed condition. There were no significant increases in plasma insulin concentrations after any of the sweetened solutions, but the apple pie resulted in significant increases in plasma insulin concentrations after both the one and three minute exposures. The taste of sweetened solutions alone, therefore, did not stimulate cephalic-phase insulin release.

Abdallah et al. (1997) conducted a randomized, double-blind, placebo-controlled study to evaluate cephalic-phase insulin release with oral exposure (sucking on the tongue) to three different tablets, tablets containing 3 g of sucrose, 18 mg of aspartame plus 3 g of polydextrose (a non-sweet carbohydrate), or 3 g of polydextrose as a placebo. Plasma glucose, insulin, and glucagon concentrations were not changed after aspartame or sucrose tablets, suggesting that sweet taste alone was not sufficient for eliciting cephalic-phase insulin release.

Coulston and Gori (2002) conducted a thorough review of aspartame research and concluded the results of these studies demonstrated no cephalic-phase insulin release induced by aspartame. The authors also noted the sweetener "does not affect the glycemic response in normal or diabetic subjects. In addition, aspartame does not affect metabolic control when administered to diabetic subjects daily for up to 18 weeks" (Coulston and Gori 2002, s81).

Low-calorie sweeteners have not been found to trigger the second phase of insulin release, which accounts for the majority of total insulin released in response to carbohydrate consumption.

Rodin (1990) subsequently conducted research to investigate her theory that low-calorie sweeteners stimulate appetite. In a study of 24 normal and overweight subjects, caloric intake and insulin levels were measured following a preload solution of fructose, glucose or aspartame, or water. Overall, there was no significant difference in intake following the aspartame, glucose or water preloads. Aspartame also had no effect on blood glucose or insulin levels. Rodin (1990) concluded: "The data on aspartame showed that, overall, an aspartame-sweetened drink does not lead to significantly greater subsequent food intake than does a preload of plain water. Thus aspartame, at least under present conditions, does not appear to have the stimulating effect on food intake that Blundell et al. suggested" (p. 434).

Additional recent clinical studies further refute the insulin theory by showing no effect on insulin levels following aspartame and/or saccharin administration. Researchers Carlson and Shah (1989) examined the effects of encapsulated aspartame or its constituent amino acids, as well as solutions of aspartame, on levels of blood glucose, insulin, and other hormones in 16 normal-weight adults. The researchers found no significant changes in blood levels of the hormones monitored. Interestingly, all tests resulted in slightly lowered insulin levels followed by a rise in blood glucose levels back to baseline, which the authors noted may represent normal post-meal alterations in these subjects. Carlson and Shah (1989) concluded: "The present results, showing no effect of aspartame or its constituent amino acids on serum glucose or insulin in normal subjects, confirm and extend several previous reports demonstrating no change in these measures in diabetic and normal subjects after oral administration of aspartame" (p. 431). They added: "These findings confirm in humans the previously reported lack of endocrine disturbance in rats given aspartame" (p. 431).

Horwitz et al. (1988) evaluated the effects of low-calorie sweeteners on blood glucose and insulin levels, which were measured in 12 normal subjects and 10 subjects with non-insulin-dependent diabetes mellitus. Subjects were given one of three fruit-flavored drinks at weekly intervals: one unsweetened, one sweetened with 400 mg of aspartame, and one sweetened with 135 mg of saccharin. The aspartame and saccharin drinks were comparable in sweetness and contained approximately the same amount of sweetener found in one liter of a sugar-free soft drink. Fasting blood glucose and insulin levels were measured before and for three hours following the test. The researchers found no significant effect on glucose levels at any time in either the normal or diabetic group. For insulin values, the only treatment effect was in normal subjects 15 minutes following the aspartame test, when insulin levels rose slightly relative to values found after the saccharin-sweetened or the unsweetened beverages.

The authors noted: "The magnitude of (this) difference was small and unlikely to be of physiological importance in the absence of differences in glucose levels." The researchers concluded: "[I]ngestion of aspartame- or saccharin-sweetened beverages by fasting subjects, with or without diabetes, did not affect blood glucose homeostasis" (Horwitz et al. 1988, p. 230).

Dhingra et al. (2007) published research alleging an association between impaired insulin response and consumption of low-calorie sweetened soft drinks. The study was an observational study, however, and did not find a causal relationship between diet soft drink consumption and impaired insulin response. The study authors noted, "Individuals with greater intake of soft drinks also have a dietary pattern characterized by greater intake of calories and saturated and trans fats, lower consumption of fiber and dairy products and a sedentary life" (p. 485) Such behaviors are known factors related to weight gain, heart disease and metabolic syndrome. Further, the American Heart Association released a statement in regard to this study and noted: "Since this is an observational study, it is important to note that the study does not show that soft drinks cause risk factors for heart disease. Diet soda can be a good option to replace caloric beverages that do not contain important vitamins and minerals. The American Heart Association supports dietary patterns that include low-calorie beverages like water, diet soft drinks, and fat-free or low-fat milk … " (AHA 2007).

Another study by Lutsey et al. (2008) alleged an adverse association between diet soda and metabolic syndrome. The researchers analyzed data from the Atherosclerosis Risk in Communities study. The authors noted: "Although prospective study designs establish temporal sequence, it is possible that reverse causality or residual confounding may explain this finding [diet soda], especially because consumption of diet soda is higher among diabetics than among nondiabetics" (Lutsey et al. 2008, p. 759). The study authors offer no definitive conclusions or reasons as to why diet soda would increase the risk of metabolic syndrome. Neither the study by Lutsey et al. (2008) nor Dhingra (2007) show causality as the studies are observational in nature. Also, because some participants in the Lutsey et al. study began the study with metabolic syndrome or developed it over the course of the study, it is impossible to determine if diet soda consumption is related to an increased risk of developing metabolic syndrome or if participants used diet soda to help manage their weight, diabetes, and so forth. The Lutsey et al. study did not control for weight gain, which is related to the development of metabolic syndrome, nor did they exclude overweight individuals from the study.

Grotz et al. (2003) conducted a randomized, double-blind, placebo-controlled study to evaluate the effect of sucralose on glycemic control in individuals with type 2 diabetes. Hundred and twenty eight adults, most defined as obese, completed the 13-week study in which they randomly received either a placebo or 667 mg encapsulated sucralose on a daily basis. The level of sucralose administered was equivalent to 7.5 mg/kg/day, approximately three times the estimated maximum intake. The study revealed no significant differences between the sucralose and placebo groups in fasting plasma glucose levels and no clinically significant differences in safety effects between the two groups. The researchers concluded: "In sum, sucralose-sweetened foods and beverages may be useful tools in the dietary management of individuals with, or at risk of, diabetes" (Grotz et al. 2003, p. 1612).

Renwick and Molinary (2010) published a review article that analyzed numerous studies to evaluate the metabolic effects of low-calorie sweeteners and, specifically, how these ingredients might affect intestinal sweet-taste receptors (e.g., GLP-1). Renwick and Molinary investigated the claim of a synergistic effect between sucralose and glucose, which reportedly triggered the release of GLP-1 in short-term studies involving animal and human cell lines, as well as genetically-modified laboratory mice. The article found no evidence to support this claim and the researchers concluded: "There is no consistent evidence that low-energy sweeteners increase appetite or subsequent food intake, cause insulin release or affect blood pressure in normal subjects" (Renwick and Molinary 2010, p. 1415).

Epidemiology

Epidemiological research also has investigated the benefits of low-calorie sweeteners for weight control. Stellman and Garfinkel (1986) noted the effect of low-calorie sweeteners on self-reported, one-year weight change in a select sub-sample of women enrolled in a prospective mortality study. Analysis was confined to white women aged 50 to 69 with no history of diabetes, heart disease or cancer, and no major change in diet in the past 10 years. According to some media reports, the results showed that use of low-calorie sweeteners may lead to weight gain. The researchers concluded that the study's long-term users of low-calorie sweeteners were more likely to gain weight than non-users (an average difference of less than two pounds) and that the rate of weight gain among users was significantly greater than in non-users. In contrast, the researchers found that among the most obese group, more users than non-users lost at least 10 pounds. Also, they left unexplained the finding that low-calorie sweetener users in the two heaviest weight groups lost more weight than non-users. Interestingly, the authors themselves have noted that artificial sweetener users who lost weight, or whose weight did not increase, could have gained weight had they not consumed the sweeteners (Stellman and Garfinkel 1988).

Lavin et al. (1994) have criticized this study for both methodological flaws in experimental design and statistical analysis. Review of this research discloses flaws in the methodology of the study, which raise serious questions concerning the reported interpretations with respect to weight gain. First of all, the study was designed to investigate cancer incidence, not obesity. Also, the researchers themselves recognized that a beneficial effect, if one exists, might best be demonstrated among short-term users of low-calorie sweeteners or among persons whose low-calorie sweetener consumption is coupled with major changes in dietary behavior, groups deliberately excluded from the analysis of Lavin et al. (1994). Exercise also was excluded from the analysis. By excluding women with major changes in diet and/or a significant level of exercise, the researchers appear to have made the assumption that mere consumption of low-calorie sweeteners will guarantee weight loss or prevent weight gain.

In addition, the subjects' actual consumption of low-calorie sweeteners, as well as their reported current weight and weight one year prior to the study, were not clinically measured but were based on self-reported questionnaires. Stellman and Garfinkel (1988) acknowledged the difficulty of conducting well-controlled efficacy studies, and commented that the high cost of this type of research makes future large-scale clinical efficacy studies unlikely.

In the 1950s, prior to the widespread popularity of low-calorie sweeteners, McCann et al. (1956) studied 247 obese individuals who had participated in weight reduction programs, as well as 100 diabetic patients. In a three-year follow-up study, the obese individuals were questioned about their use of low-calorie sweeteners. Based on the questionnaires, the researchers found no significant difference in weight loss between users and non-users in either the obese or diabetic groups. They noted, however: "it is well to remember that it is the individual who is losing weight, not the group." They added, "(T)he use of non-calorie sweeteners may have a place in the diets of some obese individuals" (McCann et al. 1956, p. 329).

Parham and Parham (1980) conducted a smaller observational study investigating the physiological benefits of low-calorie sweeteners. The researchers investigated the effect of saccharin use on sugar consumption in college students and found that use of saccharin resulted in reduced caloric intake. The researchers concluded that, using the criteria of lowered sugar or calorie intake, "saccharin is mildly beneficial to healthy consumers who are watching their weight or trying to limit their sugar intake for other reasons" (Parham and Parham 1980, p. 563).

Researchers Smith and Heybach (1988) compared aspartame consumption patterns and caloric intakes of 1500 women, utilizing data from a USDA food and nutrition survey. Twenty-five percent of the women questioned had consumed aspartame on the survey day. The mean caloric intake for aspartame users was a significant 165 calories/day less than for non-users; for women aged 35–50 the difference was 215 calories.

Fowler et al. (2005) presented a paper at the 65th Scientific Session of the American Diabetes Association alleging a relationship between the increase in "light product," specifically diet soda, usage and overweight adults. This unpublished epidemiology study raised more questions than it answered. The study, evaluated seven to eight years of data on 1550 Mexican American and non-Hispanic white Americans aged 25–64 years. Approximately 620 participants were involved in the study. The researchers' conclusions are, in some instances, contradictory to the data presented. The presentation even questioned the studies relevance to diet soft drinks stating: "A study of this kind does not prove that diet soda causes obesity." There was no information available as to whether individuals became overweight before or after they began drinking diet soda (Fowler et al. 2005).

A second study by Fowler et al. (2008), published in the journal *Obesity,* alleges a relationship between "artificially sweetened beverages" and weight gain. However, the authors concluded: "There may be no causal relationship between AS [artificial sweetener] use and weight gain. Individuals seeking to lose weight often switch to ASs in order to reduce their caloric intake. AS might therefore simply be a marker already on weight-gain trajectories, which continued despite their switching to ASs" (Fowler et al. 2008, p. 1898).

Nettleton et al. (2009) analyzed associations between diet soda consumption and risk of incident metabolic syndrome, its components, and type 2 diabetes, utilizing a food frequency questionnaire. Results allege a link between diet soda and components of metabolic syndrome, however, after excluding participants with any existing component of metabolic syndrome, no significant association between diet soda and metabolic syndrome was found. The study authors note: "We are cautious not to conclude causality between diet soda and the diabetic or pre-diabetic condition. The possibility of confounding by other dietary and lifestyle/behavioral factors cannot be excluded from these observational studies" (Nettleton et al. 2009, p. 692).

Animal Research

The effects of low-calorie sweetener consumption on short-term caloric intake also have been studied in animals. Animal research conducted by Tordoff and Friedman (1989a, 1989b, 1989c, 1989d) investigated the hypothesis that consumption of saccharin may hamper attempts to restrict food intake. In the study, rats reportedly were given either a glucose solution, saccharin solution or nothing, followed by a paired flavored food. The rats were found to prefer food paired with a sweet drink regardless of the sweetener used. Rats consuming saccharin reportedly increased their short-term food intake (compared to a maintenance diet), while those consuming glucose did not significantly alter calorie consumption. No long-term effects were observed.

Tordoff and Friedman have noted that a component of the increased food intake in rats consuming saccharin is a learned response. They have offered an additional explanation for their findings, stating that the increase in feeding might be explained as a balancing of solid intake and fluid intake by the rats consuming saccharin. While these are interesting hypotheses, of significance is the fact that the rats given saccharin did not demonstrate any long-term effects, that is, they did not gain weight.

Research conducted by Porikos in rats, contradicts claims that low-calorie sweeteners are ineffective for weight control. In addition, Porikos' animal research confirms her clinical findings that low-calorie sweeteners are effective in helping reduce calorie intake (Porikos and Pi-Sunyer 1984; Porikos and Koopmans 1988). Her research in rats found that long-term covert calorie dilution with aspartame and saccharin led to weight loss (Porikos and Koopmans 1988). In this study, 81 rats were studied for the effects of the consumption of sucrose versus aspartame and saccharin on body weight. The control group received a standard diet of laboratory chow and water. A second group received ad libitum access to an 11% sucrose solution, and a third group received a comparably sweet solution of aspartame and saccharin. After eight weeks of the 16-week study, half of the animals in the sucrose and low-calorie sweetener groups were given the other sweetened beverage.

The results showed that the rats ingesting aspartame and saccharin for the full 16 weeks weighed the same as the controls at the end of the experiment, while those consuming the sucrose solution weighed 26% more than the rats consuming low-calorie sweeteners. Rats that switched from sucrose to low-calorie sweeteners lost a significant amount of weight compared to rats that kept consuming sucrose. Those that switched from low-calorie sweeteners to sucrose showed a weight gain of 24%. Dr. Porikos concluded: "The results show that substitution of artificial sweeteners for sugars prevents weight gain and promotes weight loss in rats" (Porikos and Koopmans 1988, p. 12).

Davidson and Swithers (2004), on the basis of a small rat study, hypothesized that the use of low-calorie sweeteners might be related to the increased incidence of obesity. Two groups of weanling rats, 10 per group, were given overnight access to 50 ml of grape or cherry flavored sweet solutions along with laboratory chow ad libitum. One group received the flavored solutions sweetened with either 10% sucrose or 10% glucose. The other group received the flavored solutions sweetened with either 10% sucrose or 0.3% saccharin. The authors note that sweet taste was a reliable predictor of calories for the first group but not the second. After 10 days of this "training," all rats were given one day with just laboratory chow before testing began. The animals were then food deprived overnight before being offered 4 g of a sweet, high-calorie, chocolate-flavored premeal. Following the premeal, the rats were given laboratory chow for one hour. The rats given the two caloric solutions reportedly ate significantly less laboratory chow than the group given one caloric and one non-caloric solution. The authors conclude that the group given the two caloric solutions was better able than the group given one caloric and one non-caloric solution "to anticipate the caloric consequences of eating the sweet premeal and thus was better able to compensate for the calories contained in that meal by reducing subsequent intake of laboratory chow" (p. 934). No food or beverage intake data were given and no animal weights either before or after the study were provided. This study is purely speculative and the hypothesis is inconsistent with the results of other animal and human studies.

A second study conducted by Swithers and Davidson (2008) again hypothesized, "experiences that reduce the validity of sweet taste as a predictor of the caloric or nutritive consequences of eating may contribute to deficits in the regulation of energy by reducing the ability of sweet-tasting foods that contain calories to evoke physiological responses that underlie tight regulation" (p. 161). Swithers and Davidson conducted various experiments with rats. Eight rats in the sweet predictive group received plain, unsweetened yogurt for three days each week followed by yogurt sweetened with 20% glucose for three days. Nine rats in the sweet non-predictive group received unsweetened yogurt for three days followed by yogurt sweetened with 0.3% saccharin for another three days. A third group (the control group with 10 rats) received yogurt sweetened with 20% glucose on the same three days that the sweet non-predictive group received sweetened yogurt. After five weeks, the researchers found that during weeks two, three and five, the rats in the sweet non-predictive group gained more weight than did rats in either the sweet predictive or sweet predictive control groups. Rats in the sweet non-predictive group also had greater adiposity than those in the sweet predictive and sweet predictive control groups.

In the second experiment, the researchers evaluated, "… whether or not the strength of the predictive relationship between sweet taste and calories could influence the strength of caloric compensation" (Swithers and Davidson 2008, 164). Eleven rats in the sweet predictive group received plain, unsweetened yogurt for seven of the 14 days and yogurt sweetened with 20% glucose for the other seven days. Nine rats in the sweet non-predictive group received plain, unsweetened yogurt for seven days and yogurt sweetened with 0.3% saccharin for the other seven days. Following the two weeks of yogurt consumption, the rats were given one day of rat chow and water. Half of the rats in each group were then given a premeal of 5 g of Chocolate Ensure Plus. Rat chow was then given to all the rats and food intake was measured at one, two, four and twenty-four hours. After measuring intake, rats received rat chow again, which was then removed overnight and the premeal conditions were reversed. The researchers found that the sweet non-predictive rats ate more calories and gained more weight than did the sweet predictive rats.

The hypotheses by Swithers and Davidson are not new and have been tested in humans over the past 15 years. These studies have shown that low-calorie sweeteners and the products that contain them do not cause people to eat more, increase cravings for sweetness or cause weight gain. Generally, clinical studies with humans follow animal studies. The study by Swithers and Davidson went backwards and findings in animal (e.g., rat) studies are not necessarily applicable to humans. Further, the study conducted by Swithers and Davidson uses a very small sample size—11 rats, or less, per group. A recent study based on rats, showed that any flavor associated with a lack of calories led to overeating, even salt. However, that effect was observed only in very young rats (4 weeks) and disappeared 4 weeks later. Further, rats like the taste of saccharin and it is often used as a reward in rat studies so it is not surprising that rats would eat more of something sweetened with saccharin. On the other hand, rats do not like the taste of aspartame, which questions this study's applicability to other low-calorie sweeteners (Sclafani and Abrams 1986). These researchers report: "The results indicate that aspartame is not very palatable to rats and suggest that it has little or no sweet, that is, sucrose-like, taste to rats as it does to humans" (Sclafani and Abrams 1986, p. 253).

In a subsequent article, the Swithers et al. (2010) repeated allegations that the consumption of low-calorie sweeteners could result in a "dissociation between the sweet taste cues and caloric consequences" that also could "lead to a decrease in the ability of sweet tastes to evoke physiological responses that serve to regulate energy balance" (p. 55). No new research is presented; the article is essentially a review of the authors' earlier work (including what is discussed above), which is contrary to the significant body of science on the subject.

Another small rat study by Pierce et al. (2007) alleges that the consumption of low-calorie foods may lead to increased episodes of overeating in human children. The study, using 16 obesity prone and 16 lean juvenile rats, found that rats given a premeal sweetened with saccharin consumed more laboratory chow during meal times. From this observation, the researchers theorize that children who consume products containing low-calorie sweeteners may be desensitized to satiety signals and therefore be more likely to overeat during meals. This rat study has been criticized because of the assertion that the research findings could apply to the eating behaviors of children. Further, the vast body of scientific literature does not support the conclusions of Pierce and colleagues. Studies evaluating children, for example, Rodearmel et al. (2007) and Ludwig et al. (2001) discussed above, found that the use of low-calorie products not only aids in weight loss, but can also assist with weight control. Within the context of rising obesity rates among children (currently 32% of US children aged 6–19 years are overweight or at risk of being overweight) taking steps to assure appropriate caloric intake is important and products with low-calorie sweeteners are effective tools for reducing calories (Flegal et al. 2010; Ogden et al. 2010).

Conclusions

Pertinent to a review of scientific literature is the provocative issue of bias in research reporting, which can lead to misleading representation of data. Two recent studies address this issue. In the first, Cope and Allison (2010) demonstrated that a certain "white hat" bias may exist in obesity research (this bias refers to the tendency to distort research findings if such distortion is perceived to serve good ends). The researchers examined ways in which scientists writing new research papers referenced two studies reporting effects of sugar-sweetened beverages on body weight. An analysis of 206 papers that cited the two studies found the study results were accurately reported less than one-third of the time. More than two-thirds of the papers reportedly overstated evidence that reducing sugar-sweetened drink consumption reduced weight or obesity. The researchers also reported that data was more likely to be published when it showed statistically significant outcomes, and when data did not show sugar-sweetened drinks to have the "desired" outcome, it was less likely to be published.

In a related study, Thomas et al. (2008) analyzed the association of funding source and the quality of reporting (QR) of long-term clinical obesity trials. Of 63 long-term weight loss trials reviewed, 67% of which were industry supported, the average QR score was significantly higher (i.e., better reporting) for industry-funded studies of overweight or obese individuals than for non-industry-funded studies. After limiting the analysis to only nondrug studies that were both relatively large and long term, the statistical significance decreased with the much smaller sample size, but the QR score was still higher for industry-funded research. The researchers noted, "For the scientific process to proceed effectively, it is important that all studies, both industry-funded and not, be reported with the highest quality possible … because it is through the comprehension of published research reports that the scientific community at large can judge the merits and import of the findings" (Thoms et al. 2008, p. 1533).

Low-calorie sweeteners and the products containing them provide sweetness and "good taste" without the calories of their full calorie counterparts. The studies discussed above demonstrate that when sucrose is covertly replaced by low-calorie sweeteners non-dieting obese and normal weight individuals incompletely compensate for the calorie reduction. In other words, they consume fewer calories.

Numerous studies have demonstrated that low-calorie sweeteners do not increase hunger, appetite or food intake. Further, multidisciplinary weight control programs that include the use of reduced-calorie foods and beverages sweetened with low-calorie sweeteners may facilitate weight loss and weight maintenance.

Also of interest, research (Sigman-Grant and Hsieh 2005) was undertaken to determine the impact of reduced calorie foods and beverages (i.e., products sweetened with low-calorie sweeteners) on the micro- and macronutrient composition of the diets of American adults (20 years old and older) who provided two days of dietary recall. The data was the combined 1994–96 Continuing Survey of Food Intakes by Individuals ($n = 9323$) and the accompanying Diet, Health and Knowledge Survey ($n = 5649$). Reduced calorie product users reported consuming significantly less total and saturated fat, cholesterol, energy, and added sugars, while having significantly higher intakes of vitamins (e.g., vitamins A, E, and folate) and minerals (e.g., calcium, iron and zinc) from their foods. These users also reported consuming higher amounts of dark green and yellow vegetables.

The use of low-calorie sweeteners in place of sugar can result in products significantly reduced in calories when compared with their traditional counterparts. In light of the current obesity epidemic, it is important that consumers have available a wide variety of good tasting, reduced-calorie products as tools to assist to them in addressing their calorie goals.

References

Abdallah, L., Chabert, M., and Louis-Sylvestre, J. 1997. Cephalic phase responses to sweet taste. *American Journal of Clinical Nutrition* 65:737–743.

Allon, N. 1973. The stigma of overweight in everyday life. In *Obesity in perspective*, G.A. Bray (Ed.), 83–110. Washington, D.C.: National Institutes of Health.

Ambrus, J.L., Ambrus, C.M., Shields, R., Mink, I.B., and Cleveland, C. 1976. Effect of galactose and sugar substitutes on blood insulin levels in normal and obese individuals. *Journal of Medicine* 7:429–438.

American Dental Association. 1998. Position Statement. Role of Sugar Free Foods and Medications in Maintaining Good Oral Health. Adopted October 1998. http://www.ada.org/1874.aspx. Accessed November 6, 2010.

American Diabetes Association. 2002. Position statement of the American diabetes association. Evidence-based nutrition principles and recommendations for the treatment and prevention of diabetes and related complications. *Diabetes Care* 25(1):202–212.

American Dietetic Association. 2004. Position of the American dietetic association: Use of nutritive and non-nutritive sweeteners. *Journal of the American Dietetic Association* 104(2):255–275.

American Heart Association. 2007. Soft drink consumption and risk of developing cardiometabolic risk factors and the metabolic syndrome in middle-aged adults in the community. http://www.americanheart.org/presenter.jhtml?identifier=3050553. Accessed October 2, 2007.

Anderson, G.H, Saravis, S., Schacher, R., Zlotkin, S., and Leiter, L.A. 1989. Aspartame: Effect on lunch-time food intake, appetite and hedonic response in children. *Appetite* 13:93–103.

Anton, S.D., Martin, C.K., Hongmei, H., Coulon, S., Cefalu, W.T., Geiselman, P., and Williamson, D.A. 2010. Effects of stevia, aspartame, and sucrose on food intake, satiety, and postprandial glucose and insulin levels. *Appetite* 55:37–43.

Appleton, K.M., and Blundell, J.E. 2007. Habitual high and low consumers of artificially-sweetened beverages: Effects of sweet taste and energy on short-term appetite. *Physiology & Behavior* 92(3):479–486.

Beidler, L.M. 1975. The biological and cultural role of sweeteners. In *Sweeteners: Issues and uncertainties*, 11–18. Washington, DC: National Academy of Sciences.

Bellisle, F., and Drewnowski, A. 2007. Intense sweeteners, energy intake and the control of body weight. *European Journal of Clinical Nutrition* 61:691–700.

Berryman, G.H., Hazel, G.R., Taylor, J.D., Sanders, P., and Weinberg, M.S. 1968. A case of safety of cyclamate and cyclamate-saccharin combinations. *American Journal of Clinical Nutrition* 21:673–687.

Berthoud, H.R., Bereiter, D.A., Trimble, E.R., Siegel, E.G., and Jeanrenaudet, B. 1981. Cephalic phase, reflex insulin secretion neuroanatomical and physiological characterization. *Diabetologia* 20:393–401.

Birch, L.L, McPhee, L., and Sullivan, S. 1989. Children's food intake following drinks sweetened with sucrose or aspartame: time course effects. *Physiology & Behavior* 45:387–395.

Black, R.M., Tanaka, P., Leiter, L.A., and Anderson, G.H. 1991. Soft drinks with aspartame: Effect on subjective hunger, food selection, and food intake of young adult males. *Physiology & Behavior* 49:803–810.

Black, R.M., Leiter, L.A., and Anderson, G.H. 1993. Consuming aspartame with and without taste: differential effects on appetite and food intake of young adult males. *Physiology & Behavior* 53:459–466.

Blackburn, G.L., Kanders, B.S., Lavin, P.T., Keller, S.D., and Whatley, J. 1997. The effect of aspartame as part of a multidisciplinary weight-control program on short- and long-term control of body weight. *American Journal of Clinical Nutrition* 65:409–418.

Blundell, J.E., and Hill, A.J. 1986. Paradoxical effects of an intense sweetener (aspartame) on appetite. *Lancet* 1:1092–1093.

Blundell, J.E., and Hill, A.J. 1987. Artificial sweeteners and the control of appetite. In *Implications for the eating disorders in future of predictive safety evaluation*, A.N. Worden, D.V. Parke, and J. Marks (Eds.), 244–257. Boston: MTP Press.

Brown, R.J., De Banate, M.A., and Rother, K.I. 2010. Artificial sweeteners: A systematic review of metabolic effects in youth. *International Journal of Pediatric Obesity* 5(4):305–312.

Brownell, K.D. 1984. The psychology and physiology of obesity: Implications for screening and treatment. *Journal of the American Dietetic Association* 84:408–413.

Brownell, K.D. and Wadden, T.A. 1992. Etiology and treatment of obesity: Understanding a serious, prevalent and refractory disorder. *Journal of Consulting and Clinical Psychology* 60(4):505–517.

Bruce, D.G., Strolien, L.H., Furler, S.M., and Chisholm, D.J. 1987. Cephalic phase metabolic responses in normal weight adults. *Metabolism* 36(8):721–725.

Canty, D.J., and Chan, M.M. 1991. Effects of consumption of caloric vs noncaloric sweet drinks on indices of hunger and food consumption in normal adults. *American Journal of Clinical Nutrition* 53:1159–1164.

Carlson, H.E., and Shah, J.H. 1989. Aspartame and its constituent amino acids: Effects on prolactin, cortisol, growth hormone, insulin, and glucose in normal humans. *American Journal of Clinical Nutrition* 49:427–432.

Colagiuri, H.E., and Shah, J.H. 1989. Metabolic effects of adding sucrose and aspartame to the diet of subjects with noninsulin-dependent diabetes mellitus. *American Journal of Clinical Nutrition* 50:474–478.

Cope, M.B., and Allison, D.B. 2010. White hat bias: Examples of its presence in obesity research and a call for renewed commitment to faithfulness in research reporting. *International Journal of Obesity* 34:84–88.

Coulston, F., and Gori, G.B. 2002. Aspartame: Review of safety. *Regulatory Toxicology and Pharmacology* 35(2):78–81.

Davidson, T.L., and Swithers, S.E. 2004 A short communication on a Pavlovian approach to the problem of obesity. *International Journal of Obesity* 28:933–935.

de la Hunty, A., Gibson, S., and Ashwell, M. 2006. A review of the effectiveness of aspartame in helping with weight control. *British Nutrition Foundation Nutrition Bulletin* 31:115–128.

Dhingra, R., Sullivan, L., Jacques, P.F., Wang, T.J., Fox, C.S., Meigs, J.B., D'Agostino, R.B., Gaziano, J.M., and Vasan, R.S. 2007. Soft drink consumption and risk of developing cardiometabolic risk factors and the metabolic syndrome in middle-aged adults in the community. *Circulation* 116(5):480–488.

Drewnowski, A.1995. Intense sweeteners and the control of appetite. *Nutrition Reviews* 53:1–7.

Drewnowski, A., Massien, C., Louis-Sylvestre, J., Fricker, J., Chapelot, D., and Apfelbaum, M. 1994a. Comparing the effects of aspartame and sucrose on motivational ratings, taste preferences, and energy intakes in humans. *American Journal of Clinical Nutrition* 59:338–345.

Drewnowski, A., Massien, C., Louis-Sylvestre, J., Fricker, J., Chapelot, D., and Apfelbaum, M. 1994b. The effects of aspartame versus sucrose on motivational ratings, taste preferences, and energy intakes in obese and lean women. *International Journal of Obesity* 18:570–578.

Flegal, K., Carroll, M., Ogden, C., and Curtin, L. 2010. Prevalence and trends in obesity among US adults, 1999–2008. *Journal of the American Medical Association* 303(3):235–241.

Fowler, S.P., Williams, K., Hunt, K.J., Resendez, R.G., Hazuda, H.P., and Stern, M.P. 2005. Diet soft drink consumption is associated with increased incidence of overweight and obesity in the San Antonio Heart Study. San Antonio, TX: Annual Meeting of the American Diabetes Association.

Fowler, S.P., Williams, K., Resendez, R.G., Hunt, K.J., Hazuda, H.P., and Stern, M.P. 2008. Fueling the obesity epidemic? Artificially sweetened beverage use and long-term weight gain. *Obesity* 16:1894–1900.

The Gallup Organization, Inc. Gallup Study of Children's Nutrition & Eating Habits. 2010. Available at http://www.multisponsor.com/cat/Food%20Beverage%20and%20Nutrition/Children%20and%20Mothers/Sweetener%20Preferences%20of%20Mothers%20MS10074.pdf.

Ganong, W.F. 1973. *Review of medical physiology*, pp. 259. Los Altos, CA: Lange Medical Publications.

Goldfine, I.D., Ryan, W.G., and Schwartz, T.B. 1969. The effect of glucola, diet cola and water ingestion on blood glucose and plasma insulin. *Proceedings of the Society for Experimental Biological Medicine* 131:329–330.

Grotz, V.L., Henry, R.R., McGill, J.B., Prince, M.J., Shamoon, H., Trout, J.R., and Pi-Sunyer, F.X. 2003. Lack of effect of sucralose on glucose homeostasis in subjects with type 2 diabetes. *Journal of the American Dietetic Association* 103(12):1607–1612.

Härtel, B., Graubaum, H.J., and Schneider, B. 1993. Effect of sweetener solutions on insulin secretion and blood glucose levels. *Ernahrungs-Umschau* 40:152–155.

Hill, J.O., Wyatt, H.R., Reed, G.W., and Peters, J.C. 2003. Obesity and the environment: Where do we go from here? *Science* 299(5608):853–855.

Horwitz, D.L., McLane, M., and Kobe, P. 1988. Response to single dose of aspartame or saccharin by NIDDM patients. *Diabetes Care* 11(3):230–234.

Kanders, B.S., Blackburn, G.L., Lavin, P.T., and Kienholz, M. 1996. Evaluation of weight control. In *The clinical evaluation of a food additive: Assessment of aspartame,* C. Tschanz, H.H. Butchko, W.W. Stargel, and F.N. Kotsonis (Eds.), 289–299. Boca Raton: CRC Press.

Kanders, B.S., Lavin, P.T., Kowalchuk, M.B., Greenberg, I., and Blackburn, G.L. 1988. An evaluation of the effect of aspartame on weight loss. *Appetite* 11(suppl 1):73–84.

Lavin, P.T., Sanders, P.G., Mackey, M.A., and Kotsonis, F.N. 1994. Intense sweeteners use and weight change among women: A critique of the Stellman and Garfinkel study. *Journal of the American College of Nutrition* 13:102–105.

Leon, A., Hunninghake, D., Bell, C., Rassin, D.K., and Tephly, T.R. 1989. Safety of long-term large doses of aspartame. *Archives of Internal Medicine* 149:2318–2324.

Louis-Sylvestre, J. 1976. Preabsorptive insulin release and hypoglycemia in rats. *American Journal of Physiology* 230:56–60.

Ludwig, D., Peterson, K.E., and Gortmaker, S.L. 2001. Relation between consumption of sugar-sweetened drinks and childhood obesity: A prospective, observational analysis. *Lancet* 357(9255):505–508.

Lutsey, P., Steffen, L.M., and Stevens, J. 2008. Dietary intake and the development of the metabolic syndrome. The atherosclerosis risk in communities study. *Circulation* 117:754–761.

Magnuson, B. 2007. Aspartame: a safety evaluation based on current use levels, regulations, toxicological and epidemiological studies. *Critical Reviews in Toxicology* 37:629–727.

Maller, O., and Desor, J.A. 1973. Effect of taste on ingestion by human newborns. In *Fourth symposium on oral sensation and perception,* J.W. Weiffenbach (Ed.), 61–68. Bethesda, MD: National Institute of Dental Research, National Institutes of Health.

Mattes, R. 1990. Effects of aspartame and sucrose on hunger and energy intake in humans. *Physiology & Behavior* 47:1037–1044.

Mattes, R., and Popkin B. 2008. Nonnutritive sweetener consumption in humans: Effects on appetite and food intake and their putative mechanisms. *American Journal of Clinical Nutrition* 89:1–14.

McCann, M.B., Trulson, M.F., and Stulb, S.C. 1956. Non-caloric sweeteners and weight reduction. *Journal of the American Dietetic Association* 32:327–330.

Morris, D.H., Mance, M.J., Bell, K.J., and Ward, A. 1989. Weight loss and consumption of artificial sweeteners among overweight women and men (abstract). *Journal of the American Dietetic Association* 9, Supplement A-94.

Nettleton, J., Lutsey, P., Wang, Y., Lima, J., Michos, E., and Jacobs, D. 2009. Diet soda intake and risk of incident metabolic syndrome and type 2 diabetes in the multi-ethnic study of atherosclerosis. *Diabetes Care* 32:688–694.

Ogden, C., Carroll, M., Curtin, L., Lamb, M., and Flegal, K. 2010. Prevalence of high body mass index in US children and adolescents, 2007–2008. *Journal of the American Medical Association* 303(3):242–249.

Okuno, G., Kawakami, F., Tako, H., Kashihara, T., Shibamoto, S., Yamazaki, T., Yamamoto, K., and Saeki, M. 1986. Glucose tolerance, blood lipid, insulin and glucagon concentrations after single or continuous administration of aspartame in diabetics. *Diabetes Research and Clinical Practice* 2:23–27.

Parham, E.S., and Parham, A.R. 1980. Saccharin use and sugar intake by college students. *Journal of the American Dietetic Association* 86:560–563.

Pfaffmann, C. 1977. Biological and behavioral substrates of the sweet tooth. In *Taste and development, The genesis of sweet preference*, J.W. Weiffenbach (Ed.), 3–24. U.S. Department of Health, Education and Welfare.

Phelan, S., Lang, W., Jordan, D., and Wing, R.R. 2009. Use of artificial sweeteners and fat-modified foods in weight loss maintainers and always-normal weight individuals. *International Journal of Obesity* 33:1183–1190.

Pierce, W.D., Heth, C.D., Owczarczyk, J.C., Russell, J.C., and Proctor, S.D. 2007. Overeating by young obesity-prone and lean rats caused by tastes associated with low energy foods. *Obesity* 15(8):1969–1979.

Porikos, K.P., Booth, G., and Van Itallie, T.B. 1977. Effect of covert nutritive dilution on the spontaneous intake of obese individuals: a pilot study. *American Journal of Clinical Nutrition* 30:1638–1644.

Porikos, K.P., Hesser, M.F., and Van Itallie T.B. 1982. Caloric regulation in normal weight men maintained on a palatable diet of conventional foods. *Physiology & Behavior* 29:293–300.

Porikos, K.P., and Koopmans, H.S. 1988. The effect of non-nutritive sweeteners on body weight in rats. *Appetite* 11:12–15.

Porikos, K.P., and Pi-Sunyer, F.X. 1984. Regulation of food intake in human obesity: studies with caloric dilution and exercise. *Clinical and Endocrinological Metabolism* 13:547–561.

Porikos, K.P., and Van Itallie, T.B. 1984. Efficacy of low-calorie sweeteners in reducing food intake: Studies with aspartame. In *Aspartame: Physiology and Biochemistry*, L.D. Stegink and L.J. Filer, Jr., (Eds.), 273–286. New York: Marcel Dekker, Inc.

Rabin, A., Vasilaras, T.H., Møller, A.C., and Astrup, A. 2002. Sucrose compared with artificial sweeteners: Different effects on ad libitum food intake and body weight after 10 wk of supplementation in overweight. *American Journal of Clinical Nutrition* 76:721–729.

Remington, D. March 19, 1985. In how diet drinks may make you gain weight. *National Enquirer.*

Renwick, A.G. 1994. Intense sweeteners, food intake, and the weight of a body of evidence. *Physiology & Behavior* 55:139–143.

Renwick, A.G., and Molinary, S.V. 2010. Sweet-taste receptors, low-energy sweeteners, glucose absorption and insulin release. *British Journal of Nutrition* 104(10):1415–1420.

Rodearmel, S., Wyatt, H.R., Stroebele, N., Smith, S.M., Ogden, L.G., and Hill, J.O. 2007. Small changes in dietary sugar and physical activity as an approach to preventing excessive weight gain: The America on the Move Family Study. *Pediatrics* 120(4):869–879.

Rodin, J. 1984a. In taming the hunger hormone. *American Health*, January/February 43–47.

Rodin, J. 1984b. A sense of control. *Psychology Today*, December, 38–45.

Rodin, J. 1990. Comparative effects of fructose, aspartame, glucose, and water preloads on calorie and macronutrient intake. *American Journal of Clinical Nutrition* 51:428–435.

Rogers, P.J., and Blundell, J.E. 1989. Separating the actions of sweetness and calories: Effects of saccharin and carbohydrates on hunger and food intake in human subjects. *Physiology & Behavior* 45:1093–1099.

Rogers, P.J., Carlyle, J., Hill, A.J., and Blundell, J.E. 1988. Uncoupling sweet taste and calories: Comparison of the effects of glucose and three intense sweeteners on hunger and food intake. *Physiology & Behavior* 43:547–552.

Rogers, P.J., Fleming, H.C., and Blundell, J.E. 1990. Aspartame ingested without tasting inhibits hunger and food intake. *Physiology & Behavior* 47:1239–1243.

Rogers, P.J., Keedwell, P., and Blundell, J. 1991. Further analysis of the short-term inhibition of food intake in humans by the dipeptide l-aspartyl-phenylalanine methyl ester (aspartame). *Physiology & Behavior* 49:739–743.

Rolls, B.J. 1987. Sweetness and satiety. In *Sweetness*, J. Dobbing (Ed.), 161–173. London: Springer-Verlag.

Rolls, B.J. 1991. Effects of intense sweeteners on hunger, food intake, and body weight: A review. *American Journal of Clinical Nutrition* 53.872–878.

Rolls, B.J., Kim, S., and Fedoroff, I.C. 1990. Effects of drinks sweetened with sucrose or aspartame on hunger, thirst and food intake in men. *Physiology & Behavior* 48:19–26.

Rolls, B.J., Laster, L.J., and Summerfelt, A. 1989. Hunger and food intake following consumption of low-calorie foods. *Appetite* 13:115–127.

Rolls, B.J., and Shide, D.J. 1996. Evaluation of hunger, food intake, and body weight. In *The clinical evaluation of a food additive: Assessment of aspartame*, C. Tschanz, H.H. Butchko, W.W. Stargel, and F.N. Kotsonis (Eds.), 275–287. Boca Raton: CRC Press.

Ryan-Harshman, M., Leiter, L.A., and Anderson, G.H. 1987. Phenylalanine and aspartame fail to alter feeding behavior, mood and arousal in men. *Physiology & Behavior* 39:247–253.

Sclafani, A., and Abrams, M. 1986. Rats show only a weak preference for the artificial sweetener aspartame. *Physiology & Behavior* 37(2):253–256.

Shigeta, H., Yoshida, T., Nakai, M., Mori, H., Kano, Y., Nishioka, H., Kajiyama et al. 1985. Effects of aspartame on diabetic rats and diabetic patients. *Journal of Nutrition Science Vitaminology* 31:533–540.

Sigman-Grant, M., and Hsieh, G. 2005. Reported use of reduced sugars foods and beverages reflect high quality diets. *Journal of Food Science* 70(1):S42–S46.

Smith, J.L., and Heybach, J.P. 1988. Evidence for the lower intake of calories and carbohydrate by 19–50 year old female aspartame users from the continuing survey of food intakes by individuals (CSFII 85). *Federation of American Societies for Experimental Biology* (abstract), 2 (5) (abst. No. 5238).

Steiner, J.E. 1973. The gustofacial response. In *Fourth symposium on oral sensation and perception*, J.F. Bosma (Ed.), 145–159. Bethesda, MD: National Institutes of Health.

Stellman, S.D., and Garfinkel, L. 1986. Artificial sweetener use and one-year weight change among women. *Preventive Medicine* 15:195–202.

Stellman, S.D., and Garfinkel, L. 1988. Patterns of artificial sweetener use and weight change in an American Cancer Society prospective study. *Appetite* 11(suppl):85–91.

Swithers S.E., and Davidson, T.L. 2008. A role for sweet taste: calorie predictive relations in energy regulation by rats. *Behavioral Neuroscience* 122(1):161–173.

Swithers, S.S., Martin, A.A., and Davidson, T.L. 2010. High-intensity sweeteners and energy balance. *Physiology & Behavior* 100(1):55–62.

Teff, K.L., Devine, J., and Engelman, K. 1995. Sweet taste: Effect on cephalic phase insulin release in men. *Physiology & Behavior* 57:1089–1095.

Thomas, O., Thabane, L., Douketis, J., Chu, R., Westfall, A.O., and Allison, D.B. 2008. Industry funding and the reporting quality of large long-term weight loss trials. *International Journal of Obesity* 32:1531–1536.

Tordoff, M.G., and Alleva, A.M. 1990a. Effect of drinking soda sweetened with aspartame or high-fructose corn syrup on food intake and body weight. *American Journal of Clinical Nutrition* 51:963–969.

Tordoff, M.G., and Alleva, A.M. 1990b. Oral stimulation with aspartame increases hunger. *Physiology & Behavior* 47:555–559.

Tordoff, M.G., and Friedman, M.I. 1989a. Drinking saccharin increases food intake and preference —I. Comparison with other drinks. *Appetite* 12:1–10.

Tordoff, M.G., and Friedman, M.I. 1989b. Drinking saccharin increases food intake and preference—II. Hydration factors. *Appetite* 12:11–21.

Tordoff, M.G., and Friedman, M.I. 1989c. Drinking saccharin increases food intake and preference—III. Sensory and associative factors. *Appetite* 12:23–36.

Tordoff, M.G. and Friedman, M.I. 1989d. Drinking saccharin increases food intake and preference—IV. Cephalic phase and metabolic factors. *Appetite* 12:37–56.

Von Borstel, R.W. 1985. Metabolic and physiologic effects of sweeteners. *Clinical Nutrition* 4:215–220.

Wilson, J.F. 2000. Lunch eating behavior of preschool children: effects of age, gender, and type of beverage served. *Physiology & Behavior* 70:27–33.

Index